Sources and Studies
in the History of Mathematics and Physical Sciences

Managing Editor
J.Z. Buchwald

Associate Editors
J.L. Berggren and J. Lützen

Advisory Board
C. Fraser, T. Sauer, A. Shapiro

For further volumes:
http://www.springer.com/series/4142

The *Liber mahameleth*

A 12th-century mathematical treatise

Part Two

Translation,

Glossary

Jacques Sesiano

Jacques Sesiano
Département de Mathématiques
Ecole polytechnique fédérale
Lausanne, Switzerland

ISSN 2196-8810   ISSN 2196-8829 (electronic)
ISBN 978-3-319-03939-8   ISBN 978-3-319-03940-4 (eBook)
DOI 10.1007/978-3-319-03940-4
Springer Cham Heidelberg New York Dordrecht London

Library of Congress Control Number: 2014930575

© Springer International Publishing Switzerland 2014
This work is subject to copyright. All rights are reserved by the Publisher, whether the whole or part
of the material is concerned, specifically the rights of translation, reprinting, reuse of illustrations,
recitation, broadcasting, reproduction on microfilms or in any other physical way, and transmission
or information storage and retrieval, electronic adaptation, computer software, or by similar or
dissimilar methodology now known or hereafter developed. Exempted from this legal reservation
are brief excerpts in connection with reviews or scholarly analysis or material supplied specifically
for the purpose of being entered and executed on a computer system, for exclusive use by the
purchaser of the work. Duplication of this publication or parts thereof is permitted only under the
provisions of the Copyright Law of the Publisher's location, in its current version, and permission
for use must always be obtained from Springer. Permissions for use may be obtained through
RightsLink at the Copyright Clearance Center. Violations are liable to prosecution under the
respective Copyright Law.
The use of general descriptive names, registered names, trademarks, service marks, etc. in this
publication does not imply, even in the absence of a specific statement, that such names are exempt
from the relevant protective laws and regulations and therefore free for general use.
While the advice and information in this book are believed to be true and accurate at the date of
publication, neither the authors nor the editors nor the publisher can accept any legal responsibility
for any errors or omissions that may be made. The publisher makes no warranty, express or implied,
with respect to the material contained herein.

Printed on acid-free paper

Springer is part of Springer Science+Business Media (www.springer.com)

Table of contents
Part Two

Translation 579

Beginning of the Book (on) mahameleth 581

(Chapter A–I) on Numbers 581

Chapter (A–II) on the Premisses necessary for practical arithmetic . . 588

Chapter (A–III) on Multiplication 605

    Multiplication of digits by themselves and by other digits 607

    Chapter on the attribution of a note 609

    Chapter on the multiplication of integral numbers according to note, excepting the composite ones 613

    Chapter on the science of multiplying differences according to rules 622

    Multiplication of composites from a digit and an article 623

    Multiplication of thousands among themselves 626

    Chapter on taking fractions of thousands repeated 635

Chapter (A–IV) on Division 639

    Chapter on another way of dividing 642

    Chapter on the denominations 646

Chapter (A–V) on the Multiplication of fractions 655

    Chapter on the multiplication of a fraction by an integer 655

    Multiplication of a fraction by a fraction 662

    Chapter on the conversion of fractions into other fractions 667

    Chapter on the multiplication of a fraction by an integer and a fraction 673

    Chapter on irregular fractions, which are discussed among mathematicians 683

Chapter (A–VI) on the Addition of fractions to fractions 694

    Again on the same topic 700

    Again on addition 702

    Chapter on amounts in addition 703

    Another chapter 707

Chapter (A–VII) on Subtracting 711

    Problems on subtracting 717

    Again on subtracting 719

    Chapter on amounts in subtracting 720

vi   Part Two: Translation, Glossary

Chapter (A–VIII) on the Division of fractions, with or without integers 726

     Chapter on denominating fractions from others, with or without integers 729

     Again. Another chapter. On dividing larger by smaller 730

     Chapter on dividing, otherwise 733

     Again. Rules for the multiplication, division, addition and subtraction of fractions, in a more concise form than above 743

     Again on division 744

Chapter (A–IX) on the Determination of roots, and on their multiplication, division, subtraction, addition, and other related subjects . . . 748

     Chapter on the determination of roots 748

     Multiplication of roots 751

     Chapter on the addition of roots 753

     Chapter on the subtraction of roots 754

     Chapter on the division of roots 756

     Chapter on the multiplication of roots of roots 760

     Chapter on the addition of roots of roots 763

     Chapter on the subtraction of roots of roots 765

     Chapter on the division of roots of roots 766

     Again on roots 770

Beginning of the Second part 777

Chapter (B–I) on Buying and selling 780

     Chapter involving unknowns in buying and selling 788

     Again. Another chapter on the same topic, with things 792

     Again. Another chapter involving unknowns in buying and selling 796

     Another chapter, on modii at different prices 811

Chapter (B–II) on Profits 817

     Chapter on profits wherein that which is sold or bought is named 820

     Another chapter on profit 830

     Again. Chapter on unknown profits 837

Chapter (B–III) on Profit in partnership 843

Chapter (B–IV) on Division according to portions 848

Chapter (B–V) on Masses 852

     Again, on something else 853

Chapter (B–VI) on Drapery 856

     Again on the same topic, but otherwise 859

LIBER MAHAMELETH

Chapter (B–VII) on Linens	863
Again on the same topic, but otherwise	867
Again, on something else	870
Chapter (B–VIII) on Grinding	871
Again on the same topic	878
Again on the same topic, but another chapter involving increase	878
Chapter (B–IX) on Boiling must	882
Chapter (B–X) on Borrowing	892
Chapter (B–XI) on Hiring	895
Chapter involving unknowns in hiring for things	896
Again on the same topic, but otherwise	920
Again on the same topic, but otherwise	922
Again on the same topic, but otherwise	935
Chapter (B–XII) on the Diversity of workers' wages	943
Again on the same topic, but otherwise	948
Chapter (B–XIII) on Hiring carriers	952
Chapter (B–XIV) on Hiring stone-cutters	967
Again on the same topic, but otherwise	971
Chapter on another topic	975
Chapter (B–XV) on the Consumption of oil by lamps	979
Again on the same topic	984
Chapter (B–XVI) on the Consumption by animals	988
Chapter on unknown numbers of animals	994
Chapter on another topic	997
Chapter (B–XVII) on the Consumption of bread by men	999
Again on the same topic, but otherwise: wherein are involved measures of different regions	1003
Chapter (B–XVIII) on the Exchange of morabitini	1006
Again on the same topic	1010
Again on the same topic	1016
Chapter (B–XIX) on Cisterns	1039
Again on the same topic	1040
Again	1044
Chapter (B–XX) on Ladders	1046
Again, on another topic	1055
Again	1060
Other topic	1062

viii   PART TWO: TRANSLATION, GLOSSARY

Again 1062

Again. Science of finding the height of a tower or a tree 1065

Again, (B–XXI) on Another topic 1066

Chapter (B–XXII) on Messengers 1068

Chapter (B–XXIII) on Another topic 1071

Glossary 1075

Translation

In our translation we have made use of various signs. Any additions made solely for the convenience of reading appear in round brackets. Square brackets denote, as in the Latin text, presumed interpolations; and these we have written in italics for further identification. Angular brackets are, as in the Latin text, for lacunae; but in the translated text they have been retained only if of some importance for the history of the text, such as when the omission renders the extant text unclear and thus may explain the occurrence of a reader's intervention. The footnotes are mostly concerned with textual explanations or references and only rarely with mathematical questions since these are fully dealt with in our commentary.

Beginning of the Book (on) mahameleth

(Chapter A–I) on Numbers

Of all things that exist, some are created by man, others not. Of those which are not created by man, some are subject to motion, others not, such as God and an angel. Now, of those which are subject to motion, some cannot exist without motion [*and matter*][1], such as humanity and squareness, others may exist without it. Of those which cannot exist without motion, some can neither exist nor be conceived of apart from some specific matter, such as humanity, others may be conceived of apart from some specific matter even if they can only come into being through matter, such as squareness. Those which are mixed with motion and may exist without it are for instance unity, number, causality, and (others) of this kind.

Thus number belongs to those which are considered from each of the two points of view, in themselves and in matter: number is considered in itself when one has in mind only its nature, or its property [*by itself*][2] of being even or odd and other such properties taught in the Arithmetic of Nicomachos; but is considered in matter when [*as in what follows*][3] one has it in mind [*like three or four*][4] as a support for human uses such as multiplying, dividing, and other such uses taught in the Arithmetic of al-Khwārizmī and in (works on) mahameleth. The study of number considered by itself[5] is called theoretical or speculative; that involving matter, practical or active. Since it belongs to arithmetical science to deal with number from both points of view, there is accordingly a theoretical and a practical arithmetic.

The kinds of practical arithmetic are numerous: there is the science of collecting numbers, that of separating them, the science of commercial

---

[1] As will be seen, the first pages of the text contain numerous interpolations (originating with early readers). These may mostly be ignored as adding nothing to the sense.

[2] 'by itself' (*per se*) presumably to correct the previous 'in itself' (*in se*); see note 5.

[3] Indeed, the subject of this treatise.

[4] Refers to 'even or odd' two lines above.

[5] The text has *per se*, whereas in a similar passage (by Gundissalinus) we find *in se*, as above in our text.

582   Part Two: Translation, Glossary

transactions, the science of determining unknowns by means of numbers, and many others.[6] The science which teaches collecting numbers consists of those of adding, doubling, multiplying, while that which teaches separating numbers consists of those of subtracting, halving, dividing; but the science of determining the root of a number is included in both, since finding a root involves collecting and separating, because of multiplying and subtracting.[7] The science of commercial transactions comprises those of selling and buying, borrowing and lending, hiring and offering services, spending and saving, and many others which will be dealt with in what follows.[8] The science of determining unknown quantities by means of numbers occurs both in the above kinds of commercial transaction and in the determination of the weight, depth or volume of objects when their length or breadth is known, or inversely.[9]

[*A number may be integral or fractional.*[10] *An integral number may be a digit, an article, a limit, a composite. The digits are the first numbers, made up of units only, as are all those from one*[11] *to nine.*]

Unit is not a number, but the origin and first element of number; for any number is made up of units and breaks up into units, while unit itself is not divisible. [*If we were to suppose that it can be divided, it would follow that it does not exist* [*if now we divide, it must exist, for we say 'one part', 'two parts'*][12]; *then if it were divisible it would exist and not exist at the same time, which is impossible. Now any number may be divided, for number is what is made up of units; thus, unit is not a number.*][13] [*But unit is not a number and a numbered thing is not a number either. For when we say 'three', 'ten', 'a hundred', 'a thousand', we mean pure numbers, whereas when we want to mean numbered things, we have their number attached immediately next to them, saying 'three men', 'ten horses', 'a hundred sextarii', 'a thousand nummi', and so forth.*[14] *Thus a numbered*

---

[6] This corresponds to the two parts of our treatise: on arithmetical operations (Book A), on concrete applications (Book B). These two aspects of practical arithmetic are developed in what follows.

[7] Such are the operations of practical arithmetic. Note that (unlike in many mediaeval treatises) doubling and halving will not be taught in the *Liber mahameleth*. The characteristic of root extraction alluded to refers to the proper algorithm for extracting roots, omitted from the *Liber mahameleth*, which teaches only approximating square roots (Chapter A–IX).

[8] Subjects to be treated in Book B.

[9] But already Book A has a few algebra-like treatments (A.159–169, A.184–186, A.204–214). Determination of areas, weights and volumes is found in B.V to B.VII, B.XIV, B.XIX, B.XX.

[10] Irrelevant. We are in the domain of natural numbers.

[11] Excluded.

[12] Gloss to the gloss.

[13] What follows does not seem to originate with the same reader.

[14] 'sextarius' and 'nummus': we shall keep the Latin names throughout

LIBER MAHAMELETH

thing is not a number.] The first number made up of units is two, which is therefore the very first and the smallest. A unit added to two yields three. A unit added to three yields four. Adding always a unit in this way, number will increase to infinity.

Therefore each number could not be designated by an individual name.[15] For in any language the instruments of speech are fixed and determined, and their modulations, by means of which the articulated voice is formed, are naturally limited; therefore [*the symbols of the letters among all the nations and*][16] also their combinations by means of preposition and postposition for the purpose of representing the names of all things are multifarious yet limited. For this reason, since numbers are infinite, each single one could not have received a name, nor should; above all, people, who use numbers in almost everything, would be exceedingly hindered if the need to count compelled them to have always at hand, for their calculations, an infinite quantity of numeral names. Therefore it became indispensable to set within the infinite sequence of numbers certain boundaries and to designate them by means of a few names in order that man not be compelled when counting always to advance by new adjunctions of both numbers and names. And since it was impossible for all numbers to have names though for some it was necessary, and since they had to be multiplied among themselves,[17] the numbers were therefore arranged by orders or places.[18]

[*Each order contains nine numbers, except the first.*][19] The first order was established from one to nine; it comprises nine names, is called the order of the units, or of the digits, and its initial element, or limit, is the unit.[20] After the manner of this first order, a second order was also established, containing nine numbers, namely from ten to a hundred[21].

---

for this capacity measure and this coin.

[15] What follows appears more or less verbatim in the additional material found at the end of Johannes Hispalensis's *Liber algorismi* (pp. 127–128 in Boncompagni's 1857 *editio princeps*).

[16] Obvious interpolation: we are still only concerned with verbal expression.

[17] The multiplication table, the main tool of reckoning, must contain a restricted number of products, thus of entries, in order to be known 'by heart and readily', as the author will tell us before presenting it (below, p. 607).

[18] Here ends the passage from the *Liber algorismi*, which goes on to give the twelve names whereby all natural numbers are expressed (at least in Arabic): those of the digits from 1 to 9, then of 10, 100, 1000.

[19] Former marginal gloss summarizing what follows.

[20] All orders are said to comprise nine *numbers*, whereas the first comprises nine *names*. Indeed, 1 is not a number (see gloss) and each element of its order has a name of its own (note 18).

[21] When indicating the elements of an order, the text is inconsistent: sometimes the last element mentioned belongs to the next order, as here, sometimes not, as further below.

584 PART TWO: TRANSLATION, GLOSSARY

Its initial element, or limit, is ten, the doubling and multiplying of which generates all numbers of its order —just as before doubling and multiplying the unit generated the numbers of the first order [22]— and this order is called the order of the tens, or of the articles. The articles are all decuples of the digits, in succession from ten to ninety; for ten is the decuple of one, twenty the decuple of two, thirty the decuple of three, and likewise for the others in succession up to ninety. The third order was established from a hundred to a thousand. Its initial element, or limit, is a hundred, the doubling and multiplying of which generates all numbers of its order, after the manner of the first and the second orders, namely two hundred, three hundred, and so on up to nine hundred. This order is called the order of the hundreds. The fourth order was established from a thousand to nine thousand; its initial element, or limit, is a thousand, the doubling and multiplying of which by the first digits [23] generates the remaining numbers of its order. This order is called the order of the thousands. [*From that order the following orders began to be repeated, and this is the beginning of the repetition.*] The fifth order was established from ten thousand to ninety thousand.

In this way these orders of numbers increase to infinity, the subsequent being always decuples of the preceding. For as ten is the decuple of one [*twenty the decuple of two, and so on up to ninety*], likewise a hundred, which is the limit of the hundreds, is the decuple of the limit of the tens; then a thousand, which is the limit of the thousands, is the decuple of the limit of the hundreds. And likewise the fifth limit is the decuple of the fourth, since ten thousand is the decuple of a thousand [*and twenty thousand the decuple of two thousand, thirty thousand the decuple of three thousand, and so on in succession up to ninety thousand.[24] Next follows, in the sixth place [25], the limit of the hundred thousands, decuple of the fifth; for as a hundred thousand is the decuple of ten thousand, likewise two hundred thousand is the decuple of twenty thousand, three hundred thousand the decuple of thirty thousand, and so on in succession up to nine hundred thousand. Next follows, in the seventh place, the limit of the thousand times thousands, decuple of the sixth, in succession up to nine times a thousand times thousand. In the eighth place follows the limit of the articles of a thousand times thousand, such as ten times a thousand times thousand, or twenty times, thirty times, forty times a thousand times thousand, and so follow in succession the decuples of the preceding numbers up to ninety times a thousand times thousand. In the ninth place follows the limit of the hundred thousand times thousands, which are the decuples of*]

---

[22] 'doubling and multiplying', here and below: remember that multiplication by 2 was often considered as a special case.

[23] 'first digits'. We shall see that there are higher digits, namely those corresponding to the first nine multiples of a power 10^{4k}.

[24] The glossator apparently took the word 'limit' (*limes*) to designate the order itself with its nine elements (as in the *Liber algorismi*, p. 27, where *limes* is indeed synonymous with *ordo*).

[25] This refers to the numbers being written down.

LIBER MAHAMELETH

the preceding numbers up to nine hundred times a thousand times thousand. In the tenth place follows the limit of the thousand times thousand times thousands, which are the decuples of the preceding numbers up to nine times a thousand times thousand times thousand][26].

This may go on to infinity: you put as the subsequent the decuples of the preceding, in analogy to the former, and you begin after each third place always as before; whence, just as the first limit is the unit, the second ten, the third a hundred, so the next limit will be a thousand, the second ten times a thousand, the third a hundred times a thousand; likewise the first (next limit) will be a thousand times thousand, the second ten times a thousand times thousand, the third a hundred times a thousand times thousand; likewise again the first (next limit) will be a thousand times thousand times thousand, the second ten times a thousand times thousand times thousand, the third a hundred times a thousand times thousand times thousand; and so on without end, always beginning anew, after every third place, with the digits and through the articles arriving at the hundreds. [*For as the first place is that of the digits of units, the fourth will be that of the digits of thousands; whence, just as in the first place, that of the digits, you say 'one', 'two', 'three' and so on in succession up to 'nine', so also in the fourth place, that of the thousands, you will say 'one thousand', 'two thousand', 'three thousand' and so on in succession up to nine, counting 'thousand' with the first digits. And, just as the second place is that of the articles for tens, the fifth place will be that of the articles for thousands; whence, just as you say in the second place 'ten', 'twenty', 'thirty' and so on in succession up to 'ninety', so you will say in the fifth place 'ten thousand', 'twenty thousand', 'thirty thousand', and so on up to ninety, counting 'thousand' with the first articles. And, just as in the third place you say 'one hundred', 'two hundred', 'three hundred' and so on up to 'nine hundred', you will say in the sixth place 'one hundred thousand', 'two hundred thousand', 'three hundred thousand', and so on in succession, counting 'hundred thousand' with the first digits up to nine hundred thousand*[27]. *And so on without end for all the others: after every third place the following will always be the place of units, the second of tens and the third of hundreds, and (the numbers of) this first place will be counted with the digits, (those of) the second with the articles and (those of) the third with the hundreds.*] Therefore the first elements, or limits, of only four orders have their own names, namely one, ten, hundred, thousand, whereas the first elements of the others are derived from these and are arranged like them. The figure below will help make this clearer.[28]

---

[26] An interpolation probably prompted by the subsequent figure which goes on to the tenth place.

[27] Rather: *milia enumerando per primos centenos usque nongenta*, that is, counting 'thousand' with the first hundreds up to nine hundred.

[28] The columns in the figure represent the nine elements of the first ten orders. (A similar figure, with nine columns, is found in the *Liber algorismi*, p. 133.) This is thus our first encounter with the nine (significant) digits,

ten	9	8	7	6	5	4	3	2	1
1	1	1	1	1	1	1	1	1	1
2	2	2	2	2	2	2	2	2	2
3	3	3	3	3	3	3	3	3	3
4	4	4	4	4	4	4	4	4	4
5	5	5	5	5	5	5	5	5	5
6	6	6	6	6	6	6	6	6	6
7	7	7	7	7	7	7	7	7	7
8	8	8	8	8	8	8	8	8	8
9	9	9	9	9	9	9	9	9	9
of the thousand times thousand times thousands	of the hundred thousand times thousands	of the ten thousand times thousands	of the thousand times thousands	of the hundreds of thousands	of the tens of thousands	of the thousands	of the hundreds	of the tens	order of the digits

(The first ten orders)

[*Numbers are called composite when they are interposed between those above, or limits. They are always composed of a digit and an article or a limit, such as twelve, twenty-two, two hundred, two thousand, and the like.*][29]

[*To each order was attributed a note, or a mark, to distinguish it. The note of the first order was then taken as the unit; for since the first order is that of the units and the first in respect to the other orders, so it was appropriate enough to put the unit as its note (since it is) the first and the origin of the other notes. The note of the second order was taken as two, that of the third as three, and the note of each order is likewise as distant from one as the corresponding order is distant from the first. And since the note of the first order was one, so the note of any order is larger than the note of the preceding by one. For the purpose of the note is to show to which order belongs any given number; thus if the note of the order is ten, the number will be in the tenth order, if eleven then the number will be in*

---

not otherwise described. It appears thus that only nine (or ten) digits are needed to write numbers, just as twelve words were needed to name them.

[29] We find here four interpolations. This first one paraphrases a passage at the beginning of A–III while the second does the same for the section introducing the concept of 'note' in A–III (just before the beginning of the problems). The latter is prompted by the figure, so also the other two, which comment on it.

the eleventh order.]

[*Digits, tens, hundreds, thousands, ten times thousand, a hundred times thousand, a thousand times thousand, ten times a thousand times thousand, a hundred times a thousand times thousand, a thousand times a thousand times thousand, and so on to infinity.*]

[*The note of the units or digits is one; the note of the tens, two; the note of the hundreds, three; the note of the thousands, four; the note of 'ten times thousand', five; the note of 'hundred times thousand', six; the note of 'thousand times thousand', seven; the note of 'ten times thousand times thousand', eight; the note of 'hundred times thousand times thousand', nine; the note of 'thousand times thousand times thousand', ten. And conversely:[30] the name of one is one; the name of two, 'ten'; the name of three, 'hundred'; the name of four, 'thousand'; the name of five, 'ten thousand'; the name of six, 'hundred thousand'; the name of seven, 'thousand times thousand'; the name of eight, 'ten times thousand times thousand'; the name of nine, 'hundred times thousand times thousand'; the name of ten, 'thousand times thousand times thousand'; and so on to infinity. We shall give the rules for determining the note when the number is known, or the number when the note is known, farther below.[31]*]

Since we have in mind to demonstrate what we shall assert on multiplication, division and the other kinds mentioned before[32], we want to present as a preliminary some (propositions) necessary for the demonstration of what follows. The reader made thus acquainted with them will have greater ease with these demonstrations.[33] These premisses are the following.

---

[30] Before, the top of the columns was set in relation to the bottom (the names); it is now the other way round.

[31] If this last sentence originates with the author, so should the second of the four interpolations. In any event the concept of note will be dealt with in detail later, and introduced as if mentioned for the first time (pp. 609–610).

[32] See beginning of the treatise ('kinds of practical arithmetic', p. 581).

[33] All the demonstrations in both Book A and Book B are based only on propositions from Euclid's *Elements* and the subsequent premisses.

Chapter (A–II) on the Premisses necessary for practical arithmetic

(P_1) For any quantity of numbers exceeding one another by a same difference which are in even quantity, the sum of the two extremes is the same as the sum of any two intermediate terms of which one is as distant from the first as the other is from the last; the same will hold if they are in odd quantity but, besides, the sum of the first and the last will be the same as twice the middle one.[34]

For instance, let there be numbers A, BG, DH, $ÇK$, TQ, LM, in even quantity and exceeding one another by the same difference [*let it be PG*]. Then I say that the sum of the first, namely A, and the last, namely LM, is the same as the sum of any two among them of which one is as distant from the first as the other is from the last, namely BG and TQ or DH and $ÇK$. Let the excess of BG over A be PG, that of DH over BG be the quantity FH, that of $ÇK$ over DH be the quantity CK, that of TQ over $ÇK$ be the quantity ZQ, and let that of LM over TQ be the quantity RM, all these differences between them being equal. Then I say that the sum of A and LM is the same as the sum of BG and TQ. This is proved as follows. Drawing MN equal to A will indeed show that RN is equal to BG: A being equal to BP and to MN, BP is equal to MN; but PG is equal to RM, thus RN is equal to BG. It will also be shown that RL is equal to TQ, for that by which LM exceeds TQ is RM. So the whole LN is equal to the sum of [*each of*] BG and TQ. Now the whole LN is equal to A and LM[35]. Therefore A with LM are altogether equal to BG with TQ [*adding to TQ a line equal to BG, as before to LM MN equal to A*]. It will likewise be shown that it is equal to DH with $ÇK$.

Let them also be in odd quantity, that is, let them be A, BG, DH, $ÇK$, TQ. It will be shown as previously that the sum of A and TQ is the same as BG with $ÇK$. I say also that the sum of A and TQ is the same as twice DH. This is proved as follows. We draw QU equal to A, take from TQ a part equal to A, say TE, and take from DH a part equal to A, say DS. It is thus clear that there are in QE four of the excesses and in SH, two. So QE is twice SH. It is also clear that TE and QU is twice DS. So the whole TU is twice the whole DH. Now the whole TU is equal

---

[34] Same assertion in the *Liber algorismi*, pp. 96 & 97. All three manuscripts $\mathcal{A}$, $\mathcal{B}$, $\mathcal{C}$ contain a reader's numerical example. (The reader of $\mathcal{C}$ just reproduced that of the *Liber algorismi*, which precedes the *Liber mahameleth* in this MS).

[35] 'A and LM': the sum of A and LM. Usual expression.

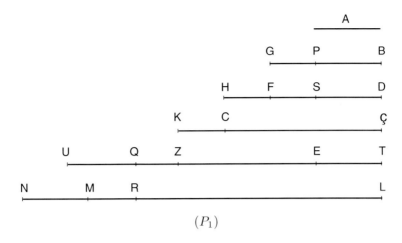

(P_1)

to the sum of A and TQ. So the sum of A and TQ is twice DH. This is what we intended to demonstrate.

$(\boldsymbol{P_2})$ For any three numbers the product of the first into the second multiplied by the third is equal to the product of the third into the second multiplied by the first.[36]

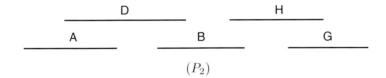

(P_2)

For instance, let there be three numbers, designated by A, B, G. Then I say that the product of A into B multiplied by G is equal to the product of G into B multiplied by A. Let the product of A into B be D and the product of G into B be H. Then I say that the product of A into H is equal to the product of G into D. This is proved as follows. The product of A into B is D [*by hypothesis*] and the product of G into B is H. Thus the products of A, G into a certain number are D and H. Therefore the ratio of the products is the same as the ratio of the multipliers, as Euclid said in the seventh book [*18th theorem*][37]. Thus the ratio of A to G is the same as the ratio of D to H. These are therefore four proportional quantities. Thus the product of A into H is equal to the product of G into D, as Euclid said [*in the 19th of the seventh*][38]. This is what we intended to demonstrate.

$(\boldsymbol{P_3})$ For any four numbers the result of multiplying the product of the first into the second by the product of the third into the fourth is equal

---

[36] Associativity and commutativity both involved.
[37] *Elements* VII.18.
[38] *Elements* VII.19.

590 PART TWO: TRANSLATION, GLOSSARY

to the result of multiplying the product of the first into the third by the product of the second into the fourth.

(a) For instance, let there be four numbers, designated respectively by A, B, G, D. Then I say that the result of multiplying the product of A into B by the product of G into D is equal to the result of multiplying the product of A into G by the product of B into D. Let the product of A into B be H, that of G into D, Z, that of A into G, K, that of B into D, T. Then I say that the product of H into Z is equal to the product of K into T. This is proved as follows. The product of A into B is H [by hypothesis] and that of B into D is T [and is conversely the same [39]]. Thus the products of A, D into B are H and T. Therefore the ratio of H to T is the same as the ratio of A to D [by the 18th of the seventh][40]. Likewise too, [it will be shown that] the products of A, D into G are K and Z.[41] Thus the ratio of K to Z is the same as the ratio of A to D [by the same][42]. But the ratio of A to D was already the same as the ratio of H to T. Thus the ratio of H to T is the same as the ratio of K to Z. These are therefore four proportional numbers [H, T, K, Z]. Then the product of H into Z is equal to the product of K into T [by the nineteenth of the seventh][43]. It is therefore clear that for any four numbers the result of multiplying the product of the first into the second by the product of the third into the fourth is equal to the result of multiplying the product of the first into the third by the product of the second into the fourth. This is what we intended to demonstrate.[44]

$(P_3\,a\text{--}b)$

(b) This proposition will also be demonstrated using a statement by Abū Kāmil in the third part of his book on algebra, which is the following:

---

[39] Meaning that the product of D into B is T (which makes B, here too, the second factor in the product).

[40] Elements VII.18.

[41] By hypothesis. So this interpolation is meaningless.

[42] Elements VII.18.

[43] Elements VII.19. This gloss, found only in MS $\mathcal{C}$, differs from the others in expression.

[44] The same figure will serve for the subsequent demonstration (whence the inclusion of Q).

(P_3') If any two numbers are individually divided by some number and the quotients are multiplied, then the product is equal to the result of dividing the product of the two dividends by the product of the two divisors.

For instance[45], multiplying A by B produces H [*by hypothesis*]; then dividing H by B will give A [*by this rule: When a number is multiplied by another, if the product is divided by one of them the result will be the other. Thus if four is multiplied by three, or inversely, the result is twelve; then by whichever of them twelve is divided the result will be the other. The same is true for all.*][46]. Likewise, multiplying G by D produces Z; then dividing Z by D will give G. Therefore we shall have, as the result of dividing H by B, A and, as the result of dividing Z by D, G. Hence dividing the product of H into Z by the product of B into D will give the product of A into G [*as shown above*][47]. Let then the product of H into Z be Q. We know also[48] that multiplying B by D produces T and that multiplying A by G produces K. Thus dividing Q by T will give K. Then the product of K into T will be Q. But the product of H into Z is also Q. Therefore the product of H into Z is equal to the product of K into T. This is what we intended to demonstrate.

(c) Having thus again demonstrated the proposition in this way [*about the problem proposed by Abū Kāmil, which is necessary to us*][49], I shall present a proof of Abū Kāmil's assertion, but much simpler than the proof he has himself presented.

A	K	D
B	T	H
G	Q	Z

$$(P_3\,c)$$

For instance, let the division of A by B give G, the division of D by H, Z, and let the multiplication of A by D produce K, the multiplication of B by H, T, the multiplication of G by Z, Q. Then I say that the division of K by T will give Q. This is proved as follows. Multiplying G by B produces A and multiplying G by Z, Q. Thus[50] the ratio of A to Q is the same as the ratio of B to Z. Likewise too, multiplying B by H produces T and multiplying Z by H, D. Thus the ratio of B to Z is

---

[45] Rather: As we know from *a*.

[46] In MS $\mathcal{C}$ this has been left in the margin, as with many interpolations.

[47] Rather: as stated above (P_3').

[48] By hypothesis in *a*.

[49] Former marginal gloss indicating what this section is about. See p. 598, note 78.

[50] Note, in this demonstration, the absence of interpolated references to (respectively) *Elements* VII. 17, 18 and 19.

592 PART TWO: TRANSLATION, GLOSSARY

the same as the ratio of T to D. But the ratio of B to Z was already the same as the ratio of A to Q. So the ratio of A to Q is the same as the ratio of T to D. Therefore the product of A into D is equal to the product of Q into T. Since multiplying A by D produces K, multiplying Q by T produces K. Thus dividing K by T will give Q. This is what we intended to demonstrate.

(P_4) When a number is divided by another and the quotient is divided by a third number, then the result is equal to the result of dividing the first by the product of the second into the third.

A

B

G

D

H

(P_4)

For instance, let the division of A by B give G and the division of G by D, H. Then I say that dividing A by the product of B into D will give H. This is proved as follows. The division of G by D gives H [*by hypothesis*]; then the multiplication of H by D will produce G [*by this rule: For any two numbers when one is divided by the other and the result is multiplied by the divisor the product will be the dividend*][51]. The division of A by B gives G; then, the multiplication of G by B will produce A. Thus multiplying H by D and the product by B will produce A. But we have already established that for any three numbers the result of multiplying the first by the second and the product by the third is equal to the result of multiplying the third by the second and the product by the first [*by the foregoing*][52]. Thus the result of multiplying H by D and the product by B is equal to the result of multiplying B by D and the product by H. Since multiplying H by D and the product by B gives A, multiplying B by D and the product by H gives A. Therefore dividing A by the product of B into D will give H. This is what we wanted to demonstrate.

(P_5) When a number is divided by another and the quotient is multiplied by a third number, then the result is equal to the result of dividing the

---

[51] Similar gloss in P_3', so the glossator must be the same. This rule will be stated, this time by the author, at the beginning of the chapter on division of integers (p. 639). It is stated and demonstrated in Abū Kāmil's *Algebra*, fol. 19^r–19^v of the Arabic text, lines 915–930 of the printed Latin translation.

[52] Above, P_2.

product of the dividend into the multiplier by the divisor.[53] [*You are to understand that the product from the (first) multiplication, of the quotient from the first division, is equal to the quotient from the second division, of the product from the second multiplication.*][54]

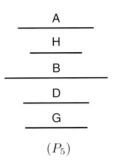

(P_5)

For instance, let the division of A by B give G and the multiplication of G by D produce H. Then I say that multiplying A by D and dividing the product by B will give H. The proof of this is the following. The division of A by B gives G; then the multiplication of G by B will produce A. But the multiplication of G by D produces H. So the multiplication of G by B and by D produces A and H. Thus the ratio of B to D is the same as the ratio of A to H [*by the 17th of the seventh Book*][55]. These are therefore four proportional numbers. Thus the product of A into D is equal to the product of B into H [*by the 18th of the same*][56]. So multiplying A by D and dividing the product by B will give H. This is what we wanted to demonstrate.

(P_6) When there are six numbers such that as the first is to the second so the third is to the fourth and as the fifth is to the second so the sixth is to the fourth, then the excess of the first over the fifth [*or the excess of the*

---

[53] This reading is found in MSS $\mathcal{AB}$ and in the margin of $\mathcal{C}$, which has in the text: 'When a number is divided (...) is equal to the product of the first into the third the result being divided by the second.' Although $\mathcal{C}$ has a generally very reliable text, we have adopted the reading of $\mathcal{A}$ and $\mathcal{B}$, in keeping with a quotation of P_5 found later in the text (B.257b, p. 958 below).

[54] The meaning of this gloss will become apparent when the proposition is written down as $\frac{a}{b} \cdot c = \frac{a \cdot c}{b}$: there is one multiplication and one division on each side, those of the left side being referred to as the 'first' and those of the right as the 'second'.

[55] *Elements* VII.17.

[56] Actually *Elements* VII.19. This gloss might however have been intended to modify the previous one (the only difference between VII.17 and VII.18 being that in one case the common factor is the multiplicand and in the other the multiplier). In PE_1 VII.18 is cited instead of VII.17.

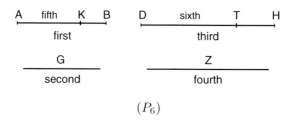

(P_6)

fifth over it] will be to the second as the excess of the third over the sixth [*or the excess of the sixth over it*] is to the fourth.

For instance, let there be six numbers AB, G, DH, Z, AK, DT and let the ratio of the first, which is AB, to the second, which is G, be the same as the ratio of the third, which is DH, to the fourth, which is Z, and let the ratio of the fifth, which is AK, to G, which is the second, be the same as the ratio of the sixth, which is DT, to the fourth, which is Z. Then I say that the excess of the first over the fifth, and let it be KB[57], will be to the second, which is G, as the excess of the third over the sixth, and let it be TH[58], is to the fourth, which is Z. The proof of this is the following. The ratio of AK to G is the same as the ratio of DT to Z; by inversion, the ratio of G to AK will be the same as the ratio of Z to DT [*by the 16th of the fifth*][59]. Therefore we have that the ratio of AB to G is the same as the ratio of DH to Z and the ratio of G to AK is the same as the ratio of Z to DT. Thus, by proportion of equality[60], the ratio of AB to AK will be the same as the ratio of DH to DT. By separation [*and conversion*], the ratio of BK to AK will be the same as the ratio of HT to TD. Since[61] the ratio of AK to G is the same as the ratio of DT to Z, then, by proportion of equality, the ratio of BK to G will be the same as the ratio of TH to Z. This is what we wanted to demonstrate.

Using the foregoing[62] we shall also demonstrate (the following proposition):

(P_7) When any two different[63] numbers are divided by some number, then the excess of the division of the greater over the division of the lesser is equal to the result of dividing the excess of the greater number over the other by the divisor.

For instance, let there be two different numbers AB and AK. [*And let the excess of the first over the second be KB.*] Let AB be divided by

---

[57] *Sic*, instead of 'which is KB'.
[58] *Sic*, instead of 'which is TH'.
[59] What *Elements* V.16 actually asserts is that if $a : b = c : d$, then $a : c = b : d$ ('alternation').
[60] *secundum proportionem equalitatis* (usually rendered today by 'ex aequali' = δι' ἴσου), see *Elements* V.22 (and def. 17) & VII.14.
[61] By hypothesis, not by what just precedes as the interpolator thought.
[62] The coming proof is based on P_6.
[63] 𝒜ℬ (and 𝒞 in the margin) have 'unequal'. But 'different' occurs below and in a later quotation (see p. 718).

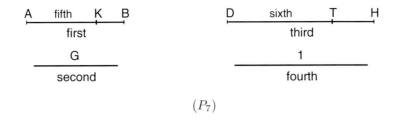

(P_7)

G to give DH, and AK be divided by G to give DT. The excess of AB over AK is KB and the excess of DH over DT is TH. Then I say that dividing KB, which is the difference between the two dividends, by G will give TH, which is the difference between the two quotients. This is proved as follows. Since AB divided by G gives DH, multiplying DH by G will produce AB. Thus G measures[64] AB as many times as there are units in DH. But the unit divides DH as many times as the unit is in DH. Therefore the ratio of one to DH is the same as the ratio of G to AB. By inversion, the ratio of AB to G will be the same as the ratio of DH to one. It will likewise be shown that the ratio of AK to G is the same as the ratio of DT to one. Therefore the ratio of AB, which is the first, to G, which is the second, is the same as the ratio of DH, which is the third, to one, which is the fourth, and the ratio of AK, which is the fifth, to G, which is the second, is the same as the ratio of DT, which is the sixth, to one, which is the fourth.[65] Then the ratio of KB, which[66] is the excess of the first over the fifth, to G, which is the second, is the same as the ratio of HT, which is the excess of the third over the sixth, to one, which is the fourth. So the ratio of KB to G is the same as the ratio of TH to one. By inversion, the ratio of one to TH will be the same as the ratio of G to KB.[67] Thus one divides TH as many times as G divides KB. But one divides TH as many times as one is in TH. So G divides KB as many times as one is in TH. Therefore multiplying G by TH will produce KB. Thus dividing KB by G gives TH. This is what we wanted to demonstrate.

This proposition, too, involves the operation of subtraction.[68] I shall now present another proposition, similar to it, which [likewise] involves the operation of addition, in which [also] appears a statement made by Euclid

---

[64] *Numerare* means 'being an exact divisor', like Euclid's μετρεῖν (whence our translation). What follows seems verbose (same below, and in P_8 & A.260a); but it is perfectly Euclidean (see *Elements* VII.16–17 or X.5–6) and is found in Abū Kāmil's *Algebra* as well (fol. 19^v–20^r, 27^r–27^v, 64^r–64^v).

[65] Conditions for applying P_6.

[66] *que* instead of (as before) *qui*. Such variations are common, sometimes used to clarify the text (differenciating quantities).

[67] Again verbosity, this time to pass from $KB : G = TH : 1$ (ratio) to $KB : G = TH$ (quotient).

[68] Both P_6 and P_7 involved excesses, thus subtraction.

in the fifth book, namely the following:[69]

(P'_7) When the ratio of the first to the second is the same as the ratio of the third to the fourth and the ratio of the fifth to the second is the same as the ratio of the sixth to the fourth, then the ratio of the sum of the first and the fifth to the second will be the same as the ratio of the sum of the third and the sixth to the fourth.

We shall omit the proof of this statement, since Euclid has given it, and turn to our purpose of proving the proposition involving the operation of addition and in which the above statement appears.

(P_8) When any two numbers are divided by some number, then the sum of the quotients is equal to the result of dividing the sum of the two numbers by the divisor.

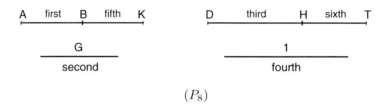

(P_8)

For instance, let AB be divided by G to give DH, and BK be divided by G to give HT. Then I say that dividing the whole AK by G gives the whole DT. The proof of this is the following. Dividing AB by G gives DH. Thus multiplying DH by G will produce AB. So G measures AB as many times as one is in DH. But one measures DH as many times as one is in DH. Thus the ratio of one to DH is the same as the ratio of G to AB. By inversion, the ratio of AB to G will be the same as the ratio of DH to one.[70] It will likewise be shown that the ratio of BK to G is the same as the ratio of HT to one. Therefore the ratio of AB, which is the first, to G, which is the second, is the same as the ratio of DH, which is the third, to one, which is the fourth, and the ratio of BK, which is the fifth, to G, which is the second, is the same as the ratio of HT, which is the sixth, to one, which is the fourth.[71] Then the ratio of the whole AK, which is the first and the fifth, to G, which is the second, is the same as the ratio of the whole DT, which is the third and the sixth, to one, which is the fourth. By inversion, the ratio of one to TD will be the same as the ratio of G to AK. Since this is so, multiplying DT by G will produce AK, as we have established in the previous proposition [*analogous to this one*][72]. Thus dividing AK by G will give DT. This is what we wanted to demonstrate.

---

[69] P'_7, the analogue of P_6 for the addition, is *Elements* V.24. It will serve to prove P_8, analogous to P_7 for the addition.

[70] Thus $AB : G = DH$ has become $AB : G = DH : 1$.

[71] Conditions for applying P'_7.

[72] End of P_7. But the whole reference may be an addition.

(P_9) When a number is divided by another and yet another by the quotient, then the result of this last division is equal to the result of dividing the product of the second dividend into the first divisor by the first dividend.

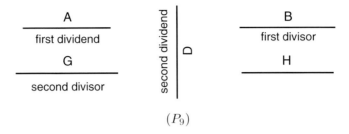

(P_9)

For instance, let A be divided by B to give G and D be divided by G to give H. Then I say that multiplying D, which is the second dividend, by B, which is the first divisor, and dividing the product by A, which is the first dividend, will give H, which is the result of dividing D by G. The proof of this is the following. A divided by B gives G; thus multiplying G by B will produce A. [*Now the product of G into B is equal to the product of B into G.*][73] Again, D divided by G gives H; thus multiplying H by G will produce D. So multiplying G by B and by H produces A and D. Therefore[74] the ratio of H to B is the same as the ratio of D to A. Thus the product of H into A is equal to the product of D into B [*as Euclid has established in the seventh book [19th theorem]*[75], *saying: For any four proportional numbers the multiplication of the first by the fourth produces the same as the second by the third*][76]. Since the product of D into B is equal to the product of H into A, then, obviously, multiplying D by B and dividing the product by A will give H. This is what we wanted to demonstrate.

[*Here end the premisses*][77]

We have thus completed, with the help of God, the necessary premisses, not in fact taken from Euclid's treatise but useful for those wishing to acquire the science of mahameleth with demonstrations. In order to prove them, however, much is used from Euclid's treatise [*what is neces-*

---

[73] Perhaps by the same glossator who applied commutativity for the same purpose in $P_3 a$.

[74] *Elements* VII.17.

[75] *Elements* VII.19.

[76] Since this theorem has been used several times before, there is no need to quote it.

[77] There now follows an adaptation of *Elements* II.1–10 to numbers. That this is a continuation of the same chapter becomes apparent from what the author says, not only when concluding this second section but also in A.200, where it is explicitly stated that P_7 belongs to the *first* section on premisses.

sary to us[78]], for on that basis are discovered the proofs of this science. We deemed it appropriate to add next what Euclid stated in the second book, in order to explain with respect to numbers what he himself explained with respect to lines. It will be necessary for their proof to use certain propositions from the seventh book, for Euclid only spoke about numbers in the seventh book and the two following ones.[79] [*For this reason Euclid should first be read and known thoroughly before embarking upon the present treatise on mahameleth.*][80]

(**PE_1**) If there are two numbers and one of them is divided into any quantity of parts, then the product of the two numbers is equal to the sum of the products of the undivided number into each part of the divided number.

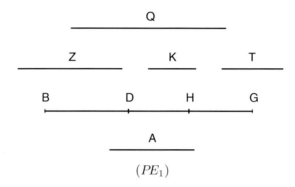

(PE_1)

For instance, let there be two numbers, A and BG, and let one of them, say BG, be divided into parts, say BD, DH, HG. Then I say that the product of A into BG is equal to the sum of the products of A into BD, DH, HG. The proof of this is the following. Let A be multiplied by BG to produce Q, and let the multiplication of A by BD produce Z, the multiplication of A by DH, K, and the multiplication of A by HG, T. Since multiplying A by BG produces Q and multiplying the same by BD produces Z, the ratio of Z to Q is the same as the ratio of BD to BG [by *the 18th of the seventh*][81]. Since multiplying A by DH produces K and multiplying the same by BG produces Q, the ratio of DH to BG is the

---

[78] Former marginal gloss indicating the subject of this introductory section. See p. 591, note 49.

[79] Indeed, Books VII–IX (often called today the 'arithmetical Books') are specifically devoted to the arithmetical properties of (natural) numbers. But Euclid's proofs in Book II concerned segments of straight line and did not use propositions from the arithmetical Books. Without this remark by the author, the reader might have wondered why propositions from Book II depended for their proofs on theorems from a later Book. Although, as a matter of fact, only VII.17 and V.24 = P'_7 are used, both in PE_1.

[80] The necessity of knowing the *Elements* (in particular Book VII) was already apparent in the first set of premisses.

[81] *Elements* VII.18 (or, rather, VII.17).

same as the ratio of K to Q. Therefore the ratio of DB to BG is the same as the ratio of Z to Q, while the ratio of DH to BG is the same as the ratio of K to Q. Thus the ratio of BH to BG is the same as the ratio of Z and K to Q, as Euclid said in the fifth.[82] Likewise too, since multiplying A by HG produces T and multiplying A by BG produces Q, the ratio of HG to BG is the same as the ratio of T to Q. Therefore we have that the ratio of BH to BG is the same as the ratio of Z and K to Q, while the ratio of HG to BG is the same as the ratio of T to Q. So[83] the ratio of the sum of BD, DH, HG to BG is the same as the ratio of the sum of Z, K, T to Q. But BD, DH, HG are (together) equal to BG; thus Z, K, T are (together) equal to Q. This is what we wanted to demonstrate.

($\boldsymbol{PE_2}$) When a number is divided into any quantity of parts, multiplying it by itself is the same as multiplying it by all its parts.

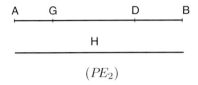

(PE_2)

For instance, let the number AB be divided into parts, say AG, GD, DB. Then I say that the product of AB into itself is the same as the products of AB into AG, GD, DB. The proof of this is the following. Let us assume another number equal to AB, say H. [*Now it has been established that the product of AB into itself is equal to the product of it into H.*][84] Now the product of AB into H is equal to the products of AG into H, GD into H, and DB into H, as we have established just before,[85] while H is equal to AB. Then the product of AB into itself is equal to the products of AB into AG, GD, DB. This is what we wanted to demonstrate.

($\boldsymbol{PE_3}$) When a number is divided into two parts, the product of the whole into one of its parts [*that in which the whole is multiplied*][86] is equal to the products of this part into itself and the two parts into one another.

(PE_3)

---

[82] *Elements* V.24 = P'_7 above.
[83] Again *Elements* V.24 = P'_7.
[84] Gloss to the next sentence.
[85] Above, PE_1.
[86] Refers to the subsequent 'this part'.

For instance, let the number AB be divided into two parts, say AG and GB. Then I say that the product of AB into GB is equal to the sum of the products of GB into itself and into AG. The proof of this is the following. Let H be assumed equal to GB. [And it has been shown that the product of AB into BG is equal to the product of H into AB.][87] Now [88] the product of H into AB is equal to the products of H into AG and into GB, and the product of H into GB is equal to the product of GB into itself. Therefore the product of AB into BG is equal to the products of AG into GB and GB into itself. This is what we wanted to demonstrate.

(**PE$_4$**) When a number is divided into two parts, then the product of the whole into itself is equal to the products of each of the parts into itself and twice their product.

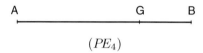

(PE_4)

For instance, let the number AB be divided into two parts, say AG and GB. Then I say that the product of AB into itself is equal to the sum of the products of AG into itself, GB into itself, and AG into GB, twice. The proof of this is the following.[89] The product of AB into itself is equal to the product of AB into AG and the product of AB into GB, taken together. Now the product of AB into AG is equal to the product of AG into itself and the product of AG into GB, while the product of AB into BG is equal to the products of BG into itself and BG into AG. [Now the product of BG into AG is equal to the product of AG into GB.][90] Therefore the product of AB into itself is equal to the products of AG into itself, GB into itself, and AG into GB, twice. This is what we wanted to demonstrate.

(**PE$_5$**) When a number is divided into two equal parts and two unequal parts, then the product of one equal part into itself is equal to the product of the unequal parts into one another and the product of the excess of an equal part over the smaller unequal part into itself.

(PE_5)

---

[87] As in PE_2, this is misplaced: it should have been inserted after the next sentence.

[88] PE_1.

[89] Relies on PE_1 and PE_2.

[90] Again (see notes 39 & 73) putting the factors in order since we have once $AG \cdot GB$ and once $GB \cdot AG$.

For instance, let the number AB be divided into two equal parts, say at the point G, and into two unequal parts, say at the point D. Then I say that the product of GB into itself is equal to the product of AD into DB and the product of GD into itself. The proof of this is the following. The product of GB into itself is equal to the products of GB into GD and GB into DB [*by the second of this*][91]. Now the product of GB into DB is equal to the product of AG into DB. Thus the product of GB into itself is equal to the products of GD into GB and AG into DB. But [92] the product of GB into GD is equal to the products of GD into itself and GD into DB. Then the product of GB into itself is equal to the products of AG into DB, GD into DB, and GD into itself. Now the products of AG into DB and GD into DB are equal to the product of AD into DB. Therefore the product of GB into itself is equal to the products of AD into DB and GD into itself. This is what we wanted to demonstrate.

(PE_6) When a number is divided into two equal parts and another number is added to it, then the product of the half with the added number into itself is equal to the products of the initial number with the number added into the added number and the half into itself.

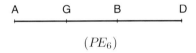

(PE_6)

For instance, let the number AB be bisected at the point G, and let next the number BD be added to it. Then I say that the product of GD into itself is equal to the products of AD into DB and GB into itself. The proof of this is the following.[93] The product of GD into itself is equal to the products of GD into GB and into DB. Now the product of GD into GB is equal to the products of GB into itself and into BD. Thus the product of GD into itself is equal to the product of GD into DB, the product of GB into DB, and the product of GB into itself. Since GB is equal to AG, the product of GD into itself is equal to the products of AG into DB, GD into DB, and GB into itself. Now the products of AG into DB and GD into DB are equal to the product of AD into DB. Therefore the product of GD into itself is equal to the product of AD into DB and the product of GB [*which is half the number*] into itself. This is what we wanted to demonstrate.

(PE_7) When a number is divided into two parts, then the product of the whole number into itself and the product of one of the parts into itself are equal to twice the product of the whole number into the same part and the product of the other part into itself.

---

[91] PE_2.
[92] PE_1.
[93] Relies on PE_1 and PE_2.

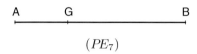

(PE_7)

For instance, let the number AB be divided into two parts at the point G. Then I say that the products of AB into itself and GB into itself are equal to the products of AB into GB, twice, and AG into itself. The proof of this is the following. The product of AB into itself is equal to the product of AG into itself, the product of GB into itself, and twice the product of AG into GB.[94] Let the product of GB into itself be (added in) common. Then the products of AB into itself and GB into itself will be equal to the products of AG into itself, GB into itself, twice, and AG into GB, twice. But the products of GB into itself, twice, and AG into GB, twice, are equal to the product of AB into BG, twice [*for the products of GB into itself, once, and AG into GB, once, are equal to the product of AB into GB, once*][95]. So the products of AB into itself and BG into itself are equal to the products of AB into BG, twice, and AG into itself. This is what we wanted to demonstrate.

(**PE_8**) When a number is divided into two parts and a number equal to one of the parts is added to it, then the product of the whole number, (thus) with the added number, into itself will be equal to the product of the initial number into the added number, four times, and the product of the other part into itself.

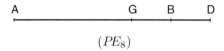

(PE_8)

For instance, let the number AB be divided into two parts at the point G, and let another number equal to GB, say BD, be added to it. Then I say that the product of AD into itself is equal to the products of AB into BD, four times, and AG into itself. The proof of this is the following. The product of AD into itself is equal to the products of AB into itself, BD into itself, and AB into BD, twice.[96] Now the products of AB into itself and BD into itself are equal to the products of AB into BD, twice, and AG into itself.[97] Therefore the product of AD into itself is equal to the products of AB into BD, four times, and AG into itself. This is what we wanted to demonstrate.

(**PE_9**) When a number is divided into two equal and two unequal parts, then the products of each of the two unequal parts into itself are equal to

---

[94] PE_4.

[95] Here interpolating in the premises, of which there is already less, appears to come to an end.

[96] PE_4.

[97] By PE_7 and considering that $BD = GB$.

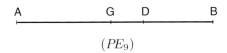

(PE_9)

twice the product of one equal part into itself and twice the product of the excess of one equal part over the smaller unequal part into itself.

For instance, let the number AB be divided into two equal parts at the point G and into two unequal parts at the point D. Then I say that the products of AD into itself and DB into itself are equal to the products of AG into itself, twice, and GD into itself, twice. The proof of this is the following. The square[98] of AB is equal to twice the square of AG and twice the square of GB.[99] Now twice the square of GB is equal to twice the square of GD, twice the square of DB, and four times the product of GD into DB; for the square of GB is equal to the square of GD, the square of DB, and twice the product of GD into DB.[100] Thus the square of AB is equal to twice the square of AG, twice the square of GD, twice the square of DB, and four times the product of GD into DB. Since[101] the square of AB is equal to the square of AD, the square of DB, and twice the product of AD into DB, so the square of AD, the square of DB, and twice the product of AD into DB are equal to the products of AG into itself, twice, GD into itself, twice, DB into itself, twice, and four times the product of GD into DB. After taking from four times the product of GD into DB twice this product and adding it to twice the square of DB, the result will be equal to twice the product of GB into BD. But twice the product of GB into BD is equal to twice the product of AG into BD. Thus the products of AD into itself, DB into itself, and AD into DB, twice, become equal to the products of AG into itself, twice, GD into itself, twice, the product of AG into DB, twice, and the product of GD into DB, twice. Now the products of AG into DB, twice, and GD into DB, twice, are equal to the product of AD into DB, twice. Removing the common (term), which is twice the product of AD into DB, will leave the products of AD into itself and DB into itself equal to the products of AG into itself, twice, and GD into itself, twice. This is what we wanted to demonstrate.

(**PE_{10}**) When a number is divided into two equal parts and another number is added to it, then the product of the whole number, (thus) with the added number, into itself and the product of the added number into itself are together twice the product of the half into itself and the product of the half, with the added number, into itself.

For instance, let the number AB be bisected at the point G, and let another number, say BD, be added to it. Then I say that the products of AD into itself and BD into itself are twice the sum of the products of

---

[98] First (provisional) use of this term.
[99] PE_4, particular case.
[100] PE_4; the latter explanation may be an addition.
[101] PE_4 again.

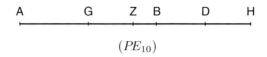

(PE_{10})

AG into itself and GD into itself. The proof of this is the following. Let DH be equal to BD and GZ be equal to BD. So AG is half of AB and GZ is half of BH. Thus the whole AZ is half of the whole AH. Therefore the number AH is divided into two equal parts at the point Z and into two unequal parts at the point D. Thus[102] the squares of AD and DH are twice the squares of AZ and ZD. But AZ is equal to GD; for AG is equal to GB and BD equal to GZ, thus AZ is equal to GD. And AG is also equal to ZD; for GZ is equal to BD, and let ZB be (added in) common, so GB will be equal to ZD; but since GB is equal to AG, so AG is equal to ZD. And DH is also equal to BD. Therefore the products of AD into itself and BD into itself are twice the products of AG into itself and GD into itself. This is what we wanted to demonstrate.

Having presented the premisses deemed necessary for the demonstration of what follows and explained with respect to numbers what Euclid explained with respect to lines, we shall return to our object, which is to deal with the aforementioned kinds of the science of practical arithmetic;[103] and we shall prove the subjects expounded in this treatise on mahameleth by means of incontrovertible demonstrations.[104]

---

[102] PE_9.

[103] These are the 'kinds' of practical arithmetic referred to in the introduction to the treatise (pp. 581–582) and mentioned again at the end of A–I (p. 587).

[104] As implied at the end of A–I, knowledge of the *Elements* and the premisses will suffice for all the proofs.

⟨Chapter (A–III) on Multiplication⟩

Now, among all these kinds of the science of (practical) arithmetic, those dealing with the collection of numbers come first. For nothing may be separated unless it has been collected.[105] That is why the kinds involving the collection of numbers must precede those involving separation. Now, of the former, addition should come first: for in any duplication and multiplication[106] there is an addition, but not the opposite. Therefore addition should be dealt with first. However, since the Arabs mostly begin with the multiplication of numbers, we too, following their example, shall start with it.[107]

You should know that a number[108] is sometimes considered by itself, without reference to another, in which case it is called an integer; sometimes in relation to another, that is, as a part or parts[109], in which case it is called a fraction; sometimes as a part or parts of an aliquot part, in which case it is called a fraction of a fraction. An integer may be a digit, or a limit, or an article, or a composite. A limit may belong to the digits, as one does[110], to the tens or primary articles[111], as ten does, to the hundreds, as a hundred does, to the thousands, as a thousand does; in this way the limits grow indefinitely, each always the decuple of the preceding. A composite number may be composed of a digit and an article, as twenty-two is, or of a digit and a limit, as one hundred and eight is, or of an article and a limit, as one hundred and thirty is. A limit may be simple, as are the first four, namely one, ten, a hundred, a thousand, or composite, as is ten thousand. [*A composite limit may be composed of the second and the fourth limit, as ten thousand is, of the third and the fourth, as a hundred thousand is, of the fourth doubled[112] or repeated more times, as a thousand times thousand or a thousand times thousand times thousand is, and so on successively to infinity.*] A fraction may be, as already said, a fraction of an integer or a fraction of a fraction.

---

[105] We cannot decompose what is not already composite.

[106] Doubling is often distinguished from multiplying (see notes 7 & 22).

[107] *maiores Arabum a multiplicatione numerorum incipiunt*, the text says. But addition of integers is not treated in the extant text of the *Liber mahameleth* (nor is subtraction).

[108] As in most cases, natural numbers are meant.

[109] Thus aliquot fraction (unit fraction) or non aliquot fraction.

[110] It is in fact the limit of the digits.

[111] The other articles are of the form $a \cdot 10^{4k}$, with $k \geq 1$ and a a digit.

[112] The *name* is 'doubled'.

606   Part Two: Translation, Glossary

These three (categories), namely integer, fraction, fraction of a fraction, multiplied among themselves necessarily involve twenty-eight cases [113] of multiplication. For we may have: a single by itself [114], which gives three cases [115]; a single by another single, which gives three more cases; a single by a pair of singles, which gives nine cases; a single by a triad of singles, which gives three cases; a pair by a pair, which gives six cases; a pair by a triad, which gives three cases; a triad by a triad, which gives one (case). All these cases add up to twenty-eight.

A single is multiplied by the same single when an integer is multiplied by an integer, or a fraction by a fraction, or a fraction of a fraction by a fraction of a fraction. A single is multiplied by another single when an integer is multiplied by a fraction or by a fraction of a fraction, or a fraction by a fraction of a fraction. A single is multiplied by a pair of singles when an integer is multiplied by an integer with a fraction, or by a fraction with a fraction of a fraction [116], or by an integer with a fraction of a fraction; or a fraction by an integer and a fraction, or by an integer and a fraction of a fraction, or by a fraction and a fraction of a fraction; or a fraction of a fraction by an integer and a fraction, or by an integer and a fraction of a fraction, or by a fraction and a fraction of a fraction. A single is multiplied by a triad of singles when an integer is multiplied by an integer with a fraction and a fraction of a fraction, or a fraction by an integer with a fraction and a fraction of a fraction, or a fraction of a fraction by an integer with a fraction and a fraction of a fraction. A pair is multiplied by a pair when an integer with a fraction is multiplied by an integer with a fraction, or by a fraction with a fraction of a fraction, or by an integer with a fraction of a fraction; or a fraction with a fraction of a fraction by a fraction with a fraction of a fraction, or by an integer with a fraction of a fraction; or an integer and a fraction of a fraction is multiplied by an integer and a fraction of a fraction. A pair is multiplied by a triad when an integer with a fraction is multiplied by an integer with a fraction and a fraction of a fraction, or when a fraction with a fraction of a fraction is multiplied by an integer with a fraction and a fraction of a fraction, or when an integer with a fraction of a fraction is multiplied by an integer with a fraction and a fraction of a fraction. A triad is multiplied by a triad once, when an integer with a fraction and a fraction of a fraction is multiplied by an integer with a fraction and a fraction of a fraction. You can also find just as many cases of division, addition and subtraction, which will be treated below.[117]

---

[113] *species*.

[114] That is, by a term of the same category. This will be developed below.

[115] *modi*.

[116] By 'a fraction with (or: and) a fraction of a fraction' is meant a compound fraction, of the form $\frac{p}{q} + \frac{k}{l}\frac{1}{q} = \frac{1}{q}\left(p + \frac{k}{l}\right)$ with p, q, k, l natural.

[117] The present chapter A–III treats the multiplication of integers and A–IV their division. Chapters A–V, VI, VII, VIII deal with the multiplication, addition, subtraction and division of fractional expressions; they will not,

LIBER MAHAMELETH

Since an integer may be a digit, or an article, or a limit, or a composite, the multiplication of an integer by an integer will necessarily involve multiplication of a single of such integers by the same single or by other single integers [*or by pairs, or by triads, or by tetrads; or pairs by pairs, by triads, or by tetrads; or triads by triads or by tetrads; or tetrads by tetrads. For we may have multiplication of a digit by a digit, of an article by an article, of a limit by a limit, of a composite by a composite; or of a digit by an article and a limit, or of an article by a limit and a composite, or of a composite, by a composite; or of an article and a limit by an article and a limit, or of a limit and a composite by a limit and an article or a composite.*] All these cases will be fully treated below.[118]

When a digit is multiplied by a digit, this may produce a digit only, or a ten only, or a digit with one or more tens, or a number of tens. Anyone who does not know their multiplication by heart and readily will never be able to fully master the science of multiplying numbers. Therefore, by way of an introduction, we shall start by setting out the multiplication of every digit by itself or by another; whoever wishes to proceed further must first strive to commit these multiplications to memory.[119]

MULTIPLICATION OF DIGITS BY THEMSELVES AND BY OTHER DIGITS

— First for one. One multiplied by one produces simply one; one multiplied by two produces simply two; one multiplied by three, simply three. And so for all; for any number which is multiplied by one [*or one into it*][120] is neither doubled nor increased.

— For two. Two multiplied by two is four; two by three, six; two by four, eight; two by five, ten; two by six, twelve; two by seven, fourteen; two by eight, sixteen; two by nine, eighteen; two by ten, twenty.

---

though, exemplify each of these cases. An allusion to the various cases concerning addition is found after problem A.154 (p. 701).

[118] Here an early reader, inspired by the enumeration of the 28 cases, has attempted to do the same with the four types of integers just seen. Had he paid attention to what was said at the beginning of this same chapter, he would have realized that the combinations he lists are merely examples of composite numbers. (A similar enumeration in the *Liber algorismi*, p. 118, is equally unsatisfactory.) What matters here is whether both multiplying factors have just the first figure significant, as do non-composite integers (digits, articles, limits), or not, like composite integers; for in the first case the operation reduces to a multiplication of digits. Accordingly, after the multiplication table for the single digits, we shall first be taught the multiplication of non-composite integers (A.9–21, A.30–31, A.33–37) and then that of composites (A.22–29, A.32, A.38–41, A.44–45). A.1–8, A.42–43 are complementary sections.

[119] Same advice in the *Liber algorismi*, p. 38.

[120] Evidence of this (?) early reader's concern with commutativity has already been seen (notes 39, 73, 90).

608   PART TWO: TRANSLATION, GLOSSARY

— For three. Three multiplied by three is nine; three by four, twelve; three by five, fifteen; three by six, eighteen; three by seven, twenty-one; three by eight, twenty-four; three by nine, twenty-seven; three by ten, thirty.

— For four. Four multiplied by four is sixteen; by five, twenty; by six, twenty-four; by seven, twenty-eight; four by eight, thirty-two; four by nine, thirty-six; and by ten, forty.

— For five. Five multiplied by five is twenty-five; by six, thirty; by seven, thirty-five; by eight, forty; by nine, forty-five; by ten, fifty.

— For six. Six multiplied by six is thirty-six; by seven, forty-two; by eight, forty-eight; six by nine, fifty-four; six by ten, sixty.

— For seven. Seven multiplied by seven is forty-nine; by eight, fifty-six; by nine, sixty-three; by ten, seventy.

— For eight. Eight multiplied by eight is sixty-four; by nine, seventy-two; by ten, eighty.

— For nine. Nine multiplied by nine is eighty-one; by ten, ninety.

— For ten. Ten multiplied by ten produces a hundred.

RULES FOR DETERMINING THE PRODUCTS OF DIGITS INTO THEMSELVES OR INTO OTHER DIGITS.

FIRST, INTO THEMSELVES.

(*i*) The multiplication of any digit by itself produces the same amount as the multiplication of its immediate neighbours, plus one.[121]

(*i'*) The multiplication of any digit by itself produces the same amount as the (multiplication of its) two (immediate) neighbours, the neighbours of the neighbours, (and so on) till unity, plus the multiplication of the differences between the middle term and the extremes. [*For multiplying five by itself produces as much as multiplying four by six and adding the product of the differences, which are both one; or as much as multiplying three by seven and adding the product of the differences, which are both two; or as much as multiplying two by eight and adding the product of the differences, which are both three; and so on till unity.*]

(*ii*) The multiplication of any digit by itself produces the same amount as the multiplication of any two terms distant from it by an equal ratio. [*For four multiplied by itself produces as much as two multiplied by eight, which are distant from four by the same ratio, namely the double ratio*[122].][123]

(*iii*) Again, the multiplication of any digit by itself produces the same amount as the multiplication of any two parts of it by themselves, plus twice one of them by the other.

---

[121] This and the following rule, and subsequent example, are found in the *Liber algorismi*, p. 104 (see also *ibid.*, pp. 93-94).

[122] The 'double ratio' is the ratio 2 : 1, to be distinguished from the 'duplicate ratio', which is the square of its amount. See, e.g., Heath's *Euclid*, II, pp. 132-133.

[123] This and the following rule are found in the *Liber algorismi*, p. 104 (no example).

LIBER MAHAMELETH 609

(*iv*) Again, the multiplication of any digit by itself produces the amount of ten times its denomination less the multiplication of its difference from ten by the digit itself. [*For six multiplied by itself will be (first) said to produce sixty, which is the decuple of its denomination; but the difference between six and ten is four, which when multiplied by this six produces twenty-four; subtracting that from sixty leaves thirty-six, and so much is the result of multiplying six by itself.*][124]

MULTIPLICATION OF DIGITS BY OTHER DIGITS.

(*v*) Again, the multiplication by one digit of another produces the same as when the first multiplies the limit and the result is diminished by the product of the difference between the multiplicand and the limit into the multiplier.[125]

(*vi*) Again, the multiplication of any digit by another digit produces as much as its multiplication by all the parts of the second. [*For three multiplied by four produces as much as when it is multiplied by two and two, which are its parts, and the two products are added.*][126]

[*RULES FOR THE MULTIPLICATION OF DIGITS BY ARTICLES, LIMITS AND COMPOSITES.*][127]

Since, however, our aim here is to explain the rules for multiplying numbers according to note, we shall start by expounding the subject of note. For this is the basis of the science of multiplying numbers, whether large or small, this science being in turn the foundation of the whole science of arithmetic.

CHAPTER ON THE ATTRIBUTION OF A NOTE

You are to know that the origin of number is one, which doubled becomes two, which is the first and smallest number; next, adding one to two makes three, and so, adding one successively, number grows indefinitely.[128] But since it was necessary to multiply numbers by themselves, they were arranged in orders[129] and to each order was attributed a distinguishing note. The first order was established from one to nine; one was put as its

---

[124] Same rule and example in the *Liber algorismi*, p. 97.

[125] Same rule with an example ($5 \cdot 7$) in the *Liber algorismi*, pp. 97-98, and without example p. 104. Since here there is none, the reader of MS C has added one.

[126] There is no real need for an example since this was proved earlier (above, Premiss PE_1). The rule alone is found in the *Liber algorismi*, p. 104.

[127] (Not in MS C.) There are no such rules here. The same subheading appears elsewhere (p. 619).

[128] This is known from A–I (p. 583).

[129] Otherwise there cannot be a (limited) multiplication table. Again, we have seen that in A–I (note 17, p. 583).

note. If something other than one had been put, this would have been admissible;[130] but one was more appropriate, for the first order is that of the units. The second order was established from ten to ninety; two was put as its note. The third order was established from a hundred to nine hundred; three was put as its note. The fourth order was likewise established from one thousand to nine thousand; its note was designated by four. And since one was put as the note of the first order and the note of any order is larger by one than the note of the previous order, so the note of any order is as distant from one as the order itself is from the first; thus the meaning of the note is to indicate in which order a number is. For if the note is ten the number will be in the tenth order, if eleven the number will be in the eleventh order.

This being so, I shall look for a rule whereby, once a number is given [131], its order may be determined, that is, when it is asked 'in which order is this or that number?', how I may know whether it belongs to the sixth order, or to the seventh, or to some other order;[132] and inversely (I shall look for) a rule whereby once the note is known the (order of the) number may be determined.[133]

I say thus that the first order was attributed to the ones, the second to the tens, the third to the hundreds, and, after the hundreds, which is the third order, were passed, the fourth was attributed to the thousands, the fifth to the tens of thousands, the sixth to the hundreds of thousands. After the sixth order was passed, to the first repetition was added another, by saying 'thousand times thousand', which is the seventh order. After three other orders were passed, one repetition was added. After three more orders were passed, one repetition was added. In this way one repetition was added to the previous repetitions each time after three orders.[134]

RULE FOR DETERMINING THE NOTE WHEN THE NUMBER IS KNOWN.[135]

(A.1) Someone asks: What is the note of a thousand times thousand repeated four times, that is, in which order is it?

We know that after three orders one repetition is always added. Therefore, we shall multiply the number of repetitions by three; in this case, four by three, which is twelve. This is the number of all orders preceding the order of a thousand times thousand repeated four times. Adding one to it makes thirteen. Thus the place of a thousand times thousand repeated four times is the thirteenth; its note is thirteen.

---

[130] For instance, in our terms, k for a power 10^k, thus 0 for the first order. The suggestions found later in a gloss (note 181, p. 619) are fanciful.

[131] That is, in words.

[132] Rule explained and applied in A.1 & A.3–5.

[133] A.2 & A.6–8.

[134] Thus the first three orders are followed by a certain number of 'repetitions' (*iterationes*), which is the number of times 'thousand' occurs in the verbal expression.

[135] Similar examples in the *Liber algorismi*, p. 123.

LIBER MAHAMELETH   611

(A.1′) If it were ten thousand times thousand repeated four times, you would add two to twelve.

(A.1″) If it were a hundred thousand times thousand repeated four times, you would add three to twelve.

CONVERSE CASE WITH RESPECT TO THE PREVIOUS ONE, THAT IS: RULE FOR DETERMINING THE (ORDER OF THE) NUMBER WHEN THE NOTE IS KNOWN.

(A.2) Also by means of the foregoing [136] it will be shown that, when it is asked 'of which number is fourteen the note?', this is found by reversing the preceding rule. Namely, we divide fourteen by three; this gives four and the remainder is two, which is the note of the tens. Thus we shall say that fourteen is the note of ten thousand times thousand repeated four times.

(A.2′) If the remainder of the division were one, we would say that the divided number is the note of a thousand times thousand repeated four times.

(A.2″) If this (given) quantity were divisible by three, that is, if it were fifteen, or some other number divisible by three, we would say that it is the note of a hundred thousand times thousand repeated four times, or repeated a number of times equal to the quotient of the division less one. For instance, when fifteen is divided by three, the result is five. Subtracting one from it leaves four. Thus fifteen is the note of a hundred thousand times thousand repeated four times.

Likewise for all.

HOW TO DETERMINE THE PLACE OF ANY GIVEN NUMBER. [137]

(A.3) You want to know to which place belongs three thousand repeated five times.

Always multiply the number of repetitions by three. In this case, the result is fifteen. Add to it the note of three, namely one, thus making sixteen, which designates the required place. The given number therefore belongs to the sixteenth place.

(A.4) Similarly, you want to know to which place belongs ten thousand repeated eight times.

Multiply the number of repetitions, which is eight, by three; the result is twenty-four. Adding to it the note of the tens, namely two, makes twenty-six. The given number therefore belongs to the twenty-sixth place.

AGAIN.

(A.5) You want to know to which place belongs a hundred thousand repeated six times.

---

[136] Instructions preceding A.1. The same numerical examples are taken.

[137] That is, the highest decimal place when the number is written. Since the note indicates this place, we are to calculate the note.

612 PART TWO: TRANSLATION, GLOSSARY

Multiply the number of repetitions, which is six, by three; the result is eighteen. Adding to it the note [*of the place*] of the hundreds, which is three [*for a hundred belongs to the third place*], makes twenty-one. The given number therefore belongs to the twenty-first place.

(*i*) You will be able to know the place of any given number in the same way. For you will always multiply the number of repetitions by three and add to the product the number of the place to which belongs the number appended to the repetition. The required place will be denominated from the resulting number.[138]

The reason for this is clear from the first chapter of the present treatise, on the composition of number [139]. For, in numeration, any repetition comprises three places: for units, tens, hundreds. Therefore, in numeration, any repetition comes into being after a third place, inasmuch as after the first, for the units, the second, for the tens, the third, for the hundreds, follows the fourth, which is the first of the next repetition [*thus every place for the units has two places before itself, those for the tens and the hundreds*]; so when you multiply the quantity of repetitions by three, all the places preceding the required place are included. Then if the number belongs to the units of a repetition, add one to the product; if it belongs to the tens, add two; if it belongs to the hundreds, add three.

Understand this and consider ⟨other such instances accordingly⟩[140].

CONVERSE CASE WITH RESPECT TO THE PREVIOUS ONE, THAT IS: HOW TO DETERMINE THE NUMBER BELONGING TO A GIVEN PLACE.

(**A.6**) You want to know to which number belongs the note eleven.[141]

Divide eleven by three; this gives three and the remainder is two, which means tens, while the three resulting from the division is the number of repetitions. So you will say that eleven is the note of the tens of thousands repeated thrice.

AGAIN.

(**A.7**) You want to know to which number belongs the note thirteen.

Divide thirteen by three; this gives four and the remainder is one; this is distinctive of the units of thousands repeated four times. Thus thirteen is the note of the thousands repeated four times.

(**A.8**) You want to know to which number belongs the note eighteen.

Divide eighteen by three; this gives five and the remainder is three, which three is distinctive of the hundreds. Thus eighteen is the note of the hundred thousands repeated five times.

---

[138] If 'five' is the resulting number, 'fifth' will denominate the place. Similar rule in the *Liber algorismi*, p. 123.

[139] A–I, second section (summarized before A.1).

[140] Lacuna filled according to the variant reading (Latin text, p. 36).

[141] Same example in the *Liber algorismi*, p. 125.

LIBER MAHAMELETH
613

(*ii*) You will proceed likewise, in all instances, to determine the number corresponding to a given note. That is, you will always divide the known note by three. The integral part of the result will be the number of repetitions. The remainder, if it is one, will indicate units, of thousands repeated as many times as is the result of the division; if the remainder is two, it will indicate tens, of thousands repeated as many times as is the result of the division. But if the note is divisible by three, after taking away three from it divide the remainder by three; the three taken away will mean a hundred, of thousands repeated as many times as are units in the result from dividing the remainder. Understand this and bear it in mind.[142]

CHAPTER ON THE MULTIPLICATION OF INTEGRAL NUMBERS
ACCORDING TO NOTE, EXCEPTING THE COMPOSITE ONES [143]

If you want to multiply integral non-composite numbers according to note (do the following). Consider the position of each number in its order, that is, whether it is the second, the third, and so on. Multiply the numbers denominating their positions,[144] and keep the product in mind. Next, add the notes of the two numbers and subtract one from the sum; find out to which order corresponds the note of the remainder. Put then for the number corresponding to the note[145] as much as is the digit, if any[146], in the number kept in mind, and put for the number following the note as much as the article, if any.

(**A.9**) For instance, you want to multiply three by forty.

Consider the position of each number in its order. Now three is third in the order of the digits and forty, fourth in the order of the articles. Then multiply the numbers denominating their positions, that is, three by four; the result is twelve. Keep this in mind.[147] Next, add the notes of the numbers; they are one and two, so the result is three. Subtract one from it; this leaves two. Since two is the note of the tens, put as many times ten as is the digit in the number kept in mind, and put for the number following this note [*which is a hundred*] as much as the article. Now the digits are, in the number kept in mind, which is twelve, two; put as many times ten, which gives twenty. There is just once an article, which is ten[148]; put then as much for the number which follows this note, that is, a hundred. Thus the multiplication of three by forty produces a hundred and twenty.

(**A.10**) Similarly if you want to multiply an article by an article, for example thirty by fifty.

---

[142] Similar rule in the *Liber algorismi*, p. 125.

[143] A.9–13. (A.14–21 are of the same kind, but the product is calculated without using the note.)

[144] Their ranks within the order.

[145] That is: in the place indicated by the note.

[146] That is, if the digit is significant (not zero).

[147] Usual way (since Mesopotamian times) of designating intermediate results. See our *Introduction to the History of Algebra*, p. 5.

[148] Rather: There is just once ten in the article. Same expression in A.10.

614   PART TWO: TRANSLATION, GLOSSARY

Multiplying the numbers of their positions, that is, three by five, produces fifteen. Next, adding their notes, which are two and two, makes four. Subtracting one from this leaves three, which is the note of the hundreds. Now there are in fifteen a digit and an article. The number of units in the digit will give the number of hundreds, thus five. Since there is just once an article, there will be as many times for the number following the note, which is a thousand. Therefore the multiplication of thirty by fifty produces one thousand five hundred.

(A.11) Similarly if you want to multiply an article by one of the hundreds, for example twenty by five hundred.

Now twenty is second in its order, and five hundred, fifth. So multiply the numbers of their positions, that is, two by five, thus producing ten. Next, add the notes of the two numbers, which are two and three, making five. Subtracting one from it leaves four; this is the note of the thousands. Therefore the multiplication of twenty by five hundred produces ten thousand.

Likewise for the others.

NEXT: MULTIPLICATION OF REPEATED THOUSANDS ACCORDING TO NOTE.

(A.12) You want to multiply a hundred thousand repeated three times by five thousand repeated four times.

Consider a hundred thousand repeated three times as if it were one [*for it is within its order as is one in its order, that is, the first*]. Consider five thousand repeated four times as if it were five [*for five thousand repeated four times is in its order as is five in its order, that is, the fifth number*][149]. Next, multiply one by five; this produces five. Keep it in one hand.[150] Next, take the note of five thousand repeated four times, which is thirteen, and add it to the note of a hundred thousand thrice repeated, which is twelve; this makes twenty-five. Removing one from it leaves twenty-four. Then find out of which number twenty-four is the note as explained before[151]; you will find, using what has been shown above, that twenty-four is the note of a hundred thousand repeated seven times. Take for each unit of the digit found before, in this case five, a hundred thousand repeated seven times. So the result of the multiplication is five hundred thousand repeated seven times.

(A.13) You want to multiply eight thousand repeated four times by four hundred thousand repeated seven times.

Consider eight thousand repeated four times as if it were eight, and four hundred thousand repeated seven times as if it were four. Then mul-

---

[149] MS *C* has both bracketed sentences in the margin, so they are probably interpolations.
[150] Retain it for the time being. See also note 152.
[151] A.2 & A.6–8 (and rule *ii* following A.8).

LIBER MAHAMELETH

tiply eight by four; the result is thirty-two. Keep it in your hand.[152] Next, add the note of eight thousand repeated four times, which is thirteen, to the note of four hundred thousand repeated seven times, which is twenty-four, thus making thirty-seven. Subtract one from this, and find out of which number the remainder is the note. You will find that thirty-six is the note of a hundred thousand repeated eleven times. Now take for each unit of the digit found before a hundred thousand repeated eleven times, and take as many thousand times thousand repeated twelve times as there are tens in the article. But the result found above was thirty-two. Since there are in the digit two units and in the article thrice ten, the product of the given numbers is three thousand repeated twelve times and two hundred thousand repeated eleven times.

CHAPTER EXPLAINING THE REASON WHY ONE IS SUBTRACTED FROM THE NOTE.

We have explained preliminarily how to determine the note of numbers and the rule for multiplying numbers according to note.[153] I shall now explain why one is subtracted from the sum of the notes of the two multiplied numbers and why the remainder is the note of their product. That is, (I shall first explain) that, when each of the multiplied numbers is the first of the numbers belonging to its order [*as one and ten are the first in their orders*], their product will also be the first in its order.[154] Then I shall explain how to proceed when this is not so: that is, when each of the multiplied numbers is not the first in its order, but the second, the third, the fourth, and so on, I shall likewise show what their product is. Next, I shall also explain how to proceed to multiply any number whatsoever by itself, be it the first in its order, the second, the third, and so on. And I shall provide all my assertions with incontrovertible proofs.[155]

(*a*) I shall take the first number of each order, that is, one for the first order, that of the units, ten for the order of the tens, a hundred for the hundreds, a thousand for the thousands, and likewise the first number for every single order as far as I want. Let one be A, ten, B, a hundred,

---

[152] A single hand can signify a number with one or two figures.

[153] A.1–8 and A.9–13 respectively.

[154] In other words: the product of two limits is a limit.

[155] These, then, are the first of the proofs announced just before A–II (p. 587) and at the end of it (p. 604). What the author will do here is give the proof of what has already been applied in the previous problems. First he will calculate the note of the product of two limits as the sum of their notes minus one (for we need only consider multiplication of the first elements of the corresponding orders); this is part a of the proof. Next, as a glossator will remark below, he will examine the multiplication of numbers occupying some other place in their respective orders, which is part b. Finally, he will examine (in our section c) the multiplication of a number by itself. This clearly indicates that the above heading, which applies only to the first part, is a later addition, formerly a marginal indication of the content.

616　　　　Part Two:　Translation, Glossary

G, a thousand, D; let H be ten thousand, Z, a hundred thousand, K, a thousand times thousand, T, ten thousand times thousand, Q, a hundred thousand times thousand, L, a thousand times thousand times thousand. It is then evident that these numbers are, from one to the last, in continued proportion.[156] Thus the ratio of one, which is A, to B is the same as the ratio of B to G, the same as the ratio of G to D, the same as the ratio of D to H, the same as the ratio of H to Z, and so on with the others to L. It is also evident that if any of these numbers is multiplied by any other of them, their product will be one of these, or another subsequent but which will be in the same continued proportion; that is, it will be the first of its order; for when some numbers starting with the unit are in continued proportion, then the smaller divides the larger according to a number from among these proportional numbers, as Euclid said in the ninth book [*12th theorem*].[157] [*Now we know* [158] *that the note of A is one, the note of B, two, the note of G, three, and likewise the note of each order exceeds the preceding note by one.*][159] Thus, if any of these numbers is multiplied by any other of them, say if D is multiplied by K and M is produced, we know that this M is one of the above numbers or some other but in the same proportion [*as is A to B*]. Now we do not know to which order the product belongs; but once its note has been determined we shall know to which order it belongs as shown above.[160]　So I say that the multiplication of D by K produces M.[161] Then K measures M as many times as one is in D. But one measures D as many times as one is in it.[162] So the ratio of one, which is A, to D is the same as the ratio of K to M. Thus it is that K and M bear to one another the same ratio as A to D. Since between A and D two numbers are interposed, all in the same proportion, there will be between K and M two consecutive numbers in the same proportion, as Euclid said [*in the eighth of the eighth*][163]. This being so, it is necessary that D be at the same distance from A as M is from K. Now we have said above[164] that the notes of (any two consecutive) such orders exceed one another by one, while we have taken two numbers which are at an equal distance from the extremes; then the sum of the extremes is equal to

---

[156]　They form a geometric progression.

[157]　Actually *Elements* IX.11.

[158]　Rule for attributing the notes (pp. 609–610).

[159]　This statement is, at best, misplaced.

[160]　A.2 and rule *ii*, p. 613. The misplaced statement might belong here.

[161]　According to the above statements, $D \cdot K$ ($=\ 10^3 \cdot 10^6$) will produce L ($= 10^9$), a discrepancy noticed by the reader of $\mathcal{C}$ (Latin text, note to line 993). This error vitiates the calculation below, though it does not invalidate the proof.

[162]　See above, notes 64 & 67, p. 595.

[163]　*Elements* VIII.8. Here the inappropriateness of the figure becomes evident.

[164]　In the general theory about notes. See also the misplaced statement above (note 159).

the sum of these two numbers, as we have established in the premisses.[165] Therefore the sum of the notes of D and K is equal to the sum of the notes of A and M. Since adding the notes of D and K makes eleven, adding the notes of A and M makes eleven. Subtracting from eleven the note of A, which is one, leaves ten, which is the note of M; thus M is the first number belonging to the tenth order. This is what we wanted to demonstrate.

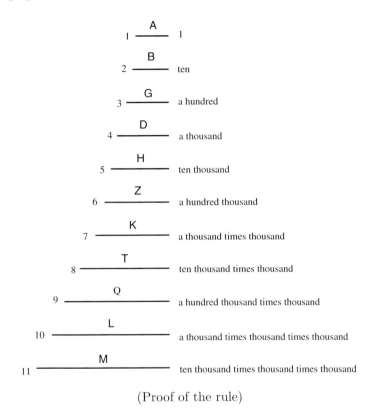

(Proof of the rule)

[*He has taught above the multiplication of these limits, as in the figure appended. He now teaches the multiplication of those which are contained within the orders of these limits, (thus) in the second, third or fourth of the ranks*[166], *and so on, which are called double, triple or quadruple of the limits, like thirty of ten, three thousand of a thousand, and similarly for the others.*]

(**b**) It is then clear from the preceding rules[167] that if you want to multiply the triple of D by the quadruple of K [like[168] *three thousand by four*

---

[165] Premiss P_1. The text is confused: 'two numbers' refers once to D and K, once to their notes. The original text might have been reworked.
[166] *gradus* instead of the usual *locus*.
[167] A.9–11 (used below).
[168] The actual values of D and K are irrelevant.

618   PART TWO: TRANSLATION, GLOSSARY

thousand times thousand], you will do the following. We know [169] that the multiplication of D by K produces M. Thus, the quadruple of K multiplied by the triple of D will produce twelve times M [*for the ratio of any two composite numbers* [170] *is compounded of the two ratios of their sides* [*by the fifth of the eighth*][171]].[172] Now twelve times M is a unit in the following order and twice M [*for the first number of any order is the decuple of the (first of the) preceding one*].

[*MULTIPLICATION OF ANY LIMIT BY ITSELF.*][173]

(*c*) It is also clear from the foregoing that if you want to multiply any of the above numbers by itself, you will double its note, subtract one from the result, and the remainder will be the note of the square of this number.

[*MULTIPLICATION OF ARTICLES BY THEMSELVES.*][174]

If you want to multiply its triple by its triple, the result will be, as shown before [175], nine times the square of this number [*for the ratio of any two squares is the duplicate ratio of their sides* [*by the 9th of the eighth*][176]].

[*MULTIPLICATION OF ARTICLES BY OTHER ARTICLES HAVING THE SAME LIMIT.*][177]

If you want to multiply its double by its triple, the result will be the sextuple of its square. If its quintuple by its triple, the result will be fifteen times its square; this is one, or the first, in the order following the order of the square and five times the square.

[*AGAIN, ACCORDING TO ANOTHER AUTHOR WHO PUTS IT MORE SUCCINCTLY. THE REASON WHY ONE IS SUBTRACTED FROM THE NOTES OF THE TWO MULTIPLIED NUMBERS IS THE FOLLOWING.*[178]

To multiply a number by a number means repeating the multiplicand as many times as there are units in the multiplier. So if you want to multiply a hundred by ten, you will just repeat a hundred as many times as there

---

[169] As assumed in *a*.

[170] 'Composite' in the sense of 'not prime' (*Elements* VII, def. 13). Here 'plane' would be more appropriate.

[171] *Elements* VIII.5.

[172] Rather: according to P_3.

[173] We now consider the multiplication of a number by itself, first when it is the initial element of an order, then when not, as already announced (p. 615). This insertion in the form of subheadings of what were once marginal glosses is misleading.

[174] This subheading is found only in MS $\mathcal{A}$. What follows are numerical examples for *c* & *b*.

[175] In the multiplication table and by P_3.

[176] *Elements* VIII.11. Presumably same glossator as in *b*.

[177] This subheading is found in all three MSS.

[178] This whole section is an early interpolation, which itself contains readers' glosses.

LIBER MAHAMELETH   **619**

are units in ten, thus ten times since there are ten units in ten. Then the ratio of one to ten is the same as the ratio of a hundred to the required quantity. These are therefore four proportional numbers. Thus the product of the first, which is one, into the fourth, which is the required quantity, is equal to the product of the second, which is ten, into the third, which is a hundred [by the 19th of the seventh][179]*. Since the product of the first into the fourth is the same as the product of the second into the third, it immediately follows that the note of one product is equal to the note of the other. [This is not yet proved.]*[180]* Thus the sum of the notes of the first and the fourth is equal to the sum of the notes of the second and the third [for the note of the product has been taken from the notes of the two factors]. Since it is clear that the sum of the notes of the first and the fourth equals the sum of the notes of the second and the third, then subtracting the note of the first, which is one, from the notes of the second and the third will leave the note of the fourth, which is required. Thus you must proceed as follows: always subtract one from the notes of the two factors, and the remainder will be the note of the product. This is what we wanted to demonstrate.*

It is therefore also clear that if two were the note of the digits and the note of each order exceeded the previous note by two you would subtract from the notes of the two factors two, which is the note of the digits.[181] *If three were the note of the digits and the note of each order exceeded the previous note by three, you would also subtract from the notes of the two factors three, which is the note of the digits. We subtract the note of the first only so as to be left with the note of the fourth, required quantity. If you were to multiply a number by itself, you would double its note and subtract the note of the digits from the result.]*

Consider other such instances likewise, and you will succeed.

AGAIN, IN ANOTHER WAY. MULTIPLICATION OF INTEGERS AMONG THEM-SELVES WITHOUT NOTE.[182]

RULES FOR THE MULTIPLICATION OF DIGITS BY ARTICLES, LIMITS AND COMPOSITES.[183]

(*i*) If you want to multiply a digit by one of the articles which go up to

---

[179] *Elements* VII.19.

[180] This gloss refers to what follows.

[181] The possibility of taking another series of numbers for the successive notes was alluded to in the section on attributing a note. See p. 610 (note 130).

[182] The *Liber algorismi* has, on pp. 120–122, the subsequent rules *i–viii* and examples A.14, A.16, A.17, A.18, A.20. (The example corresponding to A.15 is different, and those to A.19 and A.21 differ by the number of repetitions.)

[183] This same heading occurred before (p. 609) but was followed by the theory of note. Here we should in fact have 'Rules for the multiplication of digits by articles, hundreds, thousands'.

a hundred,[184] you multiply their figures[185]; the number of units in the resulting digit will give the number of tens, and the number of tens in the article will give the number of hundreds.

(A.14) For instance, you want to multiply seven by seventy.

Multiply seven by seven; the result is forty-nine. Now in forty, which is the article, there are four times ten; thus as many hundreds, which is four hundred. In the digit, which is nine, there are nine times the unit; thus as many tens, which is ninety. Therefore the multiplication of seven by seventy produces four hundred and ninety.

It will always be like that when a digit is multiplied by one of the articles which go up to a hundred.

(*ii*) If you want to multiply a digit by one of the hundreds, multiply their figures; the number of units in the resulting digit will give the number of hundreds, and the number of tens in the article will give the number of thousands.

(A.15) For instance, you want to multiply seven by three hundred.

Multiply seven by three; the result is twenty-one. Now there is twice ten in the article, which is twenty, thus as many thousands. There is once the unit in the digit, which is one, thus as many hundreds. Therefore the multiplication of seven by three hundred produces two thousand one hundred.

It will always be like that when a digit is multiplied by one of the hundreds.

(*iii*) If you want to multiply a digit by one of the thousands, or by thousands however often repeated, multiply their figures; put the digit, if any, in the place of the multiplier, and the article in the place following.

(A.16) For instance, you want to multiply six by thirty thousand.

Multiply six by three; the result is eighteen. Put the digit, which is eight, in the place of the multiplier, namely the fifth[186], and the article, which is ten[187], in the place following. This gives a hundred and eighty thousand.

And so in all similar instances.

Multiplication of articles by themselves, by others, and by hundreds and thousands.

---

[184] Not the articles of the higher orders.

[185] The 'figures' (*figure*) are the nine significant digits (thus not including zero).

[186] The MSS have here 'namely six', a change certainly due to an early reader's taking 'namely' to refer to 'multiplier'.

[187] Rather: 'and the number of tens in the article, which is one'. Same in A.19.

LIBER MAHAMELETH

(*iv*) When you multiply an article by itself or by another article among those which are from ten to a hundred [188], multiply their figures; the number of units in the resulting digit will give the number of hundreds, and the number of tens in the article will give the number of thousands.

(**A.17**) For instance, you want to multiply thirty by seventy.

Multiply three by seven; the result is twenty-one. Now there is twice ten in the article, which is twenty, thus as many thousands. There is one unit in the digit, which is one, thus as many hundreds. Therefore thirty multiplied by seventy produces two thousand one hundred.

Likewise in all instances of this kind.

(*v*) When you multiply one of the articles which go from ten up to a hundred by one of the hundreds which go up to a thousand [189], multiply their figures; the number of units in the resulting digit will give the number of thousands, and the number of tens in the article will give the number of ten thousands.

(**A.18**) For instance, you multiply thirty by five hundred.

Multiply three by five; the result is fifteen. There is only once ten in the article, thus it will be ten thousand. There is in the digit five times the unit, thus as many thousands. Therefore the multiplication of the given numbers produces fifteen thousand.

You must proceed likewise in all such instances.

(*vi*) When you multiply one of the articles by one of the thousands or thousands however often repeated, multiply their figures. Put the digit, if any, in the second place counted from the multiplier and the article in the third. [190]

(**A.19**) For instance, you multiply thirty by four thousand.

Multiply their figures, that is, three by four; the result it twelve. Put two, which is the digit, in the second place from the multiplier; this is the fifth, that of ten thousands. Put the article, which is ten, in the third place from the multiplier; here this is the sixth, that of hundred thousands. The result is a hundred and twenty thousand.

MULTIPLICATION OF HUNDREDS AMONG THEMSELVES AND BY THOUSANDS.

(*vii*) When you multiply among themselves the hundreds which go up to a thousand, multiply their figures. The number of units in the resulting digit will give the number of ten thousands, and the number of tens in the article will give the number of hundred thousands.

(**A.20**) For instance, you multiply three hundred by five hundred.

---

[188] Thus the articles of the second order.

[189] Not the hundreds of higher orders.

[190] 'the second place counted from the multiplier': the next place (that of the multiplier being counted too).

622 PART TWO: TRANSLATION, GLOSSARY

Multiply three by five; the result is fifteen. There is five times the unit in the digit, thus there are as many ten thousands; there is once ten in the article, thus there are as many hundred thousands. Therefore the multiplication of three hundred by five hundred produces a hundred and fifty thousand.

Likewise in all instances of this kind.

(*viii*) When you multiply one of the hundreds by one of the thousands, or by one of the however often repeated thousands, multiply their figures. Put the digit, if any, in the third place from the multiplier, and the article in the fourth from it.

(A.21) For instance, you multiply two hundred by five thousand times thousand.

Multiply two by five. Since the result is an article only [191], it should be put in the fourth place from the multiplier five thousand times thousand, thus here in the tenth place; this will produce a thousand times thousand times thousand.[192]

CHAPTER ON THE SCIENCE OF MULTIPLYING
DIFFERENCES ACCORDING TO RULES [193]

[(*i′*) *When you multiply a digit by an article, there will as many tens as there are units in the resulting digit, and as many hundreds as there are tens in the article.*[194]

(*ii′*) *When you multiply a digit by one of the hundreds, there will be as many hundreds as there are units in the digit, and as many thousands as there are tens in the article.*

(*iv′*) *When you multiply an article by an article, there will be as many hundreds as there are units in the digit, and as many thousands as there are tens in the article.*

---

[191] Meaning: without significant digit.

[192] Further such examples are A.30–37.

[193] We shall now be taught the rules of signs. However, first comes an interpolation repeating (in MSS $\mathcal{B}$ and $\mathcal{C}$ only) some rules already learnt. This repetition of known rules by an early reader was no doubt due to his misunderstanding the heading: difference (*differentia*) can also mean the decimal place a figure occupies, as was the case before. But here it is used in the common sense of 'difference', and multiplication of subtracted quantities will indeed appear from A.23 on. As a 13th-century arithmetical text observes (MS Caen 14, fol. 58^v), *differentia duplex dicitur in hac arte: dicitur enim differentia numerus ille secundum quem alius numerus differt ab alio; (...) item differentia dicitur spatium illud in quo continetur aliqua species figurarum.* See also Tropfke's *Geschichte*, I, p. 14 (ed. 1930). In addition, *differentia* can mean a *term* in a mathematical expression.

[194] Better, here and below: in the resulting article.

(*v′*) When you multiply an article by one of the hundreds, there will be as many thousands as there are units in the digit, and as many ten thousands as there are tens in the article.

(*vii′*) When you multiply hundreds among themselves, there will be as many ten thousands as there are units in the digit, and as many hundred thousands as there are tens in the article.

In (using) these rules observe the following: multiply together the figures of the numbers and count the digits and articles of the products, if any, as tens, or hundreds, or thousands as explained before.]

(*ix*) The multiplication of added by added always produces added.

(*x*) The multiplication of added by subtracted always produces subtracted and the multiplication of subtracted by added, subtracted.

(*xi*) The multiplication of subtracted by subtracted always produces added.

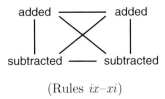

(Rules *ix–xi*)

MULTIPLICATION OF COMPOSITES FROM A DIGIT AND AN ARTICLE[195]

(**A.22**) You want to multiply thirteen by fourteen.

(***a***) Multiply ten by ten, which produces a hundred. Next, three by ten, which produces thirty. Next, four by ten, which produces forty. Next, four by three, which produces twelve. Then add all that in your hand.[196] [*Addition of a thousand ⟨and a hundred⟩ is performed in one hand, of a digit and an article in the other.*]

(***b***) Or otherwise. Multiply here[197] ten by ten, thus producing a hundred. Add after that four and three, which makes seven; multiplying it by the article, which is here ten, produces seventy. Next, multiply three by four, which produces twelve. Adding all that will make the quantity you require. This rule is applicable only when the article is the same in both numbers, as here.[198]

MULTIPLICATION OF A COMPOSITE FROM A DIGIT AND AN ARTICLE BY AN ARTICLE ONLY.

(**A.23**) You want to multiply thirty-eight by forty.

---

[195] The subject of A.22–29 is the multiplication of *various* kinds of composites.

[196] Namely 30, 40, 12, which a single hand can signify.

[197] Since this is a particular case (see remark at the end).

[198] Thus three products instead of four occur. A similar example in the *Liber algorismi*, p. 117.

624   PART TWO: TRANSLATION, GLOSSARY

(*a*) Either multiply them as explained before, that is, eight by forty and thirty by forty.[199]

(*b*) Or, since you know that thirty-eight is forty minus two, multiply forty minus two by forty, in the following way. Multiply forty by forty, thus producing one thousand six hundred. Next, multiply two, which is subtracted, by forty; this produces eighty, subtracted. Subtract it from the additive one thousand six hundred[200]; this leaves one thousand five hundred and twenty, and this is the result of the multiplication.

MULTIPLICATION OF COMPOSITES FROM DIFFERENT DIGITS AND ARTICLES.

(**A.24**) You want to multiply seventy-nine by thirty-two.

Perform the multiplication as if it were eighty minus one by thirty-two. That is, multiply eighty by thirty-two, thus producing two thousand five hundred and sixty. Subtract from this the product of one, subtracted, by thirty-two, that is, thirty-two, subtracted; the remainder is two thousand five hundred and twenty-eight.

AGAIN.[201]

(**A.25**) You want to multiply seventy-nine by fifty-eight.

Perform the multiplication as if it were eighty minus one by sixty minus two. That is, multiply eighty by sixty, thus producing four thousand eight hundred. Next, multiply one, subtracted, by sixty; the result is sixty, subtracted. Next, multiply two, subtracted, by eighty; the result is a hundred and sixty, subtracted. Adding this to sixty, subtracted, makes two hundred and twenty, subtracted. Subtracting this from four thousand eight hundred leaves four thousand five hundred and eighty. Next, multiply one, subtracted, by two, subtracted, thus producing two, added, and add it to the preceding quantity. The result of the multiplication of the given numbers is four thousand five hundred and eighty-two.

MULTIPLICATION OF COMPOSITES FROM DIFFERENT LIMITS AND ARTICLES.[202]

(**A.26**) You want to multiply three hundred and twenty by six hundred and forty.

---

[199] A.14, A.17 and rules *i*, *iv*.

[200] Since we are dealing here with subtracted quantities and not negative numbers, subtracting them will mean subtracting their absolute values. Note that subtraction of integers is not taught in the extant text (nor is addition).

[201] The two factors are now represented as differences.

[202] It should be 'from different hundreds and (different) articles'. In the subsequent subheadings, 'limit' is taken to mean one of the hundreds (*centeni*), just as 'article' is one of the tens. We have encountered this ambiguity before, in an interpolated passage (note 24, p. 584). Remember that most of the subheadings do not originate with the author.

LIBER MAHAMELETH

Multiplying three hundred by six hundred produces a hundred and eighty thousand. Keep it in one hand.[203] Next, multiply three hundred by forty; the result is twelve thousand. Next, multiply six hundred by twenty; the result is twelve thousand. Next, multiply forty by twenty; the result is eight hundred. Adding them all makes two hundred thousand and four thousand eight hundred, and this is the result of the multiplication.

MULTIPLICATION OF A COMPOSITE BY ITSELF, BUT COMPOSED OF THE SAME LIMIT AND THE SAME ARTICLE.[204]

(**A.27**) You want to multiply three hundred and ninety by three hundred and ninety.

(*a*) Either multiply them as before.

(*b*) Or perform the multiplication as if it were four hundred minus ten by four hundred minus ten, as follows. Multiply four hundred by four hundred, thus producing a hundred and sixty thousand. Next, multiply ten, subtracted, by four hundred, twice, thus producing eight thousand, subtracted. Subtracting it from a hundred and sixty thousand leaves a hundred and fifty-two thousand. Next, multiply ten, subtracted, by ten, subtracted, thus producing a hundred, added. Adding it to the preceding quantity will give as the total of the whole multiplication a hundred and fifty-two thousand one hundred, which is the result.

MULTIPLICATION OF COMPOSITES FROM A LIMIT, AN ARTICLE AND A DIGIT.

(**A.28**) You want to multiply seven hundred and twenty-eight by four hundred and sixty-four.

In this and similar problems it is necessary to write down the resulting numbers, for the hands cannot hold them all.[205] Therefore, when you want to multiply them, set out the problem in two rows.[206] Next, multiply each number in one row by each number in the other.[207] That is, multiply four hundred by seven hundred, producing two hundred and eighty thousand; next, four hundred by twenty-eight, producing eleven thousand two hundred. Write down all that, each number being put with the number of its kind. Next, sixty-four by seven hundred, producing forty-four thousand eight hundred; next, twenty-eight by sixty-four, as explained in the multiplication of digits and tens among themselves,[208] and add these quantities

---

[203] Rather: Keep 18 in one hand.

[204] 'But': it could be another kind of composite number.

[205] Remember (note 152, p. 615) that each hand can signify one number expressed by at most two significant figures of consecutive orders.

[206] *pone questionem in duobus ordinibus*, namely multiplicand above, and corresponding orders of the two numbers aligned vertically. See below (or the *Liber algorismi*, p. 38).

[207] 'number', not '(single) figure', according to the multiplications seen before.

[208] A.24–25.

626   PART TWO: TRANSLATION, GLOSSARY

in your hands,[209] making one thousand seven hundred and ninety-two. Write this down, each number being with the number of its kind as said before. Next, add all that, that is:[210] first, the digits with the digits, then the tens with the tens, after that the hundreds with the hundreds, finally the thousands with the thousands [*and then the tens of thousands*].

You will proceed accordingly, adding and multiplying, in all such instances.

[*MULTIPLICATION OF A LIMIT BY A COMPOSITE FROM AN ARTICLE AND A DIGIT.*][211]

(**A.29**) You want to multiply nine hundred and ninety-nine by themselves.

(*a*) Either multiply this as before.

(*b*) Or multiply this as if it were a thousand minus one by a thousand minus one, in the following way. Multiply one thousand by one thousand, thus producing a thousand times thousand. Next, multiply one, subtracted, by a thousand, thus producing one thousand, subtracted; next, multiply a thousand by one, subtracted, thus producing one thousand, subtracted; adding this to the preceding one thousand subtracted makes two thousand, subtracted,[212] and subtracting this from a thousand times thousand leaves nine hundred thousand and ninety-eight thousand. Next, multiply one, subtracted, by one, subtracted, thus producing one, added, and add it to the preceding quantity. The quantity produced by the whole multiplication will be nine hundred thousand and ninety-eight thousand and one, and this is what you require.

MULTIPLICATION OF THOUSANDS AMONG THEMSELVES

(**A.30**) If you want to multiply a thousand by a thousand, say 'thousand times thousand'. But if you want to multiply a thousand times thousand by a thousand, say 'thousand times thousand times thousand', thrice. You will always proceed likewise in multiplying thousands among themselves [*or by other numbers, thus a thousand times thousand by a hundred produces a hundred thousand times thousand; the same for all instances of this kind*]; for the multiplication of simple thousands[213] will always produce as much as the addition or repetition of their names.[214]

(**A.31**) You now want to multiply six thousand times thousand times thousand by seven thousand times thousand.

---

[209] As said in A.22*a*.

[210] All we shall be taught about addition of integers is here and in A.40.

[211] The glossator took the coming problem to be $900 \cdot 99$ because of an inappropriate formulation (*inter se* in the MSS instead of *in se*). He was obviously not following, at least not in this case, the computations.

[212] Use is not made here of the simplification seen in A.22*b* & A.27.

[213] That is, not combined with other terms.

[214] We shall often meet with 'addition' in the sense of appending. Note that simply repeating the name 'thousand' works in Arabic, not in Latin.

LIBER MAHAMELETH **627**

Multiply six by seven, thus producing forty-two. Adding to it the repetition disregarded, that is, the number of times 'thousand' is repeated, makes forty-two thousand repeated five times.

(A.32) You now want to multiply a hundred thousand times thousand thrice repeated and twenty thousand and six by fifty thousand times thousand, repeated twice, and a hundred thousand.

You will do the following. Place the multiplicand on one side and the multiplier on the other, as we said before[215], in the following way:

A hundred thousand thrice repeated	Twenty thousand	Six
Fifty thousand times thousand, repeated twice	A hundred thousand.	

Then multiply fifty thousand times thousand, repeated twice, by six; this produces three hundred thousand times thousand.[216] Next, multiply fifty thousand times thousand by twenty thousand; this produces a thousand times thousand four times repeated. After that, multiply fifty thousand times thousand by a hundred thousand times thousand thrice repeated; this produces five thousand times thousand repeated six times. Next, multiply a hundred thousand by a hundred thousand thrice repeated; this produces ten thousand repeated five times. Next, multiply a hundred thousand by twenty thousand; this produces two thousand thrice repeated. Next, multiply six by a hundred thousand; this produces six hundred thousand. Next, add all that; the quantity resulting from the addition will be the product of the given numbers, namely: five thousand times thousand six times repeated and ten thousand five times repeated and a thousand times thousand four times repeated and two thousand thrice repeated and three hundred thousand times thousand and six hundred thousand.

You will proceed in the same way for all such figures[217]. That is, you will multiply each number in each row by the individual numbers in the other row and add all the results; the sum is the result of the whole multiplication.

The proof of this is the following. Let a hundred thousand times thousand thrice repeated be AB, twenty thousand, BG, and six, GD; next, let fifty thousand times thousand, repeated twice, be HZ, and a hundred thousand, ZK. I want to know how to multiply AD by HK. I say thus that the result of multiplying AD by HK is equal to the products of HZ into AB, HZ into BG, HZ into GD, with the products of ZK into AB, ZK into BG, ZK into GD. The proof of this is the following. We know

---

[215] A.28. The two 'rows' or lines (*ordines*, note 206; Arabic *saṭrān*) have become 'sides' (*latera*). (*latus* and *ordo* have the same meaning in the *Liber algorismi*, p. 56.)

[216] As in A.28, each multiplication corresponds to cases already seen.

[217] *omnes figure consimiles*. Here *figura* means the frame above.

[*by the first of the second of Euclid*][218] that the product of HK into AD is equal to the products of HZ into AD and ZK into AD. Now the product of HZ into AD is equal to the products of HZ into AB, HZ into BG, and HZ into GD. It will be shown likewise that the product of ZK into AD is equal to the products of ZK into AB, ZK into BG, and ZK into GD. Thus the product of HK into AD is equal to the products of HZ into AB, HZ into BG, HZ into GD, ZK into AB, ZK into BG, and ZK into GD. This is what we wanted to demonstrate.

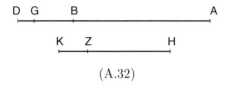

(A.32)

You will in similar instances resort to a proof like this one.

(A.33) You now want to multiply four thousand by six thousand.[219]

Disregarding in both factors the name 'thousand' leaves four and six; multiplying them produces twenty-four. Next, multiply a thousand by a thousand; this will produce as much as the result of adding together their names, that is, a thousand times thousand.[220] Appending this to twenty-four gives twenty-four thousand times thousand, and this is the quantity you require.

(A.34) You want to multiply seven thousand by two thousand.

Multiply two by seven; this produces fourteen. Appending to it the names of the thousands gives fourteen thousand times thousand.

(A.35) You want to multiply ten thousand by ten thousand.

Disregarding the two names 'thousand', there remains the multiplication of ten by ten, which produces a hundred. Appending to this the two names of the thousands gives a hundred times thousand times thousand, and this is the quantity produced by the multiplication.

(A.36) You want to multiply six thousand by forty thousand.

Disregard the repeated name. Next, multiply forty by six; this produces two hundred and forty. Appending to each [221] the name 'thousand' repeated gives as the result two hundred thousand times thousand and forty thousand times thousand.

(A.37) You want to multiply a hundred thousand times thousand thrice repeated by five thousand times thousand repeated four times.

---

[218] Thus by PE_1.
[219] A.33–36 would better fit between A.30 and A.31.
[220] As seen in A.30.
[221] There are two names of numerals, which was not the case before (except in A.32, misplaced). Same in A.38.

(*a*) Either proceed as we have taught before.

(*b*) Or proceed as follows.[222] We know that a hundred thousand times thousand thrice repeated is produced by multiplying a hundred by a thousand times thousand thrice repeated, and that five thousand times thousand repeated four times is produced by multiplying five by a thousand times thousand repeated four times. Therefore we are to multiply a hundred by a thousand times thousand thrice repeated, five by a thousand times thousand repeated four times and then the two products. But the result of multiplying the product of a hundred into a thousand times thousand thrice repeated by the product of five into a thousand times thousand repeated four times is equal to the result of multiplying the product of five into a hundred by the product of a thousand times thousand thrice repeated into a thousand times thousand repeated four times.[223] But a thousand times thousand thrice repeated and a thousand times thousand repeated four times is a thousand times thousand repeated seven times,[224] while five multiplied by a hundred is five hundred. So multiply five hundred by a thousand times thousand repeated seven times, that is, append (to five hundred) their names, which gives five hundred thousand times thousand repeated seven times, and this is what you wanted.

(**A.38**) You want to multiply fifteen thousand by fifteen thousand.

Multiply fifteen by fifteen; this produces two hundred and twenty-five. Appending to each the two 'thousand' gives two hundred thousand times thousand and twenty-five thousand times thousand.

MULTIPLICATION OF THOUSANDS AND HUNDREDS BY THOUSANDS AND HUNDREDS.[225]

(**A.39**) You want to multiply six thousand four hundred by three thousand eight hundred.

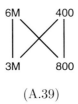

(A.39)

Multiply six thousand by three thousand, in the following way. Multiply six by three; this produces eighteen. Append to it (each) repeated 'thousand'; this gives eighteen thousand times thousand. Next, multiply three thousand by four hundred, as follows. Multiply three by four

---

[222] This is in fact the proof of the procedure already seen. Similar situation in A.51.

[223] Premiss P_3.

[224] A.30.

[225] In fact the real difference occurs in the previous problem, with the multiplication of numbers having each two significant figures.

630 PART TWO: TRANSLATION, GLOSSARY

hundred, thus producing one thousand two hundred. Append to this the 'thousand' disregarded; this gives a thousand times thousand and two hundred thousand. Next, multiply six thousand by eight hundred, as follows. Multiply six by eight hundred, thus producing four thousand eight hundred. Append to this the 'thousand' disregarded; this gives four thousand times thousand and eight hundred thousand. Next, multiply four hundred by eight hundred, thus producing three hundred thousand and twenty thousand. Adding this to the above results will make the product of the multiplication of the given numbers.

MULTIPLICATION OF THOUSANDS, HUNDREDS, TENS AND DIGITS BY THOUSANDS, HUNDREDS, TENS AND DIGITS.

(A.40) You want to multiply six thousand four hundred and sixty-eight by four thousand five hundred and sixty-four.

$$\begin{matrix} 6 \\ \text{M} \end{matrix} \quad 468$$

$$\begin{matrix} 4 \\ \text{M} \end{matrix} \quad 564$$

(A.40, first figure)

Multiply six thousand by four thousand as we have taught before;[226] this produces twenty-four thousand times thousand.[227] Then multiply four thousand by four hundred; this produces a thousand times thousand and six hundred thousand. Then multiply four thousand by sixty-eight; this produces two hundred thousand and seventy-two thousand. Then multiply five hundred by six thousand; this produces three thousand times thousand. Next, multiply five hundred by four hundred; this produces two hundred thousand. Next, multiply five hundred by sixty-eight; this produces thirty-four thousand. Next, multiply sixty-four by six thousand; this produces three hundred thousand and eighty-four thousand. Next, multiply sixty-four by four hundred; this produces twenty-five thousand six hundred. Then multiply sixty-four by sixty-eight as explained before;[228] this produces four thousand three hundred and fifty-two. Add all these together, in the following way: each one with the number of its kind; but wherever a digit results it must be left in the same place, whereas the article is always moved to the next place.[229] The resulting quantity is the product of the multiplication.

---

[226] A.33.

[227] In the attached figure (p. 631), we find the result in the upper row. The other rows do not represent intermediate results, for they contain only significant digits, which are thus inserted from the top down wherever there is an empty place.

[228] A.22a or A.24–25.

[229] *differentia*, thus decimal place.

LIBER MAHAMELETH

2	9	5	1	9	9	5	2
2	4	6	7	2	6	5	2
	1	2	3	4	3		
	3	2	8	4			
		3	2	5			
				4			

(A.40, second figure)

MULTIPLICATION OF REPEATED THOUSANDS AMONG THEMSELVES.

(**A.41**) You want to multiply three hundred thousand times thousand and forty thousand by four hundred thousand times thousand and five thousand times thousand times thousand, thrice.[230]

Multiply three hundred thousand times thousand by five thousand times thousand (times) thousand, thrice, in the following way. Multiply three hundred by five; this produces one thousand five hundred. Append to it the repetition of 'thousand' which you have disregarded; this gives the quantity a thousand times thousand repeated six times, and five hundred thousand times, thousand times repeated five times. Then multiply three hundred thousand times thousand by four hundred thousand times thousand, in the following way. Multiply three hundred by four hundred; this produces a hundred thousand and twenty thousand. Appending to each the repetition gives a hundred thousand times repeated five times and twenty thousand times repeated five times. Then multiply forty thousand by five thousand times thrice repeated; this produces two hundred thousand repeated four times. Next, multiply forty thousand by four hundred thousand times thousand; this produces six thousand times thousand repeated four times, and ten thousand times, thousand repeated four times.[231] Add all these together. The resulting quantity is the product of the multiplication of the quantities proposed.

SUBTRACTION OF REPEATED THOUSANDS AMONG THEMSELVES.[232]

(**A.42**) For instance, you want to subtract one from a thousand times thousand times thousand.

---

[230] Note how this verbal expression departs from the norm: the smaller term in the multiplier precedes the larger. (The same occurs again towards the end of the present problem; see also A.75.) We also find 'thousand times repeated' (*repetitum milies*) instead of the usual 'thousand repeated' (*repetitum mille*).

[231] We would expect: sixteen thousand times thousand repeated four times (*sexdecim milies milia quater iterata*). The two terms are again inverted.

[232] This heading is somewhat inappropriate (unless we consider the coming problem as a preliminary).

632 PART TWO: TRANSLATION, GLOSSARY

Split up a thousand times thousand (times) thousand until you reach a number from which you may subtract one[233], in the following way. Split up a thousand times thousand times thousand, thrice repeated, into nine hundred thousand times thousand and a hundred thousand times thousand. Then split up a hundred thousand times thousand into ninety-nine thousand times thousand and a thousand times thousand. Then split up a thousand times thousand into nine hundred thousand and a hundred thousand. Next, split up a hundred thousand into ninety-nine thousand and a thousand. Next, split up a thousand into nine hundred and a hundred. Then split up a hundred into ninety and ten. Subtract now one, the given number, from ten; nine remains by itself while all other numbers are left as they stand in their places.[234] Thus the remainder of the subtraction is nine hundred thousand times thousand and ninety-nine thousand times thousand and nine hundred thousand and ninety-nine thousand and nine hundred and ninety-nine. Such is the quantity resulting.

(A.43) You want to subtract twenty thousand times thousand from thirty thousand times thousand repeated five times.

Take from thirty thousand times thousand repeated five times one thousand times thousand repeated five times; this leaves twenty-nine thousand times thousand repeated five times. Then split up one thousand times thousand repeated five times into nine hundred thousand times thousand repeated four times and a hundred thousand repeated four times. Then split up a hundred thousand (times) thousand repeated four times into ninety-nine thousand times thousand repeated four times and a thousand times thousand repeated four times. Then split up this thousand times thousand repeated four times into nine hundred thousand times thousand thrice repeated and a hundred thousand thrice repeated. Next, split up this hundred thousand times thousand thrice repeated into ninety-nine thousand times thousand thrice repeated and a thousand times thousand thrice repeated. Next, split up a thousand times thousand thrice repeated into nine hundred thousand times thousand and a hundred thousand times thousand. From this hundred thousand times thousand subtract the given subtrahend twenty thousand times thousand; this leaves eighty thousand times thousand, each other number being left as it stands. Add them to eighty thousand times thousand. The remainder of the subtraction of twenty thousand times thousand from thirty thousand times thousand repeated five times will be the following quantity: twenty-nine thousand times thousand repeated five times and nine hundred thousand times thousand repeated four times and ninety-nine thousand repeated four times and nine hundred thousand thrice repeated and ninety-nine thousand thrice repeated and nine hundred thousand times thousand and eighty thousand times thousand. This is the quantity remaining from the subtraction.

You will proceed likewise in all instances of this kind.

---

[233] That is, a number which renders the subtraction straightforward.

[234] 'in their places' might be a later addition.

This procedure[235] is very useful for some other problems which are about the multiplication of repeated thousands. For it so happens that in such problems the multiplication is more easily performed by using this procedure, as already explained in the differences[236].

[*AGAIN, ACCORDING TO ANOTHER AUTHOR WHO PUTS IT MORE SUCCINCTLY.*][237]

(**A.44**) For instance, someone says: Multiply nine thousand nine hundred and ninety-nine by itself.

You may very well reach the result by multiplying this as shown above, but it will be extremely lengthy. Therefore you will proceed as follows. You know[238] that adding one to ninety-nine makes a hundred, adding a hundred to nine hundred makes a thousand, adding a thousand to nine thousand makes ten thousand. Thus this nine thousand nine hundred and ninety-nine is ten thousand minus one. [*So if you want to multiply them* [*for: ten thousand minus one by ten thousand minus one*] *it will be easier.*][239] Then multiply ten thousand minus one by ten thousand minus one in the following way. Multiply ten thousand by ten thousand, thus producing a hundred thousand times thousand. Then multiply one subtracted by ten thousand, twice, which produces twenty thousand subtracted; subtract it[240] from a hundred thousand times thousand as we have taught before in subtracting[241]; this leaves ninety-nine thousand times thousand and nine hundred thousand and eighty thousand. Next, multiply one subtracted by one subtracted, which produces one added. Add it to the above quantity. Therefore the quantity produced by the multiplication is ninety-nine thousand times thousand and nine hundred thousand and eighty thousand and one. This is what you wanted to know. [*Or: subtract from a hundred thousand times thousand twice the product of one into ten thousand and add to the remainder the product of one into itself; this will give the required quantity.*][242]

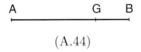

(A.44)

---

[235] *capitulum*. The subtraction taught in A.42–43.

[236] A.23–25, A.27, A.29. The coming problems A.44–45 apply this to the multiplication of higher numbers, first for a number multiplied by itself then for a pair of different numbers.

[237] Irrelevant here; see p. 618 & note 178.

[238] Splitting up as taught in A.42.

[239] If the whole is not a gloss (MS $\mathcal{C}$ has it in the margin), at least the interjection must be interpolated (presumably to clarify the 'them').

[240] That is, its amount (see note 200, p. 624).

[241] Previous problem.

[242] Summary of the previous computations (and convenient for the subsequent proof).

634   Part Two: Translation, Glossary

The proof of this is the following. Let ten thousand be AB and one, BG. We want to know the result of multiplying AG by itself. Now we know that the products of AB into itself and BG into itself are equal to the products of AB into BG, twice, and AG into itself [*by the 7th of the second*][243]. Therefore multiply AB by itself, add to the result the product of GB into itself, and subtract from the sum twice the product of AB into BG; the remainder will be the product of AG into itself. This is why we multiply ten thousand, which is AB, by itself, then subtract from the product twice the product of one into ten thousand, which is twice the product of AB into BG, and add to the remainder the product of GB into itself. The sum is the result of the multiplication.

(A.45) You want to multiply forty-eight thousand times thousand repeated six times and nine hundred thousand repeated five times and ninety-nine thousand repeated five times and nine hundred thousand repeated four times and ninety-nine thousand repeated four times and nine hundred thousand thrice repeated and sixty thousand thrice repeated by twenty thousand repeated four times and nine hundred thousand thrice repeated and ninety-nine thousand thrice repeated and nine hundred thousand twice repeated and ninety-nine thousand twice repeated and nine hundred thousand and eighty thousand.

If you want to multiply them as shown before, it will be hard work.[244] But we know from what has been said above that forty-eight thousand times thousand repeated six times and nine hundred thousand times thousand repeated five times with all the other numbers in the row of the multiplicand is forty-nine thousand times thousand repeated six times minus forty thousand times thousand thrice repeated, while all the numbers in the row of the multiplier are twenty-one thousand times thousand repeated four times minus twenty thousand. Thus multiply forty-nine thousand times thousand repeated six times minus forty thousand times thousand thrice repeated by twenty-one thousand times thousand repeated four times minus twenty thousand, in the following way. Multiply twenty-one thousand times thousand repeated four times by forty-nine thousand times thousand repeated six times, and keep the product in mind. Subtract from this product both the product of twenty thousand into forty-nine thousand times thousand repeated six times and the product of forty thousand times thousand thrice repeated into twenty-one thousand times thousand repeated four times. Add to the remainder the product of twenty thousand into forty thousand times thousand thrice repeated [*for the multiplication of added by subtracted produces subtracted and the multiplication of subtracted by subtracted produces added*][245]. The result is the quantity you require, namely

---

[243] The square of a difference between two quantities is inferred from *Elements* II.7 = Premiss PE_7.

[244] That is, if we do not use the representation as a difference.

[245] The same interpolator will consider that the proof below demonstrates these two rules.

a thousand times thousand repeated eleven times and twenty-eight thousand times thousand repeated ten times and nine hundred thousand times thousand repeated nine times and ninety-nine thousand times thousand repeated nine times and nine hundred thousand times thousand repeated eight times and ninety-eight thousand times thousand repeated eight times and a hundred thousand times thousand repeated seven times and eighty thousand times thousand repeated seven times and eight hundred thousand times thousand repeated four times.[246]

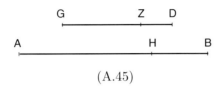

(A.45)

The proof of this is the following. Let forty-nine thousand times thousand repeated six times be AB, and what is subtracted from it, namely forty thousand times thousand thrice repeated, HB; let twenty-one thousand times thousand repeated four times be GD, and what is subtracted from it, namely twenty thousand, ZD. We want to know the result of multiplying AH by GZ. Now we know [247] that the product of AB into GD is equal to the products of AH into GZ and AH into ZD, and the products of HB into GZ and HB into ZD. Let the product of HB into ZD be (added in) common. Then the products of AB into GD and HB into ZD are equal to the products of AH into GZ, AH into ZD, HB into ZG, and HB into ZD, twice. Now the products of HB into GZ and HB into ZD, once, are equal to the product of HB into GD, while the products of AH into ZD and HB into ZD, once, are equal to the product of ZD into AB. So the products of AB into GD and HB into ZD are equal to the products of AH into GZ, AB into ZD and GD into HB. Therefore when you want to know the result of multiplying AH by GZ, and AB, GD, ZD, HB are all known, you will multiply AB by GD and add to the result the product of HB subtracted into ZD subtracted [*it is therefore clear that [the result of] multiplying subtracted by subtracted produces added*], next subtract from the sum the products of ZD into AB and HB into GD; the remainder will be the product of AH into GZ [*thus it is now also clear that [the result of] multiplying subtracted by added produces subtracted*]. This is what we wanted to demonstrate.

CHAPTER ON TAKING FRACTIONS OF THOUSANDS REPEATED

(A.46) You want to know what a third of a thousand times thousand thrice repeated is.

---

[246] The intermediate results have been added in MS $\mathcal{B}$ by its reader (see Latin text, p. 63). In his two upper rows, GZ and AH must be inverted.

[247] Proof in A.32, or PE_1.

636 PART TWO: TRANSLATION, GLOSSARY

Take a third of one thousand,[248] which is three hundred and thirty-three and a third, and multiply this by the remainder of the repetition, which is a thousand times thousand, or append to it the remainder of the repetition, that is, a thousand times thousand, twice (repeated);[249] it will give three hundred thousand times thousand and thirty-three thousand times thousand and a third of a thousand times thousand. Next, take a third of a thousand times thousand, as follows. Append to a third of a thousand, which is three hundred and thirty-three and a third, the above single 'thousand' which remains [or *multiply this third by a thousand, which amounts to the same*]. This gives three hundred thousand and thirty-three thousand, and a third of a thousand, which is three hundred and thirty-three and a third. Thus, a third of a thousand times thousand times thousand, thrice repeated, is three hundred thousand times thousand and thirty-three thousand times thousand and three hundred thousand and thirty-three thousand and three hundred and thirty-three and a third. This is what you wanted to know.

The proof of this is the following. We know[250] that a thousand times thousand thrice repeated results from multiplying a thousand by a thousand times thousand. You want therefore to take a third of the product of a thousand into a thousand times thousand; that is, you want to divide this product by three. Now it amounts to the same to multiply a thousand by a thousand times thousand and divide the product by three as to divide a thousand by three and multiply the quotient by a thousand times thousand, repeated twice, as said previously in the chapter on premisses[251]. It is because of this that you take a third of one thousand and multiply this result by the remainder of the repetition. The proof for taking a third of a thousand times thousand is similar. This is what we wanted to demonstrate.

(A.47) You want to know what a ninth of a thousand times thousand thrice repeated is.

Take a ninth of one thousand, which is a hundred and eleven and a ninth. Append to it the remaining repetition, that is, a thousand times thousand, twice. The result will be a hundred thousand times thousand and eleven thousand times thousand and a ninth of a thousand times thousand. You will find a ninth of a thousand times thousand by the same rule; it is a hundred thousand and eleven thousand and a ninth of a thousand. A ninth of a thousand being a hundred and eleven and a ninth, a ninth of a thousand times thousand thrice repeated is a hundred thousand times thousand and eleven thousand times thousand and a hundred thousand and eleven thousand and a hundred and eleven and a ninth.

---

[248] Even though the subject of division is still to come.

[249] With repeated thousands, 'multiplying by' and 'appending' mean the same, as seen in A.30 (and A.33).

[250] A.30.

[251] Premiss P_5.

LIBER MAHAMELETH

637

AGAIN.[252]

(A.48) You want to know what five sixths of ten thousand times thousand thrice repeated is.

Take five sixths of ten, which is eight and a third, and append to it the whole repetition; for you did not take anything of it.[253] The result will be eight thousand times thousand thrice repeated and a third of a thousand times thousand times thousand, thrice repeated. Find a third of a thousand times thousand thrice repeated according to the rule explained above[254], and add it to the previous quantity. So five sixths of ten thousand times thousand thrice repeated is eight thousand times thousand thrice repeated and three hundred thousand times thousand, twice (repeated), and thirty-three thousand times thousand, twice, and three hundred thousand and thirty-three thousand and three hundred and thirty-three and a third. This is the quantity you require.

(A.49) You want to know what three fifths of eight thousand times thousand is.

Append to three fifths of eight, which is four and four fifths, the repetition; this will give four thousand times thousand and four fifths of a thousand times thousand. Now four fifths of a thousand times thousand is eight hundred thousand. Therefore three fifths of eight thousand times thousand is four thousand times thousand and eight hundred thousand. This is the quantity you require.

(A.50) You want to know what five eighths of a hundred thousand times thousand repeated four times is.

Leaving out the repetition, there remains a hundred. Append to five eighths of it, which is sixty-two and a half, the whole repetition; for you did not take anything of it. This will give sixty-two thousand repeated four times and half a thousand times thousand repeated four times. Now this half is five hundred thousand thrice repeated. Therefore five eighths of a hundred thousand times thousand repeated four times is sixty-two thousand repeated four times and five hundred thousand thrice repeated.

(A.51) You want to know what five sevenths of twenty thousand times thousand thrice repeated is.

What you want here just amounts to taking five sevenths of the product of twenty into a thousand times thousand thrice repeated; but this is the same as taking five sevenths of twenty and multiplying the result by a thousand times thousand thrice repeated.[255] So take five sevenths of twenty

---

[252] We no longer consider unitary fractions.

[253] In the previous problems, unlike here (and in the subsequent problems), the initial division applied to part of the repetition.

[254] A.46.

[255] This in fact justifies the procedure seen in A.46–50. A.37b is a similar case.

and multiply the result by a thousand times thousand thrice repeated. Continue then as shown before[256], and the result will be what you wanted.

Consider other such instances likewise, and you will succeed.

---

[256] Previous problems.

Chapter (A–IV) on Division

Anyone dividing a number by a number has one of two purposes. Either his purpose is to know what will be attributed to one, namely when he divides one thing by another of a different kind; for example, if he divides ten nummi by five men, his purpose is just to know what will be attributed to one of them. Or his purpose is to know what the ratio of the two numbers is, namely of the dividend to the divisor, when he divides one thing by another of the same kind; for example, if he has to divide twenty sextarii by ten sextarii he just wants to know the ratio of the twenty sextarii to the ten. But the way of proceeding is the same in both cases.

(*i*) You must know that multiplying the result of the division by the divisor gives the dividend.

For example, you want to divide ten nummi by five men.[257] [*His purpose here is just to know what will be attributed to one of them.*] What is assigned to one is two nummi. Then what is due to each of them is two nummi, and all of them are five. So when you multiply two by five, ten will be produced, which is the given quantity to be divided by five. Therefore if the result of the division is multiplied by the divisor, the dividend will be produced according to this (first) purpose.

The same holds also for the other purpose. For example, you want to divide ten sextarii by four sextarii. The result is two and a half [*for ten is twice four plus a half (of four)*]. So if you multiply four by two and a half, ten will be produced.

Therefore multiplying the result of the division by the divisor will always give the dividend, for both purposes. Thus the way of proceeding is the same for both.

(*ii*) It must be known that in both (types of) division either larger is divided by smaller, and this is, properly speaking, division; or smaller by larger, and this is called denomination; or equal by equal, in which case the result is just one.[258]

[*In the division of a larger number by a smaller, these rules will have to be observed.*][259]

(**A.52**) You want to divide twenty by four.

---

[257] Above example.

[258] This last case is alluded to later on (A.229, p. 729), so it may belong to the original text.

[259] The intention may have been to make up for the lack of a heading here, namely 'On division of larger by smaller' (which is the subject of A.52–71, A.72–83 being about denomination).

640   PART TWO: TRANSLATION, GLOSSARY

(*a*) Look for the number which multiplying four produces twenty;[260] this is five. Such is the result of the division.

(*b*) Or denominate one from four;[261] this gives a fourth. Such a fraction, that is, a fourth, being taken of twenty, which gives five, is the result of the division.

The proof of this is clear. For the ratio of one to the divisor is the same as the ratio of the required quantity to the dividend.[262] Thus denominating one from the divisor and taking such a fraction of the dividend gives the result of the division.

(*c*) The verification in this case is as follows. Multiplying five by four produces twenty. Thus the dividend appears again. For multiplying the result of a division by the divisor gives the dividend, as has already been said.[263]

(**A.53**) You want to divide sixty by eight.

You will do the following. Look for the number which multiplying eight produces sixty. Or which multiplying eight produces a number smaller than sixty but with a difference from sixty of less than eight. Such is seven: when multiplied by eight it produces fifty-six, with a difference from sixty of four. Denominating this from eight gives a half. Adding it to seven makes seven and a half, and this is the result of the division.

(*iii*) When you want to divide a number by a number, observe whether both are divisible by some fraction[264]; then take this fraction of both numbers and divide the first result by the second; the quotient is the result of the division of the two numbers.[265]

(**A.54**) For instance, you want to divide twenty-four by eight.

Consider which same fraction divides both of them; a fourth does[266]. So divide a fourth of twenty-four, which is six, by a fourth of eight, which is two. This gives three, which also results from dividing twenty-four by eight.

DIVISION OF REPEATED THOUSANDS AMONG THEMSELVES.

(**A.55**) You want to divide a hundred thousand times thousand repeated five times by fifteen thousand times thousand, twice repeated.

---

[260] In accordance with rule *i*.

[261] Denominating is used here and in the subsequent problem, although it will be taught later on.

[262] As inferred from rule *i*.

[263] Rule *i* above.

[264] Thus if the same part can be taken of them both, that is, if they have a common divisor (not necessarily the largest, see A.54).

[265] Proof in A.55.

[266] We would say: Consider which same number divides both of them; four does. (The Latin text reads: *Vide que fractio communis numerat eos; scilicet, quarta.*)

LIBER MAHAMELETH 641

You will do the following. Subtract two from five; this leaves three.[267] Next, divide a hundred thousand times thousand, thrice, by fifteen. That is, divide a hundred by fifteen, which gives six and two thirds, and multiply it by a thousand times thousand thrice repeated. This produces six thousand times thousand thrice repeated and two thirds of a thousand times thousand thrice repeated. Add then two thirds of a thousand times thousand thrice repeated to the six thousand times thousand thrice repeated. The quantity produced is the result of the division.

The proof of this is the following. We know[268] that fifteen thousand times thousand results from multiplying fifteen by a thousand times thousand and that a hundred thousand times thousand repeated five times results from multiplying a hundred thousand times thousand thrice repeated by a thousand times thousand. We have then that the multiplication of fifteen by a thousand times thousand produces fifteen thousand times thousand and the multiplication of a hundred thousand times thousand thrice repeated by a thousand times thousand produces a hundred thousand times thousand repeated five times. These are then two numbers which when multiplied by the same number produce two numbers. Therefore the ratio of the products is the same as the ratio of the multiplicands, as Euclid said [*in the 18th of the seventh*][269]. So the ratio of a hundred thousand times thousand repeated five times to fifteen thousand times thousand is the same as the ratio of a hundred thousand times thousand thrice repeated to fifteen. Therefore the result of dividing a hundred thousand times thousand repeated five times by fifteen thousand times thousand is equal to the result of dividing a hundred thousand times thousand thrice repeated by fifteen. [*For ratio is (like) division.*][270] So dividing a hundred thousand times thousand thrice repeated by fifteen will give what you require. Now dividing a hundred thousand times thousand thrice repeated by fifteen is the same as multiplying a hundred by a thousand times thousand thrice repeated and dividing the product by fifteen. And multiplying a hundred by a thousand times thousand thrice repeated and dividing the product by fifteen is the same as dividing a hundred by fifteen and multiplying the result by a thousand times thousand thrice repeated.[271] This is what we wanted to demonstrate.

(**A.56**) You want to divide a hundred thousand times thousand by twelve.

You will do the following. Take a hundred by itself, without the repetition, and divide it by twelve. This gives eight and a third. Appending the repetition gives eight thousand times thousand and a third of a thousand times thousand. Find a third of a thousand times thousand as explained

---

[267] Thus removing the common repetition (rule *iii*).

[268] As seen in A.37*b*.

[269] *Elements* VII.18.

[270] As said in the introduction to this chapter (quantities of the same kind).

[271] Premiss P_5.

642   Part Two: Translation, Glossary

before[272], and add it to eight thousand times thousand. The result of the division is therefore eight thousand times thousand and three hundred thousand and thirty-three thousand three hundred and thirty-three and a third.

AGAIN.

(A.57) You want to divide fifty thousand times thousand repeated four times by eight thousand.

Disregarding the thousand in the eight and as much in the fifty there remains the division of fifty thousand times thousand thrice repeated by eight. Proceed as explained before.[273] That is, take fifty by itself, without the repetition, and divide it by eight; the result is six and a fourth. Appending the repetition disregarded[274] gives six thousand times thousand thrice repeated and a fourth of a thousand times thousand thrice repeated. But a fourth of a thousand thrice repeated is two hundred thousand times thousand and fifty thousand times thousand. Adding it to the previous quantity will give the result of the division.

AGAIN.

(A.58) You want to divide eight thousand times thousand thrice repeated by four hundred.

(a) Divide eight with 'thousand' only once by four hundred; this gives twenty. Appending to it the remainder of the repetition, that is, 'thousand' twice, gives twenty thousand times thousand. This is the result of the division.

(b) You may also take one 'thousand' from eight thousand times thousand thrice repeated and divide it by four hundred; this gives two and a half. Multiplying this by what remains, that is, eight thousand times thousand, produces twenty thousand times thousand, and this is the result of the division.

(A.59) You want to divide six thousand times thousand repeated four times by two hundred thousand times thousand.

Disregarding the whole repetition in the divisor and as much in the dividend, there will remain the division of six thousand times thousand, twice, by two hundred. Perform the division as explained before:[275] either divide six with 'thousand' taken once, that is, six thousand, by two hundred and append to the result the remaining repetition; or divide one of its 'thousand's by two hundred and multiply the quotient by six with the remainder of the repetition, the product being the result of the division.

CHAPTER ON ANOTHER WAY OF DIVIDING

---

[272] A.46.

[273] Previous problem.

[274] Not the common repetition initially disregarded.

[275] Previous problem.

LIBER MAHAMELETH   **643**

(**A.60**) For instance, you want to know how many money-bags correspond to twenty thousand times thousand morabitini.

A money-bag has a capacity of five hundred morabitini.

(**a**) You will do the following. Divide twenty thousand times thousand by five hundred as shown before in the chapter on division;[276] the result is forty thousand, which is the number of money-bags in question.

(**b**) Or otherwise:[277]

(*i*) Always disregard one 'thousand' in the repetition, and double the remainder; this will give what you require.

In this case, disregarding one 'thousand' in twenty thousand times thousand leaves twenty thousand, which doubled becomes forty thousand. This is the quantity you require.

We have proceeded in this way for the following reason. However many repeated thousands you want to divide by five hundred, you will take out one 'thousand', divide it by five hundred, and multiply the result by what remains of the number.[278] Now dividing one thousand by five hundred gives two, and multiplying this two by the remainder produces twice the remainder. [*The proof of this is the following.*[279] *We know that twenty thousand times thousand results from multiplying a thousand by twenty thousand. Thus we are to multiply a thousand by twenty thousand and divide the result by five hundred. This is just the same as dividing a thousand by five hundred and multiplying the quotient by twenty thousand. Since dividing the thousands by five hundred gives two, you will multiply two by twenty ⟨thousand⟩; this produces forty ⟨thousand⟩, which is the number of money-bags. This is what we wanted to demonstrate.*]

(**A.61**) Someone now asks: How many money-bags correspond to six hundred thousand times thousand, repeated four times, morabitini?

(**a**) Either perform the division by five hundred, which is the capacity of one money-bag, as shown before.[280]

(**b**) Or remove one 'thousand' in the repetition, thus leaving six hundred thousand thrice repeated. Twice this is a thousand times thousand repeated four times and two hundred thousand times thousand thrice repeated, and this is what you wanted to know.

CONVERSELY.

(**A.62**) Someone asks: In two hundred thousand money-bags how many morabitini are there?

There are five hundred morabitini in one money-bag.[281]

---

[276] Previous section of this chapter.

[277] A (specific) practical rule will be given each time.

[278] This procedure has been applied before (A.58–59).

[279] Superfluous, and somewhat carelessly expressed.

[280] A.58.

[281] Perhaps an interpolation: this has been stated twice already.

644   PART TWO: TRANSLATION, GLOSSARY

(*a*) Multiply two hundred thousand by five hundred; the result is one hundred thousand times thousand, and this is what you wanted.

(*b*) Or, inverting the previous procedure:[282]

(*i'*) Append once 'thousand' to half the given number.

Half the two hundred thousand is a hundred thousand. Appending to it one 'thousand' gives one hundred thousand times thousand, and this is what you wanted to know.

We have proceeded thus because five hundred is half a thousand. Then it is like asking that two hundred thousand be multiplied by half a thousand. Therefore multiply a half by two hundred, thus producing a hundred, and append to it the repetition which was with the half and with the two hundred;[283] this will give what you require.

(**A.63**) Someone asks: How many morabitini are there in three thousand times thousand money-bags?

(*a*) Multiply this by five hundred as we have taught before [284]; this will produce a thousand times thousand times thousand, thrice, and five hundred thousand times thousand, twice. This is what you wanted.

(*b*) You may do it in the other way. Take half of three thousand times thousand, which is a thousand times thousand and five hundred thousand. Appending to it once 'thousand' gives as the quantity a thousand times thousand times thousand and five hundred thousand times thousand, and this is what you wanted to know.

CHAPTER ON TREASURIES.

(**A.64**) Someone asks: To how many treasuries correspond ten thousand times thousand times thousand, thrice repeated, morabitini?

A treasury has a capacity of one thousand times thousand morabitini.

(*a*) Therefore divide this, that is, ten thousand thrice repeated, by a thousand times thousand as we have taught in the chapter on division [285]; the result will be ten thousand treasuries.

(*ii*) Now dividing a number by a thousand times thousand always means discarding from it twice 'thousand', what is left being the result of the division.

(*b*) (Whence) otherwise: Remove from the (given) number of morabitini twice 'thousand'; what is left is the number of treasuries to which they correspond.

(**A.65**) Someone asks: To how many treasuries correspond a hundred thousand times thousand, repeated four times, morabitini?

---

[282] Namely rule *i* in A.60*b*.
[283] One repetition with each.
[284] A.21.
[285] (If need be) A.55–59.

LIBER MAHAMELETH 645

Disregard 'thousand' twice in the given number; this leaves a hundred thousand times thousand, which is the quantity of treasuries to which the given number corresponds.

CONVERSE OF THAT.

(**A.66**) Someone asks: How many morabitini are there in ten thousand treasuries?

(*a*) Multiply the given number by the quantity of morabitini in one treasury, which is a thousand times thousand; the result is ten thousand times thousand thrice repeated. This is the quantity of morabitini contained in ten thousand treasuries.

(*ii'*) Now multiplying a number by a thousand times thousand always amounts to appending to it twice 'thousand'.

(*b*) Thus always add to the number of treasuries twice 'thousand', and this will be the quantity of morabitini which are in the treasuries.

(**A.67**) For instance someone may say: How many morabitini are there in a hundred thousand times thousand treasuries?

Append to the given number 'thousand' twice repeated; the result is a hundred thousand repeated four times, which is the quantity of morabitini contained in a hundred thousand times thousand treasuries.

ON THE NUMBER OF MONEY-BAGS IN TREASURIES.

(**A.68**) Someone asks: How many money-bags are there in three hundred treasuries?

(*a*) You already know that there are a thousand times thousand morabitini in each treasury.[286] But a thousand times thousand morabitini correspond to two thousand money-bags.[287] Therefore there are two thousand money-bags in one treasury. Thus always multiply the number of treasuries by two thousand; the result will be the quantity of money-bags in these treasuries. [*In the present problem there are six hundred thousand money-bags.*]

(*iii*) Now the multiplication of a number by two thousand always means doubling the number and appending 'thousand' once to the result.

(*b*) You may thus double here the number of treasuries and append 'thousand' to the result; this will give the quantity of money-bags which are in these treasuries.

(**A.69**) For instance someone asks: How many money-bags are there in six hundred thousand treasuries?

Double six hundred thousand. This gives a thousand times thousand and two hundred thousand. Append to it 'thousand' once. The result is a thousand times thousand times thousand and two hundred thousand times thousand, and this is the quantity of money-bags which are in the given treasuries.

---

[286] A.64.

[287] A.60*b* (rule *i*).

646 PART TWO: TRANSLATION, GLOSSARY

CONVERSE OF THIS.

(**A.70**) Someone asks: How many treasuries correspond to six hundred thousand money-bags?

(*a*) Divide the given number by the quantity of money-bags contained in one treasury, which is two thousand; this gives three hundred treasuries.

(*iii'*) Now the division of a number by two thousand always means removing one repetition from it and taking half the remainder.

(*b*) You may therefore remove here one repetition from the number of money-bags, and half the remainder will be what you wanted to know.

(**A.71**) For instance someone may ask: To how many treasuries correspond a hundred thousand times thousand money-bags?

Remove from it one repetition, thus leaving a hundred thousand. Half of it, which is fifty thousand, is the quantity of treasuries corresponding to a hundred thousand times thousand money-bags.

In order that you may better understand what has been said about money-bags, treasuries and morabitini, I have put below a figure. When you want to convert one of these into another, place one finger on what you want to convert and another finger on that into which you want to convert. Then bring down the upper finger vertically, move simultaneously the other finger towards it horizontally. In the place where they meet is indicated the rule by which you are to proceed.

CHAPTER ON THE DENOMINATIONS

Let there first be set forth some obvious facts which will enable you to denominate one number from another. They are the following.

(*i*) Any number[288] having no half does not have a fourth, sixth, eighth or tenth either.

(*ii*) Any number having no third does not have a sixth or a ninth either.

(*iii*) Any number having no fourth does not have an eighth either.

(*iv*) Any number having no fifth does not have a tenth either.

(*v*) Any number having a tenth has also a fifth and a half.

(*vi*) Any number having a ninth has a third.

(*vii*) Any number having an eighth has a fourth and a half.

(*viii*) Any number having a sixth has a third and a half.

(*ix*) Any number having a fourth has a half.

(*x*) No odd number has a fractional part denominated from an even number.[289]

---

[288] Throughout: integer, of which we examine the divisibility by the first integers from 2 to 10.

[289] That is, a fractional part taking its name from an even number (including a half, not denominated from two, either in Arabic or in Latin).

LIBER MAHAMELETH

What must be converted

Morabitini	Money-bags	Treasuries	O	
Remove from the number of morabitini two repetitions, and what remains is the number of treasuries	Remove from the number of money-bags one repetition, and half the remainder will be the number of treasuries	O	Treasuries	**O t h e r k i n d i n t o w h i c h i s c o n v e r t e d**
Remove from the number of morabitini one repetition and double the remainder, and the result is what you wanted to know	O	Double the number of treasuries and add to the result one repetition, and the result is the number of money-bags which are in these treasuries	Money-bags	
O	Add one repetition to half the number of money-bags, and the result is the number of morabitini	Add two repetitions to the number of treasuries, and the result is the number of morabitini which are in them	Morabitini	

(Summary of the rules in A.60–71)

(*xi*) Any number divisible by nine has a ninth; if it is even, it has a sixth and a third [*and the other parts of an even*], if odd a ninth and a third only. If after (division by) nine the remainder is six and the number is even, it will have a sixth and a third, if the remainder is three it will have a third.

(*xii*) Any number divisible by eight has an eighth, a fourth and a half.[290] If there remains something after (division by) eight, it will not have an eighth; if four remains, it will have a fourth.

(*xiii*) No number has a seventh unless seven divides it.

(*xiv*) Likewise, none has a sixth unless six divides it. If after (division by) six three remains, it will have a third.

(*xv*) None has a fifth unless five divides it.

(*xvi*) No number has a fourth unless four divides it.

(*xvii*) None has a third unless three divides it.

---

[290] Rule *vii*.

648 PART TWO: TRANSLATION, GLOSSARY

(*xviii*) A number which does not have a fractional part denominated from any of the digits up to one[291] does not have a fractional part except those denominated from the composite odd numbers which only the unit divides, such as eleven, thirteen, seventeen and so on.[292]

(*xix*) To find out whether a number has a tenth. Consider whether there is in it a digit or not.[293] For any number which does not have a digit has a tenth, whereas if it does it cannot have a tenth.

(*xx*) To find out whether a number has a ninth. Take one from each article or limit which is in it.[294] If the sum of the units taken, together with the digit, if any, is divisible by nine, the given number will have a ninth, otherwise not.

(**A.72**) For instance, you want to know if a hundred and fifty-four has a ninth.

Take one from this limit, which is a hundred. Take one from each article which is in fifty, thus you will take five. This added together with the unit taken from a hundred and the digit, which is four, makes ten. Since ten is not divisible by nine, the given number does not have a ninth.

(*xxi*) To find out whether a number has an eighth. Take two from each ten and four from each one hundred. If the sum of what has been taken, plus the digit, if any, is divisible by eight, the number will have an eighth, otherwise not.

(**A.73**) For instance, you want to know if two hundred and sixty-four has an eighth.

Taking four from each one hundred, you will take eight from two hundred. Taking two from each ten, you will take twelve from sixty. These added together with the previous eight and the digit found here, that is, four, make twenty-four. Since division of it by eight gives three, the proposed number has an eighth. From a thousand, however many times it is found there, you will not take anything because every thousand has an eighth: any thousand is divided by eight a hundred and twenty-five times [*since eight times a hundred and twenty-five is one thousand*].

(*xxii*) To find out whether a number has a seventh. Take three from each ten, two from each one hundred, six from each one thousand, four from each ten times thousand; take five from each one hundred times thousand, one from each one thousand times thousand, and so on: take one from each one thousand repeated an even number of times, three from each ten times

---

[291] Here too excluded (see notes 11 & 21, pp. 582–583).

[292] Composite primes, thus 'composite' in the sense seen at the beginning of A–III (p. 605).

[293] As usual, only significant digits (not zero) are considered.

[294] 'from each article or limit': from each ten in the article and each multiple of a limit. Similar ambiguity in A.72, A.76 and rules *xxiii*, *xxv*, *xxvi*. Here a limit is a power 10^k, $k \geq 2$, so also in A.76 and rules *xxiii* & *xxvi*.

LIBER MAHAMELETH

thousand repeated an even number of times, two from each one hundred times thousand repeated an even number of times; but take six from each one thousand repeated an odd number of times, four from each ten times thousand repeated an odd number of times, five from each one hundred times thousand repeated an odd number of times. Add these quantities together with the digit found there; if the sum is divisible by seven, the number has a seventh, otherwise not.

(A.74) For instance, you want to know if two thousand three hundred and forty-eight has a seventh.

Taking three from each ten, you will take, from forty, twelve; taking two from each one hundred, you will take, from three hundred, six; taking six from each one thousand, you will take, from two thousand, twelve. Add this together with the previous twelve and six, and the digit found there, namely eight, making thirty-eight. Since this is not divisible by seven, the given number does not have a seventh.

AGAIN.

(A.75) You want to know if thirty thousand times thousand and four hundred thousand times thousand and four thousand times thousand times thousand has a seventh.[295]

Taking three from each ten times thousand repeated an even number of times, you will take nine from thirty thousand times thousand; taking two from each one hundred times thousand repeated an even number of times, you will take, from four hundred thousand times thousand, eight; taking six from each one thousand repeated an odd number of times, you will take, from four thousand times thousand times thousand, twenty-four. Adding together all you have taken makes forty-one. Since this is not divisible by seven, the given number does not have a seventh.

(*xxiii*) To find out whether a number has a sixth. Take four from each article or limit, and add together what you have taken, including the digit, if any. If the sum is divisible by six, the number will have a sixth, otherwise not.

(A.76) For instance, you want to know if two thousand three hundred and twenty-four has a sixth.

Take from each article four, thus, from twenty, eight; take from each limit four, thus, from two thousand, eight, and, from three hundred, twelve. Adding together all of them, including the digit, makes thirty-two. Since this is not divisible by six, the given number does not have a sixth.

(*xxiv*) To find out whether a number has a fifth. Consider whether there is a digit other than five in this number. For every number having a digit other than five does not have a fifth; all others do.[296]

---

[295] We have already met with the expression of a number arranged by increasing powers of 10 (A.41).

[296] That is: all those ending with 5 or 0 (not a digit) are divisible by 5.

650   Part Two: Translation, Glossary

(*xxv*) To find out whether a number has a fourth. Take two from each article and add together what you have taken, including the digit, if any. If the sum is divisible by four, the number will have a fourth, otherwise not. From one hundred and beyond [*that is, articles*], you will not take anything since all [*articles of hundreds and beyond*] do have a fourth.

(*xxvi*) To find out whether a number has a third, you will proceed as you did for a ninth.[297] You will take one from each article or limit, and you will add together all you have taken, including the digit, if any. If the sum is divisible by three, the number will have a third, otherwise not.

(*xxvii*) To find out whether a number has a half. Consider whether it is even or odd. For every even and no odd has a half.

   [*In all these denominations we understand by 'parts' integral numbers only.*][298]

(*xxviii*) When it is established that a number has a tenth[299] and you want to know what its tenth is, move this number one place to the right; the number resulting from the move will be a tenth of the number before the move.

(**A.77**) For instance, you ask: What is a tenth of one thousand two hundred?

   Move one thousand two hundred back one place; it becomes one hundred and twenty, which is a tenth of one thousand two hundred.

AGAIN.[300]

(**A.78**) You want to know what a tenth of one hundred and twenty is.

   Move it back one place; this gives twelve, which is a tenth of one hundred and twenty. [*Since this twelve has a digit, it cannot have, according to the previous rule,*[301] *a tenth. But since nine dividing twelve leaves a remainder of three, and twelve is even, it has a half, a third, a fourth and a sixth.*[302]]

   One can find in the same way a tenth of all those numbers which can be moved back one place.

(*xxix*) You want to know what a fifth of some number is. Consider whether there is a digit in this number or not.[303] If there is none, move it, as said before, back one place, and you will find its tenth; double it, and you will

---

[297] Rule *xx*.

[298] This information should have appeared (if at all) at the beginning of the section.

[299] By rule *xix*.

[300] Actually, it continues with the previous result.

[301] Rule *xix* again.

[302] An early reader wanted to apply rule *xi*. Note that an even number of the form $9t + 3$, thus with t odd, will not have a fourth if $t = 4m + 3$.

[303] 'or not': that is, it ends with 0.

LIBER MAHAMELETH

have its fifth. If there is the digit five, disregard this digit and find a tenth of the remainder by the above rule; doubling the result and adding one will give its fifth.

(A.79) For instance, you want to know what a fifth of twenty-five is.

Disregard the digit, which is five. After moving the remainder back one place, you will find its tenth, which is two. Doubling two gives four, and adding one to it gives five; this is its fifth.

The same in all other instances.

(xxx) When you want to know about the ninth, the eighth, the seventh and the others,[304] divide the number of which you ask for the fraction by the number from which this fraction is denominated; the result is the required fraction of the proposed number.

(xxxi) When you want to denominate a number from a number, put first the number to be denominated, and underneath it the number from which you want to denominate. Consider what (integral) parts the latter has, starting with a tenth and ending with a half.[305] Put, underneath the number from which you want to denominate, the quantity of the first part you find.[306] Next, consider what parts this last quantity has, starting with a tenth and up to a half.[307] You will write the quantity of the first part you find underneath the quantity of which it is a part. You will do the same for this quantity and for the following ones until you reach unity. This being done, divide the number to be originally denominated by the quantity of the fraction immediately following; the quotient will be the numerator of the fraction in question[308]. After that, if there is a remainder from the division, divide it by the quantity of the fraction immediately following; the quotient will be the numerator of the fraction in question.[309] If something still remains, you will proceed in the same way until nothing remains. You will then add all the fractions, and the sum is what you are looking for.

(A.80) For instance: Denominate a thousand eight hundred and thirty-six from five thousand and forty.

First, put the number to be denominated, and underneath it the denominating number, in this way:

One thousand eight hundred and thirty-six

Five thousand and forty.

---

[304] Not including 5.

[305] It is supposed that these are the only parts the divisor has.

[306] We are to distinguish here between 'quantity' and 'part' or 'fraction', the former being the (integral) result of the division by the latter's denominator.

[307] The divisor may be divisible by 10 more than once.

[308] Literally: 'the number of that fraction by which you divide'.

[309] Which will be a fraction of a fraction.

652   PART TWO: TRANSLATION, GLOSSARY

Next, consider what parts the denominating number, which is the lower one, has, starting with a tenth and up to a half. By the previous rule[310], you will find a tenth of it, which is five hundred and four. Put it underneath the denominating number. Next, consider what parts this tenth has, starting with a tenth and up to a half; you will find according to the foregoing[311] that it has a ninth, which is fifty-six. Put it underneath five hundred and four. Next, consider what parts this ninth has; you will find an eighth of it, which is seven. Put it underneath fifty-six. Next, put a seventh of seven, which is one, underneath seven. This being done, divide the number to be originally denominated by the quantity of the fraction immediately following, that is, by five hundred and four; this will give three, thus three tenths, and there will remain three hundred and twenty-four. Divide this by the quantity of the fraction immediately following, that is, by fifty-six; this will give five, thus five ninths of one tenth, and there remains forty-four. Divide it by the quantity of the fraction following, that is, seven; this will give six, thus six eighths of one ninth of one tenth, and there remains two. Denominate it from seven, which gives two sevenths (thus two sevenths) of one eighth of one ninth of one tenth. Next, add all these parts; this will give three tenths and five ninths of one tenth and six eighths of one ninth of one tenth and two sevenths of an eighth of a ninth of a tenth. Such is the part that the number to be denominated is of the denominating number. The attached figure represents it.

	1836	to be denominated
	5040	denominating
three tenths	504	tenth
five ninths	56	ninth
six eighths	7	eighth
two sevenths	1	seventh

$$(A.80)$$

(*xxxi'*) For the succession of these fractions you may put the larger denominators first, even though they correspond to smaller fractions. For a tenth of one seventh is the same as a seventh of one tenth, and an eighth of a ninth is the same as a ninth of an eighth [*for from seven times ten and from ten times seven the same results, and so in similar cases, as Euclid has said*]. [*When something remains from the division and you do not know which part it is of the dividing number, you will by the above rule denominate it from the dividing number. For you will put what remains to be denominated first, underneath it the denominating, that is, the dividing; next, you will consider by the previous rules which parts the dividing has, starting with a tenth and up to a half; the part you find first you will place*

---

[310] Above, rule *xxviii*.

[311] Above, rule *xx* & *xxx*.

LIBER MAHAMELETH

underneath the dividing, and then you will continue with the other steps as put above.][312]

All these[313] are to be considered in the division called 'denomination', that is, when smaller is divided by larger. In order to make it clearer we append the following examples.[314]

(**A.81**) For instance, you want to denominate one from twelve.

You know[315] that twelve results from the multiplication of three by four; thus three is a fourth of twelve, and four, a third of it. Since one is a third of three, one is a third of a fourth of twelve. Again, you also know that twelve results from the multiplication of six by two; thus six is a half of twelve, and two, its sixth. Since one is a half of two, one is a half of a sixth of twelve [*or a sixth of its half*].

(**A.82**) You want to denominate one from thirteen.

You know that thirteen does not result from the multiplication of any number.[316] Thus one is its thirteenth part.[317]

(**A.83**) Similarly[318] if you want to denominate one from fourteen.

You know that fourteen results from the multiplication of seven by two; thus seven is its half, and two, its seventh. Since one is a half of two, one is a half of a seventh of fourteen.

You will likewise obtain the result for other digits.[319]

(**A.84**) You want to denominate one from a thousand.

You know that a hundred is a tenth of a thousand, ten a tenth of a hundred, one a tenth of ten. Thus, say that one is a tenth of a tenth of a tenth, thrice repeated, of a thousand. When you want to denominate one from a thousand times thousand, say that one is a tenth of a tenth six times repeated. Indeed, a thousand is a tenth of a tenth of a tenth of a thousand times thousand, whilst one is a tenth of a tenth of a tenth of a thousand; thus one is a tenth of a tenth, six times repeated, of a thousand times thousand. One will be, of a thousand times thousand times thousand, thrice, a tenth of a tenth repeated nine times. One will be, of a thousand times thousand repeated four times, a tenth of a tenth repeated twelve times. And always according to this reasoning: as many times you repeat

---

[312] Repeats (in part) rule *xxx*.

[313] Rules *i–xxxi'*.

[314] First with 1 in the numerator (A.81–85), then with other integers (A.86–89). In the coming examples factorization of the given divisors is straightforward.

[315] From the multiplication table. Same in A.83, A.84, A.87.

[316] Remember that the unit is not a number (p. 582).

[317] '*tredecima pars*' (here the Arabic would need a circumlocution). This example illustrates rule *xviii*.

[318] This refers to A.81.

[319] Digits other than 1 in the dividend.

654   PART TWO: TRANSLATION, GLOSSARY

'thousand', as many times will you repeat, (once) for each repetition, 'tenth of a tenth of a tenth', thrice.[320]

(A.85) You want to denominate one from eighty thousand.

You will do the following. Take eighty by itself and denominate one from eighty; this is an eighth of a tenth. Append to it 'of a tenth' as many times as required by each 'thousand'. Since there is here 'thousand' only once, you will say: an eighth of a tenth of a tenth repeated four times. Such is the ratio one has to eighty thousand.

(A.86) You want to denominate twelve from twenty-seven.

You know that twenty-seven results from the multiplication of three by nine. Thus three is a ninth of twenty-seven [*and nine, its third*][321]. Now twelve is four times three. Therefore twelve is four ninths of twenty-seven.

(A.87) You want to denominate fourteen from forty-five.

You know that forty-five results from the multiplication of nine by five. Thus five is a ninth of forty-five. But fourteen is twice five plus four fifths of it. Therefore fourteen is two ninths of forty-five and four fifths of a ninth of it.

Similarly in all other instances, whether with articles or composites[322].

CHAPTER ON DENOMINATING THOUSANDS REPEATED, OR OTHER NUM-BERS, FROM THOUSANDS REPEATED.

(A.88) You want to denominate five thousand from forty thousand times thousand thrice repeated.

Disregard the 'thousand' which is with five, and disregard as much from the repetition which is with forty; the latter becomes forty thousand times thousand, from which you are to denominate five.[323] So denominate five from forty, which is an eighth. Append to it 'of a tenth' as many times as required for each 'thousand'. Since 'thousand' is said twice, it will be an eighth of a tenth of a tenth repeated six times. This is what you intended to know.

(A.89) You want to denominate four hundred from ten thousand times thousand.

Denominate four hundred from ten thousand, which is two fifths of a tenth. Append to it '(of a) tenth' as many times as required for the single remaining 'thousand', thus a tenth of a tenth, thrice. Therefore the denomination will give two fifths of a tenth of a tenth repeated four times. This is what you wanted.

Similarly in all other instances.

---

[320] Analogous to the multiplication of thousands seen in A.30.

[321] This seems irrelevant here.

[322] That is, as the dividends, which in the last two examples are no longer digits.

[323] According to rule *iii*, p. 640.

Chapter (A–V) on the Multiplication of fractions

CHAPTER ON THE MULTIPLICATION OF A FRACTION BY AN INTEGER
There are four cases.[324]

FIRST CASE. MULTIPLICATION OF SEVERAL FRACTIONS[325] BY A DIGIT.

(**A.90**) You want to multiply three fourths by seven.[326]

(*a*) You will do the following. Take of the number from which a fourth is denominated, namely four, three fourths, which is three. Multiply it by seven, thus producing twenty-one. Divide it by four; this gives five and a fourth. Such is three fourths of seven.[327]

3	7
4·	
3	7
21	
4	

(A.90*a*)

The proof of this is the following.[328] The ratio of three, the numerator of the fraction, to four, the denominator, is the same as the ratio of three

---

[324] As noted by an earlier reader at the end of this section (note to line 2335 of the Latin text) there are only three: an integer may multiply a fraction (A.90–92), a fraction of a fraction (A.93–95), or a compound fraction (A.96–98). The above statement cannot have originated with the author.

[325] 'several fractions': a non-aliquot fraction. Same in the title before A.91.

[326] We find, in this chapter and the next, illustrations which reproduce in numerical figures the statement and, at times, results of calculations as well; but see Mathematical Commentary, pp. 1200–1201.

[327] The expressions 'taking three fourths of seven' and 'multiplying three fourths by seven' are thus stated to mean the same.

[328] Consider a fractional expression $\frac{p}{q}$. We find each of its two terms designated in several ways. First, $p > 1$ is called *numerus fractionis* (A.90*c*, A.93*c* and in the rule before A.95) or *numerus fractionum* (A.90*a*, A.90*b*, A.92*b*, A.93*a–c*, A.95*b*, A.96*b*, A.97*b*). We shall translate both by 'numerator of the fraction'. The parts may be specified, as in *numerus quartarum*, which will be rendered as the 'number of fourths' (A.100*b*,

656   PART TWO: TRANSLATION, GLOSSARY

fourths of seven, the required quantity, to seven. These are therefore four proportional numbers. Thus[329] the product of the first into the fourth will be as much as the product of the second into the third. So multiplying the first, which is three, namely the numerator of the fraction, by the fourth, which is seven, and dividing the product by the denominator, which is four, will give the third, which is required. This is what we wanted to demonstrate.

Another reason for the same is the following. Any number multiplied by one remains the same.[330] Thus multiplying a fourth by one will produce nothing but this fourth, and multiplying three fourths by one will produce nothing but three fourths. Multiplying it now by seven will produce twenty-one fourths; this is three fourths of seven. We now want to know how many times one is in it. Since there are in one four fourths, dividing the twenty-one (which are) fourths by four will give what you wanted to know, which is five and a fourth.

(**b**) Or otherwise. Multiply the numerator of the fraction, which is three, by the integer, which is seven, and divide the product by four, from which a fourth is denominated; the result will be what you wanted to know.   The proof of this is clear from what precedes.[331]

(**c**) Another rule about the same. Take of the integer the same part as the denominator of the fraction, thus a fourth of this integer, and multiply the result by the numerator of the fraction, in this case three. The product will be the result of the multiplication in question.

MULTIPLICATION OF SEVERAL FRACTIONS BY A COMPOSITE.

(**A.91**) You want to multiply three fifths by forty-seven.

(**a**) You may multiply them according to the previous rules.

A.106*b*).   The quantity q, or 4 here, may be called *numerus denominationis* (A.90*a*, A.93*a*, A.94*a*, A.95*a*, A.97*a*, A.100*a*), or simply *denominatio* (A.90*c*, A.93*b* & *c*, A.94*b*, A.95*b–c*, A.96*b*, A.97*b*); for both we shall write 'denominator'. (On another translation of *numerus denominationis*, see note 345, p. 660.) Here again the parts may be specified. For the expression *numerus a quo* (or: *unde*) *denominatur quarta* we shall write 'the number from which a fourth is denominated' (A.90*a*, A.90*b*, A.92*a* & *b*, A.103*a*), while for *numerus denominationis quarte* and *numerus denominans quartam* we adopted 'denominator of a fourth' (A.93*a*, A.117*i*, A.123*a*). Note finally that *denominatio* may mean not only 'denominator' but also 'denomination', that is, the (aliquot) fraction which takes its name from the denominator: compare *denominatio undecime que est undecim* in A.106*a* and *denominationes que sunt undecima et quinta* in A.98*a* (or A.96*a*). We have translated both, *denominatio que est* and *denominatio*, by 'denominator of'. (For other examples of this ambiguity, see A.121 & A.122, A.150 & A.151.)

[329] This being in common use, we shall no longer refer to *Elements* VII.19.

[330] As already said in the multiplication table for digits (p. 607).

[331] Proof in *a*.

(**b**) Or you may take three fifths of forty-five, which is the (integral) number closest to forty-seven having a fifth without a fraction; the result is twenty-seven, which you keep in mind. Next, take three fifths of the remaining two, which is the difference between the two numbers; this is one and a fifth. Adding this to the twenty-seven kept in mind will make twenty-eight and a fifth, and such is the quantity produced.

[*Or take one fifth of forty-seven and multiply it by three. The product will be what you require.*]³³²

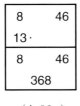

(A.91*b*)

MULTIPLICATION OF SEVERAL COMPOSITE FRACTIONS BY A COMPOSITE.³³³

(**A.92**) You want to multiply eight thirteenths by forty-six.

(**a**) You will do the following. Take eight thirteenths of the number from which a thirteenth is denominated; this is eight. Multiply it by forty-six, thus producing three hundred and sixty-eight. Dividing it by thirteen gives twenty-eight and four thirteenths, and this is what you wanted to know.

(A.92*a*)

(**b**) Or otherwise. Multiply eight, which is the numerator of the fraction, by forty-six and divide the product by the number from which a thirteenth is denominated. The result will be what you require.

(**c**) Or otherwise. Take eight thirteenths of thirty-nine, which is the closest (integral) number below forty-six having a thirteenth without a fraction; this gives twenty-four. Next, take eight thirteenths of their difference, that is, seven, in the following way: multiply eight by seven and divide the product by thirteen. This gives four and four thirteenths. Adding this to

---

³³² Not found in MSS 𝒜ℬ and indicated as superfluous by the copyist of MS 𝒞. Indeed, it corresponds to *c* in A.90, thus to *a* here.

³³³ 'Several composite fractions': the numerator is greater than 1 and the denominator an integer (prime or not) larger than 10.

658 PART TWO: TRANSLATION, GLOSSARY

twenty-four makes twenty-eight and four thirteenths, and this is what you wanted.

SECOND CASE. MULTIPLICATION OF A FRACTION OF A FRACTION BY AN INTEGER.

[*FIRST. MULTIPLICATION OF FRACTIONS OF A FRACTION OF A DIGIT BY A COMPOSITE.*][334]

(**A.93**) You want to multiply three fourths of a seventh by fifteen.

(*a*) You will do the following. Form, by multiplying the denominators of a fourth and a seventh, twenty-eight, which is the common number[335]. Next, multiply three fourths of a seventh of it, which is three, by fifteen, thus producing forty-five. Dividing this by the common[336] will give one and four sevenths and a fourth of a seventh.

The proof of this is evident from what precedes.[337] For the ratio of three fourths of a seventh of twenty-eight to twenty-eight is the same as the ratio of the required quantity, which is three fourths of one seventh of fifteen, to fifteen. Then the product of the first into the fourth will be as much as the product of the second into the third. So multiplying the first, which is three, by the fourth, which is fifteen, and dividing the product, which is forty-five, by the second will give the third, which is what we require.

Another reason explaining that we multiply the numerator of the fraction, which is three, by fifteen, thus producing forty-five, which are fourths of a seventh, just in order to know what three fourths of one seventh of fifteen are.[338] Since multiplying anything by one just produces the same, so multiplying three fourths of one seventh by one will simply reproduce the same; multiplying it now by fifteen will produce forty-five fourths of one seventh. Dividing this[339] by seven will give fourths, by four, sevenths, for all seven sevenths of a fourth are one fourth and all four fourths of a seventh are one seventh. [*Then, if they are sevenths, divide the result by seven, if they are fourths, divide it by four; this will give an integral number.*] The result is three fourths of a seventh of fifteen, and this is what we wanted to demonstrate.

---

[334] Only $\mathcal{C}$ has this subheading, and in the margin.

[335] 'common number' (*numerus communis*, see A.94*a*, A.95*a*, A.96*a*, A.101*a*) is used to designate a common denominator of two given expressions, usually just the product of all denominators. (Here one side is an integer.) This will often come to be replaced (from A.100 on) by *prelatus* (which also has another meaning, see note 493, p. 703).

[336] We adopt the mode of expression of the text.

[337] This refers to the analogous proof in A.90*a*.

[338] This other justification corresponds to A.90*a*'s 'another reason'.

[339] *Que* (that is, *partes*), although the integer forty-five is meant. This is not an unusual expression; see A.98*b*, A.104*b*, A.126*b*, A.126*c*.

LIBER MAHAMELETH

659

(*b*) Or otherwise. Always[340] multiply the numerator of the fraction, in the present case three, by the multiplier, here fifteen, and divide the result by the product of the two denominators. The result will be what you wanted.

The proof of this is the following. We had to divide forty-five by four and the quotient by seven, or conversely, first by seven and then by four, as said before[341]. But this is just the same as dividing forty-five by the product of four into seven. For dividing one number by another and the result by some third number is the same as dividing the first by the product of the two divisors. This is already clear from the premisses.[342]

(*c*) Another rule for dealing with the same is the following. Multiply the numerator of the fraction, which is three, by fifteen; this produces forty-five. Divide it by either of the denominators and in turn the quotient by the other; the quotient of the last division is the result. [*Or otherwise. Multiply the numerator of the fraction, that is, three, by fifteen; this produces forty-five, which are fourths of a seventh.*] Dividing forty-five by four gives eleven and a fourth, and dividing this by seven gives one and four sevenths and a fourth of a seventh, which is what you wanted to know.

(*d*) You may also multiply three fourths by fifteen as we have taught in the previous chapter[343]; this produces eleven and a fourth. Dividing it by seven will give what you require.

AGAIN. ANOTHER EXAMPLE, ON THE MULTIPLICATION OF FRACTIONS OF A FRACTION BY A COMPOSITE.

(**A.94**) You want to multiply three sevenths of an eleventh by thirty-six.

(*a*) Multiply the denominators, which are seven and eleven, thus producing seventy-seven; this is the common number. Multiplying three sevenths of an eleventh of it by thirty-six[344] and dividing the product by the common will give one and four elevenths and three sevenths of an eleventh, and this is what you wanted to know.

(*b*) Or otherwise. Multiply three by thirty-six and divide the product by the denominator of a seventh; the result will be elevenths. If you had divided by the denominator of an eleventh, the result would be sevenths. So divide first by whichever you want and divide the quotient by the other; the result will be what you require.

---

[340] Since antiquity, 'always' means that a general rule is given (see our *Introduction to the History of Algebra*, p. 21). Here it is the direct rule, corresponding to A.90*b*.

[341] Above, *a*, 'another reason'.

[342] Premiss P_4. This makes the previous sentence superfluous, so one or other might be an addition.

[343] A.90–92.

[344] Instead of the usual *cuius tres septimas undecime multiplica* we find *cuius undecime tres septimas multiplica*, perhaps a reader's change. We have left this on account of similar occurrences below (A.95*a* & A.97*a*).

660 PART TWO: TRANSLATION, GLOSSARY

(*c*) Or otherwise. Divide three sevenths of thirty-six by eleven. The result will be what you require.

AGAIN. ANOTHER EXAMPLE, ON THE MULTIPLICATION OF FRACTIONS OF A FRACTION OF A FRACTION BY A COMPOSITE.

If you now want to multiply a fraction of a fraction of a fraction, and any number of times repeated, by an integer, multiply the first denominator by the second, the resulting product by the third, the resulting product by the fourth, and so on to the last; the result will be the denominating number[345]. Next, multiply the numerator of the fraction by the integer and divide the product by the denominating number. This will give the required quantity.

(**A.95**) For instance,[346] you want to multiply three sevenths of a fifth of an eighth by fifty-nine.

(*a*) By multiplying the denominators of a seventh, a fifth and an eighth, you form two hundred and eighty. Multiply three sevenths of a fifth of an eighth of it, which is three, by fifty-nine, and divide the product by the common; but if the first is less than the second[347], denominate from the latter. The result will be what you require.

(*b*) Or otherwise. Multiply the numerator of the fraction, which is three, by fifty-nine, thus producing a hundred and seventy-seven. Dividing it by the denominator of a fifth gives thirty-five and two fifths, and dividing this by the denominator of a seventh gives five and two fifths of a seventh. These being eighths, the result is five eighths and two fifths of a seventh of an eighth, and this is what you wanted.   If you had divided first a hundred and seventy-seven by the denominator of a seventh, then the quotient by the denominator of a fifth, and then the result by the denominator of an eighth, this would also have been appropriate: start with the division by whichever denominator you want and end with the division by whichever you want.

---

[345] *numerus denominationis*, usually meaning 'denominator' (see note 328, pp. 655–656) is translated here as 'denominating number'. A fractional expression being given, its 'denominating number' is the product of all denominators found in it (numerous examples from A.126 on). When two fractional expressions are to be multiplied, the 'common number' is the product of their denominating numbers (but 'common number' is also used in this section, where one side is an integer). The difference between 'denominating number' and 'common number' is clearly explained in the *Liber algorismi*, p. 57: *Numerus autem denominationis est qui ex multiplicatis in se denominationibus omnium fractionum cuiuslibet lateris per se nascitur* (...). *Numerus vero communis dicitur qui ex numeris denominationum duorum laterum in se multiplicatis nascitur*.

[346] But in *a* and *b* the solution is just as before (A.93–94). The above rule (together with heading and 'for instance') might have been added later in the margin, which would explain why MS *B* has it after A.95.

[347] As is the case here.

(c) You may also divide three sevenths of fifty-nine by the denominator of a fifth[348] and the result by the denominator of an eighth.

All these ways lead to the right result. If in the fraction of a fraction there are four or more terms, do as has been said[349].

THIRD CASE. MULTIPLICATION OF A FRACTION AND A FRACTION OF A FRACTION BY AN INTEGER.

(A.96) You want to multiply five sevenths and three fourths of a seventh by ten.

(a) From the denominators of a fourth and a seventh you will form the common number, which is twenty-eight. Add its five sevenths to its three fourths of a seventh, multiply the sum by ten, and divide the product by the common. The result will be eight and a seventh and half a seventh, and this is what you wanted to know. The proof of this is the same as the preceding.[350]

(b) Or otherwise. Multiply the numerator of the fraction, that is, five and three fourths,[351] by ten, thus producing fifty-seven and a half. All being sevenths, divide by the denominator of a seventh. The result will be eight and one and a half sevenths. Such is five sevenths and three fourths of a seventh of ten, which is what you wanted.

(A.97) Suppose now you increase the (quantity of) fractions, as if you wanted to multiply three sevenths of a tenth and a fourth of a seventh of a tenth by fifty-four.

(a) You will do the following. Multiplying the denominators of a seventh, a tenth, a fourth, produces two hundred and eighty, which is the common number. Add three sevenths of a tenth of it to a fourth of a seventh of a tenth of it; this makes thirteen. Multiply this by fifty-four, thus producing seven hundred and two. Dividing this by the common gives two and five tenths and half a seventh of a tenth, and this is what you wanted.

(b) Or multiply the numerator of the fraction, which is three and a fourth, by fifty-four, thus producing a hundred and seventy-five and a half. Dividing it by the denominator of a seventh gives twenty-five and half a seventh, and dividing this by the denominator of a tenth gives two and five tenths and half a seventh of a tenth, and this is what you wanted to know.

(A.98) You now want to multiply four elevenths and a fifth of an eleventh by thirty-six.

(a) From the denominators of an eleventh and a fifth you will form the common number, which is fifty-five. Add four elevenths of it to a fifth of an eleventh of it, multiply the sum by thirty-six, and divide the product

---

[348] All three MSS have 'seventh'.

[349] In the present treatments (rather than in the preceding rule, not used).

[350] A.93a (and A.90a).

[351] The numerator of a compound fraction is always an integer with a fraction, also if there are several fractional terms (A.97b).

662   Part Two: Translation, Glossary

by the common; the quotient will be thirteen[352] and eight elevenths and a fifth of an eleventh, and this is what you wanted to know.

(**b**) Or otherwise. Multiply four and a fifth, which is the numerator of the fraction, by thirty-six, thus producing a hundred and fifty-one and a fifth. Dividing this[353] by the denominator of an eleventh will give what you require.

Multiplication of a fraction by a fraction

There are also four cases.[354]

First case.

(**A.99**) ⟨You want to multiply one fourth by one eighth.⟩[355]

When you want to multiply one fourth by one eighth, your purpose is to take of an eighth the same ratio which a fourth bears to one.[356] From this it is evident that the result of multiplying a fraction by a fraction corresponds to appending the name of one of them to the other, the former depending on the latter[357]. So when a fourth is multiplied by an eighth, the result is simply a fourth of an eighth.   The same holds for all others.[358]

[Therefore multiplying the triple of a fourth, which is three fourths, by the quintuple of an eighth, which is five eighths, must produce fifteen fourths of one eighth. [*This is the reason why when we multiply the two numerators of the fractions the result is fifteen fourths of one eighth.*] Dividing them by eight will give fourths, by four, eighths, as we have shown in the foregoing.[359]]

(**A.100**) You want to multiply three fourths by five eighths.

---

[352] The MSS have 'three'.

[353] *Quas* (referring to a hundred and fifty-one and a fifth) instead of *quos* since they are *partes*. Same in A.104*b*.

[354] This time (unlike in the previous section) we may indeed find four: fraction by fraction (A.99–101), fraction of a fraction by fraction (A.102–103), fraction of a fraction by fraction of a fraction (A.104–105), compound fraction by fraction or fraction of a fraction (A.106–107). The missing case, compound fraction by compound fraction, which requires converting fractions, will be taught in A.126–127. A further case is A.136.

[355] A similar omission in A.246.

[356] Taking a ratio of some quantity means taking of it the fraction equal to the amount of this ratio. See also A.100*b*.

[357] The use here of the verb *pendere* might correspond to the Arabic *aḍāfa*, which expresses the *status constructus* (when a noun in the genitive is appended to a governing word). We have indeed such a situation here (*quarta octave* = *rub'u thumnin*). Al-Ḥaṣṣār, whose treatise has been summarized by H. Suter, says that the new name is 'composed of' (*mu'allaf min*) the two.

[358] That is, multiplication of aliquot fractions. What now follows belongs at the end of A.100*b*.

[359] A.93–95.

LIBER MAHAMELETH   663

(*a*) Put below the corresponding side each of the denominators, thus four and eight.[360] Their multiplication produces thirty-two, which is the principal number.[361] Next, multiply three fourths of the denominator of a fourth, thus three, by five eighths of the denominator of an eighth, thus five; the result is fifteen. Denominating this from the principal will give three eighths and three fourths of an eighth, and this is what you wanted.

A
—————

B
——————————

G
————

D
—————

(A.100*a*)

The proof of this is the following. Let the denominator of a fourth be *A*, the denominator of an eighth, *B*, three fourths of the denominator of a fourth, *G*, five eighths of the denominator of an eighth, *D*. So dividing *G* by *A* will give three fourths and dividing *D* by *B*, five eighths. Now we want to multiply three fourths by five eighths. This is the same as dividing *G* by *A*, *D* by *B*, and multiplying the two quotients. This again is the same[362] as dividing the product of *G* into *D* by the product of *A* into *B*. This is what we wanted to demonstrate.

(*b*) Or multiply the number of fourths by the number of eighths, that is, five by three, thus producing fifteen. Divide it by the denominator of a fourth; this gives three and three fourths. Divide this by eight; all being eighths, they are then three eighths and three fourths of an eighth. If you had first divided by the denominator of an eighth, the result would have been fourths.

The proof of this is clear. For we know that multiplying a fraction by an integer or by another fraction is the same as taking of the multiplicand the same ratio which the multiplier bears to one [*for instance, multiplying two fifths by eight is simply the same as taking two fifths of eight, which (quantity) bears to eight the same ratio as two fifths bears to one*][363], as we said previously.

---

[360] With multiplicand written on the left and multiplier on the right. The word 'side', *latus*, then becomes virtually synonymous with 'fractional expression', which is why we shall refer to the denominators *of* one side.

[361] The 'principal number' (*prelatus*, Arabic *imām*), like the *communis* before, is thus the product of the denominators (note 335, p. 658). (In MS *C* the two designations are often found coupled.) But *prelatus* does not always mean the same as *communis*; see note 493, p. 703.

[362] By Premiss P_3' (Abū Kāmil's proposition).

[363] No need for an example here, with the fundamental procedure of re-

664 PART TWO: TRANSLATION, GLOSSARY

(**A.101**) You now want to multiply three sevenths by nine thirteenths.

(**a**) Put the denominator of a seventh below its side, and, likewise, the denominator of a thirteenth below its side. Multiply them, that is, seven by thirteen, thus producing ninety-one, which is the common number. Then multiply three sevenths of seven, which is three, by nine thirteenths of thirteen, which is nine; the product is twenty-seven. Denominating this from the common number will give three thirteenths and six sevenths of one thirteenth, and this is what you wanted to know.

(**b**) Or otherwise. Multiply three, which is the number of sevenths, by nine, which is the number of thirteenths; this produces twenty-seven. Divide this by the denominator of a seventh; the result is three and six sevenths, which are thirteenths.[364] If you had first divided twenty-seven by the denominator of a thirteenth, the result would have been two and one thirteenth, which are sevenths.

SECOND CASE. MULTIPLICATION OF A FRACTION OF A FRACTION BY A FRACTION.

(**A.102**) You want to multiply three fourths of a fifth by seven eighths.

(**a**) Multiplying the denominators of the fractions of one side, thus of a fourth and a fifth, produces twenty; put it below this side. Next, put the denominator of the other side, which is eight, below this side. Next, multiply twenty by eight; this produces a hundred and sixty, which is the principal number. Next, multiply three fourths of a fifth of twenty, which is three, by seven eighths of eight; this produces twenty-one. Denominating it from the principal number you will find an eighth and a fourth of a fifth of an eighth, and this is what you wanted to know. The proof of this is clear from the foregoing [365].

(**b**) Or otherwise. Multiply the two numerators of the fractions, thus producing twenty-one, which are fourths of a fifth of an eighth. For when a fourth of a fifth is multiplied by an eighth, the result is simply a fourth of a fifth of an eighth;[366] so when three fourths of a fifth are multiplied by seven eighths, this produces twenty-one fourths of a fifth of an eighth. Divide it by the denominator of a fourth; the quotient is five and a fourth. Divide this in turn by the denominator of a fifth; the quotient is one and a fourth of a fifth. Divide this in turn by the denominator of an eighth; the result is an eighth and a fourth of a fifth of an eighth, and this is what you wanted to know. Whether you divide twenty-one first or last by any denominator you want, this will produce the same.

Likewise for all similar instances.[367]

---

ducing a multiplication to taking a ratio already explained in A.90a (integer as multiplier) and in A.99 (fraction).

[364] No other division to be performed.

[365] A.100a.

[366] A.99 (appending names), three this time.

[367] Where reduction by division is possible, which is not the case in the coming example.

LIBER MAHAMELETH 665

(**A.103**) You want to multiply four fifths of an eleventh by one ninth.

(**a**) From the denominators of the fractions a fifth and an eleventh you form fifty-five; put it below the corresponding side. Next, for the other side, put the number from which a ninth is denominated, namely nine, below this side. Then multiply nine by fifty-five; the product is the principal number. Next, multiply four fifths of an eleventh of fifty-five, which is four, by one ninth of nine, thus producing four. Denominating it from the principal number will give what you wanted.

(**b**) Or multiply one by four, which is simply four. Divide[368] it by five, the result by nine, and the result by eleven; this will give four fifths of one ninth of one eleventh. You may also say 'four fifths of one eleventh of one ninth': you may put first the denominator of whichever fraction you want and last whichever, this produces the same.

THIRD CASE. MULTIPLICATION OF A FRACTION OF A FRACTION BY A FRACTION OF A FRACTION.

(**A.104**) You want to multiply three fourths of a fifth by seven eighths of a sixth.

(**a**) Multiplying the denominators of a fourth and a fifth produces twenty. Put it on their side below. Multiplying then, for the other side, the denominators of an eighth and a sixth produces forty-eight. Multiplying this by twenty produces nine hundred and sixty, which is the principal number. Next, multiply three fourths of a fifth of twenty, which is three, by seven eighths of a sixth of forty-eight, which is seven; the product is twenty-one. Denominating it from the principal number will give one eighth of a sixth and a fourth of a fifth of an eighth of a sixth. The proof of this is evident from the foregoing[369].

(**b**) Or otherwise. Multiply seven, the numerator of the fraction, by three; this produces twenty-one, which are fourths of a fifth of a sixth of an eighth. Divide it by whichever denominator of a fraction you want first. Dividing first by four gives five and a fourth; dividing this in turn by five gives one and a fourth of a fifth; dividing then by six and the quotient by eight will give in the end a sixth of an eighth and a fourth of a fifth of a sixth of an eighth. [*Suppose you divide first by six; this will give three and a half, which divided by four gives seven eighths, which divided by five gives seven eighths of a fifth.*][370]

(**A.105**) You want to multiply four fifths of a ninth by a seventh of an eleventh.

(**a**) Multiply as explained before in order to form the common number[371].

(**b**) Or multiply one by four, thus producing four, which is four fifths of a ninth of a seventh of an eleventh. [*Divide this by whichever denominator*

---

[368] Instead of 'denominate'. See notes 382, 464, 468, 604, 733.

[369] A.100*a*.

[370] Faulty computation.

[371] *Sic*, instead of the now usual 'principal number'.

666   PART TWO: TRANSLATION, GLOSSARY

you want first; in the end the result will be four fifths of a ninth of a seventh of an eleventh.][372] You may also say that this is four sevenths of a ninth of a fifth of an eleventh.

All these ways lead to the right result.[373] Proceed likewise however often you want to repeat the fraction of a fraction.

FOURTH CASE. MULTIPLICATION OF A FRACTION WITH A FRACTION OF IT BY A FRACTION.

(**A.106**) You want to multiply five sevenths and three fourths of a seventh by ten elevenths.

(***a***) Multiplying the denominators of the fractions a seventh and a fourth produces twenty-eight. Multiply it by the denominator of an eleventh, which is eleven, thus producing three hundred and eight. Such is the principal number. Next, add five sevenths of twenty-eight to three fourths of a seventh of it; this makes twenty-three. Multiply it[374] by ten elevenths of eleven; the result is two hundred and thirty. Denominating this from the principal will give eight elevenths and a seventh of an eleventh and half a seventh of an eleventh, and this is what you wanted. The proof of this is quite the same as the previous one and they do not differ in anything.[375]

(***b***) Or otherwise. Multiply five and three fourths, which is the number of sevenths, by ten, which is the number of elevenths; this produces fifty-seven and a half, which are sevenths of elevenths. The division by seven gives elevenths; by eleven, sevenths. Divide then by the denominator of a seventh; this gives eight and a seventh and half a seventh, which divided by the denominator of an eleventh will give eight elevenths and a seventh of an eleventh and half a seventh of an eleventh.

(***c***) Or otherwise. Multiply ten elevenths by five and three fourths as we have taught above[376]; this will produce elevenths. Dividing by seven will give what you want.

(***d***) Or otherwise. Multiply five sevenths by ten elevenths; the result is seven elevenths and a seventh of an eleventh. Next, multiply three fourths of a seventh by ten elevenths; the result is one eleventh and half a seventh of an eleventh. Adding this to the previous result will give eight elevenths and a seventh of an eleventh and half a seventh of an eleventh, and this is what you wanted.

(**A.107**) Likewise, you want to multiply three fifths of a seventh and a fourth of a fifth of a seventh by five sixths of an eighth.

(***a***) Multiplying the denominators of a fifth, a seventh and a fourth produces a hundred and forty. Put it below their side. Next, for the other

---

[372] This does not add anything. Such an irreducible case has already been encountered in A.103.

[373] Whichever denominator is taken first.

[374] *Quas*, for twenty-three are *partes*. See note 353.

[375] A.100*a*.

[376] A.92 and A.101 (although A.123 below would be more appropriate).

LIBER MAHAMELETH

side, multiply the denominators of a sixth and an eighth, thus producing forty-eight. Put it below their side. Then multiply them, and the product will be the principal number. Next, add three fifths of a seventh of a hundred and forty to a fourth of a fifth of a seventh of this same number, and multiply the sum by five sixths of an eighth of forty-eight. Denominating the product from the principal will give what you wanted.

(b) Or multiply three and a fourth by five, thus producing sixteen and a fourth, which are fifths of a sixth of a seventh of an eighth. Divide it first by whichever denominator you want, the result by another, and so on to the last. Dividing first by five gives three and a fourth[377], which are three sevenths of a sixth of an eighth and a fourth of a seventh of a sixth of an eighth. But dividing first by six, then by eight, next by five, and next by seven, gives two eighths[378] of a fifth of a seventh and four sixths of an eighth of a fifth of a seventh and a fourth of a sixth of an eighth of a fifth of a seventh. This is what you wanted.

Consider other such instances likewise.

I shall presently begin to treat the conversion of fractions into others. This is necessary for anyone who wants to perform the remaining multiplications of fractions according to other rules. I could not[379] present this before, since in the fractions treated so far some were attached to others[380]. It was therefore more appropriate to speak about the conversion of fractions here.

CHAPTER ON THE CONVERSION OF FRACTIONS INTO OTHER FRACTIONS
There are five cases.[381]

FIRST CASE. CONVERSION OF A FRACTION INTO A FRACTION.

(A.108) You want to know how many fifths three fourths are.

The meaning of this problem is that some whole is divided into four parts and then into five parts, and because of this you want to know to how many of these five parts of the whole three of the four parts correspond.

---

[377] Which is what we started with!

[378] All three MSS have 'an eighth'.

[379] Perhaps: I did not wish to (*volui* instead of *potui*).

[380] Or: depending on others (note 357). The denominators of the fractions in the result were obtained by simply putting together the individual denominators. Thus the problem of converting different denominators into a common one did not arise.

[381] Fraction into fraction (A.108–111); compound fraction into fraction (A.112–114); fraction of a fraction into fraction (A.115–116); fraction of a fraction into fraction of a fraction (A.117–120); compound fraction (several terms) into fraction of a fraction (A.121–122). Note that the final expression will contain all the required denominators plus some of the expression to be converted; thus in any event more than announced. This division into five cases is an addition anyway.

668   Part Two: Translation, Glossary

This being so, it is clear that the ratio of the three parts to the four parts forming the whole is the same as the ratio of the required number of parts to the five parts likewise forming the whole. So multiply these three by five and divide the product by the four parts. The result will be three and three fourths, which is the number of fifths found in three fourths, which are three fifths and three fourths of a fifth.

(A.109) You now want to know how many tenths three eighths are.

You will do the following. You will always multiply the numerator of the fraction to be converted, whether it is one or more, by the denominator of the fraction into which you are to convert. In this case, multiply three by ten, thus producing thirty. Divide it by eight; the quotient is three and three fourths. Dividing[382] this in turn by ten gives three tenths and three fourths of a tenth.

(A.110) You now want to know how many sixths four sevenths are.

Multiply four sevenths by six; this produces three and three sevenths, which are three sixths and three sevenths of a sixth.

(A.111) You want to know how many thirteenths one fifth is.[383]

Multiply a fifth by thirteen; this produces two and three fifths, which are two thirteenths and three fifths of a thirteenth, and this is what you wanted.[384]

$$1 \longrightarrow 13$$
$$5 \cdot$$
$$5 \longrightarrow 13$$

(A.111)

Second case. Chapter on converting a fraction and a fraction of a fraction into a fraction.

(A.112) You want to know how many elevenths five sevenths and two thirds of a seventh are.

You know[385] that the ratio of five and two thirds to seven is the same as the ratio of the required quantity to eleven. So multiply five and two thirds by eleven and divide the product by seven; the result will be what you wanted[386].

---

[382] Instead of 'denominating'; same in A.113–118, A.120–122.

[383] Note, in the figure, how 'one fifth' is represented (cf. 'a fifth' in A.118).

[384] Here, as in A.125, the last line of the figure indicates the denominators. See A.116.

[385] A.108.

[386] That is, the number of elevenths.

(A.113) Likewise if you want to know how many tenths three eighths and half an eighth are.

Multiply three eighths and half an eighth by ten, from which a tenth is denominated; this produces four and three eighths. Dividing this by ten gives four tenths and three eighths of one tenth. Such is the quantity to which three eighths and half an eighth correspond.

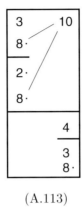

(A.113)

(A.114) You now want to know how many eighths four elevenths and a third of one eleventh are.

Multiply four and a third elevenths by eight; this produces the integer three and one eleventh and two thirds of one eleventh. Dividing this by eight gives three eighths and an eleventh of one eighth and two thirds of an eleventh of one eighth. Such is the quantity to which four elevenths and a third of an eleventh correspond.

THIRD CASE. CHAPTER ON CONVERTING A FRACTION OF A FRACTION INTO A FRACTION.

(A.115) You want to know how many sixths three fourths of one tenth are.

Multiply three fourths of a tenth by six; this produces eighteen fourths of one tenth, which are four tenths and half a tenth. Dividing this by six gives four tenths of one sixth and half a tenth of a sixth.

(A.115)

670 PART TWO: TRANSLATION, GLOSSARY

(A.116) You now want to know how many eighths three fifths of one eleventh are.

Multiply three fifths of one eleventh by eight; this produces twenty-four fifths of one eleventh, which are four elevenths and four fifths of one eleventh. Dividing this by eight gives four elevenths of one eighth and four fifths of one eleventh of one eighth.

$$
\begin{array}{ccc}
3 & \!\!\!\!\rule[0.3em]{1em}{0.4pt}\!\!\!\! & 8 \\
5\cdot & & \\
\hline
1 & & \\
11\cdot & & \\
\hline
24 & & 8
\end{array}
$$

(A.116)

FOURTH CASE. CHAPTER ON CONVERTING A FRACTION OF A FRACTION INTO A FRACTION OF A FRACTION.

(A.117) You want to know how many sixths of one tenth three sevenths of one eighth are.

(a) Multiply three sevenths of an eighth by the quantity produced by multiplying the denominators, which is sixty [*since six times ten is sixty*][387]; the result is three and a seventh and half a seventh. Dividing this by sixty gives three sixths of one tenth and a seventh of a sixth of one tenth and half a seventh of one sixth of one tenth.

$$
\begin{array}{cc}
3 & 6\cdot \\
7\cdot & \overline{} \\
\overline{} & 1 \\
1 & 10\cdot \\
8\cdot & \overline{} \\
& 60
\end{array}
$$

(A.117a)

(b) Or otherwise. The multiplication of the denominators seven and eight produces fifty-six. Next, multiplying the other denominators, of a sixth and a tenth, produces sixty. Next, multiply three by sixty and divide the product by fifty-six; the result will be what you wanted.

---

[387] A reader wanted to point out that we multiply the denominators of the fractions into which we wish to convert.

LIBER MAHAMELETH **671**

(i) You will proceed in the same way [388] for any number of fractions, be there three or more, namely as follows. Form a single number by multiplying the denominators of all the fractions to be converted, and likewise another from the others, those into which we are to convert. Next, say: 'So many or so many parts of such or such number, how many parts are they of this or that number?' Then multiply the first number, which is the numerator of the fractions to be converted, by the last, or fourth, which is the product of the denominators of the fractions into which must be converted, and divide the result by the second, which is the product of the denominators of the fractions to be converted; this will give the third, which is required.

(A.118) You now want to know how many eighths of one eleventh a fifth of one seventh is.

Multiply a fifth of a seventh by eighty-eight, which is the product of the denominators [389]; the result is two and three sevenths and three fifths of one seventh. Dividing this by eighty-eight gives two eighths of one eleventh and three sevenths of one eighth of one eleventh and three fifths of one seventh of one eighth of one eleventh.

$$\frac{5\,\cdot}{1} \qquad \frac{8\,\cdot}{1}$$
$$7\,\cdot \qquad 11\cdot$$

(A.118)

(ii) If now among the various fractions of both sides the same fraction occurs, it should be disregarded.

(A.119) For example, you ask: How many elevenths of a fifth of an eighth are five sixths of an eighth of a seventh?

You will remove the two '(of an) eighth's and say: 'How many elevenths of one fifth are five sixths of a seventh?' You will then proceed as has been shown above.[390]

(A.120) Or you might ask: How many sixths of one seventh are three fourths of one sixth?

Since both sides contain like fractions you will omit them: in this case the sixth with the sixth, leaving three fourths to be multiplied by seven. This produces five and a fourth. Dividing this by forty-two, which is the product of the denominators of a sixth and a seventh, will give five sixths of one seventh and a fourth of a sixth of one seventh.

---

[388] As in b. This rule and b occur together in the disordered text.

[389] That is, those of the fractions into which we convert.

[390] A.117–118.

$$
\begin{array}{c|c}
3 & 6\,\cdot \\
\dfrac{4\,\cdot}{1} & 1 \\
6\,\cdot & 7\,\cdot \\
\hline
\end{array}
$$

(A.120)

FIFTH CASE. CONVERTING A FRACTION OF A FRACTION AND A FRACTION OF A FRACTION OF A FRACTION INTO A FRACTION OF A FRACTION.

(**A.121**) You want to know how many sixths of one seventh three eighths of one tenth and half an eighth of one tenth are.

Multiply three eighths of a tenth and half an eighth of a tenth by forty-two, which is the product of the denominators of a sixth and a seventh; this produces the integer one and eight tenths and three eighths of one tenth. Dividing this by forty-two will give one sixth of a seventh and eight tenths of a sixth of a seventh and three eighths of a tenth of a sixth of a seventh.

$$
\begin{array}{|c|c|}
\hline
\begin{array}{c} 3 \\ 8\,\cdot \\ \hline 1 \\ 10\,\cdot \\ \text{and } 2\,\cdot \\ 8\,\cdot \\ 1 \\ 10\,\cdot \end{array} &
\begin{array}{c} 6\,\cdot \\ 1 \\ 7\,\cdot \\ \\ \\ \\ \\ \end{array} \\
\hline
\end{array}
$$

(A.121)

(**A.122**) You now want to know how many tenths of one thirteenth four sevenths of one tenth and half a seventh of a tenth are.

Disregard the two 'tenth's since they are the same (fraction), as said before,[391] and multiply four sevenths and half a seventh by thirteen, from which a thirteenth is denominated; the result is eight and two sevenths and half a seventh. Dividing this by a hundred and thirty, which is the product of the denominators ten and thirteen, will give eight tenths of one thirteenth and two sevenths of a tenth of one thirteenth and half a seventh of a tenth of a thirteenth.

---

[391] Rule *ii* above.

LIBER MAHAMELETH

$$
\begin{array}{|c|c|}
\hline
4 & 10\cdot \\
7\cdot & 1 \\
1 & 13\cdot \\
10\cdot & \\
\text{and } 2\cdot & \\
7\cdot & \\
10\cdot & \\
\hline
\end{array}
$$

(A.122)

You will proceed likewise for all other instances.

Chapter on the multiplication of a fraction by an integer and a fraction

(A.123) You want to multiply five sevenths by six and two thirds.

(a) You will do the following. Multiply the denominators of the fractions a seventh and a third, thus producing twenty-one, which is the principal. Multiply five sevenths of it, which is fifteen, by six and two thirds, and divide the product by the principal; the result will be what you wanted.

The proof of this is clear.[392] For the ratio of five sevenths of twenty-one to twenty-one is the same as the ratio of five sevenths of six and two thirds to six and two thirds. Thus multiplying five sevenths of twenty-one by six and two thirds and dividing the product by twenty-one will produce what is required.

(b) Or otherwise. Multiply five by six and two thirds and divide the product by seven; the result will be what is required. The proof of this is evident from the foregoing[393].

(c) Or otherwise. Multiply five sevenths by six, thus producing four and two sevenths. Next, multiply five sevenths by two thirds, thus producing three sevenths and a third of a seventh. Adding this to four and two sevenths will make four and five sevenths and a third of a seventh, and this is what you wanted. The proof of this is evident from the foregoing[394] [*in the chapter on the multiplication of thousands repeated*[395] [*and from the first theorem in the second book of Euclid*]].

---

[392] Analogous proofs in A.90a & A.93a.

[393] Or 'from the premisses' (Premiss P_5).

[394] Premiss PE_1 = *Elements* II.1.

[395] The expression *ex premissis* used for the reference is ambiguous since it can mean both 'from the premisses' (thus A–II) and, less precisely, 'from the foregoing' (note 393). Whence this early reader's attempt to locate what was being referred to, which he found in A.32 (though A.123 precedes A.32 in the disordered text).

674 PART TWO: TRANSLATION, GLOSSARY

CHAPTER ON THE MULTIPLICATION OF AN INTEGER AND A FRACTION BY AN INTEGER AND A FRACTION.

There are five cases of the multiplication of an integer and a fraction by an integer and a fraction.[396]

FIRST CASE.

(A.124) You want to multiply four and five eighths by nine and three fifths.

(*a*) You will do the following. Multiply the denominator of one fraction by the denominator of the other, in this case eight by five; this produces forty, which is the principal number. Next, convert the integer and the fraction of the multiplying side into its largest denominator[397], that is, into eighths, as follows. Multiply the integer of the multiplying side, that is, four, by the denominator of this side, that is, eight; this gives thirty-two. Adding to it five, which is the numerator of the fraction, makes thirty-seven. Put it separately. Next, convert the other side into fifths in the same manner. That is, multiply the integer, which is nine, by five, which is the denominator of this side, thus producing forty-five. Adding to it three, which is the numerator of the fraction, makes forty-eight. Multiply it by the thirty-seven of the other side; this produces one thousand seven hundred and seventy-six. Divide it by the common number,[398] namely forty; this gives the integer forty-four and two fifths[399], which is the result of multiplying the proposed quantities.

The proof of this is clear from what has been said in the chapter on the multiplication of a fraction by a fraction.[400] Nevertheless, I shall repeat it in order to better engrave it upon your memory. I shall also show how this proof may be used in each case of the multiplication of fractions.[401]

Let then four and five eighths be A, nine and three fifths, B, that from which an eighth is denominated, G, the denominator of a fifth, D. Let A multiplied by G produce H, and B multiplied by D produce Z.

---

[396] Integer and fraction by integer and fraction (A.124–125); integer and compound fraction by integer and fraction (A.128); integer and compound fraction by integer and compound fraction (A.129); integer with two fractions by integer with two fractions (A.130). The heading 'second case' has been attributed to the multiplication of compound fractions, without any integer (A.126–127). Furthermore, A.123 (fraction by integer and fraction) and A.137 (integer and compound fraction by compound fraction) are other such cases. Once again, it is clear that this enumeration of cases, like the one beginning with A.90, originated with an early reader.

[397] *ultimum genus sue denominationis*, thus the denominator which is of the highest kind. (The rule is expressed generally; here there is just one.) See note 419.

[398] Called 'principal number' before.

[399] The figure has the unsimplified fraction.

[400] A.100*a*.

[401] See remark following the two proofs, p. 676, and note 404.

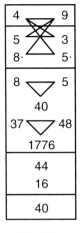

(A.124a)

Since multiplying A by G produces H, dividing H by G will give A. Since multiplying B by D produces Z, dividing Z by D will give B. Now we want to multiply A by B. This is the same as dividing H by G, Z by D, and multiplying the two quotients; this is again the same[402] as dividing the product of H into Z by the product of G into D. Therefore multiplying A by B is the same as multiplying H by Z and dividing the product by the product of G into D. This is what we wanted to demonstrate.

(A.124a, first proof)

There is also for this case another, easier proof.[403] Let namely four and five eighths be A, and nine and three fifths, B. Let A be multiplied by B and G be produced, which is the required quantity. Let next the denominator of an eighth be D and let its multiplication by four and five eighths produce H. Then A multiplied by B produces G and multiplied by D, H. Therefore the ratio of G to H is the same as the ratio of B to D. Let then the denominator of a fifth be K, and let its multiplication by nine and three fifths produce Z. Let the multiplication of D, which is

---

[402] By P'_3 (Abū Kāmil's proposition).
[403] It is simpler in the sense that it requires only well-known theorems of the *Elements* (VII.17–19).

676   PART TWO: TRANSLATION, GLOSSARY

the denominator of an eighth, by K, which is the denominator of a fifth, produce T. Then B multiplied by K produces Z and D multiplied by K produces T. Therefore the ratio of Z to T is the same as the ratio of B to D. But the ratio of B to D was the same as the ratio of G to H. So the ratio of G to H is the same as the ratio of Z to T. Thus the product of Z into H is equal to the product of G into T. Therefore when Z, which is the product of nine and three fifths into five, is multiplied by H, which is the product of four and five eighths into eight, and the result is divided by T, which is the product of the denominators of the two fractions, the quotient will be G, which is what we require. This is what we wanted to demonstrate.

$$
\begin{array}{ll}
G \;\underline{\text{what is required}} & \underline{37}\; H \\[2ex]
\qquad B\;\underline{\text{9 and 3 5}^{\text{ths}}} & \underline{8}\; D \\[2ex]
\qquad\;\; A\;\underline{\text{4 and 5 8}^{\text{ths}}} & \underline{5}\; K \\[2ex]
Z\;\underline{48} & \underline{40}\; T
\end{array}
$$

(A.124a, second proof)

Anyone who has paid careful attention to these two proofs and is fully acquainted with them may use them to prove all multiplications of fractions.[404]

(**b**) Or otherwise. You may also multiply the given quantities term by term, namely as follows.[405] You multiply four by nine, thus producing thirty-six. Next, you multiply five eighths by nine, thus producing forty-five eighths, which is the integer five and five eighths; you put the digit, which is five, with the thirty-six, whereas you put the five eighths separately. Next, you will multiply five eighths by three fifths, thus producing three eighths. Then you will multiply three fifths by four, thus producing two and two fifths. You put the two fifths and the previous three eighths with the above five eighths, but the two with the previous integer. The multiplication being completed, and the quantities associated by kind, that is, integer with integer, fraction with fraction, fractions of fractions with fractions of fractions [406], some of the fractions are converted into the others to make them all alike [407], as in this case converting two fifths into eighths, which gives three eighths and a fifth of one eighth. You then add the

---

[404] Indeed, they will be referred to in A.126a, A.128a, A.129a.

[405] Multiplying together the integers, then each integer by the fractional part of the other side, finally the two fractional parts. Thus four products altogether.

[406] There are none here; the rule is just expressed generally.

[407] That is, with the same denominator. Not necessary here.

LIBER MAHAMELETH

fractions together, namely the three eighths and the previous three eighths and five eighths, thus making one and three eighths. Then you add this unit to the previous integers, thus making forty-four. So the result of the multiplication of the given quantities is forty-four and three eighths and a fifth of one eighth.

(*c*) Or otherwise in this problem. Multiply nine and three fifths by four; this produces thirty-eight and two fifths. Put the integer and the two fifths separately. Next, multiply five eighths by nine and three fifths; this produces six. Add it to the previous integer; this makes forty-four. Therefore the result of multiplying the proposed quantities is forty-four and two fifths.

You will proceed likewise in all such instances. That is, you will multiply the integer of the multiplying side by the integer and the fractions of the other side, then multiply each fraction of the first side by the integer and the fractions of the other side. Adding them all as explained before [408] will then give the required quantity. [*This too is proved by the first theorem of the second book of Euclid.*][409]

(**A.125**) You now want to multiply seven and two fifths by eight and four elevenths.

Multiply the integer eight and four elevenths by the integer of the other side, which is seven; this produces fifty-eight and six elevenths. Next, multiply two fifths by eight and four elevenths, as follows. You multiply two fifths by eight, thus producing the integer two, which added to fifty-eight makes sixty;[410] then you convert the three, remaining from eight, and four elevenths into elevenths; this gives altogether thirty-seven elevenths. Two fifths of this is fourteen elevenths and four fifths of one eleventh. Adding to it the above six elevenths makes twenty elevenths and four fifths of one eleventh, which is the integer one and nine elevenths and four fifths of one eleventh. Therefore the quantity resulting from the multiplication of the proposed quantities is sixty-one and nine elevenths and four fifths of one eleventh. This is the quantity which is required.

SECOND CASE. MULTIPLICATION OF A FRACTION AND A FRACTION OF A FRACTION BY A FRACTION AND A FRACTION OF A FRACTION.

(**A.126**) You want to multiply five eighths and two thirds of an eighth by six sevenths and three fourths of a seventh.

---

[408] That is, after arranging the terms by kind and using conversion if need be; see *b* above.

[409] Thus by PE_1, as with the multiplication in case *b*.

[410] Although cumbersome (and already encountered in A.91*b* and A.92*c*), this computation, which consists in multiplying two fifths not by 8 but by $5 + 3$, fits in with the remainder of the text. An early reader must have attempted to modify the text, for we read here (in all three MSS): 'You multiply two fifths by eight, thus producing the integer three and a fifth, which three added to fifty-eight makes sixty-one'. (The rest is unchanged.)

678 PART TWO: TRANSLATION, GLOSSARY

7	8
2	4
5·	11·
5	11

(A.125)

(***a***) Multiplying the denominators of the fractions of the first side, thus eight by three, you form twenty-four, which is one denominating number[411]. Next, multiplying the denominators of the other side, thus seven by four, produces twenty-eight, which is likewise one denominating number. Multiply it by the other, that is, by twenty-four; this produces six hundred and seventy-two, which is the principal number by which we shall divide. Next, put five eighths and two thirds of an eighth of twenty-four, which is seventeen, separately; for the other side, put likewise six sevenths and three fourths of a seventh of twenty-eight, which is twenty-seven, separately. Multiplying this by the seventeen of the first side produces four hundred and fifty-nine. Denominating it from the principal number [*that is, six eleventh s and a fifth of an eleventh and a sixth of a fifth of an eleventh*][412] will give the result of multiplying the proposed quantities. The proof of this is clear from the two previous proofs.[413]

5	6
8·	7·
2	3
3·	4·
8·	7·
24	28
672	
17	27
459	

(A.126*a*)

(***b***) Or otherwise. Multiply five eighths and two thirds of an eighth by six sevenths as we have taught above[414]; the result is four eighths and six

---

[411] Terminology already encountered in the rule preceding A.95, p. 660.
[412] This is the result of A.127.
[413] A.124*a*.
[414] A.106.

LIBER MAHAMELETH

sevenths of an eighth. Next, multiply three fourths of a seventh by five eighths and two thirds of an eighth, as follows. Multiply three fourths of a seventh by five eighths, thus producing three eighths of a seventh and three fourths of an eighth of a seventh; next, multiply three fourths of a seventh by two thirds of an eighth, thus producing half an eighth of a seventh. Next, add them all. That is, add half an eighth of a seventh to three fourths of an eighth of a seventh; this gives an eighth of a seventh and a fourth of an eighth of a seventh. Add this to the six sevenths of an eighth, which is six eighths of a seventh [415]; this makes seven eighths of a seventh and a fourth of an eighth of a seventh. Add this to the four eighths as follows. Convert first the four eighths into eighths of a seventh, which gives twenty-eight eighths of a seventh; add it to the seven eighths of a seventh and a fourth of an eighth of a seventh; this makes thirty-five eighths of a seventh and a fourth of an eighth of a seventh. Add to it the three eighths of a seventh; this gives thirty-eight eighths of a seventh and a fourth of an eighth of a seventh. Dividing this [416] by eight will give sevenths, namely four sevenths and six eighths of a seventh and a fourth of an eighth of a seventh.

Consider likewise other such instances. [*The proof of this is obtainable from the first theorem of the second book of Euclid.*][417]

(*c*) Or otherwise. You will multiply the given number of eighths by the number of sevenths, thus five and two thirds by six and three fourths; this produces thirty-eight and a fourth, which are sevenths of eighths. Dividing this [*thirty-eight and a fourth*][418] by eight will give sevenths, and dividing it by seven will give eighths. So divide it by seven; this will give five and three sevenths and a fourth of a seventh, all of which are eighths. Therefore the result of multiplying the proposed quantities is five eighths and three sevenths of one eighth and a fourth of one seventh of one eighth.

AGAIN.

(**A.127**) You want to multiply seven elevenths and a third of an eleventh by four fifths and a fourth of a fifth.

You will do the following. Multiply seven and a third by four and a fourth; this produces thirty-one and a sixth. Divide it by five; the result will be elevenths. Therefore the multiplication of the given quantities produces six elevenths and a fifth of an eleventh and a sixth of a fifth of an eleventh.

THIRD CASE. MULTIPLICATION OF AN INTEGER AND A FRACTION AND A FRACTION OF A FRACTION BY AN INTEGER AND A FRACTION.

(**A.128**) You want to multiply two and five sevenths and two thirds of a seventh by four and three eighths.

---

[415] These are the fractions in which the result will be expressed.

[416] *Quas*, meaning: thirty-eight and a fourth. See the gloss in *c*.

[417] This should have been inserted before the previous sentence. Same glossator in A.129*b*, presumably also in A.123*c* & A.124*c*.

[418] To make clear what 'this' refers to.

7	4
11·	5·
3·	4·
11·	5·

(A.127)

(**a**) You will do the following. Multiplying seven and three, the denominators of the fractions, produces twenty-one, which is one denominating number. Next, for the other side, eight will be the denominating number. Multiplying it by twenty-one produces a hundred and sixty-eight, which is the principal number. Next, reduce the first side to the lowest kind of fraction[419], that is, to thirds of sevenths, multiplying each term by the corresponding denominating number, which is twenty-one; the result is fifty-nine. Next, convert the other side into eighths, multiplying each term by eight, which is the corresponding denominating number; the result is thirty-five. Multiplying it by fifty-nine produces two thousand and sixty-five. Divide it by the principal number; the result will be twelve and two eighths and two sevenths of one eighth and a third of a seventh of an eighth. This is the result of multiplying the proposed quantities. The proof of this is obtainable from the two above.[420]

2	4
5	3
7·	8·
2	
3·	
7·	
21	8
168	
59	35
2065	

(A.128a)

(**b**) Or otherwise. Perform the multiplication term by term, as follows.[421] Two by four is eight. Then multiply five sevenths and two thirds of a seventh by four, thus producing three and a seventh and two thirds of a

---

[419] *ultimum genus fractionum*; here, the product of the denominators.

[420] Once again, A.124a.

[421] A compound fraction counts as one 'term' (*differentia*).

LIBER MAHAMELETH

seventh. Next, multiply two by three eighths, thus producing six eighths. Then multiply five sevenths and two thirds of a seventh by three eighths, thus producing two eighths and three sevenths of an eighth. Put then each quantity together with the quantity of the same kind, that is, integer with integer, fraction with fraction, fraction of a fraction with fraction of a fraction. Next, convert the fractions (in one side) into the others, in this case a seventh and two thirds of a seventh into eighths; this gives an eighth and six sevenths of one eighth and a third of one seventh of an eighth. Next, add six sevenths of an eighth to three sevenths of an eighth; this makes one eighth and two sevenths of an eighth. Add together this eighth, the six eighths, the two eighths, and the single eighth; this makes one and two eighths. Add together one and the previous integers. The final result of multiplying the proposed quantities will be the following: twelve and two eighths and two sevenths of an eighth and a third of a seventh of an eighth.

FOURTH CASE. MULTIPLICATION OF AN INTEGER AND A FRACTION AND A FRACTION OF A FRACTION BY AN INTEGER AND A FRACTION AND A FRACTION OF A FRACTION.

(**A.129**) You want to multiply five and seven eighths and two thirds of an eighth by four and ten elevenths and half an eleventh.

(*a*) You will do the following. Multiplying the denominators of each side separately produces their respective denominating numbers, twenty-four for the first side and twenty-two for the second side. Multiplying them produces five hundred and twenty-eight, which is the principal number. Next, in order that all be reduced to the lowest kind of fraction in their respective sides, multiply five and seven eighths and two thirds of an eighth by twenty-four, thus producing a hundred and forty-three thirds of eighths; for the other side, multiply four and ten elevenths and half an eleventh by twenty-two, thus producing a hundred and nine half-elevenths. Multiplying this by the above hundred and forty-three produces fifteen thousand five hundred and eighty-seven. Divide it by the principal number, which is five hundred and twenty-eight; the result will be the required quantity. The proof of this is obtainable from the two above.[422]

(*b*) Or otherwise. Multiply term by term, as follows. Four by five is twenty. Next, multiply seven eighths and two thirds of an eighth by four, thus producing three and six eighths and two thirds of an eighth. Next, multiply ten elevenths and half an eleventh by five, thus producing the integer four and eight and a half elevenths. Next, multiply ten and a half elevenths by seven eighths and two thirds of an eighth as we have taught, that is, as follows.[423] You multiply ten and a half by seven and two thirds; this produces eighty and a half; dividing it by the denominator of an eighth gives ten and half an eighth; dividing it in turn by the denominator of the other side, namely eleven, gives ten elevenths and half an eighth

---

[422] A.124*a*.
[423] A.126*c*.

5	4
7	10
8·	11·
2	2·
3·	11·
8·	
24	22
528	
143	109
15587	

(A.129a)

of an eleventh. After the multiplication is performed, put each number together with the number of its kind, that is, integer with integer, fraction with fraction, fractions of fractions with fractions of fractions, and convert some of the fractions in order to make them all alike. Next, add them, beginning with the smallest [424] to the largest. The quantity resulting from the addition is the quantity resulting from the multiplication. [*The proof of this is obtainable from the first theorem of the second book of Euclid.*]

FIFTH CASE. MULTIPLICATION OF AN INTEGER WITH TWO FRACTIONS BY AN INTEGER WITH TWO FRACTIONS.

(**A.130**) You want to multiply six and a fifth and a third by eight and five sixths and a fourth.

(**a**) From the denominators of the fractions form the denominating numbers. Namely, multiply three by five, thus producing fifteen. Next, multiply four by six, thus producing twenty-four. Multiplying it by fifteen produces three hundred and sixty, which is the principal number. Next, multiply six and a fifth and a third by fifteen, thus producing ninety-eight. After that, multiply eight and five sixths and a fourth by twenty-four. Multiplying the result by ninety-eight and dividing the product by the principal number will give the result of the multiplication of the proposed quantities.[425]

(**b**) Or otherwise. Convert a third into fifths; this gives a fifth and two thirds of one fifth. Add it to six and a fifth, thus making six and two fifths and two thirds of one fifth. Next, convert a fourth into sixths; this gives one and a half sixths. Add it to five sixths, which makes the integer one and half a sixth, and add this integer to eight; the result is the integer nine and half a sixth. Then it is as if you were to multiply six and two fifths and two thirds of a fifth by nine and half a sixth. You will proceed here as

---

[424] The fraction with the largest denominator.

[425] The results of the calculations are found in the figure.

LIBER MAHAMELETH

6	8
5·	5
3·	6·
	4·
15	24
360	
98	218
21364	
59	
124	
360	

$(A.130a)$

has been shown before[426].

You will do the same for all cases of the above section which are about the multiplication of a fraction [*by an integer*].[427]

CHAPTER ON IRREGULAR FRACTIONS,
WHICH ARE DISCUSSED AMONG MATHEMATICIANS

(A.131) You want to multiply three fourths of five by seven.

(*a*) Four, from which a fourth is denominated, is in this case denominating number and principal.[428] Multiply three fourths of it, that is, three, by five; this produces fifteen. Multiply this fifteen by seven, thus producing a hundred and five. Dividing this by the principal, which is four, gives twenty-six and a fourth, and this is the quantity you require.

The proof of this is clear. For we had to multiply three by five and divide the product by four, thus obtaining three fourths of five, which we must in turn multiply by seven. But dividing fifteen by four and multiplying the result by seven is the same as multiplying seven by fifteen and dividing the product by four.[429] This is what we wanted to demonstrate.

(*b*) Or otherwise. Multiply five by seven, which produces thirty-five. It is then as if you were to multiply three fourths by thirty-five. You will do as we have taught above.[430]

---

[426] Presumably A.128–129.

[427] That is (disregarding the interpolation), if in A.99–103 one fraction is replaced by a sum of two fractions. See the example in A.136, which in fact belongs here.

[428] There is no other fraction in the multiplicand, and none in the multiplier.

[429] Premiss P_5.

[430] A.91.

$$
\begin{array}{|c|}
\hline
\begin{array}{ll} 3 & \quad 7 \\ 4\cdot & \\ \text{of } 5 & \end{array} \\
\hline
\begin{array}{ll} \quad 4 & \\ 15 & \quad 7 \\ \quad 105 & \end{array} \\
\hline
\end{array}
$$

$$(\text{A.131}a)$$

The proof of this is clear. For multiplying three fourths of five by seven is just the same as multiplying three fourths by five and the product by seven, which is the same as multiplying five by seven and multiplying the product by three fourths.[431] This is why we multiply five by seven and the product by three fourths, the result being what we wanted.

(**c**) Or otherwise. Proceed according to the wording of the problem. That is, multiply three fourths of five, which is three and three fourths, by seven, as follows. Multiply three by seven, thus producing twenty-one. Add it to the product of three fourths into seven, which is five and a fourth, thus making twenty-six and a fourth. This is what is required.

[*Or otherwise. Multiply three fourths of five, which is three and three fourths, by seven; the result will be the quantity you require.*][432]

$$
\begin{array}{|c|}
\hline
\begin{array}{ll} \dfrac{3}{} & \quad 7 \\[2pt] 3 & \\ 4\cdot & \end{array} \\
\hline
\end{array}
$$

$$(\text{A.131}c)$$

(**d**) Or otherwise. Multiply three fourths of seven by five; the result will be the quantity you require.

(**A.132**) You want to multiply four fifths of six and a third by eight.

(**a**) From the denominators of the fractions you will form the denominating number, namely fifteen, which will be at the same time denominating number and principal.[433] Multiply four fifths of it, thus twelve, by six and a third; this produces seventy-six. Multiplying it by eight [434] and dividing

---

[431] Premiss P_2.

[432] Reformulation of *c* to match *d*. This does not appear in MS $\mathcal{A}$.

[433] The multiplier is an integer.

[434] Result in the figure only. Same in A.133, A.134a, A.135a.

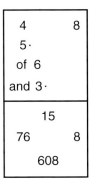

(A.132a)

the result by the principal number, namely fifteen, will give the quantity you require.

(**b**) Or otherwise. Multiply four fifths by six and a third. That is, multiply four by six and a third and divide the product by five; the result is five and a third of a fifth, which is four fifths of six and a third. Multiplying five and a third of a fifth by eight will give the quantity you require.

(**c**) Or otherwise. Multiply four fifths of eight by six and a third; the result will be the quantity you require.

(**d**) Or otherwise. Multiply six and a third by eight; four fifths of the result will be the quantity you require.

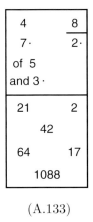

(A.133)

(**A.133**) You now want to multiply four sevenths of five and a third by eight and a half.

Form the denominating number from the denominators of a seventh and a third; this is twenty-one. Next, multiply two, which is the denominating number of the other side, by twenty-one, thus producing forty-two, which is the principal number. Multiply four sevenths of twenty-one, which is twelve, by five and a third; this produces sixty-four. Next, multiply eight

686 PART TWO: TRANSLATION, GLOSSARY

and a half by two, which is the denominating number of this side; this produces seventeen. Multiplying this by sixty-four and dividing the product by the principal number will give the quantity you require. Likewise in this case, you will proceed according to all the rules which have been given just above.[435]

(A.134) You now want to multiply two thirds of four by five eighths of seven.

(a) The multiplication of three and eight, which are in this case the denominating numbers, produces twenty-four, which is the principal number. Multiply two thirds of three, which is two, by four; this produces eight. Next, multiply five eighths of eight, which is five, by seven; this produces thirty-five. Multiplying this by eight and dividing the product by the principal number will give the quantity you require.

2	5
3·	8·
of 4	of 7
3	8
	24	
8	35
	280	

(A.134*a*)

The proof of this is the following. We know that multiplying two thirds of four by five eighths of seven is the same as multiplying two thirds by four, five eighths by seven, and then the first product by the second. But multiplying two thirds by four is nothing but taking two thirds of the denominator of a third, which is three, thus two, multiplying it by four and dividing the product by three;[436] so multiply two by four, thus producing eight, and dividing then eight by three will give two thirds of four[437]. Likewise for the multiplication of five eighths by seven: this means taking five eighths of the denominator of an eighth, which is eight, thus five, multiplying it by seven and dividing the product by eight; since the product of five into seven is thirty-five, dividing thirty-five by eight will give five eighths of seven. Then it is clear that multiplying two thirds of four by five eighths of seven is the same as dividing eight by three, thirty-five by eight, and multiplying the two quotients. But this is the same as dividing the product of eight into thirty-five by the product of three into eight, as

---

[435] In A.132.

[436] A.90*a*.

[437] All MSS have 'by four' and 'of three'.

LIBER MAHAMELETH

has been established in the chapter on premisses[438]. So multiplying two thirds of four by five eighths of seven is the same as dividing the product of eight into thirty-five by the product of three into eight. This is what we wanted to demonstrate.

(**b**) Or otherwise. Multiply two thirds of four, which is two and two thirds, by five eighths of seven, which is four and three eighths; the result will be the quantity you require.

(**c**) Or otherwise. Multiply four by seven, thus producing twenty-eight. Next, multiply two thirds by five eighths, thus producing three eighths and a third of an eighth. Multiplying this by twenty-eight will give the quantity you require.

The proof of this is the following. Multiplying two thirds of four by five eighths of seven is the same as multiplying two thirds by four, five eighths by seven, and the first product by the second. Now this is the same as multiplying two thirds by five eighths, four by seven, and the product by the product, as we have established in the beginning[439]. Therefore multiplying two thirds of four by five eighths of seven is the same as multiplying two thirds by five eighths, four by seven, and the product by the product. This is what we wanted to demonstrate.

2	2
3 ·	7·
of 5	of 6
and 4 ·	and 2 ·
12	14
168	
42	26
1092	

(A.135a)

(**A.135**) You now want to multiply two thirds of five and a fourth by two sevenths of six and a half.

(**a**) Form the denominating number by multiplying the denominators of one side, which are three and four; this is twelve. Next, form likewise the denominating number from the seventh and the half of the other side; this is fourteen. Multiply it by twelve; this produces a hundred and sixty-eight, which is the principal number. Next, multiply two sevenths of fourteen, which is four, by six and a half, thus producing twenty-six. Next, multiply

---

[438] Premiss P'_3 (Abū Kāmil's proposition).

[439] Premiss P_3.

two thirds of twelve, which is eight, by five and a fourth, thus producing forty-two. Multiplying this by twenty-six and dividing the result by the principal number will give the quantity you require. The proof of this is the same as the previous one[440].

(**b**) Or otherwise. Take two thirds of five and a fourth, which is three and a half. Next, take two sevenths of six and a half, which is one and six sevenths. You will then multiply them as you multiply an integer with a fraction by an integer and a fraction[441].

(**c**) Or otherwise. Multiply two thirds of six and a half, which is four and a third, by two sevenths of five and a fourth, which is one and a half; the result will be the quantity you require.

(**d**) Or otherwise. Multiply two thirds by two sevenths, thus producing a seventh and a third of a seventh. Next, multiply five and a fourth by six and a half, thus producing thirty-four and an eighth. Multiply these two products; the result will be six and three sevenths and half a seventh, and this is what you wanted.

(**A.135′**) This can also take many other ways than the one above.[442] Thus if you want to multiply two thirds of five, and a fourth of one, by two sevenths of six, and half of one.

(**a**) Multiply in this case two thirds of the denominating number, which is twelve, thus eight, by five; this produces forty. Add then a fourth of this denominating number, thus three, to the forty, making forty-three. Next, for the other side, multiply two sevenths of fourteen, which is four, by six; this produces twenty-four. Next, add half of fourteen, which is seven, to the twenty-four, thus making thirty-one. Multiplying this by forty-three and dividing the result by the principal number will give the quantity you require.

(**b**) Or otherwise. Two thirds of five, which is three and a third, becomes, with the addition of a fourth, three and three sixths and half a sixth. Likewise, two sevenths of six, which is one and five sevenths, becomes, with the addition of a half, two and a seventh and half a seventh. Then it is as if you were to multiply three and three sixths and half a sixth by two and a seventh and half a seventh. You will perform the multiplication according to the rule for fractions with integers.[443]

There are more other ways[444] occurring with fractions; I shall speak about them and explain how to deal with them.[445]

---

[440] A.134*a*.

[441] A.124.

[442] The figure with the data remains essentially the same, only the interpretation is different.

[443] A.129 (integers with compound fractions).

[444] That is, cases of ambiguity in the formulation.

[445] After A.137.

LIBER MAHAMELETH

689

You must know that anyone who has understood perfectly what has been said about the multiplication of fractions and has paid full attention to their proofs will be able to solve with ease any problem which might arise involving a multiplication of fractions.[446]

(A.136) Thus someone asks you to multiply two ninths and two sevenths by ten elevenths.[447]

(*a*) The solution of this problem will also be shown by means of what has been said about the multiplication of fractions,[448] as follows. You must reduce one of the two fractions to the kind of the other as shown before, in the chapter on the conversion of fractions into others[449]. Adding them makes a fraction and a fraction of a fraction. Multiply them by ten elevenths as shown above in the chapter about this kind of multiplication[450].

(*b*) Or otherwise. Multiply two ninths by ten elevenths as shown above[451]. Next, multiply two sevenths by ten elevenths. Add the result to the product of two ninths into ten elevenths. The sum will be what you wanted.

(A.137) You now want to multiply five eighths and two thirds of an eighth by six and two sevenths and three fourths of a seventh.[452]

You will do the following. Multiply five eighths and two thirds of an eighth by six as shown above[453], and keep in mind the result. Next, multiply five eighths and two thirds of an eighth by two sevenths and three fourths of a seventh. Add the result to that kept in mind before. The sum is what you wanted.

[*Anyone who understands what has been said about the conversion of fractions and is acquainted with their proofs*[454] *will easily solve the above problems, as well as many others which have not been stated here, and will know how to prove them all, without needing any further explanation.*]

There can be many other problems involving the multiplication of fractions which are subject to several interpretations.[455] We have deemed it appropriate to place them, along with their treatments and their various

---

[446] Since there is no explicit mention of the conversion of fractions an early reader enlarges upon this statement below (after A.137).

[447] This problem (and the preceding observation) should follow the remark placed after A.130 (p. 683).

[448] A.99–107. But here we need conversion of fractions.

[449] A.108–122.

[450] A.106.

[451] A.100–101.

[452] This example of the multiplication of a compound fraction by an integer with a compound fraction belongs to the group A.123–125, A.128–130.

[453] A.96.

[454] The proofs for the multiplication.

[455] We have seen one example in A.135–135′.

690   PART TWO: TRANSLATION, GLOSSARY

meanings, at the end of (the chapter on) multiplication.[456] Such are the following.

ANOTHER CHAPTER ON THE SAME TOPIC.[457]

(**A.138**) You want to multiply a fourth of five and two fifths of six by a tenth of three and an eighth of four.

This can take two ways.

(*a*) The first is when you want to add one fourth of five to two fifths of six and multiply the result by the sum of a tenth of three and an eighth of four.

If you want to solve that, you will do the following. By multiplying the denominators of the fractions a fourth and a fifth, form the denominating number; this is twenty. Next, for the other side, form likewise from a tenth and an eighth the denominating number; this is eighty. Multiply this by the twenty of the opposite side. This will produce the principal number, namely one thousand six hundred. Next, multiply a fourth of twenty, which is five, by five, thus producing twenty-five. Next, multiply two fifths of twenty, which is eight, by six, thus producing forty-eight. Add it to twenty-five; this makes seventy-three. Next, for the other side, multiply a tenth of eighty, namely eight, by three, thus producing twenty-four. Next, multiply an eighth of eighty, namely ten, by four, thus producing forty. Add it to twenty-four; this makes sixty-four. Multiplying this by seventy-three and dividing the result by the principal will give the required quantity.

1	1
4 ·	10 ·
of 5	of 3
and 2	and 1
5 ·	8 ·
of 6	of 4
20 80	
1600	
73 64	

(A.138*a*)

(*b*) The second way is when you want to add two fifths of six to five and take a fourth of the sum, then add an eighth of four to three and multiply a tenth of the sum by the former result.

If you want to solve that, you will do the following. Having taken a fourth of twenty, which is the denominating number, you multiply it by

---

[456] The place of A.140–144, which we have put after them, is uncertain.

[457] That is: ambiguity of problems involving irregular fractions.

five, thus producing twenty-five. After that, having taken a fourth of two fifths of twenty, which is two, you multiply it by six, thus producing twelve. Adding this to twenty-five makes thirty-seven. Next, after taking a tenth of the denominating number for the other side, which is eighty, multiply it by three, thus producing twenty-four. Next, multiply a tenth of an eighth of eighty, which is one, by four, thus producing four. Add it to twenty-four; this makes twenty-eight. Multiplying this by the thirty-seven of the other side and dividing the product by the principal will give the quantity which is the result.

(**A.139**) Someone proposes multiplying two thirds of seven and a half and two fifths of six and a third by two sevenths of four and a tenth and three fourths of nine and a ninth.

2 3· of 7	2 7· of 4
and 2·	and 10·
and 2 5· of 6	and 3 4· of 9
and 3·	and 9·

(A.139)

This can take four ways.

(**a**) Two of them are those we have spoken about just before.[458] That is, either adding two thirds of seven and a half to two fifths of six and a third, then adding two sevenths of four and a tenth to three fourths of nine and a ninth, and multiplying this latter sum by the former. The rest is then as we have taught before.

(**b**) Or by adding two fifths of six and a third to seven and a half, and taking two thirds of the sum, then adding three fourths of nine[459] and a ninth to four and a tenth and multiplying two sevenths of this sum by the previous result. How to perform the rest has already been shown by us.

(**c**) The third way is the following. You add a half to two thirds of seven, you add a third to two fifths of six, and you form one sum out of the two; next, for the other side, you add a tenth to two sevenths of four, you add a ninth to three fourths of nine, and you make one sum out of the two; you multiply this sum by the sum of the other side. The rest is then as we have taught just before.

---

[458] A.138.

[459] The MSS have 'of six'.

692   PART TWO: TRANSLATION, GLOSSARY

(*d*) The fourth way is the following. You add, to two thirds of seven, a half and two fifths of six and a third; then, for the other side, you add, to two sevenths of four, a tenth [*of nine* [460] *and a ninth*][461] and three fourths of nine [462] and a ninth. You multiply this sum by the first one. The rest is then as we have already taught.

So all the rules about these are already clear to anyone who has understood what has been said previously about fractions. [*This can also take ways other than these.*][463]

(**A.140**) You want to multiply three fourths of four fifths of five sixths of six sevenths of seven eighths by four fifths of five sixths of six sevenths of seven eighths of eight ninths of nine tenths.

If you want to do this by forming the denominating numbers from the denominators of the fractions, it will be long. There is another treatment, which is the following. Multiply the number from which an eighth is denominated by the number from which a tenth is denominated; this produces eighty. Put it as the principal. Next, take seven eighths of eight, which is seven; take, of six sevenths of this, which is six, its five sixths, which is five; take four fifths of this, which is four; take three fourths of this, which is three, and keep it in mind. After that, take nine tenths of ten, which is nine; take eight ninths of this, which is eight; take seven eighths of this, which is seven; take six sevenths of this, which is six; take five sixths of this, which is five; take four fifths of this, which is four. Multiply this four by the three kept in mind; the result is twelve. Dividing [464] it by eighty gives a tenth and half a tenth, and this is what you wanted.

(**A.141**) You want to multiply three fourths minus one sixth by five and a third.[465]

You will do the following. Multiplying the denominators four and six produces twenty-four. Take three fourths of it, which is eighteen. Subtract from it a sixth of twenty-four, which is four; this leaves fourteen. Multiply it by five and a third, and divide the product by twenty-four. The result is what you wanted to know.

(**A.142**) You want to multiply three fourths minus a third less an eighth by seven and a half.

You will do the following. Multiplying the denominators four and three produces twelve. Take three fourths of it, which is nine. Next take a third of twelve, which is four. Subtract from it an eighth of twelve, which is one

---

[460] The MSS have 'of six'.

[461] A reader's attempt to clarify the text.

[462] The MSS have 'of six'.

[463] In one sense, this is true. But since this statement contradicts what was initially indicated ('four ways'), it must be an addition.

[464] Rather: Denominating.

[465] A.141–144 are found only in MS $\mathcal{B}$, where they follow A.140.

LIBER MAHAMELETH 693

and a half; this leaves two and a half. Subtract it from nine; this leaves six and a half. Multiply it by seven and a half; this produces forty-eight and three fourths. Divide it by twelve. The result is four and half an eighth, and this is what you require.

(**A.143**) You want to multiply three fourths of five sixths of seven and a fifth by nine tenths of four, minus half an eighth of nine and three fifths.

You will do the following. Take three fourths of five sixths of seven and a fifth, which is four and a half. Multiply it by nine tenths of four minus half an eighth of nine and three fifths, which is three. The result is thirteen and a half, and this is what you wanted to know.

(**A.144**) You want to multiply nine tenths minus the product of half a seventh into eleven by the product of five ninths of one fourth of three into a fifth of twelve.

You will do the following. Look for a number having a tenth and half a seventh; such is seventy. Multiply half a seventh of it, which is five, by eleven, and subtract the result from nine tenths of seventy, which is sixty-three; this leaves eight. Then multiply a fifth of twelve, which is two and two fifths, by three; this produces seven and a fifth. Of a fourth of it, which is one and four fifths, take five ninths, which is one; multiply it by eight, and denominate the product from seventy. The result is a tenth and a seventh of a tenth, and this is what you wanted to know.

Chapter (A–VI) on the Addition of fractions to fractions[466]

There are five cases of addition of a fraction to a fraction.[467]

FIRST CASE.

(**A.145**) You want to add three eighths to four fifths.

(*a*) You will do the following. Multiplying the denominators of the fractions produces forty, which is the principal number. Next, add three eighths of the principal number, thus fifteen, to four fifths of this same principal number, thus thirty-two; this makes forty-seven. Dividing it by the principal gives one and an eighth and two fifths of an eighth, and this is the result from the addition of the proposed quantities.

3	4
8 ·	5 ·
8	5
40	
15	32
47	

(A.145*a*)

(*b*) Or otherwise. Convert the fractions of one side into the others, namely four fifths into eighths; this gives six eighths and two fifths of an eighth. Add it to three eighths; this makes nine eighths, which is one and an eighth, and two fifths of an eighth. This is the quantity you require.

(**A.146**) You want to add three sevenths to ten elevenths.

(*a*) You will do the following. Multiply the denominators of the fractions, namely seven and eleven; the result is the principal number, which is seventy-seven. Add to three sevenths of it, which is thirty-three, ten

---

[466] Additions involving integers are not included. See note 476, p. 700.

[467] Fraction to fraction (A.145–146); compound fraction to fraction (A.147); compound fraction to compound fraction (A.148); fraction of a fraction to fraction of a fraction (A.149); compound fraction to fraction of a fraction (A.150).

LIBER MAHAMELETH

3	10
7·	11·
7	11
77	
33	70
103	

$$(\text{A.}146a)$$

elevenths of it, which is seventy; this makes a hundred and three. Dividing this by the principal will give the result from the addition of the proposed quantities.

(**b**) Or otherwise. Convert the three sevenths into elevenths; this gives four elevenths and five sevenths of one eleventh. Adding this to the above ten elevenths will give the result.

(**c**) Or conversely. Convert the elevenths into sevenths and add to the result three sevenths; the sum is the required quantity.

SECOND CASE. CHAPTER ON ADDING A FRACTION AND A FRACTION OF A FRACTION TO A FRACTION.

(**A.147**) You want to add three fifths and a fourth of a fifth to five eighths.

(**a**) The rule is the following. By multiplying the denominators of the fractions a fourth and a fifth you form the denominating number, which is twenty. The multiplication of it by eight produces a hundred and sixty, which is the principal number. Add three fifths and a fourth of a fifth of it, which is a hundred and four, to five eighths of it, which is a hundred; this makes two hundred and four. Dividing this by the principal will give the required quantity.

3	5
5·	8·
and 4·	
5·	
20	8
160	
104	100
204	

$$(\text{A.}147a)$$

696   PART TWO: TRANSLATION, GLOSSARY

(***b***) Or otherwise. Convert three fifths and a fourth of a fifth into eighths; this gives five eighths and a fifth of an eighth. Adding it to five eighths makes one and two eighths and a fifth of an eighth, and this is what you require.

(***c***) You may also convert the five eighths into fifths; this gives three fifths and an eighth of a fifth. Adding it to three fifths and a fourth of one fifth makes one and a fifth and three eighths of one fifth, and this is what is required.

THIRD CASE. CHAPTER ON ADDING A FRACTION AND A FRACTION OF A FRACTION TO A FRACTION AND A FRACTION OF A FRACTION.

(**A.148**) You want to add two fifths and three fourths of a fifth to two sevenths and two thirds of a seventh.

You will do the same as in the previous cases, namely the following.

(***a***) You form, by multiplying the denominators of the fractions of each side separately, the denominating number of each side; multiplying them produces the principal number, which is four hundred and twenty. Add two fifths and three fourths of a fifth of it, which is two hundred and thirty-one, to two sevenths and two thirds of a seventh of it, which is a hundred and sixty, and divide[468] the sum by the principal. The result will be the quantity you require.

2	2
5 ·	7 ·
and 3	and 2
4 ·	3 ·
5 ·	7 ·
20	21
420	
231	160
391	

(A.148*a*)

(***b***) You may also convert the two fifths and three fourths of a fifth into sevenths; this gives three sevenths and four fifths of one seventh and a fourth of a fifth of a seventh. Next, add two sevenths to three sevenths; this makes five sevenths. And add two thirds of a seventh to four fifths of a seventh and a fourth of a fifth of one seventh, converting the fractions (of one side) into the others as shown above[469]; this will give a seventh and

---

[468] Rather: denominate.
[469] A.117–122.

LIBER MAHAMELETH 697

two fifths of a seventh and three sixths of one fifth of one seventh and half a sixth of a fifth of a seventh. Adding this to the previous five sevenths makes six sevenths and two fifths of one seventh and three sixths of a fifth of a seventh and half a sixth of one fifth of one seventh, and this is the quantity you require.

FOURTH CASE. CHAPTER ON ADDING A FRACTION OF A FRACTION TO A FRACTION OF A FRACTION.

(**A.149**) You want to add five sixths of an eighth to four sevenths of an eleventh.

(*a*) You will do the following. Multiplying the denominators of the fractions of the side to be added, namely six and eight, produces forty-eight, which is the denominating number. Next, for the other side, multiplying seven by eleven produces seventy-seven, which is likewise the denominating number. Multiply it by the other, of the other side; this produces three thousand six hundred and ninety-six, which is the principal number. Next, add five sixths of an eighth of it, which is three hundred and eighty-five, to four sevenths of an eleventh of it, which is a hundred and ninety-two; this makes five hundred and seventy-seven. Denominate this number, since it is smaller, from the principal; the result will be the quantity obtained from the addition.

$$
\begin{array}{|cc|}
\hline
5 & 4 \\
6\cdot & 7\cdot \\
8\cdot & 11\cdot \\
\hline
48 & 77 \\
& 3696 \\
385 & 192 \\
& 577 \\
\hline
\end{array}
$$

(A.149*a*)

(*b*) Or otherwise. Convert the fractions into a same kind, say the sixths of an eighth into sevenths of an eleventh, using the rules seen above [470]. The result will be eight sevenths of an eleventh and a sixth of an eighth of a seventh of an eleventh. Adding to this the four sevenths of an eleventh will make the quantity which is the result.

FIFTH CASE. ADDING A FRACTION AND A FRACTION OF A FRACTION TO A FRACTION OF A FRACTION.

(**A.150**) You want to add four fifths and a third of a fifth to five sixths of an eleventh.

---

[470] A.117–118.

(a) You will do the following. Multiply the denominators of the side to be added, namely five and three, thus forming fifteen, which is the denominating number. Next, for the other side, multiply six by eleven, thus producing sixty-six, which is likewise the denominating number. Multiply this denominating number by the other, of the other side; the product will be the principal number. Adding four fifths and a third of a fifth of it to five sixths of an eleventh of it and dividing the sum by the principal will give the required quantity.

$$
\begin{array}{|cc|}
\hline
4 & 5 \\
5 \cdot & 6 \cdot \\
\cline{1-1}
\text{and } 3 \cdot & 11 \cdot \\
5 \cdot & \\
\hline
15 & 66 \\
990 & \\
\hline
\end{array}
$$

(A.150a)

(b) Or otherwise. Convert four fifths and a third of a fifth into elevenths;[471] this gives nine elevenths and two fifths of an eleventh and two thirds of a fifth of an eleventh. Next, convert two fifths of an eleventh and two thirds of a fifth of an eleventh into sixths of an eleventh; this gives three sixths of an eleventh and a fifth of a sixth of an eleventh. Add this to the above five sixths of an eleventh; this makes one eleventh and two sixths of an eleventh and a fifth of a sixth of an eleventh. Then add one eleventh to the previous nine. This will give (altogether) ten elevenths and two sixths of an eleventh and a fifth of a sixth of an eleventh, and this is the quantity which is the result.

AGAIN ON THE SAME TOPIC.

(A.151) You want to add together two ninths, three eighths, ten elevenths.

(a) You will do the following. Multiply the denominators of a ninth, an eighth, an eleventh; this produces seven hundred and ninety-two, which is the principal number. Add together its two ninths, which is a hundred and seventy-six, its three eighths, which is two hundred and ninety-seven, and its ten elevenths, which is seven hundred and twenty, and divide the sum by the principal. The result will be the required quantity.

The proof of this is the following. Put as two ninths AB, three eighths, BG, and ten elevenths, GD; let one be H, the denominator of all the fractions, which is the principal number, Z, its two ninths, which is a hundred and seventy-six, KT, its three eighths, which is two hundred and

---

[471] Conversion into sixths of elevenths will follow.

$$\frac{\begin{array}{c}2\\9\cdot\end{array}}{3}$$

$$\frac{8\cdot}{10}$$

$$\overline{11\cdot}$$

(A.151)

ninety-seven, TQ, its ten elevenths, which is seven hundred and twenty, QL. Then[472] the ratio of AB to H is the same as the ratio of KT to Z, the ratio of BG to H is the same as the ratio of TQ to Z, and the ratio of GD to H is the same as the ratio of QL to Z. It follows[473] that the ratio of AD to H is the same as the ratio of KL to Z. Thus the product of line AD into Z is equal to the product of line KL into H. Now H is one. [*But anything multiplied by one does not increase.*] So the product of line AD into Z is KL. Thus dividing KL by Z will give AD. This is what we wanted to demonstrate. This proof is valid for all the previous cases of addition of fractions.[474]

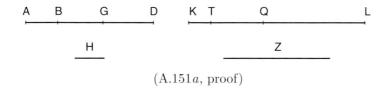

(A.151a, proof)

(***b***) Or otherwise. Convert all the fractions into the kind of whatever fraction you want, say into eighths. Converting then two ninths into eighths gives one eighth and seven ninths of one eighth, and converting ten elevenths into eighths gives seven eighths and three elevenths of an eighth. After addition of all the eighths, which makes one and three eighths, and of the seven ninths of an eighth to the three elevenths of an eighth as we have taught before[475], the total quantity will be one and four eighths and five elevenths of one ninth of one eighth. This is the quantity you required. It is preferable, though, to convert such fractions into the kind of the smallest fraction of them all, as here the eleventh.

---

[472] *Elements* VII.17, or A.90a.
[473] *Elements* V.24 (or Premiss P'_7) repeated.
[474] This last remark, although pertinent, could be a later addition. (It appears in the body of the text in MSS $\mathcal{B}$ and $\mathcal{C}$ but at the bottom of the page in $\mathcal{A}$.) A (misplaced) analogous statement at the end of A.191 is very probably a later insertion.
[475] A.149.

700   PART TWO: TRANSLATION, GLOSSARY

[*If in the addition you have also put digits, add the digits and the fractions separately. If from the addition of the fractions a digit results, add it to the previous digits. The resulting quantity will be the required sum.*][476]

AGAIN ON THE SAME TOPIC [477]

(A.152) You want to add three fifths of six to seven eighths of nine.

3	7
5·	8·
of 6	of 9

(A.152)

(*a*) You will do the following. Multiply the denominators, namely five and eight; this produces forty, which is the principal number. Multiply three fifths of it, which is twenty-four, by six, thus producing a hundred and forty-four, then multiply seven eighths of it, which is thirty-five, by nine, thus producing three hundred and fifteen. Adding this to the former one hundred and forty-four makes four hundred and fifty-nine. Dividing it by the principal will give the quantity which is the result.

The proof of this is the following. It was necessary to take three fifths of six and add them to seven eighths of nine. Now to take three fifths of six means finding any number having a fifth, say forty, multiplying three fifths of it, which is twenty-four, by six, thus producing a hundred and forty-four, then dividing a hundred and forty-four by forty; the result will be three fifths of six.[478] Taking seven eighths of nine proceeds likewise: that is, find a number having an eighth, say forty, multiply seven eighths of it, which is thirty-five, by nine, thus producing three hundred and fifteen, then divide it by forty; the result will be seven eighths of nine. Thus it was necessary to divide a hundred and forty-four by forty, to also divide three hundred and fifteen by forty, and to add the two quotients; but this is the same as adding a hundred and forty-four to three hundred and fifteen and dividing the sum by forty, as established in the chapter on premisses[479].

(*b*) Or otherwise. Add three fifths of six, which is three and three fifths, to seven eighths of nine, which is seven and seven eighths; this makes the quantity which is the result.

---

[476] It is true that we have not considered the case of adding fractions with integers; there is, however, no real need for it and in any event, according to the heading of this chapter, it was not intended to appear. The use of the word 'digit' instead of 'integer' also suggests that here we have an early reader's remark.

[477] Addition of irregular fractions.

[478] See A.90.

[479] Premiss P_8.

LIBER MAHAMELETH

2	3
5·	8·
of 4	of 6
and 2·	and 3·

(A.153)

(**A.153**) You now want to add two fifths of four and a half to three eighths of six and a third.

(*a*) You will do the following. From the denominators, as said above[480], you will form the principal number, which is two hundred and forty. You will multiply two fifths of it by four and a half, thus producing four hundred and thirty-two, then you will multiply three eighths of it by six and a third, thus producing five hundred and seventy. Adding this to the former four hundred and thirty-two and dividing the sum by the principal number will give the quantity you require. The proof of this is clear from what precedes.[481]

(*b*) Or otherwise. You will add two fifths of four and a half, which is one and four fifths, to three eighths of six and a third, which is two and three eighths. The sum is the required quantity.

(**A.154**) You now want to add two fifths of four and half of one to three eighths of six and a third of one.[482]

(*a*) You will do the following. By multiplying the denominators you will first form the principal number, which is two hundred and forty. Multiply two fifths of it by four, add to the product half the principal number, and keep in mind the result. Next, multiply three eighths of this same principal number by six, and add to the product a third of the principal number. Add the result to the quantity kept in mind, and divide their sum by the principal. The result is the quantity you require.

(*b*) Or otherwise. Add to two fifths of four half of one. Add to three eighths of six a third of one. Adding this sum to the former sum will give the quantity which is the result.

You are to know that there is for addition as many cases as for multiplication.[483] But this must be observed in all of them: you take as much of the principal number as are all the fractions in each of the two sides

---

[480] From A.145 on.

[481] A.152*a*.

[482] Another interpretation of A.153 (whence the separation in the above figure). But the Latin case-endings make the formulation unequivocal (even without the addition of 'of one').

[483] The twenty-eight cases mentioned at the beginning of A–III (see p. 606). This remark would fit better before or after A.151.

702   PART TWO: TRANSLATION, GLOSSARY

and you divide the sum by the principal; the result will be the required quantity.

AGAIN ON ADDITION[484]

(A.155) You want to add together four fifths of nine and three fourths of nine.

You will do the following. Multiplying the denominators, namely four and five,[485] you will produce twenty, which is the principal number. Next, add together four fifths of it and three fourths of it; this makes thirty-one. Multiplying this by nine and dividing the product by the principal will give the required quantity.

This rule is taken from a proportion [*which Euclid asserts in the sixth book*][486], namely that the ratio of thirty-one to twenty is the same as the ratio of the required quantity to nine. Here the third term is unknown and the division is performed by the second. Thus multiplying the first, which is thirty-one, by the fourth, which is nine, and dividing the product by the second, which is twenty, will give the unknown term, which is required.

You may also divide thirty-one by the principal and multiply the result by nine, or divide nine by the principal and multiply the result by thirty-one, and the result will be the unknown term.   All these ways lead to the right result.[487]

(A.156) You now want to add together four fifths of nine and three fourths of nine, add to the sum its half and know the resulting sum.

You will do the following. Multiplying all the denominators, that is, of a fourth, a fifth and a half, you form the principal number, which is forty. Adding together four fifths of it and three fourths of it makes sixty-two. Adding to this its half, namely thirty-one, makes ninety-three. Then the ratio of ninety-three to the principal is the same as the ratio of the required quantity to nine.[488]   Therefore multiplying ninety-three by nine and dividing the product by the principal will give what you require.   You may also divide either of the factors by the principal and multiply the result by the other factor; this will give what you require.

(A.156′) If you now wanted in this problem ⟨to add to the sum⟩ half the remainder, it would not be solvable, for after taking from nine its four fifths

---

[484] Problems 'on amounts'.

[485] Usually the denominators are taken in order of their occurrence in the formulation, but not here or in the subsequent problem.

[486] Book VI applies to plane geometry the theory of proportion taught in Book V, and Book VII applies it to numbers. In this sense VI.16 corresponds to VII.19, which infers from $a : b = c : d$ that $a \cdot d = b \cdot c$. This is perhaps what the glossator had in mind. The proof itself is just of the type already seen in A.90a and A.93a.

[487] These banal changes (see P_5) are also proposed in the next two problems and in A.159.

[488] As seen in A.155.

LIBER MAHAMELETH   **703**

and its three fourths, nothing remains.[489] Indeed, such a problem cannot be solvable unless the sum of the proposed fractions is less than one.

(**A.157**) You want to add together two fifths of nine, a fourth of it, and a third of the remainder and know the resulting sum.

You will do the following. Multiplying the denominators of a fifth, a fourth, and a third produces sixty, which is the principal number. Adding two fifths of it and a fourth of it makes thirty-nine. Subtract it from sixty; this leaves twenty-one. Add its third, namely seven, to thirty-nine; this makes forty-six. Then the ratio of this to the principal, which is sixty, is the same as the ratio of the required quantity to nine. So multiply forty-six, which is the sum, by nine and divide the product by the principal; the result will be what you require. You may also divide either of the factors by the principal and multiply the result by the other factor; the product will be what you require.

You will again do the same if there is a fraction with the integer.

(**A.158**) For instance, you want to add together five eighths of seven and a half, two thirds of the same quantity, and a fourth of the resulting sum.

You will do the following. Multiply the denominators of the fractions an eighth, [*a half*,] a third and a fourth, thus producing ninety-six[490]. Add five eighths of it to two thirds of it, and add to the sum its fourth; this makes a hundred and fifty-five. Then the ratio of this to ninety-six is the same as the ratio of the required quantity to seven and a half. So multiplying a hundred and fifty-five by seven and a half and dividing the product by ninety-six will give what is required.

Consider other such instances likewise, and you will succeed.

CHAPTER ON AMOUNTS IN ADDITION [491]

(**A.159**) Someone asks: What is the amount of which the addition of a third and a fourth makes ten?[492]

You will solve it as follows. Multiply the denominators of the parts a third and a fourth; this produces twelve. Add a third and a fourth of this number; this makes seven. Let it be in this case the principal[493]. Then it is clear that the ratio of seven to twelve is the same as the ratio of ten to the amount. So multiply the given quantity, namely ten, by the product[494],

---

[489] This is inadequately formulated. In A.203″ (similar situation) the formulation is correct.

[490] The half is irrelevant here.

[491] Problems 'on unknown amounts' (equations of the first degree).

[492] An example of the same kind in the *Liber algorismi*, p. 110.

[493] In the previous problems, the principal, which is the final divisor, was identical to the product of the denominators; here, because of the proportion set out below, the principal is the sum of the proposed fractions taken of it. See also A.204 (amounts in subtraction).

[494] The product of the denominators. See previous note.

704   Part Two: Translation, Glossary

namely twelve, and divide the product by the principal, namely seven; the result will be seventeen and a seventh. Such is the amount of which the addition of a third and a fourth makes ten. You may also divide one of the two factors, namely ten or twelve, by the principal and multiply the result by the other factor; the product will be the amount in question.

(**A.160**) Now someone asks: What is the amount of which the addition of a third and a fourth, plus two nummi, makes ten?

You will solve it as follows. Subtract two from the proposed quantity, that is, ten; this leaves eight. So eight is a third and a fourth of the amount in question. It is then like saying: 'What is the amount of which the addition of a third and a fourth makes eight?' Proceed according to the rule given above.[495]

(**A.161**) Now someone asks: What is the amount of which the addition of a third and a fourth, minus two nummi, makes ten?

You will solve it as follows. Add two, which is lacking, to the proposed quantity, namely ten; this makes twelve. So twelve is a third and a fourth of the amount. It is then like saying: 'What is the amount of which the addition of a third and a fourth makes twelve?' Solve it according to the rule given above.[496]

You will proceed likewise in all problems of this kind, adding to the proposed quantity what is lacking and subtracting what is in excess; the result is then the sum of the given fractions of the amount.

(**A.162**) Now someone asks: What is the amount of which the addition of a third, plus one nummus, and a fourth, minus three nummi, makes ten?

You will solve it as follows. Add one, added, to three, subtracted; this makes two, subtracted.[497] It is then like saying: 'What is the amount of which (the addition of) a third and a fourth, minus two nummi, makes ten?' Solve it according to the rule given above.[498]

(**A.163**) Someone asks: What is the amount of which the addition of a third, plus two nummi, a fourth, minus one nummus, and half the remainder, plus four nummi, makes ten?

You will do the following. Add two, added, to one, subtracted; this leaves one, added. It will be subtracted from[499] the remainder of the amount. Add half of this unit subtracted, that is, a half, subtracted, to

---

[495] A.159.

[496] Again, A.159.

[497] The wording here (also met with in A.208) does not mean that negative numbers are considered. On the appearance of negative numbers in the Middle Ages, see our *Introduction to the History of Algebra*, pp. 103–118.

[498] This is just A.161.

[499] Rather: in —unless by 'remainder' is now understood what remains of the amount disregarding the additive and subtractive constants. Same below.

LIBER MAHAMELETH

the four nummi; the result is three and a half nummi, added. Adding this to the nummus added before makes four and a half, added. It is then like asking: 'What is the amount of which the addition of a third, a fourth, and half the remainder, plus four and a half nummi, makes ten?' You will do the following. Subtract four and a half nummi from ten, and proceed then with the rest as we have shown before[500].

(**A.164**) Now someone asks: What is the amount of which the addition of a third, minus five nummi, a fourth, plus two nummi, and half the remainder, minus one, makes ten?

You will do the following. Add the five subtracted to the two added; this makes three, subtracted. This will be added to[501] the remainder of the amount. Add half of it, which is one and a half, added, to the given unit subtracted; this leaves a half, added. Add it to the above three, subtracted; this makes two and a half, subtracted. It is then like asking: 'What is the amount of which the addition of a third, a fourth, and half the remainder, with two and a half subtracted, makes ten?' You will do the following. Add two and a half to ten; this makes twelve and a half. Continue as has been shown before.[502]

There can be here many other, longer problems. Those who have grasped what has been said will understand them with ease [*one such is the following:*][503]

(**A.165**) Someone asks: What is the amount of which the addition of a third, a fifth, and a fourth of the remainder makes twenty?

You will do the following. Multiply the denominators of a third, a fifth, a fourth, thus producing sixty. Add a third and a fifth of it, which is thirty-two, to a fourth of the remainder; this makes thirty-nine. Then the ratio of thirty-nine to sixty is the same as the ratio of twenty to the required amount. Thus multiplying sixty by twenty and dividing the product by thirty-nine will give the required amount.

(**A.166**) Someone asks: What is the amount of which the addition of a fifth, plus two nummi, and half the remainder, plus four nummi, makes ten?

(**a**) You will do the following. You know that subtracting from something its fifth and two nummi leaves (its) four fifths minus two nummi; half of this is (its) two fifths minus one nummus. Add to it four nummi; this makes two fifths of the amount, and three nummi. It is then clear that adding together a fifth of the amount with two nummi and (its) two fifths and three nummi makes ten nummi. So add two nummi to three nummi, which makes five, next subtract this from ten, which leaves five. It is then

---

[500] As this means A.165, A.163 (and A.164) must be misplaced (though they immediately follow A.159–162 in the disordered text).

[501] Rather: in.

[502] Again, A.165.

[503] Inappropriate remark (only in MS $\mathcal{A}$, same hand).

706   PART TWO: TRANSLATION, GLOSSARY

like asking: 'What is the amount of which three fifths is five?' Proceed as has been shown before.[504]

(**b**) Or otherwise. You know that adding together its fifth and two nummi, these two nummi are added to[505] this sum but subtracted from[506] the remainder of the amount, so that from half the remainder[507] one nummus is subtracted. But we already had four nummi added. Making up for the nummus subtracted leaves three added. Adding this to the other two added makes five added. It is then like asking: 'What is the amount of which the addition of a fifth and half the remainder, plus five nummi, makes ten?' Proceed as I have taught above[508], and this will give what you require.

(**A.167**) Someone asks: What is the amount of which the sum of a third and a fourth, when multiplied by itself, produces forty-nine?

You will do the following. Take the root[509] of forty-nine. This, namely seven, is a third and a fourth of the amount. It is then like asking: 'What is the amount of which the addition of a third and a fourth makes seven?' Proceed as we have taught above[510].

(**A.168**) Someone asks: What is the amount of which the sum of a third and a fourth, when it is multiplied by itself and the product divided by the amount, gives four and half a sixth?

You will do the following. Add a third and a fourth; this makes three sixths and half a sixth. Always divide one by this[511]; the result is one and five sevenths. Multiplying this by itself, and the product by four and half a sixth, will give the amount required.

The proof of this is the following. We know[512] that multiplying three sixths and half a sixth of the amount by itself is the same as multiplying this amount by four and half a sixth. Then the ratio of the amount to (its) three and a half sixths is the same as the ratio of (its) three and a half sixths to four and half a sixth. So the ratio of the amount to four and half a sixth is the same as the duplicate ratio[513] of one[514] to three and a half sixths. Now the ratio of one[515] to three and a half sixths is one and five

---

[504] This is an odd reference (in a similar situation, in A.213, the result is given directly). The reader of MS *C* has performed the calculation here.

[505] Rather: in.

[506] Rather: in.

[507] See above note.

[508] Presumably A.165.

[509] First mention (apart from the general introduction) of roots, a subject to be treated in A–IX.

[510] A.159.

[511] Division of 1 by a fraction will be taught in A–VIII (A.215–220).

[512] From the data and the previous computation.

[513] *Elements* V, def. 9.

[514] 'of the amount' in the MSS.

[515] 'of the amount' in the MSS.

LIBER MAHAMELETH

707

sevenths.[516] Therefore the ratio of the amount to four and half a sixth is two[517] and six sevenths and four sevenths of a seventh. So multiply four and half a sixth by two and six sevenths and four sevenths of a seventh; this produces twelve, and such is the amount. This is what we wanted to demonstrate.

CHAPTER ON THE SAME TOPIC, BUT DIFFERENT FROM THOSE PRECEDING.[518]

(A.169) There are two unequal numbers such that a third of one of them added to a fourth of the other makes ten. What is each of these numbers?

(a) This problem is [*likewise*] indeterminate.[519] You will do for it the following. Multiply the denominators of the fractions, namely three and four, thus producing twelve. Let this twelve be either of the two above numbers, say that of which a third is added. Subtract its third, which is four, from ten; this leaves six. Thus six is a fourth of the other number, which is therefore twenty-four. Adding together a third of twelve and a fourth of twenty-four makes ten, as was indeed proposed.

(b) Or otherwise, in order that the rule become more general. Divide ten into two unequal[520] parts, and multiply one of them by four and the other by three. The two products will be the two numbers required.

ANOTHER CHAPTER[521]

(A.170) You want to know the sum of all the numbers in sequence from one to twenty.

Add one to the last number, with which you ended; this makes twenty-one. Multiply it by half of twenty, which is ten; this produces two hundred and ten. Such is the quantity resulting from the addition of the given numbers.

(A.171) You want to know the sum of the numbers in sequence from nine to twenty.

Add one to the last number, with which you ended, namely twenty; this makes twenty-one. Multiply it by half of twenty, which is ten; this produces two hundred and ten. Next, subtract one from the first number, with which you started, in this case nine; this leaves eight. Multiplying its half, which is four, by nine produces thirty-six. Subtract it from two

---

[516] As calculated above.

[517] Literally (and not inappropriately): twice.

[518] There will now be two unknowns.

[519] 'likewise' is found in both MSS ($\mathcal{A}$ and $\mathcal{B}$) but makes sense only in $\mathcal{B}$, where A.169 follows a pair of (misplaced) indeterminate problems (A.273–273′). It would seem that $\mathcal{A}$ put them back in the right order without removing what must have been a reader's addition.

[520] This is not the appropriate condition.

[521] The *Liber algorismi* contains similar rules for summing consecutive natural, even, odd and square numbers (pp. 94–96).

hundred and ten. The remainder is a hundred and seventy-four, and this is the quantity resulting from the addition of the given numbers.

(A.172) You want to know the sum of the odd numbers in sequence from one to nineteen.

You will do the following. Add one to nineteen; this makes twenty. Multiply its half, which is ten, by itself; this produces a hundred, and such is the quantity resulting from the addition of the proposed odd numbers.

(A.173) You want to know the sum of the odd numbers in sequence from nine to twenty-nine.

Add one to the last number, with which you ended, in this case twenty-nine; this makes thirty. Multiply its half by itself, thus producing two hundred and twenty-five. Next, subtract one from the first number, with which you started, in this case nine; this leaves eight. Multiply its half by itself; this produces sixteen. Subtracting it from two hundred and twenty-five leaves two hundred and nine, and so much is the result of adding the proposed odd numbers.

(A.174) You want to know the sum of all the even numbers in sequence from two to twenty.

Add two to twenty; this makes twenty-two. Multiply its half, which is eleven, by half of twenty, and this will be the result of adding the proposed even numbers.

(A.175) You want to know the sum of the even numbers from ten to thirty.

Always add two to the last number, in this case thirty; this makes thirty-two. Multiply its half by half of thirty; this produces two hundred and forty. Next, subtract two from ten; this leaves eight. Multiply its half by half of ten; this produces twenty. Subtracting it from two hundred and forty leaves two hundred and twenty, and this is the result of adding the given numbers.

(A.176) You want to know the sum of all the squares from the square of one to the square of ten.

(a) You will do the following. Always add one to the last number, in this case ten; this makes eleven. Multiplying it by half of ten produces fifty-five. Next, always add to two thirds of the last number, in this case ten, a third of one; this makes seven. Multiplying it by fifty-five produces three hundred and eighty-five, and such is the result of adding the squares proposed.

(b) Or otherwise. Always add one to the last number, in this case ten, and multiply the sum by half of ten, that is, of the last number, and keep the result in mind. Next, always subtract one from the last number, in this case ten, and then always add one to two thirds of the remainder, in this case to six, thus making seven. Multiplying it by the quantity kept in mind will produce what you require.

LIBER MAHAMELETH

709

(A.177) You want to know the sum of all the squares of odd numbers from the square of one to the square of nine.

You will do the following. Add two to nine, which makes eleven. Multiply this by half of the subsequent even number[522], thus by five; this produces fifty-five. Multiply it by a third of the last odd number, in this case nine, thus by three. This produces a hundred and sixty-five, which is what you wanted to know.

(A.178) You want to know the sum of all the squares of even numbers from the square of two to the square of ten.

You will do the following. Always add two to the last number, in this case ten, which makes twelve. Multiply half of this by half of the last even number, in this case ten, thus by five; the result is thirty. Multiplying it by two thirds of the last even number, plus two thirds of one, will give what you wanted to know, in this case two hundred and twenty.

[**(A.179)** *You want to know the sum of all the squares in sequence from four to ten*[523].

Always add one to the last number, in this case ten, which makes eleven. Multiply it by half of the last even[524] *number, in this case ten, thus producing fifty-five. Multiply it by two thirds of the last even number, in this case ten, plus one third; keep in mind the result. Next, subtract one from the even number with which you started, in this case four; this leaves three. Multiply its half by four*[525], *and multiply the product by two thirds of the three remaining from the four, plus one third. Subtracting the result from what was kept in mind will leave what you wanted to know.*]

(A.180) You want to know the sum of all the cubes from the cube of one to the cube of ten.

You will do the following. Always add one to the last number, in this case ten, and multiply the result by half of ten, which produces fifty-five. Multiplying it by itself produces three thousand and twenty-five, and this is the sum you wanted to know.

(A.181) You want to know the sum of all the cubes from the cube of five to the cube of ten.

Add one to the last number, in this case ten, which makes eleven. Multiply it by half of ten, thus producing fifty-five. Multiplying this by itself produces three thousand and twenty-five. Keep it in mind. Next, subtract one from the number with which you started, in this case five, which leaves four. Multiply its half by five, and multiply the product by itself, thus producing a hundred. Subtracting it from three thousand and

---

[522] The even number just following the last term.

[523] *Sic*, instead of 'from the square of four to the square of ten'. There are numerous incongruities in this problem.

[524] This (recurring) 'even' is irrelevant.

[525] It is usually half of the even number which is taken.

710   PART TWO: TRANSLATION, GLOSSARY

twenty-five leaves two thousand nine hundred and twenty-five, and this is the sum you require.

(A.182) You want to know the result of adding all the cubes of odd numbers from one to nine.

Add one to the last number, in this case nine, which makes ten. Multiply its half by itself, and multiply the product by twice this, minus one. The result, namely one thousand two hundred and twenty-five, is what you wanted to know.

(A.183) You want to know the sum of all the cubes of even numbers from the cube of two to the cube of ten.

Always add two to the last number, in this case ten, and multiply half the sum by half of ten; this produces thirty. Multiplying it by twice this, which is sixty, produces one thousand eight hundred, and this is the sum you require.

AGAIN, ON THE SAME TOPIC.[526]

(A.184) The sum of all the numbers in sequence from one to some unknown number is fifty-five. What is this number?

You will do the following. Always double the sum, in this case fifty-five, which gives a hundred and ten. Always add a fourth of one to this, and take the root of the result; this is ten and a half. Removing the half leaves ten, and this is the number you require.

(A.185) The sum of all the odd numbers from one to some odd number is a hundred. What is this odd number?

You will do the following. Subtract one from twice the root of a hundred; the remainder is the number you require.

(A.186) The sum of all the even numbers from two to some even number is a hundred and ten. What is this even number?

You will do the following. Always multiply the sum, in this case a hundred and ten, by four; this produces four hundred and forty. Always add one to this and take the root of the result; this is twenty-one. Subtracting one from it leaves the number you require.

---

[526] Like A.167, A.184–186 involve roots, a subject to be treated later on.

Chapter (A–VII) on Subtracting

FIRST CASE. SUBTRACTION OF A FRACTION FROM A FRACTION.

(**A.187**) You want to subtract three eighths from four fifths.

(**a**) You will do the following. Multiply the denominators of an eighth and a fifth; this produces forty, which is the principal number. Subtract three eighths of it, which is fifteen, from four fifths of it, which is thirty-two, thus leaving seventeen. Denominate it from the principal number; this gives three eighths and two fifths of an eighth, and such is the remainder.

(**b**) Or otherwise. Convert the four fifths into eighths; this gives six eighths and two fifths of an eighth. Subtracting three eighths from this leaves three eighths and two fifths of an eighth, that is, the same remainder as above.[527]

(**c**) Or convert three eighths into fifths; this gives a fifth and seven eighths of one fifth. Subtracting this from four fifths leaves two fifths and an eighth of a fifth. This is three eighths and two fifths of an eighth, that is, the same as above.[528]

(**A.188**) You now want to subtract three sevenths from ten elevenths.

(**a**) You will do the following. Multiply the denominators of the fractions, namely seven and eleven; this produces seventy-seven, which is the principal number. Subtract three sevenths of it from ten elevenths of it; this leaves thirty-seven. Denominating it from the principal number gives five elevenths and two sevenths of an eleventh, and such is the remainder.

(**b**) Or otherwise. Convert the three sevenths into elevenths and subtract the result from ten elevenths. The remainder is the result of the subtraction.

(**c**) Or convert the elevenths into sevenths and subtract from the result three sevenths. The remainder is the result of the subtraction.

THIRD CASE.[529] SUBTRACTING A FRACTION AND A FRACTION OF A FRACTION FROM A FRACTION.

(**A.189**) You want to subtract three eighths and half an eighth from six sevenths.

(**a**) You will do the following. Multiply the denominators of an eighth, a half, and a seventh; this produces a hundred and twelve, which is the

---

[527] For the conversion in *c* leads to a different expression.

[528] This and a similar statement at the end of *b* may originate with an early reader. From the previous chapter we know very well that different conversions will lead to the result taking different forms.

[529] A.193 has the heading 'second case'. See p. 714 & note 542.

712 PART TWO: TRANSLATION, GLOSSARY

principal number. Subtract three and a half eighths of it, which is forty-nine, from six sevenths of it, which is ninety-six; this leaves forty-seven. Denominate it from the principal number; the result is the remainder of the subtraction.

(**b**) Or otherwise. Convert the three and a half eighths into sevenths; this gives three sevenths and half an eighth of a seventh. Subtracting it from six sevenths leaves two sevenths and seven eighths of a seventh [530] and half an eighth of a seventh, and this is the remainder of the subtraction.

FOURTH CASE. SUBTRACTING A FRACTION AND A FRACTION OF A FRACTION FROM A FRACTION AND A FRACTION OF A FRACTION.

(**A.190**) You want to subtract three fifths and two thirds of a fifth from seven and a half eighths.

(**a**) You will do the following. Multiplying the denominators, as we have taught above,[531] you will form the principal number, which is two hundred and forty. Subtract its three fifths and two thirds of a fifth of it from seven and a half eighths of it. Denominate the remainder from the principal; the result is the remainder of the subtraction.

(**b**) Or otherwise. Convert the three fifths and two thirds of a fifth into eighths; this gives five eighths and four fifths of an eighth and a third of a fifth of an eighth. Subtract it from seven eighths; this leaves an eighth and two thirds of a fifth of an eighth. Add it to the half of an eighth which was with the seven eighths. The result is the remainder of the subtraction.

SUBTRACTING A FRACTION AND A FRACTION OF A FRACTION FROM AN INTEGER AND A FRACTION AND A FRACTION OF A FRACTION.[532]

(**A.191**) You now want to subtract three fifths and two thirds of a fifth from one and four and a half elevenths.

(**a**) You will do the following. Multiplying the denominators, as said above,[533] you will form the principal number. Next, multiply one and four and a half elevenths by the principal number. Subtract from the result three fifths of this principal number and two thirds of a fifth of it, and divide the remainder by the principal. The result is what you wanted to know.

(**b**) You may also convert the three fifths and two thirds of a fifth into elevenths and subtract the result from one and four and a half elevenths, which is fifteen and a half elevenths. The remainder is what you wanted to know.

(**c**) Or subtract the three fifths and two thirds of a fifth from one; this leaves a fifth and a third of a fifth. Add it to four and a half elevenths

---

[530] No reduction: the result must be in sevenths.
[531] A.187–189, but now with four denominators.
[532] Except for A.192 (misplaced) and A.196, A.191–198 involve integers.
[533] A.190a.

as shown above for the addition.[534] The result is the remainder of the subtraction.

You will operate in all other instances in the same way.

(**A.192**) You now want to subtract two sevenths and three tenths from ten elevenths.[535]

(*a*) You will do the following. Multiply the numbers from which a seventh, a tenth and an eleventh are denominated, which produces seven hundred and seventy; put it as the principal. Add two sevenths of it to three tenths of it and subtract from ten elevenths of it the sum. Denominate the remainder from the principal. The result is what you wanted.

The proof of this is the following. Let the principal number be A, two sevenths of it, BG, three tenths of it, GD, ten elevenths of it, BH; let two sevenths be ZK, three tenths, KT, ten elevenths, the whole ZQ —therefore TQ is what is required— and let one be L. Then[536] the ratio of line ZK to L is the same as the ratio of line BG to A, while the ratio of line KT to L is the same as the ratio of line GD to A. Then[537] the ratio of line ZT to L is the same as the ratio of line BD to A. Now the ratio of the whole ZQ to L is the same as the ratio of the whole BH to A. Thus the ratio of the remainder, which is TQ, to L is the same as the ratio of DH to A, as we have taught before in the chapter on premisses[538]. Therefore the product of one[539] into DH is equal to the product of TQ, which is required, into A. Now the product of one into DH is simply DH; so the product of line TQ into A is DH. Thus dividing DH by A will give TQ. This is what we wanted to demonstrate.

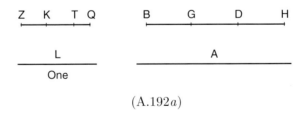

(A.192*a*)

(*b*) Or otherwise. Convert the two sevenths into elevenths as we have taught in the chapter on converting fractions[540], and convert also the three tenths into elevenths; add all that, and subtract the sum from ten elevenths. Indeed, converting two sevenths into elevenths gives three elevenths and a seventh of an eleventh, while converting three tenths into elevenths gives

---

[534] A.148.
[535] See heading to A.199.
[536] As in A.151*a*.
[537] Premiss P'_7.
[538] Premiss P_6.
[539] We would expect: L.
[540] A.108–111.

three elevenths and three tenths of one eleventh. Add then the three elevenths to the previous three elevenths, thus making six elevenths. Next, add the seventh of an eleventh to the three tenths of an eleventh, in the following way. Convert either of them into the kind of the other, for instance the seventh of an eleventh into tenths of an eleventh. It is then like asking: 'How many tenths of an eleventh is one seventh of an eleventh?' Remove both 'eleventh's.[541] Then you will say: 'How many tenths is one seventh?' So multiply one by ten and divide the product by seven; the result will be one and three sevenths, and this is one tenth of an eleventh and three sevenths of a tenth of an eleventh. Add it to three tenths of an eleventh; this makes four tenths of an eleventh and three sevenths of a tenth of one eleventh. Add this to six elevenths; this makes six elevenths and four tenths of one eleventh and three sevenths of one tenth of one eleventh. Subtract this from ten elevenths. That is, subtract six elevenths from ten elevenths, which leaves four elevenths; taking away from them one eleventh leaves three, and subtracting from this eleventh four tenths of an eleventh and three sevenths of a tenth of an eleventh leaves five tenths of an eleventh and four sevenths of a tenth of one eleventh; adding this to three elevenths makes three elevenths and five tenths of one eleventh and four sevenths of one tenth of one eleventh, and this is what you wanted.

Doing the same in all instances of this kind you will find what you require.

SECOND CASE.[542] CHAPTER ON SUBTRACTING A FRACTION FROM AN INTEGER AND A FRACTION.

(**A.193**) You want to subtract eight elevenths from one and three eighths.

(*a*) You will do the following. Multiply the denominators of the fractions, that is, eleven and eight; this produces eighty-eight, which is the principal number. Eight elevenths of it is sixty-four. Next, multiply one and three eighths by the principal number; this produces a hundred and twenty-one. Subtracting sixty-four from this leaves fifty-seven. Denominate it from the principal number. The result is the remainder of the subtraction.

(*b*) Or otherwise. Convert the eight elevenths into eighths; this gives five eighths and nine elevenths of an eighth. Subtracting this from one and three eighths, which is eleven eighths, leaves five eighths and two elevenths of an eighth.

(*c*) Or convert the three eighths into elevenths; this gives four elevenths and an eighth of an eleventh. Now add eleven elevenths, which are in one, to four elevenths and an eighth of an eleventh; this makes fifteen elevenths and an eighth of an eleventh. Subtracting eight elevenths from this leaves seven elevenths and an eighth of an eleventh, which is the remainder of the subtraction.

---

[541] Rule *ii*, p. 671.

[542] According to the disordered text. The enumeration in this section is found only in MSS $\mathcal{B}$ and $\mathcal{C}$.

(***d***) You may also subtract the eight elevenths from one, thus leaving three elevenths. Adding it to the three eighths which are in excess to one will give the remainder of the subtraction.

CHAPTER ON SUBTRACTING AN INTEGER AND A FRACTION FROM AN INTEGER AND A FRACTION.

(**A.194**) You want to subtract two and three fourths from five and eight ninths.

(***a***) You will do the following. Multiply the denominators of a fourth and a ninth, thus producing thirty-six. Put it as the principal. Multiply it by two and three fourths, and keep the result in mind. Next, multiply the principal by five and eight ninths. Subtracting from this product the other and dividing the remainder by the principal will give what you wanted.

The proof of this is the following. Let one be A, two and three fourths, BD, five and eight ninths, BG —therefore GD is what is required; let the principal be H; let further the product of H into two and three fourths be ZK, and the product of H into five and eight ninths be ZT. It is then clear that the ratio of BG to A is the same as the ratio of ZT to H. But [543] the ratio of BD to A is the same as the ratio of ZK to H. So [544] the ratio of the remainder GD to A is the same as the ratio of the remainder KT to H. Then the product of GD into H is equal to the product of one [545] into KT. Now the product of one into KT is simply KT. So the product of GD into H is KT. Thus dividing KT by H will give what is required, namely GD. This is what we wanted to demonstrate.

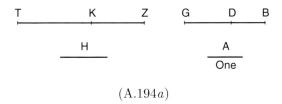

(A.194a)

(***b***) Or otherwise. Subtract two and three fourths from five, which leaves two and a fourth. Add it to eight ninths in the following way. Adding a fourth to eight ninths makes one and a ninth and a fourth of a ninth; adding this to two makes three and a ninth and a fourth of a ninth, and this is what you wanted.

SUBTRACTING A FRACTION AND A FRACTION OF A FRACTION FROM AN INTEGER AND A FRACTION.

(**A.195**) You now want to subtract four fifths and a third of a fifth from one and four thirteenths.

---

[543] We would expect: and that.
[544] Premiss P_6.
[545] As in A.192a, we would expect 'A' or 'A, which is one'.

716   PART TWO: TRANSLATION, GLOSSARY

(*a*) You will do the following. Form, by multiplying the denominators, the principal number, and continue as explained above.

(*b*) Or convert four fifths and a third of a fifth into thirteenths and subtract the result from one and four thirteenths, which is seventeen thirteenths, as has been shown shortly before.[546] The remainder is what you wanted to know.

FIFTH CASE.   CHAPTER ON SUBTRACTING A FRACTION OF A FRACTION FROM A FRACTION OF A FRACTION.

(**A.196**) You want to subtract three fifths of an eighth from five sevenths of a sixth.

(*a*) You will do the following. Multiplying the denominators of a fifth and an eighth produces forty, while from a seventh and a sixth there results forty-two. Multiplying this by the forty produces one thousand six hundred and eighty, which is the principal number. Subtract three fifths of an eighth of this, which is a hundred and twenty-six, from five sevenths of a sixth of it, which is two hundred; this leaves seventy-four.[547] Denominating it from the principal will give what you wanted.

(*b*) Or otherwise. Convert three fifths of an eighth into sevenths of a sixth; this gives three sevenths of a sixth and an eighth of a seventh of a sixth and a fifth of an eighth of a seventh of a sixth. Subtracting this from five sevenths of a sixth leaves a seventh of a sixth and six eighths of a seventh of a sixth and four fifths of an eighth of a seventh of a sixth, and this is what you wanted to know.

SUBTRACTING A FRACTION OF A FRACTION FROM AN INTEGER AND A FRACTION.

(**A.197**) You now want to subtract three fourths of a fifth from one and an eleventh.

(*a*) You will do the following. Multiplying the denominators of a fourth, a fifth and an eleventh produces the principal number, which is two hundred and twenty. Multiply one and one eleventh by it, thus producing two hundred and forty. Subtracting from this three fourths of a fifth of the principal number, that is, thirty-three, leaves two hundred and seven. Denominate it from the principal number and what you require will appear.[548]

(*b*) Or otherwise. Convert three fourths of a fifth into elevenths; the result is one eleventh and three fifths of an eleventh and a fourth of a fifth of an eleventh. Subtracting this from one and an eleventh, which is twelve elevenths, leaves ten elevenths and a fifth of an eleventh and three fourths of a fifth of an eleventh, which is what you wanted to know.

---

[546] A.193*b*.

[547] The extant text has 'sixty-four'.

[548] Turn of phrase not found elsewhere.

LIBER MAHAMELETH   717

CHAPTER ON SUBTRACTING AN INTEGER AND A FRACTION AND A FRAC-
TION OF A FRACTION FROM AN INTEGER AND A FRACTION AND A FRAC-
TION OF A FRACTION.

(A.198) You now want to subtract two and two sevenths and three fourths of a seventh from two and five eighths and two thirds of an eighth.

You will do the following. Disregard both twos. It is then like saying: 'Subtract two sevenths and three fourths of a seventh from five eighths and two thirds of an eighth.' Proceed as we have taught above[549], and this will give what you wanted to know.

CHAPTER ON SUBTRACTING SEVERAL FRACTIONS FROM OTHERS.[550]

(A.199) You want to subtract three fifths and four sevenths from one and seven eighths.

(a) You will do the following. You form the principal number by multiplying the denominators of all the fractions; this is two hundred and eighty. Add three fifths of it, which is a hundred and sixty-eight, to four sevenths of it, which is a hundred and sixty; the result is three hundred and twenty-eight. Now multiply one and seven eighths by the principal; the result is five hundred and twenty-five. Subtract three hundred and twenty-eight from this; it leaves a hundred and ninety-seven. Denominating it from the principal will give what you wanted to know.

(b) Or convert the fifths and the sevenths into eighths; the result is one and an eighth and two sevenths of an eighth and three fifths of a seventh of an eighth. Subtracting this from the given quantity one and seven eighths leaves five eighths and four sevenths of an eighth and two fifths of a seventh of an eighth. This subtraction is performed as follows. You subtract one from one, and an eighth from seven eighths; this leaves six eighths. Take away one eighth from this, thus leaving five eighths, and convert this eighth into sevenths of an eighth, which gives seven sevenths of an eighth; subtract from it two sevenths of an eighth, thus leaving five sevenths of an eighth. Take away one seventh of an eighth from this, thus leaving four sevenths of an eighth, and convert this seventh of an eighth into fifths of a seventh of an eighth, which gives five fifths of a seventh of an eighth; subtract from it three fifths of a seventh of an eighth; this leaves two fifths of a seventh of an eighth. The remainder is therefore five eighths and four sevenths of an eighth and two fifths of a seventh of an eighth, and this is what you wanted to know.

PROBLEMS ON SUBTRACTING[551]

(A.200) You want to subtract two thirds of four from five sevenths of six.

(a) You will do the following. Multiplying the denominators of a third and a seventh produces the principal number, namely twenty-one. Multiply two

---

[549] A.190.

[550] This would better apply to A.192.

[551] Rather: Problems on subtracting involving irregular fractions.

718   PART TWO: TRANSLATION, GLOSSARY

thirds of it by four, thus producing fifty-six. Next, multiply five sevenths of the principal number by six, thus producing ninety. Subtract fifty-six from this, which leaves thirty-four. Denominating it from the principal[552] will give what you require.

The proof of this is clear. Multiplying two thirds of the principal by four and dividing the product by the principal will give two thirds of four. Likewise again, multiplying five sevenths of the principal by six and dividing the product by the principal will give five sevenths of six. It is then necessary to divide the product of two thirds of the principal into four by the principal and subtract the quotient from the result of dividing the product of five sevenths of the principal into six by the principal. Now this is the same as subtracting the product of two thirds of the principal into four from the product of five sevenths of the principal into six and dividing the remainder by the principal, as we have established in the first chapter on premisses[553] [*namely: If each of two different numbers is divided by some other, the excess of one of the quotients over the other is equal to the result of dividing the excess of the greater number over the other by the divisor* [554]]. This is what we wanted to demonstrate.

(**b**) Or otherwise. Follow the wording of the problem. That is, subtract two thirds of four, which is two and two thirds, from five sevenths of six, which is four and two sevenths, in the way we have taught above.[555] This will give what you wanted.

(**A.201**) You now want to subtract two fifths of four and a half from six sevenths of eight and a third.

(**a**) You will do the following. Form, from the denominators of all the fractions, thus of a fifth, a half, a seventh and a third, the principal number; this is two hundred and ten. Multiply two fifths of this by four and a half; the result is three hundred and seventy-eight. Next, multiply six sevenths of the principal number by eight and a third; the result is one thousand five hundred. Subtracting three hundred and seventy-eight from this leaves one thousand one hundred and twenty-two. Dividing it by the principal will give the result of the subtraction. The proof of this is evident from what precedes.[556]

(**b**) Or otherwise, following the order of the wording. Take two fifths of four and a half, which is one and four fifths, and take six sevenths of eight and a third, which is seven and a seventh. It is then as if you were to subtract one and four fifths from seven and a seventh. Proceed as has been

---

[552] Rather: Dividing it by the principal.

[553] Premiss P_7. The 'second chapter on premisses' is an adaptation of Euclid's propositions II.1–10 (our PE_1–PE_{10}).

[554] Is the author likely to have repeated P_7 here but not P_6 in A.192 or A.194?

[555] A.194.

[556] A.200.

LIBER MAHAMELETH   719

explained above[557], and the result will be what you wanted to know.

(**A.201′**) You now want to subtract two fifths of four, and half of one[558], from six sevenths of eight, and a third of one.

(**a**) You will do the following. You will first form, by multiplying the denominators of all the fractions, the principal number, which is two hundred and ten. Multiply two fifths of this by four, add to the product half the principal number, and keep the sum in mind. Next, multiply six sevenths of the principal number by eight, and add to the product a third of the principal number. Subtracting from this sum the former sum and dividing the remainder by the principal will give what you require.

(**b**) Or otherwise. Add to two fifths of four half of one and keep the sum in mind. Next, add to six sevenths of eight a third of one. Subtract from this sum the former sum as we have taught before[559], and the remainder will be what you wanted to know.

AGAIN ON SUBTRACTING[560]

(**A.202**) You want to know what remains after subtracting from six its seventh and its ninth.

You will do the following. You will form, by multiplying the denominators of the fractions, the principal number, which is sixty-three. Subtract from this its seventh and its ninth; this leaves forty-seven. Multiply this by six and divide the product by the principal; the result is four and three sevenths and a third of a seventh. This is what you wanted to know.

This rule too is taken from a proportion.[561] For forty-seven is to sixty-three as the required quantity is to six. Thus the third term is unknown and the division is performed with the second.[562]

Or otherwise. Denominate forty-seven from the principal number; this gives six ninths and five sevenths of a ninth. Taking these fractions of six will give the quantity you require.

Or otherwise. Denominate six from the principal; this gives two thirds of a seventh. Thus two thirds of a seventh of forty-seven is the quantity you require.

(**A.203**) Someone asks: What quantity remains after subtracting from nine its fifth and its two sevenths and adding to the remainder its half?

---

[557] Again, A.194.

[558] The (uncorrected) text has *duas quintas de quatuor et dimidio unius*, which is the previous problem.

[559] A.194–195, A.197–198.

[560] Problems 'on known amounts'.

[561] Rather than to the proof in A.194, the 'too' may well allude to that in A.155, the first problem 'on known amounts' (in the chapter on addition). In MS $\mathcal{C}$ (disordered text), the two problems are just one leaf apart.

[562] It will be with another in the problems 'on unknown amounts'.

720 PART TWO: TRANSLATION, GLOSSARY

You will solve this as follows. You form, by multiplying the denominators of the fractions a fifth, a seventh and a half, seventy, which is the principal number. Subtracting from this principal its fifth and its two sevenths, thus thirty-four, leaves thirty-six. Adding to it, as stated, its half, which is eighteen, makes fifty-four. It is then clear[563] that the ratio of this fifty-four to seventy is the same as the ratio of the required quantity to nine. So multiply the fifty-four by nine and divide the product by the principal; the result will be what you wanted to know. Or else divide either of the factors by the principal number and multiply the quotient by the other; the result will be what you require.

(A.203′) Someone asks: What remains after subtracting from nine its fifth and its two sevenths, and subtracting from the remainder its third?

You will do the following. First you will form, from the denominators of all the fractions, the principal number, which is a hundred and five. Subtract from it its fifth and its two sevenths; the remainder is thirty-three.[564] Subtracting from this remainder its third, as said, leaves twenty-two. It is then clear that the ratio of this twenty-two to one hundred and five is the same as the ratio of the required quantity to nine. Thus multiply the twenty-two by nine and divide the product by the principal; the result will be the quantity you require. Or else divide either of the factors, namely nine or twenty-two, by the principal and multiply the result by the other. The product will be the quantity you require.

If in such problems there are more fractions, the treatment will nevertheless remain the same. Whenever in this kind of problem the sum of the fractions exceeds one, the problem will not be solvable.[565]

(A.203″) This is the case if someone asks: What remains after subtracting from eight its two thirds and its two fifths?

This problem is not admissible since two thirds and two fifths of something are more than the whole; therefore the sum of these fractions of eight is larger than eight [*and greater cannot be subtracted from lesser*]. This will become evident by taking the sum of these fractions of the principal number: the sum becomes larger than the principal number itself.[566]

CHAPTER ON AMOUNTS IN SUBTRACTING[567]

(A.204) Someone asks: What is the amount which, when its third and its fourth are subtracted, leaves ten?

You will do the following. You will form, by multiplying the denominators of a third and a fourth, twelve. Subtracting its third and its fourth

---

[563] Previous problem.
[564] *Sic*, instead of fifty-four.
[565] As already remarked in A.156′.
[566] This too may well be an addition.
[567] Problems 'on unknown amounts'.

LIBER MAHAMELETH

721

leaves five. It is then clear[568] that the ratio of five to twelve is the same as the ratio of ten to the required quantity. So multiply ten by twelve and divide the product by the principal, which is five;[569] the result is twenty-four, and this is the amount. Or divide either of the factors by the principal and multiply the result by the other; the product is the quantity of the required amount.

(**A.205**) Someone asks: What is the amount which, when its third and its fourth and half the remainder are subtracted, leaves ten?

(**a**) You will do the following. Multiply the denominators of the fractions a third, a fourth and a half, thus producing twenty-four. Subtracting from it its third and its fourth leaves ten. Subtracting from this its half leaves five. It is then clear[570] that the ratio of five to twenty-four is the same as the ratio of ten to the required quantity. So multiply ten by twenty-four and divide the product by the principal;[571] the result will be the quantity you require. Or divide either of the factors by the principal and multiply the result by the other; the product is the required quantity.

(**b**) Or otherwise. You already know that if you subtract from the amount its third, its fourth, and half the remainder, and ten remains, this ten is half the remainder after subtracting from the amount its third and its fourth. So double it; the result is twenty, which is the remainder of the amount after subtraction of its third and its fourth. It is then like saying: 'What is the amount which, when its third and its fourth are subtracted, leaves twenty?' Proceed as we have taught before[572], and the result will be what you require.

(**A.205′**) Now if you had said 'when a third of the remainder is subtracted', ten would of course be two thirds of the remainder. So add to ten its half; this gives fifteen, and such is the remainder of the amount after subtraction of its third and its fourth.

(**A.206**) Someone asks: What is the amount which, when its third, its fourth and two nummi are subtracted, leaves ten?

You will do the following. Add the two nummi to ten; the result is twelve, which is the remainder of the amount after subtraction of its third and its fourth. It is then like asking: 'What is the amount which, when its third and its fourth are subtracted, leaves twelve?' You will proceed as we have taught before.

(**A.207**) Someone asks: What is the amount which, when its third and its fourth minus two nummi are subtracted, leaves ten?

You will do the following. Subtract the two nummi from ten; this leaves eight, which is the remainder of the amount after subtraction of its

---

[568] A.202.

[569] See note 493, p. 703.

[570] A.202 & A.204.

[571] Which is 5, as in the previous problem.

[572] A.204. Same reference for A.206–208.

third and its fourth. It is then like asking: 'What is the amount which, when its third and its fourth are subtracted, leaves eight?' You will proceed as we have taught before.

You will proceed likewise with other such problems[573], always adding the additive quantities and subtracting the subtractive quantities; the result will be the remainder of the amount after subtraction of the parts proposed [*which is the opposite of what is done in addition*], as already explained.[574]

(A.208) Someone asks: What is the amount which, when its third and one nummus and its fourth minus three nummi are subtracted, leaves ten nummi?

You will do the following. Add the nummus added to the three subtracted; this makes two, subtracted[575]. It is then like asking: 'What is the amount which, when its third and its fourth minus two nummi are subtracted, leaves ten?' Subtract from ten the two nummi which are to be subtracted; this leaves eight, which is the remainder of the amount after subtraction of its third and its fourth. Continue as we have taught before.

(A.209) Someone asks: What is the amount which, when its third minus two nummi and its fourth plus three nummi are subtracted, leaves ten?

You will do the following. Add the three nummi, added, to the two subtracted; this leaves one, added. It is then like asking: 'What is the amount which, when its third, its fourth and one nummus are subtracted, leaves ten?' Proceed as we have taught before.[576]

(A.210) Someone asks: What is the amount which, when its third and two nummi, its fourth minus one nummus, and half the remainder and four nummi are subtracted, leaves ten?

You will do the following. Add two, added, to one, subtracted; this leaves one, added, to be subtracted from[577] the remainder of the amount. Add half of this nummus, which is a half, subtracted, to the given four additive nummi; the remainder is three and a half, added. Add it to the additive nummus mentioned before; the result is four and a half, added. It is then like asking: 'From an amount are subtracted its third, its fourth, and half the remainder plus four and a half nummi, and the remainder is ten; what is this amount?' Add four and a half nummi to ten, and continue as we have taught above.[578]

---

[573] That is, involving nummi added or subtracted.

[574] A.161, to which the gloss alludes.

[575] Not: negative (see note 497, p. 704). Same in A.210, A.212.

[576] A.206.

[577] Unclear expression, here and in A.212 ('added to the remainder'). See note 499, p. 704.

[578] A.205.

LIBER MAHAMELETH

(A.211) Someone asks: From an amount are subtracted its third and two nummi, its fourth minus one nummus, and half the remainder minus three nummi, and this leaves ten. How much is it?

You already know [579] that if the subtraction of half the remainder minus three nummi leaves ten, half the remainder and three nummi equal ten nummi. Thus the half alone equals seven nummi and the whole remainder, fourteen. Then you will say: 'From an amount are subtracted its third and two nummi, and its fourth minus one nummus, and the remainder is fourteen; what is this amount?' Two nummi are additive and one is subtractive; restoring [580] the latter by means of one unit of the former leaves one, added. It is then like asking: 'From an amount are subtracted its third and its fourth and one nummus and the remainder is fourteen; what is this amount?' Proceed as I have taught above.[581] The amount will be thirty-six.

Consider likewise other such instances, and you will find what you require.[582]

(A.212) Someone asks: From an amount are subtracted its third minus five nummi, its fourth and two nummi, and half the remainder minus one nummus, and this leaves ten. What is this amount?

You will do the following. Add the five subtracted to the two added; this makes three subtracted, which will be added to the remainder of the amount. Add its half, which is one and a half added, to one, subtracted, which leaves a half added. Add it to the above three, subtracted, thus making two and a half subtracted. It is then like asking: 'From an amount are subtracted its third, its fourth and half the remainder minus two and a half nummi, and this leaves ten; what is this amount?' Subtract two and a half nummi from ten, and continue as we have shown before.[583]

(A.213) Someone asks: What is the amount which, when its third and two nummi, half the remainder and five nummi, and from the last remainder two fifths of it minus one nummus are subtracted, leaves eleven?

$$\text{B} \quad \text{KT Z H} \quad\quad\quad \text{D G} \quad\quad\quad\quad \text{A}$$

(A.213)

You will do the following. Let the amount be AB, its third to be subtracted, AG, two nummi, GD, half the remainder, DH, five nummi,

---

[579] A.205b.

[580] First occurrence of *restaurare* (Arabic *jabara*).

[581] A.206.

[582] This seems out of place, for we continue with a similar problem.

[583] A.205. In this last computation we feel the limits to verbal mathematics. That is probably why, in the next problem, the treatment is illustrated by means of segments of straight lines.

724 PART TWO: TRANSLATION, GLOSSARY

HZ, two fifths of the last remainder, ZK, one nummus subtracted, KT. Thus TB will be eleven. Subtract from it KT, which is one nummus; this leaves ten as BK. This being three fifths of ZB, ZB is sixteen and two thirds. But HZ is five, so HB is twenty-one and two thirds. This being half of DB, DB is forty-three and a third. But GD is two, so BG is forty-five and a third. This being two thirds of AB, AB is sixty-eight. This is what we wanted to demonstrate[584].

(A.214) Someone asks: What is the amount which, when its third and two nummi are subtracted, and the remainder is multiplied by itself, produces the amount plus twenty-four nummi?

You will do the following. Multiply the two nummi by themselves, thus producing four. Subtract it from twenty-four; this leaves twenty. Next, find by which number two thirds must be multiplied to produce one; you will find one and a half.[585] Adding one and a half to the above four makes five and a half. Multiply its half, which is two and three fourths, by itself; the result is seven and four eighths and half an eighth. Add it to twenty; this makes twenty-seven and four eighths and half an eighth. Add its root to two and three fourths;[586] this makes eight, which is two thirds of the amount. Adding to it its half makes twelve, which is the required amount.

$$B \quad\quad G \quad D \;\; H \quad Z \quad\quad A$$

(A.214)

The proof of this is the following. Let the amount be AB, its third to be subtracted, BG, which leaves, as two thirds of the amount, AG. Subtracting from it two nummi, say GD, leaves AD. Thus the product of AD into itself is equal to the product of AG into one and a half,[587] plus twenty-four nummi. Let twice the product of AG into DG be (added in) common. So the products of AD into itself and AG into DG, twice, are equal to the products of AG into one and a half, and AG into DG, twice, plus twenty-four nummi. ⟨Now the products of AD into itself and AG into DG, twice, are equal to the products of AG into itself and DG into itself.[588] So the products of AG into itself and DG into itself are equal to the products of AG into one and a half, AG into DG, twice, plus twenty-four nummi.⟩ Now twice the product of AG into DG is equal to the product

---

[584] Thus does Abū Kāmil conclude (with a Q.E.D., as he does for a proof) computations involving a geometrical representation. See our *Introduction to the History of Algebra*, p. 163 n. 214. Other examples in A.298a, B.138, B.177, B.185c, B.351c, B.353.

[585] How to find that —if need be— will be taught below (A.216).

[586] There have already been isolated occurrences of roots in A.167 & A.184–186.

[587] Thus to the whole amount AB.

[588] *Elements* II.7 $= PE_7$.

LIBER MAHAMELETH 725

of AG into four. Thus the products of AG into four and into one and a half[589], plus twenty-four, are equal to the products of AG into itself and DG into itself. Now the product of DG into itself is four. So the product of AG into itself, plus four, is equal to the products of AG into four and into one and a half, plus twenty-four. Subtracting four from twenty-four leaves twenty. Thus the product of AG into itself is equal to the products of AG into four and into one and a half, plus twenty.[590] Therefore AG is greater than five and a half.[591] So cut from it five and a half, say AH. So the product of AG into itself is equal to the product of AG into AH, plus twenty. But[592] the product of AG into itself is equal to the product of it into AH and the product of it into HG. Thus the products of AG into AH and into HG are equal to the product of AG into AH, plus twenty. Subtracting the product of AG into AH, which is common, will leave, as the product of AG into HG, twenty.[593] So bisect AH at the point Z. Then[594] the products of AG into GH and ZH into itself will be equal to the product of ZG into itself. Now the product of AG into GH is twenty, and the product of ZH into itself is seven and a half and half an eighth. Therefore the product of ZG into itself is twenty-seven and a half and half an eighth, so ZG is five and a fourth. But AZ is two and three fourths. Thus the whole AG is eight. This is two thirds of the amount, therefore the amount is twelve. This is what we wanted to demonstrate.

---

[589] *Sic*, instead of, more conveniently, 'into five and a half' (also in what follows). Perhaps a reader became confused because of the lacuna and attempted to emend the text.

[590] Equation of the form $x^2 = bx + c$.

[591] Addition finally performed.

[592] *Elements* II.2 $= PE_2$.

[593] Difference and product of AG and HG thus known. This is our first encounter with the geometric solution of a quadratic equation.

[594] *Elements* II.6 $= PE_6$.

Chapter (A–VIII) on the Division of fractions, with or without integers

Those who divide a number by a number want to know what will be attributed to one, as we said before in the division of integers [595], whether the division is made by less or more than one. Now it is useful for those who want to divide fractions to know two things. The first is to know the number by the multiplication of which a fraction is redintegrated into one, the other is to know what ratio the unit has to an integer with a fraction. [596]

(*i*) That by which a fraction is redintegrated is like asking:

(A.215) What is the number by which a third when multiplied is redintegrated into one?

We shall answer: the multiplication of it by three does it. For there are three thirds in one, and the above third is a third of one; multiplying it by three produces one. A fourth is itself redintegrated by multiplying it by four, a half likewise by multiplying it by two, a sixth by multiplying it by six.

(A.216) Someone then asks: What is the number by which two thirds are to be multiplied to produce one?

Say that the multiplication of them by one and a half does it. For there are in one three thirds, and two thirds are as if it were two; then two, to produce three, will be multiplied by one and a half. Likewise, three fifths are redintegrated into one by multiplying them by one and two thirds; five sevenths are redintegrated into one by multiplying them by one and two fifths; and, similarly, eight elevenths by multiplying them by one and three eighths.

(A.217) Someone asks for the number by which half a sixth is to be multiplied to produce one.

Say: the multiplication of it by twelve does it. Likewise: half an eighth when multiplied by sixteen, and a fourth of a seventh when multiplied by twenty-eight.

(A.218) Someone asks: By what number are three fourths of one fifth to be multiplied to produce one?

---

[595] Beginning of A–IV (p. 639). But this was said about sharing, thus for division of quantities of different kinds. See also A.236 & A.246.

[596] Subject of A.215–220 and A.221–224 respectively. None of that appears in MS $\mathcal{A}$ and only MS $\mathcal{B}$ has this short introduction.

LIBER MAHAMELETH

Say: the multiplication of them by six and two thirds does it. This is so because there are in one twenty fourths of fifths. Now he proposed three, and three is to be multiplied by six and two thirds to produce twenty.

(**A.219**) Someone asks: Three fourths of one eleventh, by what number are they to be multiplied to produce one?

Say: the multiplication of them by fourteen and two thirds does it.

(**A.220**) Someone now asks: By what number are two and a half sevenths to be multiplied to produce one?

Say: the multiplication of them by two and four fifths does it. This is so because there are fourteen halves of a seventh in one, while two and a half sevenths of fourteen is five. Thus one is as if it were fourteen and two and a half sevenths of it is as if it were five, and five produces fourteen when multiplied by two and four fifths.

Consider all instances of this kind accordingly.

(*ii*) Now someone asks: What is the ratio of one to an integer and a fraction?[597] [*He asks this to know how many such fractions are in one.*][598] ⟨You need first to know how many such fractions are in one.⟩ Then you will denominate the result from the number and the fraction once they are reduced to this fraction.[599] This will be what you want.

(**A.221**) For instance, someone asks: What is the ratio of one to one and a half?

Say: two thirds. For one is two halves, one and a half are three, and two of three is two thirds.

(**A.222**) Someone asks: What is the ratio of one to two and a fourth?

Say: four ninths. For one is four fourths, two and a fourth are nine fourths, and four of nine is four ninths.

(**A.223**) Someone asks: What is the ratio of one to three and three elevenths?

Say: two ninths and three fourths of one ninth. For ⟨one is eleven elevenths, while three and three elevenths are thirty-six elevenths, so⟩ eleven [*elevenths*] will be denominated from thirty-six [*elevenths*].[600]

(**A.224**) Likewise if someone asks: Which part is one of two and five and a half sevenths?

---

[597] Required the *amount* of this ratio. We also have with the same meaning 'find what fraction one is of' (A.254*b*–256*b*, A.261*c*).

[598] Accounts for the subsequent lacuna.

[599] That is: the result will be denominated from the *quantity* of such fractions.

[600] What we have put in brackets is not wrong but departs from the usual expression and may have been added because of the gap.

728   PART TWO: TRANSLATION, GLOSSARY

Say: four thirteenths and two thirds of one thirteenth. Indeed, one is fourteen halves of a seventh while two and five and a half sevenths are thirty-nine halves of a seventh. So you will denominate fourteen from thirty-nine and this will be what you wanted.

Consider other instances likewise.[601]

For those who want to divide fractions by fractions it is useful to know the number by which a fraction when multiplied is redintegrated into one. The rule for this is the following:

(*iii′*) When you want to determine the number by which a fraction, or a fraction with an integer, or a fraction of a fraction, must be multiplied to produce one, divide one by it, whatever it is[602]; the result will be the number by which the fraction, or whatever else, must be multiplied to produce one.

(**A.225**) For instance, I seek the number by which two eighths are to be multiplied to produce one.

So I divide one by two eighths, which gives four. Therefore four is the number by which two eighths are to be multiplied to produce one.

Likewise for all.

Again. Another rule is the following:

(*iii″*) Multiply, by the number from which the fraction is denominated, the whole quantity you want to redintegrate into one, whether or not an integer is with the fraction, and divide by the product the number from which the fraction is denominated. The result is the number by the multiplication of which the quantity proposed will produce one.

(**A.226**) For instance, you want to redintegrate three fifths into one.

Multiply three fifths by the number from which the fraction is denominated, which is five, thus producing three. Divide five by three. The result is one and two thirds. This is the number by the multiplication of which[603] three fifths will be redintegrated into one.

AGAIN.

(**A.227**) You want to redintegrate one and a third into one.

By the previous rule, you will do the following. Multiply one and a third by the number from which a third is denominated, which is three; this produces four. Divide[604] three by four. The result is three fourths.

---

[601] As already said (note 596), the part of this chapter seen so far does not appear in MS 𝒜, probably because what follows (in all three MSS) is on the very same topic.

[602] That is: fraction, integer and fraction, fraction of a fraction (the second case requires denominating).

[603] *Et hoc est in quod* (not: *id in quod*) *si multiplicaveris* in the text. This expression occurs several times in what follows.

[604] Rather: Denominate. Same in the two following problems.

LIBER MAHAMELETH

729

This is the number by the multiplication of which one and a third will be redintegrated into one.

AGAIN.

(A.228) You want to redintegrate two and four sevenths into one.

By the previous rule, you will do the following. Multiply the whole quantity proposed, that is, two and four sevenths, by the number from which a seventh is denominated, which is seven; the result is eighteen. Divide by this the denominator, which is seven; the result is two sixths and a third of a sixth, and this is the number by the multiplication of which two and four sevenths are redintegrated into one.

AGAIN.

(A.229) You want to redintegrate two and three sevenths and half a seventh into one.

By the previous rule, you will do the following. Multiply together the numbers from which the fractions, namely a seventh and a half, are denominated, thus seven and two; the result is fourteen. Multiply two and three sevenths and half a seventh by this, thus producing thirty-five. Dividing fourteen by thirty-five gives two fifths. This is the number by the multiplication of which two and three sevenths and half a seventh are redintegrated into one.

[*It was said above that in division either larger is divided by smaller, or equal by equal, or smaller by larger, in which latter case the division is called denomination.*][605]

CHAPTER ON DENOMINATING FRACTIONS FROM OTHERS,
WITH OR WITHOUT INTEGERS [606]

(A.230) You want to denominate a fourth from a third.

You will do the following. Multiply the denominators of a fourth and a third, namely four and three; this produces twelve. Denominate its fourth, which is three, from its third, which is four. This gives three fourths, and this is what you wanted.

(A.231) You want to denominate five sixths from six sevenths.

You will do the following. Multiply the denominators of a sixth and a seventh, namely six and seven; this produces forty-two. Denominate its five sixths, which is thirty-five, from its six sevenths, which is thirty-six; this will give eight ninths and three fourths of one ninth.

(A.232) You want to denominate one from two and a half.

---

[605] Rather than an explanation of the subsequent heading, this must be a reader's reaction to the previous carelessness ('divide' instead of 'denominate', note 604). His 'above' refers to rule *ii* in the introduction to A–IV (see p. 639).

[606] In A.232–233 (misplaced?) the dividend is an integer.

730 PART TWO: TRANSLATION, GLOSSARY

The rule is the following. Multiply the number from which a half is denominated, namely two, by two and a half; the result is five. Next, multiply two by one, which produces two. Denominating two from five gives two fifths.

AGAIN.

(A.233) You want to denominate two from six and two thirds.

You will do the following. Multiply the denominator of a third, namely three, by six and two thirds; the result is twenty. Next, multiply three by two, thus producing six. Denominating six from twenty gives three tenths.

(A.234) You want to denominate one and a half from three and a third.

You will do the following. Multiply the denominators, which are two and three, thus producing six. Multiply this by three and a third; the result is twenty. Next, multiply six by one and a half; this produces nine. Denominating this from twenty gives four tenths and half a tenth, and this is what you wanted.

(A.235) You want to denominate three and three fourths from four and three tenths.

Multiply the denominators, which are four and ten, thus producing forty. Multiply this by three and three fourths; the result is a hundred and fifty. Next, multiply forty by four and three tenths; the result is a hundred and seventy-two. Denominate from it a hundred and fifty as we have taught above[607]; this will give thirty-seven and a half forty-thirds, and this is what you wanted.[608]

AGAIN. ANOTHER CHAPTER. ON DIVIDING LARGER BY SMALLER[609]

THE FIRST CHAPTER HERE IS ON THE DIVISION OF A FRACTION BY A FRACTION.

(A.236) You want to divide four fifths by three fourths.

The meaning of this problem is the following. Knowing that four fifths has been attributed to three fourths of some thing, we must infer how much will be attributed to the whole thing.

The rule is the following. Multiply the denominators of a fifth and a fourth, which are four and five; this produces twenty. Divide by three

---

[607] Presumably refers to the previous problems (rather than to A.86–87).

[608] The second hand of MS $\mathcal{A}$ corrected the answer to 'thirty-seven forty-thirds and half a forty-third'. Although such a reading is found in A.263c, this is still hardly necessary (for otherwise the text would have had *triginta septem et dimidia quadragesima tertia*).

[609] That is: Dividing larger fractions by smaller ones. We find successively: fraction by fraction (A.236–238); fraction by fraction of a fraction *or inversely with the same values* (A.239–239′); fraction by compound fraction *or inversely* (A.240–240′); compound fraction by compound fraction (A.241).

LIBER MAHAMELETH 731

fourths of it, which is fifteen, four fifths of it, which is sixteen. The result is one and a third of a fifth, and this is what is attributed to one when four fifths is attributed to three fourths.

(**A.237**) You want to divide three fifths by a third.

(**a**) Multiplying the denominators of a fifth and a third produces fifteen. Its third is five, which is the principal[610]. Next, divide three fifths of fifteen, which is nine, by the principal; the result will be what you require.

(**b**) Or else find the number by which a third is to be multiplied to produce one. This is three.[611] Multiply by three the dividend, which is three fifths; this produces one and four fifths, which is the result of the division.

(**A.238**) You now want to divide five sixths by eight elevenths.

(**a**) Do as explained above.[612]

(**b**) Or else find the number by which eight elevenths are to be multiplied to produce one. This is one and three eighths.[613] Multiplying this number by the five sixths to be divided will produce what you wanted to know.

DIVISION OF A FRACTION BY A FRACTION OF A FRACTION.[614]

(**A.239**) You now want to divide ten thirteenths by three fourths of a fifth.

(**a**) Multiplying the denominators, namely thirteen, four and five, produces two hundred and sixty. Dividing its ten thirteenths by three fourths of a fifth of it will give what you wanted.

(**b**) Or reduce each side, the dividend and the divisor, to the kind of the smallest fraction[615] occurring in the divisor, so that the divisor may become an integer. This is attained by multiplying each by twenty, which is the product of the denominators of a fourth and a fifth. Dividing the product resulting from the dividend by the product resulting from the divisor will give the quantity you require.

(**c**) Or take the number by which three fourths of a fifth are to be multiplied to produce one, namely six and two thirds, and multiply it by ten thirteenths; the result will be the quotient of the division.

(**A.239′**) You may want to invert the division, that is, to divide three fourths of a fifth by ten thirteenths.

---

[610] In the division the principal, or final divisor, is not the product of the denominators. We had the same situation in the problems on unknown amounts (see A.159, A.204).

[611] See A.215 (if need be).

[612] A.236–237*a*.

[613] 'one and three fifths' in the MSS.

[614] Includes (except in MS $\mathcal{A}$) the inverse case of the division of a fraction of a fraction by a fraction.

[615] *genus minoris fractionis*; same in A.263*b*.

732   Part Two: Translation, Glossary

(*a*) You will do what has been explained above.[616] That is, denominate thirty-nine, which is three fourths of a fifth of the product of the denominators, from two hundred, which is ten thirteenths of the product of all the denominators[617]; the resulting fraction is the quotient of the division.

(*b*) Or take the number by which ten thirteenths are to be multiplied to produce one, which is one and three tenths, and multiply it by the dividend, namely three fourths of a fifth. Do this in the following way. Multiply one and three tenths by three, thus producing three and nine tenths; divide this by four, from which a fourth is denominated, and divide the result by five, from which a fifth is denominated. The result is three fourths of a fifth and nine tenths of a fourth of a fifth, and this is the quotient you wanted to know.

Division of a fraction by a fraction and a fraction of a fraction.[618]

(**A.240**) You now want to divide four sevenths by an eighth and two thirds of an eighth.

(*a*) All the denominators, that is, of a seventh, an eighth and a third, produce a hundred and sixty-eight. Divide its four sevenths, which is sixty-four[619], by an eighth of it and two thirds of an eighth of it, which is thirty-five; the result will be the quotient you require.

(*b*) Or reduce the dividend and the divisor to thirds of an eighth by multiplying each by twenty-four, as explained before[620]. The multiplication of the dividend will produce thirteen and five sevenths, that of the divisor, five. Dividing the first by the second will give the quotient you require.

(*c*) Or take the number by which an eighth and two thirds of an eighth are to be multiplied to produce one, namely four and four fifths, and multiply it by the dividend, which is four sevenths. The product is the quotient you require.

(**A.240′**) You may want, conversely, to divide an eighth and two thirds of an eighth by four sevenths.

(*a*) Proceed as before.[621] You will end by denominating thirty-five from sixty-four.

(*b*) Another way[622] for this inverse problem is to take the number by which four sevenths are to be multiplied to produce one, namely one and three

---

[616] A.239*a*.

[617] 'all': not only those in the divisor as in A.239*b*.

[618] Includes the inverse case of the division of a fraction and a fraction of a fraction by a fraction.

[619] *Sic*, instead of ninety-six. This error occurs in both A.240 and (the related) A.240′.

[620] Above, A.239*b*.

[621] A.240*a*.

[622] Corresponding to A.240*c*.

LIBER MAHAMELETH 733

fourths, and to multiply it by the dividend, which is an eighth and two thirds of an eighth. The result will be the quotient of the division.

DIVISION OF A FRACTION AND A FRACTION OF A FRACTION BY A FRACTION AND A FRACTION OF A FRACTION.

(**A.241**) You now want to divide five eighths and three fourths of an eighth by two sevenths and half a seventh.

(**a**) You will do the following. Multiply together the denominators of a fourth and an eighth, thus producing thirty-two. Next, multiply together the denominators of a seventh and a half, thus producing fourteen. Next, multiply fourteen by thirty-two. Put [*the sum of*] two and a half sevenths of the product as the principal. Next, divide five eighths and three fourths of an eighth of this same product by the principal. The result will be what you wanted. [*The proof of this is clear from the preceding.*][623]

(**b**) Or otherwise. Take of the product of the denominators of a seventh and a half, thus fourteen, two and a half sevenths, which is five, and put it as the principal. Next, divide five eighths of fourteen and three fourths of an eighth of it, that is, ten and half an eighth, by five. The result is two and half an eighth of a fifth. We have chosen to take (into account only) the fractions of the divisor in order to have no fractions in the principal.

Treat other such instances as explained, and you will find that this is the way.

CHAPTER ON DIVIDING, OTHERWISE [624]

(**A.242**) You want to divide three fourths of six by two fifths of four.

Multiplying the denominators of a fifth and a fourth [625] produces twenty. Multiply three fourths of this, which is fifteen, by six, thus producing ninety, which is the dividend. Next, multiply two fifths of twenty, which is eight, by four; the product is thirty-two. Divide ninety by it. The result is what you require, namely two and six eighths and half an eighth.

(**A.243**) You now want to divide three fourths of three and a fifth by two fifths of two and a half.

Multiplying the denominators of a fourth and a fifth produces twenty. [*Likewise, multiplying the denominators of a fifth and a half produces ten. Now when we wanted to multiply twenty by ten in order to produce the common number, and found that all parts of ten were also in twenty,[626] we then took twenty as the common number.*] Multiply three fourths of it, namely fifteen, by three and a fifth, thus producing forty-eight. This is the

---

[623] Allusion to the proof in A.260*a*, just a few lines before A.241 in the disordered text of MSS $\mathcal{B}$ & $\mathcal{C}$ (but the words 'from the preceding' do not appear in MS $\mathcal{C}$).

[624] Division of irregular fractions.

[625] Here the order of the formulation is not respected; this is unusual. Same in A.262.

[626] The divisors of 10 are divisors of 20.

734   PART TWO: TRANSLATION, GLOSSARY

dividend. Next, multiply two fifths of twenty, which is eight, by two and a half, thus producing twenty. Divide forty-eight by this. The result is what you require, namely two and two fifths.

(A.244) You want to divide seven eighths of six by two thirds of five.

(a) You will do the following. Multiplying the denominators of an eighth and a third produces twenty-four. Multiply two thirds of this by five and put the product as the principal. Next, multiply seven eighths of twenty-four by six and divide the product by the principal. The result will be what you require.

This is proved as follows. We know that what we want is just to take seven eighths of six and divide the result by two thirds of five. But taking seven eighths of six is like taking seven eighths of twenty-four, multiplying the result by six and dividing the product by twenty-four.[627] Likewise also, taking two thirds of five is like taking two thirds of twenty-four, multiplying the result by five and dividing the product by twenty-four. Thus, it became necessary to divide a hundred and twenty-six, the product of seven eighths of twenty-four into six, by twenty-four and divide the quotient by the result of dividing eighty, the product of two thirds of twenty-four into five, by twenty-four; this will give what is required. But this is just the same as dividing a hundred and twenty-six by eighty. This is proved as follows.[628] Dividing a hundred and twenty-six by twenty-four gives some number which, when multiplied by twenty-four, produces a hundred and twenty-six. Likewise also, dividing eighty by twenty-four gives some number which, when multiplied by twenty-four, produces eighty. Thus multiplying these two quotients by twenty-four produces a hundred and twenty-six and eighty. Therefore[629] the ratio of the dividend to the divisor is the same as the ratio of a hundred and twenty-six to eighty, so that dividing the dividend by the divisor is like dividing a hundred and twenty-six by eighty. But dividing the dividend by the divisor gives what we require. Thus dividing a hundred and twenty-six by eighty will give what we require. This is what we wanted to demonstrate.

(b) You may proceed according to the wording of the problem, namely as follows. Take seven eighths of six as we have shown[630]; this gives five and a fourth. Next, take two thirds of five; this gives three and a third. Divide five and a fourth by this. The result will be one and five tenths and three fourths of a tenth, and this is what you wanted.

(A.245) You want to divide seven eighths of six and two thirds by two fifths of four and a half.

(a) You may proceed according to the wording of the problem. That is, take two fifths of four and a half, which is one and four fifths. Next, take

---

[627] As known from A.90a.

[628] Proves that the quotient of two fractions with the same denominator equals the quotient of their numerators.

[629] *Elements* VII.18.

[630] A.90.

LIBER MAHAMELETH

735

seven eighths of six and two thirds, which is five and six eighths and two thirds of an eighth. Divide this by one and four fifths, and the result will be what you wanted.

(**b**) Or otherwise. Multiply the denominators of an eighth, a third, a fifth and a half; the result is two hundred and forty. Multiply two fifths of this by four and a half, and put the product as the principal. Next, multiply seven eighths of two hundred and forty by six and two thirds and divide the product by the principal; the result will be what you wanted. The proof of this is clear from the preceding.[631]

Treat other such instances accordingly, and you will find that this is the way.

DIVISION OF AN INTEGER BY A FRACTION.

(**A.246**) ⟨You want to divide ten by a fourth.⟩

Those who want to divide ten by a fourth want to know, given that ten is attributed to a fourth of one, what will be attributed to the whole unit.[632]

(**a**) Take then of the denominator, namely four, its fourth, which is one; this is the principal by means of which the division will be made.[633] Next, multiply ten by four; the result is forty. Divide it by one,[634] and the result is what you wanted to know.

(**b**) Or otherwise. Look for the number by which a fourth is to be multiplied to produce one as explained before.[635] You will find[636] that its multiplication by four does it. Thus multiplying four by ten will give what you require.

(**A.247**) You want to divide twenty by three fourths.

Your purpose here is, given that twenty is attributed to three fourths of one, to know what will be attributed to one.

You will do the following. Multiply the number from which a fourth is denominated, namely four, by three fourths; the result is three, which is the principal. Next, multiply twenty by four; dividing the product by the principal will give what you wanted.

Or otherwise. Divide four by three and multiply the product by twenty; the quotient will be what you wanted. The proof of this is clear:

---

[631] In A.244*a*.

[632] This is repeated in A.247 (which, in MSS $\mathcal{B}$ and $\mathcal{C}$, does not follow A.246). We find a similar explanation at the beginning of the previous section (A.236).

[633] The author wants to point out the difference between (given) divisor and divisor in the computation. Same in A.248*a*, A.253*a*, A.254*a*.

[634] The method is explained generally. Same in A.249*a*.

[635] A.215, if need be.

[636] We sometimes wonder if the author is writing mechanically. See also notes 1118, 1310, 1615.

736　　　Part Two:　Translation, Glossary

four had to be multiplied by twenty and the product divided by three;[637] but this is the same as dividing four by three and multiplying the result by twenty.[638]

(**A.248**) You want to divide fifteen by four sevenths.

(*a*) You will do the following. Take four sevenths of the denominator, namely seven; this is four, which is the principal number whereby you are to divide. Next, multiply seven by the fifteen to be divided and divide the product by four. The result is what you require.

(*b*) Or otherwise. Seek the number by which four sevenths when multiplied is redintegrated into one. You will find[639] that the multiplication of it by one and three fourths does it. So multiplying this number by fifteen will produce the result of the division.

Treat other such instances in the same manner.

DIVISION OF AN INTEGER BY A FRACTION OF A FRACTION.

(**A.249**) You want to divide eight by a third of a fifth.

(*a*) You will do the following. Multiply together the denominators of a third and a fifth, thus producing fifteen. A third of a fifth of this is one, which is the principal. Next, multiply eight by fifteen; the product is a hundred and twenty. Dividing this by one, which is the principal, gives what you wanted to know.

(*b*) Or otherwise. Seek the number by which a third of a fifth is to be multiplied to produce one. You will find[640] that the multiplication of it by fifteen does it. So multiply fifteen by eight; the product is the result of the division.

(**A.250**) You now want to divide fifteen by four fifths of an eleventh.

(*a*) Multiplying the denominators of a fifth and an eleventh produces fifty-five. Four fifths of an eleventh of this is four. Dividing by this the result of multiplying fifteen by fifty-five will give the quantity you require.

(*b*) Or otherwise. Seek the number by which four fifths of an eleventh are to be multiplied to produce the whole unit. You will find[641] that the multiplication of them by thirteen and three fourths does it. Multiplying this by the fifteen to be divided will produce the result of the division.

DIVISION OF AN INTEGER BY A FRACTION AND A FRACTION OF A FRACTION.

(**A.251**) You now want to divide ten by three eighths and half an eighth.

(*a*) Multiplying the denominators of the fractions a half and an eighth produces sixteen. Three and a half eighths of this is seven, which is the

---

[637] According to what precedes.

[638] Premiss P_5.

[639] A.216, A.225–226.

[640] A.217 (if need be).

[641] A.218–219.

LIBER MAHAMELETH 737

principal. Next, multiply sixteen by the ten to be divided; the result is a hundred and sixty. Dividing this by seven will give what you require.

(*b*) Or otherwise. Consider by what three and a half eighths are to be multiplied to produce the whole unit. You will find[642] two and two sevenths. Multiplying this by the ten to be divided will give the result of the division.

(**A.252**) You now want to divide ten by five elevenths and a third of an eleventh.

(*a*) Take the result of multiplying the denominators, namely thirty-three, and multiply it by the divisor and by the dividend. Dividing the latter product by the former will give the quantity you require.

(*b*) Or otherwise. Take the number by which five elevenths and a third of an eleventh are to be multiplied to produce the whole unit, namely two and half an eighth, and multiply it by the dividend. The result will be what you require.

(**A.253**) You now want to divide eight by four fifths of an eleventh and two thirds of a fifth of an eleventh.

(*a*) Multiplying the denominators of a fifth, an eleventh and a third produces a hundred and sixty-five. Four fifths of an eleventh and two thirds of a fifth of an eleventh of this is fourteen. Such is the principal number by which we divide. Next, multiply a hundred and sixty-five by the eight to be divided and divide the product by the divisor[643]. The result is what you require.

(*b*) Or otherwise. Take the number by which the given fractions are to be multiplied to produce the whole unit, namely eleven and five and a half sevenths, and multiply it by the dividend. The result will be what you require.

CHAPTER ON THE DIVISION OF AN INTEGER BY AN INTEGER AND A FRACTION.

(**A.254**) You want to divide twenty by two and two thirds.

(*a*) Multiply the number from which a third is denominated, namely three, by the divisor, which is two and two thirds. The result is eight, which is the number by which we divide.[644] Next, multiply three by the dividend, which is twenty; this produces sixty. Dividing it by eight will give what you require.

(*b*) Or otherwise. Determine what fraction one is of two and two thirds as explained before[645]. You will find that it is three eighths. Thus three eighths of twenty, which is seven and a half, is the result of the division.

---

[642] A.220. Also used in A.252*b* (and A.253*b*).

[643] That is, the principal.

[644] Thus the 'principal', not the divisor mentioned before.

[645] A.221–223, A.227–228.

738 PART TWO: TRANSLATION, GLOSSARY

CHAPTER ON THE DIVISION OF AN INTEGER BY AN INTEGER AND A FRACTION OF A FRACTION.

(A.255) You now want to divide thirty by four and half a sixth.

(*a*) Multiplying the denominators of a half and a sixth produces twelve. Multiply it by the divisor, which is four and half a sixth; the result is forty-nine, which is the principal number. Next, multiply twelve by the dividend, which is thirty; the result is three hundred and sixty. Dividing it by the principal, which is forty-nine, will give what you require.

(*b*) Or otherwise. You need to know what fraction one is of four and half a sixth. You will find that it is a seventh and five sevenths of a seventh. Multiply this by the dividend, which is thirty, as we have taught before for the multiplication of fractions[646]. The result will be the quotient you require.

DIVISION OF AN INTEGER BY AN INTEGER AND A FRACTION AND A FRACTION OF A FRACTION.

(A.256) You now want to divide forty-five by three and four elevenths and a third of an eleventh.

(*a*) Multiplying the denominators of a third and an eleventh produces thirty-three. Multiply this by the divisor, which is three and four elevenths and a third of an eleventh; the product is a hundred and twelve, which is the principal number. Next, multiply thirty-three by the dividend; this produces one thousand four hundred and eighty-five. Dividing this by the principal will give what you require.

(*b*) Or otherwise. Find what fraction one is of the divisor. You will find[647] that it is two eighths and two sevenths of an eighth and half a seventh of an eighth. Multiplying all this by the dividend will produce the quotient you require.

CHAPTER ON THE DIVISION OF AN INTEGER AND A FRACTION BY A FRACTION.

(A.257) You want to divide thirty and two thirds by four fifths.[648]

(*a*) Multiplying the denominators of a third and a fifth produces fifteen. Four fifths of this is twelve, which is the principal number. Next, multiply fifteen by the dividend; the result is four hundred and fifty-five. Dividing this by the principal, which is twelve, will give what you wanted to know.

(*b*) Or otherwise. Reduce the divisor and the dividend to fifths by multiplying each of them by five. The divisor becomes four and the dividend, a hundred and fifty-one and two thirds. Dividing this by four gives thirty-seven and five and a half sixths, and this is what you wanted to know.

---

[646] A.96.

[647] A.224, A.229.

[648] In *a* and *b* the dividend is taken to be thirty and a third.

LIBER MAHAMELETH

(c) Or find the number by which four fifths are to be multiplied to produce one.[649] This is one and a fourth. Multiply it by thirty and two thirds in the manner we have taught for the multiplication of fractions[650]. The result will be what you wanted to know.

DIVISION OF AN INTEGER AND A FRACTION BY A FRACTION OF A FRAC-TION.

(A.258) You now want to divide twenty-three and three fourths by two thirds of a fifth.

(a) Multiplying all the denominators[651], that is, of a fourth, a third and a fifth, produces sixty. Two thirds of a fifth of this is eight, which is the principal. Next, multiply sixty by the dividend. Dividing the product by the principal will give the quotient you require.

(b) Or reduce the divisor and the dividend to thirds of a fifth, which is done by multiplying each of them by fifteen. The multiplication of the divisor produces two and the multiplication of the dividend, three hundred and fifty-six and a fourth. Dividing this by two will give the quotient you require.

(c) Or find the number by which two thirds of a fifth are to be multiplied to produce one.[652] This is seven and a half. Multiply by this the dividend, which is twenty-three and three fourths. The product is the quantity you require.

⟨DIVISION OF AN INTEGER AND A FRACTION BY A FRACTION AND A FRACTION OF A FRACTION.⟩[653]

(A.259) You now want to divide twenty-six and three fifths by four sevenths and half a seventh.

The method for dividing does not differ from those previously explained.[654]

(a) Either reduce each side to halves of sevenths; the divisor becomes nine and the dividend, three hundred and seventy-two and two fifths. Dividing this by nine will give what you wanted to know.

(b) Or find the number by which four and a half sevenths are to be multiplied to produce one.[655] This is one and five ninths. Multiply the dividend by this number. The product is the result of the division.

Proceed accordingly in other such instances.

---

[649] A.216.

[650] *eo modo quo docuimus in multiplicatione fractionum* (same expression in A.262c). Here it refers to A.124–125.

[651] 'all': unlike in *b*, where we consider only those of the divisor. See p. 732, note 617.

[652] A.217–219.

[653] This heading seems necessary.

[654] A.258 and before. Here the first way is not repeated.

[655] A.220.

740 PART TWO: TRANSLATION, GLOSSARY

CHAPTER ON THE DIVISION OF AN INTEGER AND A FRACTION BY AN INTEGER AND A FRACTION.

(A.260) You want to divide twelve and three fourths by one and two sevenths.

(*a*) You will do the following. Multiply the denominators of a fourth and a seventh; the result is twenty-eight. Multiply this by the divisor and put the product as the principal. Next, multiply twenty-eight by the dividend. Dividing the product by the principal will give what you wanted.

$$(A.260\,a)$$

This is proved as follows. Let twelve and three fourths be A, one and two sevenths, B, and twenty-eight, G. Let A divided by B give D. So multiplying D by B produces A. Therefore B measures[656] A as many times as one is in D. But one measures D as many times as one is in D. Thus the ratio of one to D is the same as the ratio of B to A. Let B be multiplied by twenty-eight, which is G, and H be produced, and let Z result from multiplying A by G. So[657] the ratio of B to A is the same as the ratio of H to Z. But the ratio of B to A is the same as the ratio of one to D. Thus the ratio of one to D is the same as the ratio of H to Z. Therefore the product of one into Z is equal to the product of D into H. The product of one into Z being simply Z, the product of D into H is Z. Therefore dividing Z by H will give D. This is what we wanted to demonstrate.

(*b*) You may also consider that taking a single denominator, namely that with the divisor, will suffice [*instead of both denominators*] to make the principal an integer, without any fraction; multiply this denominator by the divisor and let the product be the principal; multiply the denominator also by the dividend; dividing this product by the principal will give what you wanted. In the present case, you will do the following. Multiply the number from which a seventh is denominated, namely seven, by one and two sevenths, thus producing nine. Put it as the principal. Next, multiply seven by twelve and three fourths. Dividing this product by the principal will give what you wanted. The proof of this is clear from the previous one.

(A.261) You want to divide twenty and three fourths by two and a third.

---

[656] See note to P_7.

[657] *Elements* VII.18.

LIBER MAHAMELETH

741

(*a*) Form, by multiplying the denominators of a fourth and a third, twelve. Multiply by this the divisor, which is two and a third; the result is twenty-eight, which is the principal number. Next, multiply twelve by the dividend, which is twenty and three fourths, thus producing two hundred and forty-nine. Dividing this by the principal will give what you require.

(*b*) Or otherwise. Convert the dividend and the divisor into the kind of the lowest fraction occurring in the divisor,[658] that is, a third, so that the whole divisor may become an integer; this is done by multiplying both by three, from which a third is denominated. From the multiplication of the divisor results seven, which is the principal, and from the multiplication of the dividend, sixty-two and a fourth. Dividing this by the principal will give what you require.

(*c*) Or otherwise. Find what fraction one is of the divisor, which is two and a third. You will find[659] that it is three sevenths. Multiply this by the dividend, which is twenty and three fourths. The product is what you require.

CHAPTER ON THE DIVISION OF AN INTEGER AND A FRACTION BY AN INTEGER AND A FRACTION ⟨AND A FRACTION⟩[660] OF A FRACTION.

(**A.262**) You now want to divide seventeen and ten elevenths by three and seven eighths and half an eighth.

(*a*) Multiplying the denominators of a half, an eighth and an eleventh[661] produces a hundred and seventy-six. Multiply the divisor by this; the result is six hundred and ninety-three, which is the principal. Next, multiply a hundred and seventy-six by the dividend; the result is three thousand one hundred and fifty-two. Dividing this by the principal will give what you require.

(*b*) Or otherwise. Convert each of the two sides into the kind of the lowest fraction occurring in the divisor in the following way. Multiply both by sixteen, which is the product of the denominators in the divisor, thus of a half and an eighth. The multiplication of sixteen by the divisor produces sixty-three, which is here the principal number;[662] dividing by this the product of sixteen into the dividend will give the quotient you require.

(*c*) Or otherwise. Consider what fraction one is of the divisor. You will find[663] that it is two ninths and two sevenths of a ninth. Multiply it by the dividend in the manner we have taught for the multiplication of

---

[658] *genus ultime fractionis* (also: *ultimum genus fractionis*, see notes 397 & 419). Since there is only one kind of fraction, the rule is just expressed generally.

[659] A.221–223, A.227–228.

[660] See note 665.

[661] See note 625, p. 733.

[662] 'here': identical neither with that in *a* nor with the given divisor.

[663] A.224, A.229.

742   Part Two: Translation, Glossary

fractions[664]. The product is the quotient you require.

⟨Chapter on the division of an integer and a fraction by an integer and a fraction of a fraction.⟩[665]

(A.263) You want to divide twenty-five and four fifths by four and two thirds of a seventh.

(*a*) Multiply the product of all the denominators by the divisor and the dividend and divide one of the two resulting products by the other.

(*b*) Or in the second way, which is easier.[666] Convert each of the two sides into the kind of the smallest fraction occurring in the divisor. Dividing one of the two results by the other will give what you require.

(*c*) Or otherwise. Consider what fraction one is of the divisor. You will find that it is ten forty-thirds and half a forty-third. Multiplying this by the dividend will give what you require.

Treat other such instances likewise.

(A.264) The multiplication of an integer and a fraction by an integer and a fraction produces forty. What are the multiplier and the multiplicand?

You will do the following. Put as the integer with the fraction, that is, as the multiplicand or the multiplier, any number, for instance four and three eighths. Divide forty by it; this gives nine and a seventh, which is the other number. Then multiplying four and three eighths by nine and a seventh will produce forty, and this is what you wanted.

If you had put as the first number five and three fifths, you would divide forty by this, and the result for the second number would be seven and a seventh. Then multiplying five and three fifths by seven and a seventh will produce forty.

Likewise if you had put as the first number three and three fourths and had divided forty by this; the result would be ten and two thirds, which is the second number. Then multiplying three and three fourths by ten and two thirds will produce forty.

You will always do the same. That is, you will put as either of the numbers, the multiplier or the multiplicand, any number with any fraction. You will divide the proposed quantity by it, and the result will be the other.

(A.264′) Likewise if it is said that the multiplication of a fraction by an integer and a fraction produces forty.

You will do the following. You will put as the fraction any fraction, say three fourths, and divide by this the proposed quantity, which is forty. The result is fifty-three and a third, and this is the other number. Then

---

[664] A.123–129; but an example of the case 'integer and fraction by compound fraction' is not included.

[665] Heading omitted, presumably because it would have been the same as the (lacunary) previous one.

[666] 'which is easier': in comparison with *c*, not with *a*.

LIBER MAHAMELETH 743

multiplying three fourths by fifty-three and a third will produce forty, and this is what you wanted.

Consider all other such instances likewise.

AGAIN. RULES FOR THE MULTIPLICATION, DIVISION, ADDITION AND SUBTRACTION OF FRACTIONS, IN A MORE CONCISE FORM THAN ABOVE

(i) When you want to multiply any fractions by any fractions, whether with integers or not, put the fractions multiplying on one side and the fractions to be multiplied on the other.[667] Next, if there is more than one fraction on one side, multiply their denominators to the last; put the product on this same side below and call it 'quantity'[668]. Should there be only one fraction, put its denominator as this quantity. You will do the same for the other side. Next, multiply the 'quantity' on one side by the 'quantity' on the other; put the result below, between the two 'quantities', and call it 'principal'[669]. After that, multiply whatever integer and fractions, one or several, are on each side by the 'quantity' belonging to that side and put the product under the corresponding 'quantity' and call it 'intermediate result'[670]. Next, multiply together the two intermediate results and divide the product by the principal. The result is the product of the two proposed expressions.

(ii) But if you want to divide, add and subtract, you will, to form the intermediate result, multiply everything on each side not by its 'quantity' as was the case in multiplying, but by the principal, and you will put under the corresponding side the product as the intermediate result. Next, to perform the division, you will divide the intermediate result on the side of the dividend by the other intermediate result, and the quotient will be what you require [*should there be no fraction on one of the sides, then the principal will be the 'quantity' of the other side, and the intermediate result (for this side) will be its integer itself in the case of multiplication, the product of this integer into the principal in the case of division, addition and subtraction*][671]; for the addition, you will add the intermediate results and divide the sum by the principal, and the result will be what you require; for the subtraction, you will subtract one intermediate result from the other and divide the remainder by the principal, and the result will be what you require.

(A.265) You want to divide ten by seven eighths of five and a third, minus a third of two and a fourth.

---

[667] Here 'side' (*latus*) clearly implies left/right. See notes 215 & 360.

[668] What was called 'denominating number' earlier (see, for instance, A.126a, A.128a, A.129a).

[669] Or 'common number' (A.101a & note 335, p. 658).

[670] Literally 'kept (in mind)'; see note 147, p. 613. The *Liber algorismi* (p. 66; also pp. 57–58), in a similar set of rules, speaks of *numerus collectionis*.

[671] This is either an interpolation (inclusion of this particular case being, at best, of little importance) or a genuine, but misplaced statement.

744   Part Two: Translation, Glossary

You will do the following. Take seven eighths of five and a third, which is four and two thirds. Next, take a third of two and a fourth, which is three fourths. Subtract this from four and two thirds; this will leave three and five sixths and half a sixth. Dividing ten by this will give what you wanted.

(A.266) You want to divide three fourths of three fifths of nine by three tenths of one, and the quotient by two thirds of seven minus the result of the division of two and an eighth by seven eighths and half an eighth.

You will do the following. Multiply the denominators of a fourth and a fifth, thus producing twenty. Multiply three fourths of three fifths of twenty, which is nine, by nine; the result is eighty-one. Next, look for the number by which three tenths are to be multiplied to produce one.[672] You will find three and a third. Multiply it by eighty-one; the result is two hundred and seventy. Divide it by twenty; the result is thirteen and a half.[673] Next, take two thirds of seven, which is four and two thirds. After that, seek the number by which seven and a half eighths are to be multiplied to produce one.[674] You will find one and a third of a fifth. Multiply it by two and an eighth, thus producing two and a fifth and a third of a fifth. Subtracting this from four and two thirds leaves two and two fifths. Dividing by this the thirteen and a half kept in mind above will give five and five eighths, and this is what you wanted.

Again on division [675]

(A.267) You divide a hundred by[676] three, the result by seven, the result by four, the result by five and you want to know what the share of each of the five after this last division is.

You will do the following. Multiply three by seven, the product by four, and the product by five, thus producing four hundred and twenty. Denominating a hundred from this gives a seventh and two thirds of a seventh, and this is the share of each of the five.

(A.268) You want to divide ninety nummi among nine men in such a way that the second man has one nummus more than the first, the third man (one more) than the second, and so on to the last.

(*i*) You will do the following. Always subtract, in such a problem, one from the number of men, in this case from nine;[677] this leaves eight. Next, add all the numbers from one to eight, in the following manner[678]. Add one to eight, thus making nine; multiply it by half of eight, thus producing

---

[672] A.216, A.225–226.

[673] Here we would expect to see 'and keep it in mind' (see below).

[674] A.220.

[675] Namely on sharing, which is the second aspect of division (as said by the author in the introduction to A–IV, p. 639 above).

[676] Or, here and further on, 'among' (*per* having both meanings).

[677] 'Always': whatever the number of men. See note 340, p. 659.

[678] Already explained in A.170.

LIBER MAHAMELETH

745

thirty-six, which is the result of adding the numbers from one to eight. Next, subtract the thirty-six from ninety, which leaves fifty-four. Dividing it by the number of men gives six, which is the share of the first. The share of the second is seven, that of the third, eight, and so on for each to the ninth. This rule only applies if the men's shares exceed one another by one, and not in other cases.

(*ii*) In these cases[679] and in others where the difference is the same for all except between the second and the first, the rule is the following. Always subtract one from the number of men and keep the remainder in mind. Next, always subtract two from the number of men and multiply the remainder by the difference by which they[680] exceed one another. Add the result to twice the difference between the second and the first. Multiply the sum by half of what was kept in mind above. Subtract the product from the quantity of nummi to be divided and divide the remainder by the number of men. The result is the share of the first.

(**A.269**) For instance, you want to divide a hundred among eight men in such a way that the second has three nummi more than the first and the shares of all the others exceed one another by two.

Subtract one from eight; this leaves seven. Next, subtract two from eight; this leaves six. Multiply it by two, thus producing twelve. Add this to twice three; this makes eighteen. Multiply it by half of seven, thus producing sixty-three. Subtract it from a hundred; this leaves thirty-seven. Dividing it by eight gives four and five eighths, which is the share of the first.

(*iii*) But when the differences are unequal, add them all, that is, all the differences by which each man exceeds the first, and subtract the sum from the quantity of nummi to be divided. Dividing the remainder by the number of men will give the share of the first.

(**A.270**) For instance, you want to divide eighty (nummi) among five men in such a way that the second has three more than the first, the third, one more than the second, the fourth, two more than the third, the fifth, six more than the fourth.

You will do the following. The second has three more than the first but one less than the third, thus the third has four more than the first. He himself has two less than the fourth, thus the fourth has six more than the first. But he himself has six less than the fifth, who therefore has twelve more than the first. Adding this together with six, four and three makes twenty-five. Subtracting it from eighty leaves fifty-five. Dividing this by the number of men, which is five, gives eleven, which is the share of the first.

---

[679] Where the difference is a constant other than one.

[680] That is, their shares. The same below, *iii*. See also p. 943, notes 1522, 1523.

746 PART TWO: TRANSLATION, GLOSSARY

AGAIN ON THE SAME TOPIC.[681]

(A.271) You divide eighteen nummi among a certain number of men, and adding the share of each to their number makes nine. How many men are there?

You will do the following. Multiply half of nine by itself, thus producing twenty and a fourth. Subtract eighteen from this, thus leaving two and a fourth. Adding the root of this, which is one and a half, to half of nine makes six, and this is the number of men.

(A.272) Likewise, you divide forty nummi among a certain number of men, and subtracting the share of (each) one from their number leaves three. What is the number of men?

You will do the following. Multiply half of three, which is one and a half, by itself, thus producing two and a fourth. Adding it to forty makes forty-two and a fourth. Add the root of this, which is six and a half, to one and a half; this makes eight, which is the number of men.

(A.273) You divide an unknown number of nummi among an unknown number of men. Then, after adding two men, you divide again the same number of nummi among them, and the share of (each) one in the second group is the root of the share of (each) one in the first. How many nummi are there, and how many men in the two groups?

This problem is indeterminate.[682] It must be treated as follows. Put any number as the number of men in the first group, say four. Adding two to it makes six. Multiply this by itself, thus producing thirty-six. Dividing it by four gives nine, which is the number of nummi.

(A.273′) If it is said in this problem that the share of each man in the second group is thrice the root of the share of each in the first, you will do as I have explained above. Next, multiply the quantity of roots[683] by itself; multiplying this product by the quotient of the division will give the number of nummi.

(A.274) One number when divided by two leaves one, divided by three leaves one, divided by four leaves one, divided by five leaves one, divided by six leaves one, and divided by seven leaves nothing. What is this number?

You will do the following. Multiply three by four and the product by five; the result is sixty. Add one to it; this makes sixty-one. Dividing it by two, three, four, five or six will always leave one, but dividing it by seven will leave five. Now it has been stated that nothing is left after the division by seven. So subtract one from five, which leaves four. Seek then a number which, when this four is multiplied by it and one added to the product, the result is divisible by seven. Such is for instance five, or twelve —for the

---

[681] The shares are now equal. A.271–273′ involve roots, which is the subject of the next chapter.

[682] See note 519, p. 707.

[683] Namely: three.

LIBER MAHAMELETH

problem is indeterminate. Suppose you take five. Multiply it by sixty and add one to the product; the result is three hundred and one, and this is a number which, when divided by two, three, four, five and six, always leaves one but by seven, nothing. Multiplying twelve by sixty and adding one to the product will also give such a required number, namely seven hundred and twenty-one.

748   PART TWO: TRANSLATION, GLOSSARY

Chapter (A–IX) on the Determination of roots, and on their multiplication, division, subtraction, addition, and other related subjects

The previous topics having been treated, what remains is for us to deal with the determination of roots and their multiplication, division and other related subjects.[684] For this is extremely useful to know, above all to those who want to operate using algebra[685]. Abū Kāmil did indeed treat some of it, but he did not expound it plainly.[686] For our part, we shall present proofs which are clearer than his. In some of them, however, it will be necessary to make use of some of Euclid's tenth book; for the tenth is the only book where Euclid deals with roots.[687]

CHAPTER ON THE DETERMINATION OF ROOTS

The root of a number is a number from the multiplication of which by itself the other number is produced. For instance, two is the root of four, three is the root of nine, four is the root of sixteen; in this way, it is easy to find the root of other numbers which are not non-square.[688]

(i) If the number is non-square and you want to find its approximate root:[689] Seek the number closest to it, whether larger or smaller, having a rational root;[690]

– (i') if this number is larger, double its root, denominate from this the difference between them, and subtract the result from the root of the larger number; the remainder is the approximate root of the non-square number;

---

[684] Roots have already been encountered in A.167, A.184–186, A.214, A.271–273′.

[685] The text has 'using *gebra* and *muchabala*' (the two fundamental operations of algebra). Since roots are involved, what is meant here is algebra of the second degree (and biquadratic equations).

[686] 'indeed' (*enim*): since he was writing about algebra, he obviously had to deal with roots.

[687] Until now we have used for the proofs only a few theorems from Books II, V, VII–IX.

[688] At most this may mean that it is easy to set out a table of square integers from their roots. An extraction method is found both in the *Arithmetic* of al-Khwārizmī and in Johannes Hispalensis' *Liber algorismi*.

[689] We translate *surdus* by 'non-square' as being more appropriate than 'irrational'.

[690] For convenience, this number will be an integer.

LIBER MAHAMELETH 749

– (i'') if this number is smaller, denominate the difference between the non-square number and the smaller square from twice the root of the latter and add the result to the root of the smaller; the sum is the approximate root of the non-square number.[691]

(**A.275**) For instance, you want to find the root of five.

You will do the following. The number closest to five, and smaller than it, having a rational root is four. Its root is two [*for multiplying two by itself produces four*] and the remainder is one. Denominating it from twice two, which is four, gives a fourth. Adding this to two will give the approximate root, namely two and a fourth.

(ii) Now if you want to find a root even closer, you will do the following. Always multiply the number of which you seek the root by some other number having a root, as[692] five by a hundred. Having found the root of the product as I have taught before[693], divide it by the root of a hundred. The result will be a root closer than the root found before. If you want to find a root which is even closer, you will do the following. You will multiply the number of which you seek the root by a number, larger (than the previous one), having a root, such as ten thousand; for by the larger you multiply, the closer the root obtained. Next, continue as explained above.[694]

(**A.276**) For instance, you want to find the root of two.

You will do the following. Multiply two by ten thousand, thus producing twenty thousand. Divide its approximate root, found as explained above, by the root of ten thousand; the result will be the approximate root of two.

(**A.277**) You now want to find the root of fourteen.

You will do the following. Seek the closest integral number to fourteen having a root; this is sixteen. Then denominate their difference, which is two, from twice the root of sixteen, namely eight; this gives a fourth. Subtracting this from the root of sixteen, which is four, leaves three and three fourths, and this is the approximate root of fourteen.

(**A.278**) You want to know what the root of six and a fourth is.

You will do the following. Seek a number having a root and a fourth; such is four. Put its root, namely two, as the principal[695]. Next, multiply six and a fourth by four; the result is twenty-five. Divide its root, which is five, by two; the result is what you require, namely two and a half.

(**A.279**) You now want to know what the root of five and four ninths is.

---

[691] Similar rule in the *Liber algorismi*, pp. 77-78.

[692] Case of A.275.

[693] Above, i.

[694] Rule ii is explained in the *Liber algorismi*, p. 86.

[695] Thus as the divisor.

Seek a number having a root and a ninth; such is nine. Put its root, which is three, as the principal. Next, multiply nine by five and four ninths; the result is forty-nine. Divide its root, which is seven, by the root of nine, which is three[696]; the result is two and a third, and this is what you require.

(**A.280**) You want to know what the root of six eighths and an eighth of an eighth is.

You will do the following. Seek a number having a root, an eighth, and an eighth of an eighth;[697] such is sixty-four. Put its root, which is eight, as the principal. Next, multiply six eighths and an eighth of an eighth by sixty-four; the result is forty-nine. Denominate its root, which is seven, from the principal [*that is, from the root of sixty-four, which is eight*]; this gives seven eighths, and this is what you wanted to know.

(**A.281**) You want to know what the root of two and a half is.

You will do the following. Put the number from which a half is denominated, namely two, which does not have a root, as the principal. Next, multiply it by two and a half, thus producing five. Multiply this by two, thus producing ten.[698]

(*a*) Divide its root, found as I have taught above[699], which is three and a sixth, by the principal; the result will be the root[700] you seek, which is one and a half and half a sixth.[701]

[*You may also multiply two by itself, the product by two and a half, and continue as explained above.*]

(*b*) Or otherwise. Multiply ten by a hundred, thus producing a thousand. Next, multiply two by the root of a hundred, thus producing twenty. Put it as the principal. Next, divide the root of a thousand, found as I have taught above[702], which is thirty-one and five eighths, by the principal; the result, namely one and a half and half a tenth and five eighths of half a tenth, is the root of two and a half.

(**A.282**) You want to know what the root of one and three fifths is.

You will do the following. Put the number from which a fifth is denominated, namely five, as the principal. Multiply one and three fifths by it, and the product once again by five, thus producing forty.

---

[696] We would expect simply: Divide its root, which is seven, by the principal. A reader may have changed the text, as in the next problem.

[697] 'having an eighth of an eighth' would have sufficed.

[698] As a reader will remark below, it would have been simpler to say that we are to take the square of 2, the denominator, and multiply it by the radicand.

[699] Rule *i*.

[700] Here and in what follows for 'approximate root'.

[701] The MS has 'one and three fifths' (see A.282); this is corrected by the second hand (who also recomputed the result in *b*).

[702] Again, rule *i*.

LIBER MAHAMELETH **751**

(*a*) Divide its root, which is six and a third, by the principal; this gives one and a fifth and a third of a fifth, and this is the root you require.

(*b*) Or otherwise. Multiply forty by a hundred and divide the root of the product, found as I have taught above[703], by the product of five into the root of a hundred, namely fifty; the result will be what you require.

(**A.283**) You want to know what the root of three thirteenths is.

You will do the following. Put the number from which a thirteenth is denominated, namely thirteen, as the principal. Multiply its three thirteenths, which is three, by the principal; the result is thirty-nine. Denominate its root, which is six and a fourth, from thirteen; this gives six thirteenths and a fourth of one thirteenth, which is the root you require.

You should know that it is not humanly possible to find the true root[704] of a non-square number.

MULTIPLICATION OF ROOTS

(**A.284**) You want to multiply the root of ten by the root of six.

You will do the following. Multiply six by ten, thus producing sixty. Its root is the result of multiplying the root of six by the root of ten.

$$\begin{array}{ccc} \underline{G} & \Big| & \underline{A} \\ \underline{D} & H & \underline{B} \\ & \underline{Z} & \end{array}$$

(A.284)

The proof of this is the following. Let ten be A, its root, B, six, G, its root, D. We want to know the result of multiplying B by D. Let the multiplication of B by D produce H and the multiplication of A by G produce Z, which is sixty. Then I say that H is the root of Z. This is proved as follows. We know that multiplying B by itself produces A and multiplying B by D produces H. Thus[705] the ratio of B to D is the same as the ratio of A to H. Likewise also, multiplying D by B produces H and multiplying D by itself produces G. Thus the ratio of B to D is the same as the ratio of H to G. Now the ratio of B to D was already the same as the ratio of A to H. Therefore the ratio of A to H is the same as the ratio of H to G. Thus[706] the product of A into G is equal to the product

---

[703] Rule *i* above.

[704] That is, a rational (exact) value of this root.

[705] *Elements* VII.17.

[706] *Elements* VII.19.

752 PART TWO: TRANSLATION, GLOSSARY

of H into itself. Since multiplying A by G produces Z, multiplying H by itself produces Z. But Z is sixty; so H is the root of sixty. This is what we wanted to demonstrate.

(A.285) You now want to multiply eight by the root of ten.

You will do the following. You already know[707] that eight is the root of sixty-four. It is then as if you were to multiply the root of sixty-four by the root of ten. Proceeding as I have taught above[708] will give as the result the root of six hundred and forty.

(A.286) You will again do the same if you want to know of which number three roots of ten, that is, thrice the root of ten, is the root. For you already know that this is the same as three multiplied by the root of ten. Proceed therefore as I have taught above[709], and the result will be what you require.

(A.287) Likewise also if you want to multiply three roots of six by five roots of ten.

You will do the following. Find as I have taught[710] of which number three roots of six is the root; you will find that it is the root of fifty-four. Next, find also of which number five roots of ten is the root; you will find that it is the root of two hundred and fifty. It is then as if you were to multiply the root of fifty-four by the root of two hundred and fifty. Proceeding as we have taught above[711] will give the root of thirteen thousand five hundred, and this is what you wanted to know.

You are to know that when two numbers have to one another the ratio of one square to another, the product of their roots will be rational.[712]

(A.288) For instance, let the two numbers be eight and eighteen.

Their ratio is the same as the ratio of one square number to another square number[713]. Thus the product of the root of the first [*which is eight*] into the root of the other [*which is eighteen*] will be a rational number; this is twelve.

The proof of this is clear. We know[714] that two numbers which have to one another the same ratio as two squares are similar plane numbers. But the product of two similar plane numbers is a square, as Euclid said in the ninth book[715] [*thus the product of the two numbers will be a square*]; its root is therefore rational [*resulting from multiplying the root of one of*

---

[707] If need be, multiplication table in A–III (pp. 607–608).

[708] A.284.

[709] A.285.

[710] A.285–286.

[711] A.284.

[712] The proof follows the example.

[713] 'Indeed, four to nine' noted a reader of MS $\mathcal{A}$ in the margin.

[714] Converse of *Elements* VIII.26.

[715] *Elements* IX.1.

LIBER MAHAMELETH   753

these two numbers by the root of the other][716]. This is what we wanted to demonstrate.

Treat as explained other such instances which might occur, and you will find that this is the way.

CHAPTER ON THE ADDITION OF ROOTS

(*i*) You are to know that when two numbers have to one another the same ratio as two squares, the sum of their roots will be the root of a number.

This is proved as follows. We know[717] that when two numbers have to one another the same ratio as two square numbers their roots are commensurable. Thus, since these roots are commensurable, their sum will be commensurable with each of these roots.[718] But if this is so, then the sum of these roots is rational in square[719]. Therefore, when two numbers have to one another the ratio of one square number to another square number, the sum of their roots will be the root of some number. This is what we wanted to demonstrate.

(*ii*) You are also to know that, contrariwise, thus when two numbers have not to one another the same ratio as one square number to another square number, their roots can neither be added nor be the root of some number[720]. The proof of this is clear from the tenth book of Euclid.[721]

Therefore, when you want to add the roots of two numbers, consider first whether these two numbers have to one another the same ratio as one square number to another square number, in which case the roots can be added. If not, you will just repeat what you wanted to do; thus, if you wanted to add the root of two and the root of five, you would say 'the root of two and the root of five'.

(*i'*) Now when you want to know whether the ratio of one number to another number is the same as the ratio of one square number to another square number, you will multiply these two numbers. If the product is a square, they will have to one another the same ratio as two square numbers; if the product is not a square, these numbers will not have to one another the same ratio as two square numbers. The proof of this is clear from what precedes.[722]

(**A.289**) You want to add the root of two and the root of eight.

---

[716] With such a condensed proof, the presence of these interpolations is hardly surprising.

[717] *Elements* X.9.

[718] *Elements* X.15.

[719] The sum is rational 'in square' (*in potentia* $= \delta\upsilon\nu\acute{\alpha}\mu\epsilon\iota$), that is, its square is, like those of the two roots, rational.

[720] Rather: 'their roots cannot be added in such a way that the sum is the root of some number' (that is, of a rational number). This is correctly formulated below, for the subtraction.

[721] *Elements* X.9 again.

[722] Proof seen in A.288.

754 PART TWO: TRANSLATION, GLOSSARY

You will do the following. You know that the roots of these numbers can be added. For the product of two into eight is sixteen, which is a square; thus the squares of these roots have to one another the same ratio as two square numbers; therefore the roots can be added, that is, (their sum) be the root of a number. Then when you want to know of which number their sum is the root (proceed as follows). Multiply two by eight, thus producing sixteen. Keep in mind two roots of this, that is, twice its root, which is eight. Next, add two and eight, which makes ten. Add this to the two roots of sixteen kept in mind, which is eight; this makes eighteen. Therefore, the root of eighteen is the sum of the root of two and the root of eight.

$$\text{A} \qquad \text{B} \qquad\qquad\qquad \text{G}$$

(A.289)

This is proved as follows. Let the root of two be AB and the root of eight, BG. We want then to know of which number AG is the root. Now we know [723] that the product of AG into itself is equal to the products of AB into itself, BG into itself, and AB into BG, twice. Therefore, when you want to know the square of AG, you will add the square of AB, the square of BG, and twice the product of AB into BG; the sum will be the product of AG into itself.[724] Now the square of AB is two, the square of BG, eight, while taking twice the product of AB into BG is the same as multiplying the square of AB by the square of BG and taking two roots of the product, which is eight. Adding this all will give the square of AG. This is what we wanted to demonstrate.

(A.290) You now want to add the root of six to the root of ten.

You already know that these two roots cannot be added, for the ratio of six to ten is not the same as the ratio of one square number to another square number. Therefore, if you want to add them according to the rule explained, you will not succeed. For multiplying six by ten produces sixty, the root of which is irrational; thus you cannot obtain a rational result from twice this root.[725] Now if you (really) want to add the roots of six and of ten, you will have to say 'the root of the sum of sixteen and the root of two hundred and forty'. So if you want to add the root of six and the root of ten, it will be easier to say 'the root of six and the root of ten' than to say 'the root of the sum of sixteen and the root of two hundred and forty'.

Treat other such instances likewise, and you will find that this is the way.

CHAPTER ON THE SUBTRACTION OF ROOTS

---

[723] By *Elements* II.4 = PE_4.

[724] This is redundant; but see similar passages in A.291, A.298b, A.311, A.317a, A.320a.

[725] The non-square radicand is merely multiplied by a square.

756   PART TWO: TRANSLATION, GLOSSARY

Whenever such a problem occurs, consider whether the two numbers have to one another the ratio of one square number to another square number. If they do, the subtraction can be performed in such a way that the remainder is the root of some number. If not, your answer must be like the question, for doing so will be more convenient.

Treat other such instances likewise, and you will find that this is the way.

CHAPTER ON THE DIVISION OF ROOTS

(A.293) You want to divide the root of ten by the root of three.

You will do the following. Divide ten by three; the root of the result will be what you wanted.

[*This is proved as follows.*] For we know that when some number is divided by another, then taking the root of the result is the same as dividing the root of the dividend by the root of the divisor.[730] [*For instance, let A be divided by B and the result be G, and let the square of A be D, the square of B, H, the square of G, Z. Then I say that dividing D by H will give Z. This is proved as follows. A divided by B gives G; thus the ratio of one to G is the same as the ratio of B to A. So the ratio of the square of one to the square of G is the same as the ratio of the square of B to the square of A. Since the square of one is one, the square of G, Z, the square of B, H, and the square of A, D, the ratio of one to Z is the same as the ratio of H to D. Therefore dividing D by H will give Z. This is what we wanted to demonstrate.*]

$$
\begin{array}{c}
\mathrm{A} \\
\hline
\end{array}
$$

$$
\begin{array}{c}
\mathrm{D} \\
\hline
\end{array}
$$

$$
\begin{array}{c}
\mathrm{B} \\
\hline
\end{array}
$$

$$
\begin{array}{c}
\mathrm{H} \\
\hline
\end{array}
$$

$$
\begin{array}{c}
\mathrm{G} \\
\hline
\end{array}
$$

$$
\begin{array}{c}
\mathrm{Z} \\
\hline
\end{array}
$$

(A.293)

(A.294) You now want to divide ten by the root of five.

Proceed as I have taught above in the chapter on the multiplication of roots.[731] That is, consider of which number ten is the root; namely, a hundred. It is then as if you were to divide the root of a hundred by the

---

[730] This has been implicitly used in approximating roots (rule *ii*, p. 749; A.276 & A.278–283).

[731] A.285.

LIBER MAHAMELETH
757

root of five. Proceeding as explained before[732] will give as the result the root of twenty. This is what you wanted.

(**A.295**) Likewise if you want to divide the root of ten by two. You will multiply two by itself, and it will be as if you were to divide the root of ten by the root of four. You will proceed as has been explained before.

(**A.296**) Likewise also if you want to divide two roots of ten by three roots of six.

You need first to know of which number two roots of ten is the root, namely forty, and similarly of which number three roots of six is the root, namely fifty-four. It is then as if you were to divide[733] the root of forty by the root of fifty-four. Proceeding as has been explained before will give as the result the root of six ninths and two thirds of one ninth.

(**A.297**) You now want to divide the root of six and the root of ten by the root of three.

You will do the following. Divide the root of six by the root of three, then divide the root of ten by the root of three, and add the quotients. The sum will be what you require.

When such a problem occurs, find out (first) whether the two roots can be added in such a way that the sum is the root of some number. If this is the case, add them and divide the result by the divisor;[734] the quotient will be what you want. If they cannot be added, divide each by the divisor and add the quotients [*if they can be added; if not, repeat the roots as they were proposed*].

You will also proceed likewise if there are more than two roots, namely as follows. Add them all if they can be added, or only those which can be, and divide the sum. If they cannot be added, divide each separately and add the quotients [*if you can, all; if you cannot add them all, add those quotients which can be added; if they cannot be added at all, just repeat the roots proposed as they were*].

Treat other such instances likewise and you will succeed.

AGAIN ON DIVISION.[735]

(**A.298**) You want to divide ten by two and the root of three.

(*a*) You will do the following. You know[736] that the sum of two and the root of three forms a binomial. Now the product of any binomial into its apotome produces a rational result.[737] Since the apotome of two and the root of three is two minus the root of three, multiply the sum of two and

---

[732] A.293. So also in the next two problems.

[733] Rather: as if you were to denominate.

[734] Situation of A.293.

[735] The two terms are now in the divisor.

[736] From *Elements* X.36.

[737] The proof is given below.

the root of three by two minus the root of three. The result is one.[738] Let
the ten to be divided be A and the sum of two and the root of three, B. Let
A divided by B give G, which is what is required. Let two minus the root
of three be H, and let the product of two minus the root of three into two
and the root of three be D, that is, one. We have then that the division
of A by B gives G; so G multiplied by B will produce A. Likewise again, H
multiplied by B produces D. Thus the multiplication of G by B produces
A and the multiplication of H by B produces D. Therefore[739] the ratio of
A to D is the same as the ratio of G to H. Since A is the decuple of D,
G is the decuple of H. Then multiplying H by ten produces twenty minus
the root of three hundred, which is what we look for and what we wanted
to prove[740].

$$\text{A} \qquad\qquad\qquad \text{B} \qquad \text{G} \qquad\quad \text{D} \quad \text{H}$$

$$(\text{A.298}a)$$

(**b**) I shall now present a proof whereby it will be demonstrated that the
multiplication of any binomial by the corresponding apotome produces a
rational result. Let the binomial be line AB, and AG and BG the two
terms[741] of which it is composed. It is then clear that AG and GB are
in square only rational and commensurable[742]. Let the greater be AG. I
shall cut from it a part equal to BG, say GD. Then AD is the apotome
of AG and GB. Then I say that the product of AB into AD is rational.
This is proved as follows. We know that DB is bisected at the point G
and that AD is added to it. So[743] the products of AD into AB and DG
into itself are equal to the product of AG into itself. Therefore, when you
want to multiply AB by AD, multiply AG by itself and subtract from the
result the product of DG into itself. Now we know that subtracting the
product of DG into itself from the product of AG into itself leaves a rational
remainder; for the product of each of them into itself being rational, the
remainder is rational. But the remainder is equal to the product of AB
into AD. Therefore the product of AB into AD is rational. This is what
we wanted to demonstrate.

$$\text{A} \qquad\qquad\qquad \text{D} \quad \text{G} \quad \text{B}$$

$$(\text{A.298}b)$$

---

[738] We are now left with calculating $10\left(2 - \sqrt{3}\right)$, and this will be done by
means of a geometrical representation.

[739] *Elements* VII.18.

[740] Geometrical computation. See p. 724, note 584.

[741] Literally: the two names (whence 'binomial').

[742] By definition of a binomial. One of the two terms, though, may be
rational.

[743] *Elements* II.6 $= PE_6$.

LIBER MAHAMELETH

(A.299) You now want to divide the root of ten by two and the root of six.

As you already know, the apotome of two and the root of six is the root of six minus two[744]. Therefore multiplying two and the root of six by the root of six minus two will produce a rational result, namely two. Find then the ratio[745] of the root of ten to two, in the following way. Divide the root of ten by two as we have taught before;[746] the result is the root of two and a half. Multiplying then the root of two and a half by the root of six minus two will produce what is required, namely the root of fifteen minus the root of ten, and this is what you wanted. The proof of this is the same as the previous one, without any difference.[747]

(A.300) You now want to divide the root of ten by the root of two and the root of three.

You will do the following. Multiply the apotome of the root of two and the root of three, which is the root of three minus the root of two, by the root of two and the root of three; this produces one. Dividing the root of ten by one [*in order to determine the ratio of ten to one*][748] and multiplying the result by the root of three minus the root of two will produce the root of thirty minus the root of twenty, which is what is required. The proof of this is clear.[749]

AGAIN.[750]

(A.301) You want to divide ten by two minus the root of three.

You know that two minus the root of three is an apotome and that multiplying it by the corresponding binomial produces a rational result, as we have established above[751]. Now multiplying it by its binomial produces one,[752] and dividing ten by one gives ten. Multiplying this by two and the root of three produces twenty and the root of three hundred, namely what is required.

(A.302) Likewise also if you want to divide the root of ten by the root of five minus the root of three.

Multiplying the root of five minus the root of three by its binomial will produce a rational result, as we have established above, namely two. Thus divide the root of ten by two and multiply the result by the root of five and the root of three. The result will be what you require.

---

[744] Since $\sqrt{6} > 2$.

[745] That is, the amount of the ratio.

[746] A.295.

[747] The procedure here is exactly the same as in A.298a.

[748] The early reader did not realize that here the author is describing the procedure in general terms, irrespective of the numerical values.

[749] A.298.

[750] The divisor is now an apotome.

[751] A.298b. Same reference in A.302.

[752] A.298a.

760 PART TWO: TRANSLATION, GLOSSARY

You will proceed likewise in all similar instances, and the result will be what you require.

CHAPTER ON THE MULTIPLICATION OF ROOTS OF ROOTS

(A.303) You want to multiply the root of the root of seven by the root of the root of ten.

You will do the following. Multiply seven by ten, thus producing seventy. The root of the root of this is what you wanted.

The proof of this is the following. Let ten be A, its root, B, the root of B, G —thus G is the root of the root of A; let seven be D, its root, H, the root of H, Z —thus Z is the root of the root of D. Let A multiplied by D produce K, which is seventy, B by H produce T, and G by Z produce Q. Then I say that Q is the root of the root of K. This is proved as follows. We know that T is the root of K.[753] It will also be demonstrated in the same way that Q is the root of T; for G is the root of B and Z is the root of H; now multiplying B by H produces T, and multiplying G by Z produces Q; thus Q is the root of T. But T was the root of K; so Q is the root of the root of K. This is what we wanted to demonstrate.

A	B	G
D	H	Z
K	T	Q

(A.303)

(A.304) You want to multiply the root of ten by the root of the root of thirty.

You already know that the root of ten is the root of the root of a hundred. Thus multiply the root of the root of a hundred by the root of the root of thirty; the result will be the root of the root of three thousand.

(A.305) Likewise also if you want to multiply five by the root of the root of ten.

For you know [754] that five is the root of the root of six hundred and twenty-five. It is then as if you were to multiply the root of the root of six hundred and twenty-five by the root of the root of ten. Proceed according to what has been explained before;[755] the result will be the root of the root of six thousand two hundred and fifty, and this is what you wanted.

---

[753] A.284.

[754] This must simply mean that what follows is obvious, like the Arabic *qad ʿalimnā* in similar contexts (see our *Introduction to the History of Algebra*, p. 157, n. 171).

[755] A.303.

LIBER MAHAMELETH   761

AGAIN.

(**A.306**) You want to multiply three roots of the root of ten by two roots of the root of six.

You will triple the root of the root of ten in the following way. You will multiply three by the root of the root of ten as we have taught before,[756] thus producing the root of the root of eight hundred and ten. Next, you will double the root of the root of six, thus producing the root of the root of ninety-six. It is then as if you were to multiply the root of the root of eight hundred and ten by the root of the root of ninety-six. You will multiply eight hundred and ten by ninety-six, and the root of the root of the product will be what you wanted.

Note that when you multiply the root of the root of some number by the root of the root of another number, the result will always necessarily be (i) a number, or (ii', ii'') the root of a number, or (iii) the root of the root of a number.

(i, ii') When you multiply the root of the root of some number by the root of the root of another number and the ratio the roots of these numbers have to one another is the same as the ratio of a [*square number to a square number, the roots of the root are commensurable in length and the result of their multiplication will be a number. But if the ratio the roots of these numbers have to one another is the same as the ratio of a*][757] non-square number to a non-square number, the result of their multiplication will only be either (i) a number or (ii') the root of a number.

For if the ratio of the root of a number to the root of another number is the same as the ratio of a non-square number to a non-square number, the roots of the roots of these numbers will be incommensurable in length but commensurable in square. Now if two quantities are such, their multiplication will produce a result which is either rational or medial; Euclid has established this in the tenth book, saying that any area contained by two medial lines commensurable in square only will be either rational or medial[758]. To illustrate this, I shall present two examples.

— The first is the following.[759]

(**A.307**) For instance, you want to multiply the root of the root of three by the root of the root of twenty-seven.

These are two medial quantities commensurable in square only. For the ratio of the square of one of them, namely the root of three, to the square of the other, namely the root of twenty-seven, is the same as the ratio of three to nine; they are therefore commensurable in square but incommensurable in length. Now the result of their multiplication is rational, namely three.

---

[756] A.305.

[757] Added by the second hand of MS $\mathcal{A}$ in the margin. (Should be ignored.)

[758] *Elements* X.25.

[759] Example of i.

762 PART TWO: TRANSLATION, GLOSSARY

— The second example is the following.[760]

(A.308) For instance, you want to multiply the root of the root of eight by the root of the root of eighteen.

Here again[761] it is clear that these quantities are medial and commensurable in square only. For the ratio of the square of one of them to the square of the other is the same as the ratio of six to nine; they are therefore commensurable in square, since the ratio their squares have to one another is the same as the ratio of a number to a number, but incommensurable in length. The result of their multiplication is medial, namely the root of twelve.

(*iii*) When you multiply the root of the root of some number by the root of the root of another number and the ratio the two numbers[762] have to one another is not the same as the ratio of a square number to another square number, the result of their multiplication must always be the root of the root of some number.

For when the two numbers are multiplied, the result will be a non-square number, as we have already established in what precedes.[763] To illustrate this, we present hereafter such an example.

(A.309) For instance, you want to multiply the root of the root of ten by the root of the root of fifteen. Do as I have taught above[764]. The result will be the root of the root of a hundred and fifty.

(*ii″*) When you multiply the root of the root of some number by the root of the root of another number and the ratio the roots of these numbers have to one another is the same as the ratio of one square number to another square number, the result of their multiplication will always be the root of some number.

The proof of this is evident. We know that when the ratio of the root of one number to the root of another number is the same as the ratio of a square number to a square number, then the ratio of the root of the root of one of these numbers to the root of the root of the other number will be the same as the ratio of a number to a number. Thus the root of the root of one of these numbers will be commensurable with the root of the root of the other. But when one medial is commensurable with another medial, the result of their multiplication must be medial.[765] The proof of this is

---

[760] Example of *ii′*.

[761] As in the previous problem.

[762] Unlike in cases *i* and *ii′*, here we consider the ratio of the two numbers and not of their square roots.

[763] A.288 and rule *ii* in the addition of the square roots (irrationality of the mixed term).

[764] A.303.

[765] Note here the ambiguity of 'medial' for numerical quantities: one is a fourth root, the other a square root.

LIBER MAHAMELETH — 763

evident to those who read the tenth book of Euclid.[766] For the sake of greater clarity, though, we present hereafter such an example.

(**A.310**) For instance, you want to multiply the root of the root of two by the root of the root of thirty-two. Do as I have taught above.[767] The result will be the root of eight, and this is what you wanted.

CHAPTER ON THE ADDITION OF ROOTS OF ROOTS

(*i*) You are to know that when you want to add the root of the root of one number to the root of the root of another number and the ratio of the square of one of them, thus the root of one of these numbers, to the square of the other, thus the root of the other number, is the same as the ratio of a square number to a square number, then the result of their addition will always be the root of the root of some number.

The proof of this is evident. We know that when the ratio of the square of one of them to the square of the other is the same as the ratio of a square number to a square number, then the ratio of one of them to the other will be the same as the ratio of some number to another; Euclid has already established this in the tenth book[768]. Now if the ratio of one of them to the other is the same as the ratio of some number to another, they will be commensurable; this has likewise been established in the tenth of Euclid[769]. But if they are commensurable, their sum will be commensurable with each; this has likewise been already established by Euclid[770]. Since each is medial, their sum will be medial; for everything commensurable with a medial is medial, as Euclid has likewise demonstrated[771]. For the sake of greater clarity, though, I shall present hereafter an example and explain the rule for addition.

(**A.311**) For instance, you want to add the root of the root of three to the root of the root of two hundred and forty-three.

$$A \qquad B \qquad\qquad\qquad G$$

(A.311)

Let the root of the root of three be AB and the root of the root of two hundred and forty-three be BG. We want to know of which number the whole AG is the root of the root. We know[772] that the product of AG into itself is equal to the products of AB into itself, BG into itself and AB into BG, twice. So when you want to know the amount of the product of

---

[766] *Elements* X.24.

[767] A.303.

[768] *Elements* X.9.

[769] *Elements* X.6.

[770] *Elements* X.15.

[771] *Elements* X.23.

[772] *Elements* II.4 $= PE_4$.

764   PART TWO: TRANSLATION, GLOSSARY

AG into itself, you will multiply AB by itself and BG by itself, and add to their sum twice the product of AB into BG. Now we know that the square of AB is the root of three and that the square of BG is the root of two hundred and forty-three. Then add the root of three to the root of two hundred and forty-three. For these roots can certainly be added and their sum be made the root of some number.[773] Indeed, the ratio of one of them to the other is the same as the ratio of some number to another [*and they are commensurable*]; thus their sum will be the root of some number, namely of three hundred. Add now this root of three hundred to twice the product of the root of the root of three into the root of the root of two hundred and forty-three, which is the root of a hundred and eight. It is certain that the root of a hundred and eight can be added to the root of three hundred and that the result will be the root of some number, for these two roots are commensurable; this will always be so: when some quantity [*a line*] is divided into two commensurable parts, the sum of their squares will be commensurable with their product taken twice; the proof of this is evident to those who know the tenth book of Euclid.[774] So add the root of a hundred and eight to the root of three hundred; the result will be the root of seven hundred and sixty-eight. The root of this, which is the root of the root of seven hundred and sixty-eight, is what you require. This is what we wanted to demonstrate.

(*ii*) If now the ratio of the square of one of them[775] to the square of the other is the same as the ratio of a non-square number to a non-square number while their multiplication produces a square[776], then the result of their addition will be the root of the sum of some number and the root of a number, and this is a quantity producing a rational and a medial, as Euclid has demonstrated[777]. For the sake of greater clarity I shall present hereafter an example.

(**A.312**) For instance, you want to add the root of the root of three to the root of the root of twenty-seven.

Do as I have taught above,[778] namely as follows. Add the root of three

---

[773]   A.289 and preceding rule *i*.

[774]   *Elements* X.19.

[775]   As in *i*, we consider the roots of the roots of the given quantities, and form the ratio of their squares, thus of square roots.

[776]   The manuscript has 'their (*earum*, thus the fourth roots) multiplication produces a rational result'. We have changed the text to 'their (*eorum*, thus the squares of the fourth roots) multiplication produces a square', for as we have in *iii* below the two conditions $\sqrt{a} = t \cdot \sqrt{b}$ and $\sqrt{a} \cdot \sqrt{b}$ not a square, we should have here $\sqrt{a} = t \cdot \sqrt{b}$ and $\sqrt{a} \cdot \sqrt{b}$ a square.

[777]   *Elements* X.40. A line 'produces' (*potest supra*, δύναται) an area when the square on it is equivalent to this area. For quantities, it means that the first, that which 'produces' the other, is the square root of this other.

[778]   A.311.

LIBER MAHAMELETH

to the root of twenty-seven;[779] this will make the root of forty-eight. Add to it twice the product of the root of the root of three into the root of the root of twenty-seven, which is six; this will make six and the root of forty-eight. The root of this sum is what you require, which is the root of the sum of six and the root of forty-eight.

(*iii*) But if there results from their multiplication a non-square number, then the resulting sum will be the root of the sum of the root of a number and the root of a number, which is a quantity producing two medials, as Euclid has demonstrated in the tenth by an incontrovertible proof[780]. For the sake of greater clarity, though, I shall present hereafter such an example.

(**A.313**) For instance, you want to add the root of the root of eight to the root of the root of eighteen.

Add them as we have taught above.[781] That is, add the root of eight to the root of eighteen; this will make the root of fifty. Next, multiply the root of the root of eight by the root of the root of eighteen,[782] (and take the result) twice; this will produce the root of forty-eight. You are to add it to the root of fifty. But in no way can their sum become the root of some number.[783] You will therefore have to say 'the root of the sum of the root of forty-eight and the root of fifty'. This is what you wanted.

(*iv*) When you want to add the root of the root of some number to the root of the root of another number and the ratio of one of these numbers to the other is not the same as the ratio of a square number to a square number, then these roots cannot be added and their sum can only be the root of the sum of the root of some number and the root of the root of another number. [*The proof of this is evident from the tenth of Euclid.*][784] If such a problem occurs, the resulting sum will merely reproduce the disposition given in the statement.

CHAPTER ON THE SUBTRACTION OF ROOTS OF ROOTS

You are to know that subtracting roots of roots from roots of roots is analogous to adding roots of roots to roots of roots, just as we have taught above that subtracting roots from roots is analogous to adding roots together[785]. Therefore those who are acquainted with what has been said about the addition of roots of roots, and also with the application of the tenth book of Euclid thereto, will by such means be able to attain the science of subtracting roots of roots; but when subtracting roots of roots

---

[779] A.289.

[780] *Elements* X.41.

[781] A.311–312.

[782] A.308.

[783] A.290 (and preceding rule *ii*).

[784] This looks like the gloss of a reader missing the numerical examples given in the previous cases.

[785] A.291–292 and preceding rules.

766 PART TWO: TRANSLATION, GLOSSARY

they will have to consider apotomes wherever for the addition we have considered binomials.

CHAPTER ON THE DIVISION OF ROOTS OF ROOTS

(A.314) You want to divide the root of the root of ten by the root of the root of five.

You will do the following. Divide ten by five; the root of the root of the result, which is the root of the root of two, is what you wanted.

The proof of this is evident. We know that when some number is divided by another, then taking the root of the result is the same as dividing the root of the dividend by the root of the divisor[786]. Then when you want to divide the root of the root of ten by the root of the root of five, you will divide the square of the root of the root of ten, which is the root of ten, by the square of the root of the root of five, which is the root of five; the result will be the root of two, as we said before[787]; the root of this, namely the root of the root of two, is what you wanted.

(A.315) You now want to divide two roots of the root of ten by three roots of the root of eight.

You will do the following. You need first to know of which number two roots of the root of ten is the root of the root, as follows. Multiply two by the root of the root of ten as we have explained before[788]; the result will be the root of the root of a hundred and sixty. Next, you need to know of which number three roots of the root of eight is the root of the root, as follows. Multiply three by the root of the root of eight; the result will be the root of the root of six hundred and forty-eight. It is then as if you were to divide the root of the root of a hundred and sixty by the root of the root of six hundred and forty-eight. You will divide, using denomination, a hundred and sixty by six hundred and forty-eight, and the root of the root of the result will be what you wanted.

(A.316) You now want to divide the root of the root of ten by two.

You will do the following. You need first to know of which number two is the root of the root; you will find that this number is sixteen. It is then as if you were to divide[789] the root of the root of ten by the root of the root of sixteen. Proceed as we have taught before[790]. The result will be the root of the root of five eighths.

Treat other such instances likewise, and you will find that this is the way.

ON THE SAME TOPIC.[791]

---

[786] A.293.

[787] Computation above.

[788] A.305.

[789] As above, 'using denomination'.

[790] A.314, with denominating.

[791] Now two terms in the divisor.

LIBER MAHAMELETH

(A.317) You want to divide ten by two and the root of the root of three.

(a) First the following must be known. (*i*) If some number and the root of the root of a number are multiplied by the corresponding apotome, the result will be an apotome. (*ii*) If the root of a number and the root of the root of a number are multiplied by the corresponding apotome, the result will also be an apotome. (*iii*) If the root of the root of a number and the root of the root of a number are multiplied by the corresponding apotome, the result will also be an apotome. I shall present the proof for one of these cases, by means of which the other two cases may be ascertained.[792]

A D B G

(A.317*a*)

Then I say that if the root of the root of some number and the root of the root of another number are multiplied by the corresponding apotome, which is the root of the root of the larger number diminished by the root of the root of the smaller number, the result must be an apotome. So let the root of the root of the larger number be AB and the root of the root of the smaller number, BG. I shall cut from AB a part equal to BG, say DB; then AD is the apotome of the two given numbers. Then I say that the multiplication of AG by AD produces an apotome. This is proved as follows. We know [793] that the products of AG into AD and DB into itself are equal to the product of AB into itself. So if you want to multiply AD by AG, you will multiply AB by itself and subtract from the result the product of DB into itself, and this will leave the product of AG into AD. Now we know that the product of AB into itself is the root of a number and that the product of DB into itself is the root of a number. Thus the product of AD into AG is the root of a number diminished by the root of a number, which is an apotome. This is what we wanted to demonstrate.

Only in this case may the multiplication lead to the root of some number, namely when the roots of the two numbers are commensurable: for then one can be subtracted from the other so as to leave the root of a number [*then the remainder will be the root of a number, which remainder is thus the product of the root of the root of a number and the root of the root of another number into the corresponding apotome*].

In the other cases [794] the remainder can only be an apotome. For the given binomial will be either a number and the root of the root of a number or the root of a number and the root of the root of a number. Then if it is a number and the root of the root of a number, multiplying the number by itself will produce a number while multiplying the root of the root of the number by itself will produce the root of a number; now

---

[792] He chooses case *iii*.

[793] By *Elements* II.6 = PE_6 (or A.298*b*).

[794] That is, if the two roots in case *iii* are not commensurable and in the two remaining cases *i* and *ii*, explicitly mentioned in what follows.

it is possible neither to subtract the root of a number from a number nor to subtract a number from the root of a number, which is clear since they are not commensurable.[795] And if it is the root of a number and the root of the root of a number, it will likewise be impossible as well since the multiplication of the root of a number by itself can only produce a number.

(**b**) You now want to divide ten by two and the root of the root of three. You will do the following. Multiply two and the root of the root of three by two minus the root of the root of three; the result will be, as we have said before,[796] an apotome, namely four minus the root of three. Divide ten by this as we have taught before[797]; the result will be, in thirteenths, forty and the root of three hundred. Multiplying this by two minus the root of the root of three will give what you require.

(**A.318**) You want to divide ten by the root of the root of three and the root of the root of twelve.

You will do the following. Multiply the root of the root of twelve and the root of the root of three by the root of the root of twelve minus the root of the root of three; the result will be rational in square, namely the root of three[798]. Divide ten by this;[799] the result will be the root of thirty-three and a third. Multiplying it by the root of the root of twelve minus the root of the root of three will give what you require.

(**A.319**) You want to divide ten by the root of five minus the root of the root of two.

You will do the following. Multiply the root of five minus the root of the root of two by the corresponding binomial, which is the root of five and the root of the root of two; the result will be, as said before[800], an apotome, namely five minus the root of two. Divide ten by this,[801] and multiply the result by the root of five and the root of the root of two. The result will be what you require.

Treat other such instances likewise, and you will find that this is the way.

AGAIN ON THE DIVISION OF ROOTS.[802]

(**A.320**) You want to divide ten by two and the root of three and the root of ten.

(**a**) You are first to know that when three terms[803] are given in the above manner, subtracting one of them from the two others and multiplying the

---

[795] The root cannot be subtracted from the (rational) number so as to leave a single term.

[796] Case *i* in *a* above.

[797] A.301.

[798] Case *iii* above, with commensurable terms.

[799] A.294.

[800] Case *ii* above.

[801] A.301.

[802] A.320–322 would better fit after A.302.

[803] *tria nomina* (a 'trinomial').

sum of the three by the remainder will produce an apotome, or a binomial, or a rational in square.

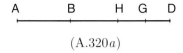

(A.320a)

Let then the trinomial be AD, its first term, AB, the second, BG, the third, GD. Then I say that subtracting one term of AD from the other two and multiplying AD by the remainder will produce an apotome, or a binomial, or a rational in square. I shall cut from AG a part equal to GD, say HG, thus leaving AH. Then I say that multiplying AH by AD will produce an apotome, or a binomial, or a rational in square. This is proved as follows. Line HD is bisected at the point G and line AH is added to it. Thus [804] the products of AH into AD and HG into itself are equal to the product of AG into itself. Therefore, if you want to multiply AH by AD, you will multiply AG by itself and subtract from the result the product of HG into itself; this will leave the product of AD into AH. But we know that multiplying AG by itself produces a first binomial; for when any binomial is multiplied by itself the result is a first binomial, as can be shown from the tenth of Euclid [805]. Now a first binomial is a number and the root of a number. We also know that multiplying HG by itself produces a number. Then subtract this number from the first binomial, which is a number and the root of a number. (*i*) If the number you subtract is smaller than the number in the first binomial, the remainder is a number and the root of a number, which is a binomial. (*ii*) If the number you subtract is larger than the number in the first binomial, the remainder will be the root of a number minus a number, which is an apotome. (*iii*) If the numbers are equal, the remainder will be the root of a number, which is a quantity rational in square. This is what we wanted to demonstrate.

(***b***) These preliminaries now being known, you want to divide ten by two and the root of three and the root of ten. You will do the following. Multiply two and the root of three and the root of ten by two and the root of three minus the root of ten; the result will be the root of forty-eight minus three, which is an apotome.[806] Dividing ten by this and multiplying the quotient by two and the root of three minus the root of ten will give what you require. [*The proof of this has already been given*.]

(**A.321**) Likewise also if you want to divide the root of ten by the root of six and the root of seven and the root of eight: you will do the same here again. That is, you will multiply the root of six and the root of seven and the root of eight by the root of six and the root of seven minus the root of eight; the result will be five and the root of a hundred and sixty-eight,

---

[804] *Elements* II.6 = PE_6.
[805] *Elements* X.54. See also below, A.323a (*i*).
[806] The subsequent interpolation refers to this.

770 PART TWO: TRANSLATION, GLOSSARY

which is a binomial. Dividing the root of ten by this as explained before[807] and multiplying the quotient by the root of six and the root of seven minus the root of eight will give what you require.

(A.322) You want to divide ten and the root of fifty by the root of two and the root of three and the root of five.

You already know[808] that dividing the sum of ten and the root of fifty by the sum of the root of two and the root of three and the root of five is the same as dividing ten by the root of two and the root of three and the root of five, and dividing the root of fifty by the root of two and the root of three and the root of five, and adding the results of these two divisions. Divide therefore ten by the root of two and the root of three and the root of five as we have taught before[809]; again, divide the root of fifty by the root of two and the root of three and the root of five;[810] and add the results of the two divisions.

Treat other such instances likewise, and you will find that this is the way.

Again on roots[811]

(A.323) You want to know what the root of the sum of eight and the root of sixty is.

(*a*) Consider, in this kind of problem, the quantity of which the root is required.

(*i*) If it is a first binomial the root found will always be a binomial. The proof of this is evident from the tenth of Euclid[812], where it is said that when some area is contained by a first binomial and a rational line, the line producing it is a binomial[813].

(*ii*) If the quantity of which the root is required is a second binomial[814], its root will be a first bimedial. For when an area is contained by a rational line and a second binomial, the line producing it [*whatever it is*] is a first bimedial.
Likewise also:

(*iii*) If it is a third binomial, its root will be a second bimedial.

(*iv*) If it is a fourth binomial, its root will be a major line.

---

[807] A.301–302.

[808] A.297.

[809] A.320–321.

[810] As taught in A.321.

[811] Taking the square root of binomials and apotomes.

[812] *Elements* X.54 (and X.55–59 for the next cases).

[813] 'producing': see p. 764, note 777.

[814] The manuscript has, like Euclid and as below, 'is contained by a second binomial and a rational line'. Remember that the root of a quantity is, geometrically, the side of a square equivalent to an area having as its sides 1 and the quantity considered.

(*v*) If it is a fifth binomial, its root will be a quantity producing a rational and a medial.

(*vi*) If it is a sixth binomial, its root will be a quantity producing two medials.

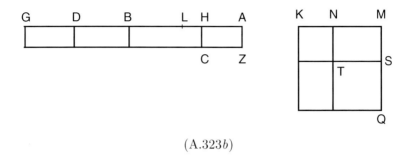

(A.323*b*)

(**b**) I shall indicate the method by which all these cases may be solved. The example will be our above one, namely: You want to know what the root of the sum of eight and the root of sixty is.

You already know that the sum of eight and the root of sixty is a first binomial: eight being greater than the root of sixty, the larger term is thus commensurable with a rational line and the smaller incommensurable with a rational line, but eight produces [*the root of*] sixty by means of the square of two, that is to say by means of the addition to sixty of the square of a line commensurable with it; so [815] the sum of eight and the root of sixty is a first binomial. Thus its root is a binomial.[816] Let therefore eight be AB, the root of sixty, BG, and line AZ, one. Then the area ZG is the sum of eight and the root of sixty. We want to know its root. Let BG be bisected at the point D. So BD is the root of fifteen. I shall now add on line AB an area equal to the square of line BD in such a way that the line will be incomplete by a square area equal to the product of HB into itself.[817] Then the product of AH into HB is fifteen, while AB is eight. So eight is divided into two parts the product of which is fifteen.[818] Let thus AB be bisected at the point L. So[819] the products of AH into HB and HL into itself are equal to the product of AL into itself. Since the product of AL into itself is sixteen and the product of AH into HB, fifteen, this leaves one as the product of HL into itself. Thus HL is one. Since AL is four, AH is three, and line HB will be five.[820] We have then that line AH is

---

[815] *Elements*, def. 1 following X.47.
[816] Above, *i*.
[817] *Elements* VI.28 (here rectangular area). The line is incomplete by the side of the square area.
[818] Sum and product of AH and HB thus known.
[819] *Elements* II.5 = PE_5.
[820] From now on, the word 'line' regularly occurs, probably because a pair of letters may also designate an area.

three and AZ, one; therefore the area ZH is three. Likewise again, line HB is five and CH is one, therefore the area CB is five. Let there now be a square area equal to the area ZH, say KT, and again another square area equal to the area CB, say TQ.[821] Thus the area KT is three and the area TQ, five. I shall now complete the area KQ. It is then clear that the area KQ is equal to the area ZG; Euclid has provided the proof of this in the tenth book[822]. Since, as we know, the area KT is three, line KN is the root of three. But the area TQ is five; so line NM, which is equal to the line TS, is the root of five. Therefore the whole line KM is [*the root of the area ZG, which area is produced by*] the root of three and the root of five [*and this is a binomial*][823]. This is what we wanted to demonstrate.

AGAIN ON THE SAME TOPIC.

(*a*) Likewise also if a problem about apotomes occurs: you will treat it in the same manner as for binomials.

(**A.324**) For instance, you want to know what the root of two hundred and twenty-five minus the root of fifty thousand is[824].

(*i*) You are to know that this[825] is a first apotome.[826] Therefore the line producing it is an apotome. For when some area is contained by a rational line and a first apotome, the line producing it is an apotome.

(*ii*) If we have a second apotome, the line producing it will be a first apotome of a medial.

(*iii*) If it is a third, the line producing it will be a second apotome of a medial.

(*iv*) If it is a fourth, the line producing it will be a minor.

(*v*) If it is a fifth, the line producing it will be the line which together with a rational produces a medial whole.

(*vi*) If it is a sixth, the line producing it will be the line which together with a medial produces a medial whole.

The proofs of all these cases have already been provided in the tenth of Euclid[827].

(*b*) I shall now explain how to find the root of two hundred and twenty-five minus the root of fifty thousand, and this will serve to investigate the other

---

[821] *Elements* II.14.

[822] *Elements* X.54.

[823] The aim here was just to provide a computation with the help of a geometrical construction. In any event both content and wording suggest that these are interpolations.

[824] The Latin wording is ambiguous but, for mathematical reasons (225 a square), it can only be understood as $\sqrt{225 - \sqrt{50\,000}}$.

[825] Refers to the quantity proposed.

[826] It obeys the same conditions as above but with a second term subtractive.

[827] *Elements* X.91–96.

cases. Let then two hundred and twenty-five be line AB, the root of fifty thousand, the part BG of it —thus AG is two hundred and twenty-five minus the root of fifty thousand— and line AZ, one. I shall construct the area ZB, and the area ZG which is two hundred and twenty-five minus the root of fifty thousand. Next, I shall bisect GB at the point D. It is then clear that GD is the root of twelve thousand five hundred. I shall then add on AB an area equal to the square of line GD in such a way that the line will be incomplete by a square area, equal to the product of AH into itself.[828] It is then clear that the product of AH into HB is twelve thousand five hundred, while AB is two hundred and twenty-five. So two hundred and twenty-five is divided into two parts the product of which is twelve thousand five hundred.[829] Proceed then as taught in algebra.[830] The result will be a hundred as HB. But AZ is one, and AZ is equal to BN. So the area NH is a hundred, and the area ZH, a hundred and twenty-five. Let there be a square area equal to the area ZH, say QK,[831] on the diagonal of which let there be an area equal to the area NH, say KT; I shall now draw the two lines PC and MX. It is then clear that the area QT is equal to the area ZG; Euclid has already demonstrated this[832]. Therefore the line producing the area ZG, thus the root of two hundred and twenty-five minus the root of fifty thousand, is CT. We want to know its value. Now it has been shown that the area KT is a hundred[833]; so line TP is ten. Since the area QK is a hundred and twenty-five, line LK, which is equal to line CP, is the root of a hundred and twenty-five. Therefore line CT is the root of a hundred and twenty-five minus ten. This is what we wanted to demonstrate.

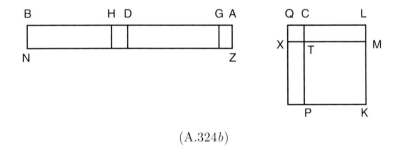

(A.324b)

You will do the same in all such instances, and you will find that this is the way.

---

[828] *Elements* VI.28 (here rectangular area).
[829] Sum and product of AH and HB thus known.
[830] The text reads *sicut docetur* (not, as later on, *docuimus* or *ostensum est*) *in agebla*. The geometrical calculation, based on *Elements* II.5, has been seen in the previous problem.
[831] *Elements* II.14.
[832] *Elements* X.91.
[833] For, by construction, the areas KT and NH are equal.

774   PART TWO: TRANSLATION, GLOSSARY

We have thus explained how to deal with the first apotome and the first binomial[834]. The procedure is the same for the other cases. I shall present only one such problem, which will serve to investigate the remaining such cases.[835]

(A.325) Someone asks: What is the root of ten and the root of a hundred and eighty?

I shall show here that the method is the same as for the first binomial, namely as follows. Multiply ten by itself, thus producing a hundred. Always take a fourth of it, which is twenty-five. Next, divide the root of a hundred and eighty into two parts the product of which is twenty-five as we have taught above,[836] that is, as follows. Take half the root of a hundred and eighty,[837] which is the root of forty-five. Multiply it by itself, thus producing forty-five. Subtract twenty-five from this, which leaves twenty. Adding the root of this, which is the root of twenty, to the root of forty-five, which is half the root of a hundred and eighty, will give the larger part, namely the root of twenty and the root of forty-five. With them, addition and subtraction are possible.[838] Thus, add the root of twenty to the root of forty-five; this makes the root of a hundred and twenty-five. Subtract also the root of twenty from the root of forty-five; this leaves the root of five. Then take the root of the root of five, and the root of the root of a hundred and twenty-five; (together) they are the root of the root of five and the root of the root of a hundred and twenty-five. This is a first bimedial, for the product of the root of the root of five into the root of the root of a hundred and twenty-five is rational, namely five.[839] And since ten and the root of a hundred and eighty is a second binomial, its root is a first bimedial.[840] We have thus illustrated Euclid's assertion.

The proof of all that has been said above is evident from what has been explained for the first binomial and does not differ in any way. It will be the same for the third binomial as well: the root of a third binomial will be found by the same method [*a second bimedial, as Euclid said*[841]]. Moreover, those who have understood the method for finding the root of a first apotome and the corresponding proof will understand the method for finding the root of a second apotome, just as we have shown the method for dealing with a second binomial from the method for a first binomial.[842] You will learn in the same way also how to deal with a third apotome.

---

[834] In the last two problems.

[835] The given expression will be a second binomial (which we are not told until the end of the problem).

[836] A.323 (pure calculation here).

[837] A.295.

[838] A.289 and preceding rule *i*.

[839] *Elements* X.37.

[840] Rule *ii*, p. 770.

[841] *Elements* X.56 (or rule *iii* in A.323).

[842] A.323 & A.325.

LIBER MAHAMELETH

But if some problem concerns the fourth, fifth and sixth binomial, your answer should be like the question. For if you were to determine its root in the way we have taught for the first, second and third binomial, the result would be, for the root of a fourth binomial, a major. Now a major is nothing but the root of the sum of a number and the root of a number, plus the root of this (first) number diminished by the root of the second. Therefore it is easier to say 'the root of the sum of a number and the root of a number' than to say 'the root of the sum of a number and the root of a number, added to the root of the (first) number diminished by the root of the other number'.

(A.326) For instance, someone asks thus: What is the root of ten and the root of eighty?

Say 'the root of the sum of ten and the root of eighty'; for this is a fourth binomial. If now we were to proceed here according to the method which we have indicated for the first, second and third binomial, the result would be the root of the sum of five and the root of five, plus the root of five diminished by the root of five. Therefore it is easier to say 'the root of the sum of ten and the root of eighty' than to say 'the root of the sum of five and the root of five, plus the root of five diminished by the root of five'.

Likewise, if someone were to state a problem regarding a fifth binomial, your answer should be like the question. For it is easier to say this, namely 'a line producing a rational and a medial'. Likewise again for the sixth binomial.

Likewise again if you were to ask about a problem concerning a fourth, fifth or sixth apotome. In all of them your answer should always be as was the question; for this is easier and nearer to hand.

Treat other such instances likewise, and you will find that this is the way.

Beginning of the Second part

We return here to the subject of four proportional numbers and what is inferred from them, although we have often operated with them before.[843]

(*i*) For instance, if there are four proportional numbers, which is to say that as the first is to the second so is the third to the fourth, then the first multiplied by the fourth produces the same result as the second by the third.[844] Of such four numbers the first and the fourth, and the second and the third, are called 'associates'. Thus,[845] generally, if any of them is unknown, it will be determined by dividing either of the other pair by the unknown number's associate and multiplying the quotient by the associate of the dividend; or by dividing the product of the other pair by the unknown number's associate.[846] Suppose, therefore, that three of them are given: if only[847] the fourth is unknown, it will be determined by multiplying the second by the third and dividing the product by the first; if only the first is unknown, it will be determined by multiplying the second by the third and dividing the product by the fourth; if only the second is unknown, it will be determined by multiplying the first by the fourth and dividing the product by the third; if the third is unknown, it will be determined by multiplying the first by the fourth and dividing the product by the second.

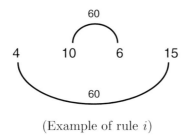

(Example of rule *i*)

To make clear what we say, let there be four proportional numbers, say four, ten, six and fifteen. The ratio of four to ten is the same as the ratio of six to fifteen. From the multiplication of four by fifteen results sixty,

---

[843] But in Book B the numbers have a concrete signification and the proportions serve to determine the unknown ones.
[844] *Elements* VII.19.
[845] What follows to the end of the paragraph (Rule of Three) is also found in the *Liber algorismi*, pp. 98–99.
[846] This is commented on in rule *ii* below.
[847] Case in which just one of the four quantities is unknown; we shall meet the case where both terms on the same side of the proportion are unknown from B.15 on (problems 'involving unknowns').

and this also results from the multiplication of six by ten. If the fourth of them is unknown, multiply the second by the third, thus producing sixty, and divide this by the first, and the result will be fifteen; if the first is unknown, divide sixty by the fourth, which is fifteen, and the result will be the first. If the third is unknown, multiply the first by the fourth, thus producing sixty, and divide this by the second, which is ten, and the result will be the third, namely six; if the second is unknown, divide sixty by the third, which is six, and the result will be the second.

(*ii*) For any three numbers, whether you multiply the first by the second and divide the product by the third or divide either of the factors by the divisor and multiply the quotient by the other factor, it amounts to the same.[848]

For instance, let there be three numbers: four, ten and six. Six multiplied by ten produces sixty, and when this is divided by four, the result is fifteen; the two numbers multiplying one another are six and ten, and the divisor is four.

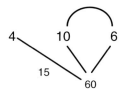

(First example of rule *ii*)

If now we divide six by four[849] and multiply the quotient by ten, the result will also be fifteen.

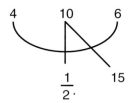

(Second example of rule *ii*)

If we divide ten by four[850] and multiply the quotient by six, this will again give the same result, namely fifteen.

Understand this, for it will be very useful in what is to follow on buying and selling, and on many other occasions.

---

[848] Premiss P_5'. But this equivalence has been noted and applied several times before, quite clearly in A.155. See also note 852.
[849] The result, $\frac{1}{2\cdot} = 1 + \frac{1}{2}$, is in the figure.
[850] The $\frac{2}{2\cdot}$ ($= 2 + \frac{1}{2}$) in the figure.

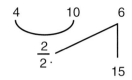

(Third example of rule *ii*)

Chapter (B–I) on Buying and selling

When, in buying and selling, it is asked about something 'what is its price?', you will do the following. Multiply the middle (term) by the last and divide the product by the first; the result will be what is required. Or divide the middle by the first and multiply the result by the last. Or divide the last by the first and multiply the result by the middle.[851] In all these ways is found the unknown which is required.[852]

But when it is asked about something 'what is my part?' or 'how much shall I have', you will do the following. Multiply the first (term) by the last and divide the product by the middle; the result will be what is required. Or divide the first by the middle and multiply the result by the last. Or divide the last by the middle and multiply the result by the first. In all these ways is found what is required.

In order to make clearer what we say, we shall put an example for each of these two cases.

— An example of the first case, that is, 'what is its price', is the following.

(B.1) For instance, someone asks: Three sextarii are given for ten nummi. What is the price of fourteen sextarii?

This is found in three ways. The ratio of the first sextarii to their price is the same as that of the second to their price. These are four proportional numbers, of which the first is three, the second, ten, the third, fourteen, while the fourth is the unknown which is required. Therefore multiply the second by the third and divide the product by the first; the result will be the fourth. Or divide either of the two factors by the first, which is the divisor, and multiply the quotient by the other factor; the result will be the fourth, which is required. That is, divide ten by three and multiply the quotient by fourteen, and the result will be the unknown. Or divide fourteen by three and multiply the result by ten.[853] The proof of this is clear from the foregoing;[854] for multiplying ten by fourteen and dividing

---

[851] According to rule *ii* above. In the proportion considered here, quantity : price = quantity : price, three terms are known; the 'last' in the text is the last given, thus the third term of the proportion here but the fourth in the next case.

[852] These 'three ways', already alluded to at the end of the above introduction, will indeed often be presented as other possible calculations.

[853] MS C has this last sentence in the margin. If it is not interpolated, at least it is misplaced and should appear at the end of the problem.

[854] P_5 or *ii* above.

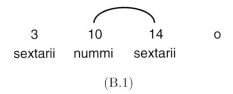

(B.1)

the product by three is the same as dividing ten by three and multiplying the quotient by fourteen.

— An example of the second case, that is, 'what is my part?' or 'how much shall I have?', is the following.

(B.2) For instance, someone says: Three sextarii are given for ten nummi. What am I to receive for sixty nummi?

The ratio of the first sextarii to their price is the same as that of the second, unknown ones, to their price, which is sixty. There are thus here four proportional numbers: the first, namely three, the second, namely ten, the third which is unknown, the fourth, namely sixty. So multiplying the first by the fourth and dividing the product by the second will give the third. Or divide either of the two factors by the divisor, which is the second, and multiply the quotient by the other factor; the result will be the third, which is required, as we said in the beginning.[855]

Understand (this) and operate in all other instances in the same way.

AGAIN. FURTHER EXAMPLES OF SELLING AND BUYING FOR THE FIRST CASE, THAT IS, 'WHAT IS ITS PRICE?', WITH FRACTIONS.[856]

(B.3) For instance, someone asks: One sextarius is given for five nummi and a third. What is the price of ten sextarii?

You will do the following. Multiplying the price of one sextarius by the quantity of sextarii, which is ten, produces fifty-three and a third. This is the price of ten sextarii, which is required.

AGAIN.[857]

(B.4) Someone asks: One sextarius is given for six nummi and a third. What is the price of ten sextarii and a fourth?

$$\frac{6}{3}. \qquad \frac{10}{4}. \qquad o$$

(B.4)

---

[855] Introduction to Book B.
[856] Fractions in the data (except B.5).
[857] This subheading would better fit after B.4 when we are no longer given the price per unit. See subheading before B.10.

782   PART TWO: TRANSLATION, GLOSSARY

You will do the following. Multiply six nummi and a third, which is the price of a sextarius, by ten sextarii and a fourth as we have taught in the multiplication of an integer and a fraction by an integer and a fraction[858]; the result will be what is required.

This is indeed how people usually proceed: they begin by stating (the price of) just one sextarius.[859]

(B.5)[860] Someone asks now: Three sextarii are given for ten nummi. What is the price of thirteen sextarii?

This is solved in three ways. The first is by multiplying the price of the three sextarii, which is ten, by the quantity of the other sextarii, which is thirteen, and dividing the product by three; the result will be what you require. Or divide ten, the price of the three sextarii, by three, and multiply the quotient by thirteen; the result will be what you require. Or divide thirteen by three and multiply the quotient by ten; the result will be what you require.

(B.6) Someone asks: Two thirds of a sextarius is given for two nummi and a fourth. What is the price of ten sextarii and a half?[861]

$$\frac{2}{3}\cdot \qquad \frac{2}{4}\cdot \qquad \mathrm{o} \qquad \frac{10}{2}\cdot$$

(B.6)

(*a*) This problem is solved in the same manner as the preceding[862], namely as follows. You may multiply two and a fourth, which is the middle term, by the last, which is ten and a half, and divide the product by the first, which is two thirds; the result will be what you require. Or divide two and a fourth, which is the middle (term), by the first, which is two thirds, and this gives three and three eighths; multiply this by ten and a half, and the result will be what you require. Or divide the last, which is ten and a half, by two thirds, which is the first, and multiply the quotient by two and a fourth; the result will be what you require.

---

[858] A.124.

[859] As a rule, in everyday life, it is the *unit price* which is given. This will not be the case in the subsequent problems, which will accordingly be solved by the 'three ways' (above, *ii*, p. 778).

[860] This problem, almost identical to B.1, and which does not really fit here (see note 856), is omitted in MS $\mathcal{A}$; but the next problem refers to it. Note, too, that B.9 is almost identical to B.2.

[861] The unknown term being the price, **o** in the figure (and in that of B.6*b*) should appear last. These figures are interpolated, by an early reader misunderstanding 'last' (see note 851). Another instance of **o** standing for the unknown in Smith, II, p. 490, footnote; see also below, p. 852 & note 1146.

[862] This must refer to B.5; see note 860.

(**b**) This problem is (also) solved in another way, namely by forming a number from the denominators [*as some have already taught in their writings*]⁸⁶³; those who do not know how to multiply fractions will operate in this manner, namely as follows⁸⁶⁴. They will multiply the denominators of the fractions in the first and the second number, which are⁸⁶⁵ a third and a fourth, thus producing twelve. They will multiply the first number by this, thus producing eight, and put it below the first number; let this now be taken as the first. Next, they will multiply the second number by twelve, thus producing twenty-seven, and put it below the middle (term); let this now be taken as the middle (term). Then it is like asking: 'When eight sextarii are given for twenty-seven nummi, what is the price of ten and a half sextarii?' Multiplying ten and a half by twenty-seven and dividing the product by eight will give what you require. Or dividing either of the two factors by the divisor and multiplying the quotient by the other factor will give the quantity you require.

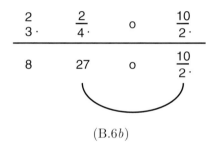

(B.6*b*)

(**c**) Or multiply the denominators of the first and the last fractions, thus producing six. Multiply by this the first and the last (terms), and put each result below the corresponding multiplier; so four will be below the first and sixty-three below the last. Then it is like asking: 'When four sextarii are given for two nummi and a fourth, what is the price of sixty-three sextarii?' Proceeding as we have taught before⁸⁶⁶ will give what you require.

[(**d**) *You may want the factors not to contain any fraction.*⁸⁶⁷ *Multiply then by the product of all the denominators, which is twenty-four, the first, the second and the third*⁸⁶⁸*, and put each of these products below the corresponding multiplier. There will then be sixteen below the first, fifty-four below the second, two hundred and fifty-two below the last. Then it is like asking: 'When sixteen sextarii are given for fifty-four, what is the price of*

---

⁸⁶³ This must be an addition: such a procedure was constantly used in Book A.
⁸⁶⁴ If this is genuine, it cannot be intended for the reader acquainted with Book A and is merely a pedagogical remark. See also B.10, end.
⁸⁶⁵ Namely the fractional parts.
⁸⁶⁶ B.1.
⁸⁶⁷ This whole treatment is erroneous.
⁸⁶⁸ Subsequently called 'the last'.

784 PART TWO: TRANSLATION, GLOSSARY

two hundred and fifty-two sextarii?' Proceeding as explained before will give what you require.]

NEXT: CHAPTER ON SELLING AND BUYING FOR THE SECOND CASE, WHICH IS 'HOW MUCH SHALL I HAVE?' OR 'WHAT IS MY PART?', WITH FRACTIONS.[869]

(**B.7**) For instance, [*someone asks: A sextarius is sold*[870] *for six nummi. How much shall I have for fifty nummi? Divide fifty by six. The result will be what you require.*

AGAIN.]^[871]

(**B.8**) Someone asks: A sextarius is given for six nummi and a half. How much is due to me for forty-four and a third?

You will do the following. Divide forty-four and a third by six and a half as we have taught above in the division of fractions.[872] The result will be six sextarii and ten thirteenths and two thirds[873] of a thirteenth, and this is the number of sextarii due to you[874].

AGAIN.

(**B.9**) Someone asks: Three modii are given for ten nummi. How much is due to me for sixty-four nummi?

This problem is solved in three ways.[875] You may multiply the first by the last, thus producing a hundred and ninety-two, and divide this by the middle; this will give nineteen modii and a fifth. Or divide three by ten and multiply the quotient by sixty-four; the result will be what you require. Or divide sixty-four by ten and multiply the quotient by three; the result will be what is required. [*Their proofs are like those preceding.*]^[876]

AGAIN.

(**B.10**) Someone asks: A modius and two thirds are given for ten nummi and a half.[877] How much is due to me for thirty-four nummi and four fifths?

(***a***) You will do the following. Either multiply the first by the last and divide the product by the middle, which is ten and a half; the result will

---

[869] In B.2 the data were integers. This is also the case in B.7 and B.9.

[870] Instead of the usual 'given'.

[871] Only MS *C* has this rather suspect passage.

[872] A.260–261.

[873] The manuscripts have 'a third'.

[874] Before, 'you' meant the reader, now it refers to the asker. (See also notes 1019, 1020, 1022, 1077, 1278.)

[875] Unlike B.7–8, where there was just one term in the numerator since the other was 1.

[876] The equivalence of these three ways was explained in the introduction to Book B (rule *ii*, p. 778).

[877] Here as elsewhere (A.210, A.212, B.43, B.46, B.120) we translate *obolus* by 'a half'. An obol is worth half a nummus.

LIBER MAHAMELETH

be what is required. Or divide either of the factors by the middle, that is, the divisor, and multiply the quotient by the other; the result will be what is required.

(**b**) Or multiply the denominator of the fraction of the middle term by the denominator of the fraction of the first or the last. Multiply this product by the middle, and put the result below it. Next, multiply the product of the denominators by the number with the fraction the denominator of which was used for the product, which is either the first or the last, and put the result below it. Next, you will proceed according to the rules for 'how much shall I have?' or 'what is my part?' —which is the same— and the result will be what you require. In such a case, the divisor and one of the two factors will be integers, and it will be easier for those who are not acquainted with the multiplication and the division of fractions.[878]

(**B.11**) Someone asks: Five sextarii and a half are given for seven nummi and a third. What is the price of ten sextarii and a fifth?[879]

(**a**) You will do the following. Multiply the denominators of a half, a third and a fifth, thus producing thirty. Multiply this by five and a half, and put the product as the principal. Next, multiply thirty by seven and a third, then the product by ten and a fifth, and divide the result by the principal; this will give what you wanted.

This is proved as follows. We know that the ratio of five sextarii and a half to seven nummi and a third is the same as the ratio of ten and a fifth to the required quantity. Now when five and a half and seven and a third are multiplied by some number, the ratio of the two products will be the same as the ratio of the two multiplicands.[880] Since the product of five and a half into thirty is a hundred and sixty-five, and the product of seven and a third into thirty is two hundred and twenty, the ratio of a hundred and sixty-five to two hundred and twenty is the same as the ratio of five and a half to seven and a third. But the ratio of five and a half to seven and a third is the same as the ratio of ten and a fifth to the required quantity. Thus the ratio of a hundred and sixty-five to two hundred and twenty is the same as the ratio of ten and a fifth to the required quantity. Therefore multiplying ten and a fifth by two hundred and twenty and dividing the product by a hundred and sixty-five will give the required quantity. This is what we wanted to demonstrate.

(**b**) Or otherwise. Multiply ten and a fifth by seven and a third and divide the product by five and a half; the result will be what you require. [*The proof of this is clear from the preceding one.*][881]

---

[878] Although generally a fraction will remain in one of the terms (and in the result). For the proof of this procedure, see B.13. See also note 864.

[879] B.11–12 belong to the first case, and should thus more appropriately follow B.6.

[880] *Elements* VII.17.

[881] This is just the plain proportion.

LIBER MAHAMELETH

You want to subtract the root of some number from the root of another number.

(i) If these numbers have to one another the same ratio as one square number to another square number, the subtraction can take place in such a way that the remainder after subtraction is the root of some number.

(ii) But if these numbers have not to one another the same ratio as one square number to another square number, the subtraction cannot take place in the aforesaid manner. The proof of this is clear.[726]

(**A.291**) You want to subtract the root of eight from the root of eighteen.

You know that this can be done, for eight is to eighteen as one square number is to another square number.[727] You will then do the following. Multiply eight by eighteen, thus producing a hundred and forty-four. Two roots of this, or twice the root, is twenty-four; keep it in mind. Next, add eight to eighteen; this makes twenty-six. Subtracting twenty-four from this leaves two. The root of two is therefore the remainder after subtracting the root of eight from the root of eighteen.

$$A \qquad G \qquad\qquad B$$

(A.291)

This is proved as follows. Let the root of eighteen be AB and the root of eight, BG. I want to know of which number AG is the root. We know [728] that the products of AB into itself and BG into itself are equal to the products of AB into BG, twice, and AG into itself. Thus subtracting twice the product of AB into BG, which is twenty-four, from the products of AB into itself and BG into itself, which are twenty-six, will leave two as the product of AG into itself. This is what we wanted to demonstrate.

(**A.292**) You now want to subtract the root of six from the root of ten.

Here there is no way of arriving at a remainder which is the root of some number [as Euclid has demonstrated][729]. You may want, though, to proceed according to the above rule. You will multiply six by ten, thus producing sixty. Subtracting two roots of this, which is the root of two hundred and forty, from sixteen and taking the root of the remainder will give in the end the root of the remainder from sixteen after subtracting from it the root of two hundred and forty. Now it is easier to say 'the root of ten minus the root of six' than 'the root of the remainder from sixteen after subtracting from it the root of two hundred and forty'.

---

[726] From the case of addition.

[727] A.288.

[728] By *Elements* II.7 = PE_7.

[729] This must be a misplaced gloss, originally referring to the use of *Elements* II.7 in the demonstration of A.291.

786   PART TWO: TRANSLATION, GLOSSARY

(*c*) You may now want (just) the principal[882] to be without fraction. Multiply the denominator of the fraction which is with the first number by this first number, and put the result as the principal. Next, multiply this same denominator by the second number, then the result by the third, and divide the product by the principal; the result will be what you require.   The proof of this is evident from what has been said in the beginning.[883]

(*d*) Or divide either of the two factors by the divisor and multiply the quotient by the other; the result will be what you require.[884]

(**B.12**) Someone asks: Five eighths of a sextarius is given for three fourths of a nummus. What is the price of ten elevenths of a sextarius?

(*a*) You will do the following. Multiply the denominators of an eighth, a fourth and an eleventh, thus producing three hundred and fifty-two. Put five eighths of this as the principal. Next, multiply three fourths of this same number by ten elevenths and divide the product by the principal; the result will be what you wanted.

(*b*) Or otherwise. Take five eighths of the number from which an eighth is denominated, thus of eight, and put it as the principal. Next, multiply three fourths of eight by ten elevenths and divide the product by the principal; the result will be what you wanted.

(*c*) Or otherwise. Multiply three fourths by ten elevenths and divide the product by five eighths; the result will be what you wanted.   Or otherwise. Divide either of the two factors by the divisor and multiply the quotient by the other; the product will be what you wanted.

The proofs for all these are the same as those for the foregoing.[885] Consider other such instances likewise, and you will find that this is the way.

AGAIN.

(**B.13**) Someone asks: Two sextarii and a third are given for seven and a half (nummi). How much shall I have for six and three sevenths?

(*a*) You will do the following. Multiply the denominators of a third, a half and a seventh, thus producing forty-two. Multiply this by seven and a half, and put the product as the principal. Next, multiply forty-two by two and a third, then the result by six and three sevenths, and divide the product by the principal; the result will be what you wanted.

The proof of this is the following. We know that the ratio of two and a third to seven and a half is the same as the ratio of the required quantity to six and three sevenths. But if you multiply two and a third and seven and a half by some number, the ratio of the two products will be the same as

---

[882] First occurrence of this term in Book B.

[883] Above, *a*.

[884] This would be expected at the end of *b*. Compare with B.12*c*, B.13*b*.

[885] See B.11 and introduction to Book B.

LIBER MAHAMELETH   787

the ratio of two and a third to seven and a half.[886] Since the product of two
and a third into forty-two is ninety-eight and the product of seven and a
half into forty-two is three hundred and fifteen, the ratio of two and a third
to seven and a half is the same as the ratio of ninety-eight to three hundred
and fifteen. But the ratio of two and a third to seven and a half is the same
as the ratio of the required quantity to six and three sevenths. Thus the
ratio of ninety-eight to three hundred and fifteen is the same as the ratio
of the required quantity to six and three sevenths. Therefore multiplying
ninety-eight by six and three sevenths and dividing the product by three
hundred and fifteen will give the required quantity.

(**b**) Or otherwise. Multiply two and a third by six and three sevenths and
divide the product by seven and a half; the result will be what you require.
Or divide either of the two factors by the divisor and multiply the quotient
by the other.

(**c**) You may want the principal to be without fraction. Then multiply
the denominator of a half, namely two, by seven and a half, and put the
product as the principal. Next, multiply this same two by two and a third,
then the result by six and three sevenths, and divide the product by the
principal; the result will be what you require.

The proof of all these will be evident from the foregoing to those who
think about them.[887]

AGAIN.

(**B.14**) Someone asks: Four fifths of a sextarius is given for two thirds of
a nummus. How much shall I have for seven eighths of a nummus?

(**a**) You will do the following. Multiply the denominators of a fifth, a
third and an eighth, thus producing a hundred and twenty. Put two thirds
of this, which is eighty, as the principal. Next, multiply four fifths of a
hundred and twenty, which is ninety-six, by seven eighths, thus producing
eighty-four. Dividing this by the principal will give one and half a tenth,
which is what is required.

(**b**) Or otherwise. Take of the denominator of the fraction which is with
the middle term, namely three, two thirds, which is two, and make it[888]
the principal. Next, multiply seven eighths by three, then the result by four
fifths, and divide the product by two; the result will be what you wanted.

The proof of these two ways is evident from what precedes.[889]

(**c**) Or otherwise. Multiply four fifths by seven eighths and divide the
product by two thirds; the result will be what you wanted. Or divide

---

[886] Again, *Elements* VII.17.

[887] All that is known from the previous problems, and B.13 is simply
analogous to B.11, but with the *quantity* being required instead of the
price. In particular, *c* is proved as *a* is.

[888] *eas*, thus the 'two thirds of three'.

[889] B.13*a*.

788   PART TWO: TRANSLATION, GLOSSARY

either of these two factors by the divisor and multiply the quotient by the other; the result will be what you require.

The proof of these ways is evident from the foregoing.[890]

To sum up all these explanations. When a known quantity of sextarii is given for a known quantity of nummi and it is asked how much is due for some (other) known quantity of sextarii, this problem is called 'what is the price?' In this case multiply the middle number by the third and divide the product by the first; the result will be what you require. If there is a fraction with the first number, multiply them both by the denominator of this fraction and put the result as the principal; next, multiply the middle number with its fraction, if any, otherwise itself, by the denominator of the first number's fraction and multiply the result by the third number with its fraction, if any, otherwise by the number itself, and divide the product by the principal; the result will be what you require. But if it is asked: a known quantity of sextarii is given for a known quantity of nummi, and what is obtained for a known quantity of nummi is required, this is the problem 'what shall I have?' In this case multiply the first number by the last and divide the product by the middle; the result will be what you require. If there are fractions with the middle number, multiply their denominator, or the number formed from their denominators, by the middle number with its fraction [*if any*][891], and put the product as the principal. Next, multiply this same denominator by the first number, with its fraction if any, then multiply the result by the last number and divide the product by the principal; the result will be what you require. Note that in the problem 'what is the price?' the fourth is unknown, whereas in the problem 'how much shall I have?' or 'what is my part?' —which is the same— the third is unknown.[892]

Treat all other such instances like this, and you will succeed.

CHAPTER INVOLVING UNKNOWNS IN BUYING AND SELLING [893]

Three cases are found here: where the two (terms) are unknown but their sum is known; where the two are unknown but their difference is known; and where the two are unknown but their product is known.[894]

— An example of the first case is the following.

(B.15) For instance, someone asks: Three modii are given for thirteen

---

[890] Introduction to Book B. This statement may well have originated from a reader's addition.

[891] Meaningless addition.

[892] Keeping the proportion in the form quantity : price = quantity : price.

[893] 'involving unknowns': *de ignoto*. An early reader's gloss, now incorporated in the text of $\mathcal{B}$, mentions that the help of a teacher is needed.

[894] The unknown terms are the last two. The first two terms of the proportion, which set the price, are always given.

LIBER MAHAMELETH 789

nummi. What is the quantity of modii, bought at the price set,[895] which, when added to their price, makes sixty?

It is asked here for the quantity of modii and their price.[896]

You will do in this case the following. Add three to thirteen, thus making sixteen, which will be here the principal.[897] If you want to know the quantity of modii, multiply three by sixty and divide the product by the principal; the result will be the quantity of modii. If you want to know their price, multiply the price of the first quantity, which is thirteen, by sixty and divide the product by the principal; the result will be the amount of the price of the unknown quantity of modii[898].

The reason for our multiplying and dividing in this manner is the following. The ratio of the three modii to sixteen, which is the sum of the modii and their price, is the same as (the ratio) of the unknown modii to sixty, which is the sum of the unknown quantity of modii and its price. Thus the third is unknown. Therefore multiply the first, which is three, by the fourth, which is sixty, and divide the product by the second, which is sixteen; the result will be the third. Likewise also, the ratio of the thirteen of the price of the modii to sixteen, which is the sum of the modii and their price, is the same as (the ratio) of the price of the unknown quantity of modii to sixty, which is the sum of the unknown quantity of modii and its price. The third being unknown, multiply and divide as just said, and the result will be what is required.

Or take of sixty the same fraction which thirteen is of sixteen, and this will be the price of the unknown quantity of modii. Or divide sixty by sixteen and multiply the result by thirteen; the product will be the price of the unknown quantity of modii. These two ways consist in first dividing either of the two factors by the divisor and then multiplying the quotient by the other.[899]

— An example of the second case is the following.

(B.16) For instance, someone asks: Three modii are given for thirteen nummi. What is the quantity of modii, bought at the price set, which when subtracted from their price leaves sixty?

It is asked here for the unknown quantity of modii and their price.

You will do in this case the following. Subtracting the three modii from thirteen leaves ten, which is the principal. If you want to know the unknown quantity of modii[900], multiply three by the given sixty and divide

---

[895] Remember that (here as in this whole chapter) there is no distinction between selling price and buying price.

[896] This was not quite explicit in the formulation. Same addition in the next problem.

[897] 'here': before it resulted from a multiplication.

[898] Literally (here and recurrently): 'the unknown modii'.

[899] Thus they are the banal 'other ways' seen before (notes 852, 876).

[900] Literally (and often below): 'the quantity of unknown modii'.

790 PART TWO: TRANSLATION, GLOSSARY

the product by the principal; the result will be the unknown quantity of modii. If you want to know the amount of the price, multiply the price of the modii, namely thirteen, by sixty and divide the product by the principal; the result will be the amount of the price of the unknown quantity of modii.

The proportion here is evident. The ratio of the three modii to the remaining ten is the same as the ratio of the unknown quantity of modii to the remaining sixty. And the ratio of thirteen, which is the price of the modii, to the remaining ten is the same as the ratio of the price of the unknown quantity of modii to the remaining sixty.

You may also divide either of the two factors by the divisor, which is the principal, and multiply the quotient by the other, as has been said before[901]; the result will be what you require.

— An example of the third case is the following.

(**B.17**) For instance, someone asks: Three modii are given for eight nummi. What is the quantity of modii, bought at the price set, which when multiplied by their price produces two hundred and sixteen?

You will do the following. If you want to know the unknown quantity of modii, multiply three, the quantity of the first, by two hundred and sixteen and divide the product by eight; the result will be eighty-one. The root of this, namely nine, is the unknown[902] quantity of modii. If you want to know the amount of their price, multiply the price of the first modii, namely eight, by two hundred and sixteen and divide the product by three; the result will be five hundred and seventy-six. The root of this, which is twenty-four, is the price of the unknown quantity of modii.[903]

The problem may be complicated by the fact that the number of which you require the root does not have one[904]; nevertheless its approximate root, found as has been shown before[905], will be taken as, respectively, the unknown quantity of modii and the amount of their price. The same holds for all other instances.

(**B.18**) Someone asks: Four sextarii are given for nine nummi. What is the unknown quantity of sextarii, bought at the price set, of which the root added to the root of their price makes seven and a half?

(*a*) You will do the following. Add the root of four, which is two, to the root of nine, which is three, thus making five. Dividing seven and a half

---

[901] For the last time at the end of B.15.

[902] The text often continues to qualify the quantity as 'unknown' even when it has been determined.

[903] MSS *B* and *C* contain here the treatment of a problem which would be formulated as follows: *Cum quatuor modii dentur pro novem nummis, tunc quot sunt modii empti ad idem forum ex quorum radice radicis multiplicata in radicem radicis sui pretii proveniunt tredecim et dimidium?*

[904] That is, does not have a rational (expressible) root.

[905] Beginning of A–IX (above, pp. 748–749).

LIBER MAHAMELETH

791

by this gives one and a half. Multiply this result by two, namely the root of four; this produces three, which is the root of the unknown quantity of sextarii; therefore the unknown quantity of sextarii is nine. Next, multiply the result of the division, namely one and a half, by three, namely the root of the nine representing the amount of the price[906]; the result is four and a half. Multiplying this by itself produces twenty and a fourth, and so much is the price.

(b) Or otherwise. Divide nine by four, which gives two and a fourth. Add one to its root,[907] thus making two and a half. Divide seven and a half by this, which gives three. Multiplying it by itself produces nine, and so many are the sextarii.

(c) Or otherwise. Double seven and a half; this gives fifteen. Next, multiply seven and a half by itself, thus producing fifty-six and a fourth. Next, divide nine by four, and subtract one from the quotient; this leaves one and a fourth. Divide fifteen by this; the result is twelve. Next, divide fifty-six and a fourth by one and a fourth; the result is forty-five. Next, multiply half of twelve by itself, thus producing thirty-six. Add this to forty-five, subtract from the root of the sum half of twelve, and multiply the remainder by itself; the product will be the quantity of sextarii. If you want to know the price, multiply the quantity of sextarii by the two and a fourth resulting from dividing nine by four; the product is the price of the sextarii.

(**B.19**) Someone asks: Four sextarii are given for nine nummi. What is the unknown quantity of sextarii, bought at the price set, of which the root subtracted from the root of their price leaves one and a half?

(a) You will do the following. Subtract the root of four, which is two, from the root of nine, which is three, thus leaving one. Divide one and a half by this; the result is one and a half.[908] Multiplying this by two produces three, which is the root of the unknown quantity of sextarii; therefore the unknown quantity of sextarii is nine. Next, multiply one and a half, which resulted from the division[909], by three; this produces four and a half, which is the root of their price; therefore the price is twenty and a fourth.

(b) Or otherwise. Divide nine by four, and from the root of the quotient always subtract one; this will leave a half. Divide one and a half by this and multiply the quotient by itself; the result will be nine, and this is the quantity of sextarii.

(c) Or otherwise. Double one and a half, which gives three. Next, multiply one and a half by itself, thus producing two and a fourth. Next, divide nine

---

[906] A welcome clarification since nine is also the quantity of sextarii just found.

[907] Better: 'Always add one to its root', this being a general rule and thus independent of the present data. (See also B.19b.)

[908] The author describes the formula, so the actual values are irrelevant. See note 634, p. 735.

[909] Not that of the data.

792   Part Two: Translation, Glossary

by four, which gives two and a fourth, and subtract one from this, thus leaving one and a fourth. Divide three by this; the result will be two and two fifths. Next, divide also by one and a fourth the product of one and a half into itself; this will give one and four fifths. Next, multiply by itself half of two and two fifths, that is, one and a fifth, add the result to one and four fifths, and add the root of the sum to one and a fifth. The result is the root of the quantity of sextarii. Multiplying it by itself will produce the quantity of sextarii. If you want to know the price of the sextarii, multiply the quantity of sextarii by the result of dividing nine by four; the product is the price.

(**B.20**) Four sextarii are given for nine nummi. What is the quantity of sextarii, bought at the price set, of which the root when multiplied by the root of their price produces twenty-four?

(*a*) You will do the following. Multiply the root of four by the root of nine, thus producing six. Divide twenty-four by this and multiply the quotient by four; the result is the quantity of sextarii, namely sixteen. Again, multiply the quotient by nine; the result will be the price, which is thirty-six.

(*b*) Or otherwise. Divide nine by four, and divide twenty-four by the root of the quotient, that is, by one and a half. The result is the quantity of sextarii.

(*c*) Or otherwise. Multiply twenty-four by itself, thus producing five hundred and seventy-six. Then it is like saying: 'Four sextarii are given for nine nummi; what is the quantity of sextarii which, when multiplied by its price, produces five hundred and seventy-six?' Do as we have taught above,[910] and the result will be what you wanted.

> Again. Another chapter on the same topic, with things

(**B.21**) Someone asks: Three modii are given for ten nummi and one thing, and this thing is the price of one modius. What is the price of this thing?

(*a*) The way of solving here is the following. You must know the price of three modii according to the sale of one modius for one thing; you will find three things. But their price was ten nummi and one thing. Then three things equal ten nummi and one thing.[911] But this thing equals one of these three things;[912] and two things are left, which are equivalent to ten nummi. Thus one thing is worth five nummi. This is what you require.

(*b*) Or otherwise. Subtract one modius from the three modii, thus leaving two. Next, subtract one thing from ten nummi and a thing, thus leaving ten nummi. Next, denominate from the two the one modius subtracted[913], which gives a half. Thus half of ten, which is five, is the price of this thing.

---

[910] B.17.

[911] Equation of the form $a_1 x + b_1 = a_2 x$, with x the price of the thing.

[912] A necessary clarification since this is the first time the reader encounters such 'things'.

[913] The MSS have *emptum*, sold, instead of *demptum*, subtracted.

LIBER MAHAMELETH

793

(**B.22**) Someone asks thus: Four modii are given for twenty nummi and two things, while one and a half modii are given for two things and three nummi. What is the value of this thing?

(**a**) You will do the following. Find, according to the selling price of one and a half modii for two things and three nummi, what is due for four modii as follows. Look for the number which multiplying one and a half produces four in the way we have taught before.[914] This is two and two thirds. Multiplying this by two things and three nummi produces five things and a third of a thing, and eight nummi. This is the price of four modii. But their price was already twenty nummi and two things. Thus five things and a third of a thing and eight nummi are equivalent to twenty nummi and two things.[915] Subtract two things from five things and a third of a thing, and subtract eight nummi from twenty nummi; this leaves twelve nummi equivalent to three things and a third of a thing. Thus the thing is worth three nummi and three fifths. This is what you require.

(**b**) Or otherwise. Subtract the one and a half modii from the four modii, thus leaving two and a half. Next, subtract two things and three nummi from twenty nummi and two things, thus leaving seventeen nummi. Next, denominate one and a half from two and a half; this gives three fifths. Next, subtract from three fifths of seventeen, which is ten and a fifth, three nummi; this leaves seven and a fifth, and so much are the two things worth. Thus one thing is worth three and three fifths. But this rule is valid only when the same number of things occurs in each price.[916]

By verifying these two problems you will find that it is as we say.[917]

(**B.23**) Someone asks: Eight modii are sold for twenty nummi and one thing, and two modii are bought[918] for one thing minus a nummus. What is the price of this thing?

(**a**) You will do the following. Subtract two modii from eight, thus leaving six. Next, subtract one thing minus a nummus from twenty nummi and one thing, thus leaving twenty-one nummi since the nummus subtracted will be added to the twenty. Next, denominate two modii from six modii; this gives a third. After that, add to a third of twenty-one, which is seven, the nummus which was subtracted; this makes eight, and so much is the price of the thing. But if there were two things instead of one, eight would be the price of the two things.[919]

(**b**) Or otherwise. You know that the ratio of eight to two corresponds to four times; therefore twenty and a thing is four times a thing minus

---

[914] A.254.

[915] Equation of the form $a_1 x + b_1 = a_2 x + b_2$.

[916] Only then does it have the advantage of eliminating all the things on one side.

[917] In order to check that the solution to B.21 and B.22 found by the new, algebraic method is correct. See also B.26.

[918] At the selling price.

[919] Case where we have m things instead of one. Same remark in B.27–28.

794   PART TWO: TRANSLATION, GLOSSARY

one nummus, for the ratio of the quantities of modii is the same as the ratio of the prices. So multiply one thing minus one nummus by four, thus producing four things minus four nummi, which equal twenty and one thing.[920] Completing what is subtracted and removing what is repeated[921] will leave twenty-four, which equal three things. Therefore the thing is worth eight, and this is what you wanted to know.

(B.24) Six sextarii[922] are given for ten nummi and a thing and someone receives two sextarii for a thing. What is this thing worth?

You will do the following. Subtract two sextarii from six, thus leaving four. Next, subtract a thing from ten nummi and a thing, thus leaving ten. Next, denominate two sextarii from four; this fraction, that is, a half, taken of ten, which gives five, is what the thing is worth.

(B.25) Six sextarii are given for ten nummi and a thing and someone receives two sextarii for a thing and one nummus. What is the thing worth?

Subtract two from six, thus leaving four. Next, subtract a thing and a nummus from ten nummi and a thing, thus leaving nine nummi. Next, denominate two sextarii from four, and take such a fraction, that is, a half, of nine, which gives four and a half. Subtracting one nummus from this leaves three and a half, and such is the price of the thing.

(B.26) Someone asks: Three modii are given for twenty nummi and one thing, and half a modius is bought for two thirds of this thing minus two nummi. What is this thing worth?

You will do the following. Say to yourself: 'If half a modius is given for two thirds of a thing minus two nummi, what is the price of three modii?' You will find that it is four things minus twelve nummi. Now twenty nummi and one thing was already the price of three modii. So the two prices are equivalent.[923] Subtract then the thing on one side from the four things on the other side, thus leaving three things, and add the twelve nummi subtracted on one side to the twenty nummi added on the other side, thus making thirty-two. So three things are equivalent to thirty-two nummi. Therefore the thing is worth ten nummi and two thirds. By verifying this you will find that it is as we say.

(B.27) Someone asks: Six modii are sold for ten nummi minus one thing, and two modii are bought for one thing. What is this thing worth?

(a) You will do the following. Add two to six, thus making eight. Next, add a thing to ten minus a thing, which makes ten. After that, denominate two modii from eight modii; this gives a fourth. The price of the thing is

---

[920] Equation of the form $a_1 x + b_1 = a_2 x - b_2$.

[921] *comple ergo quod est demptum et deme quod est iteratum*. First occurrence of the *two* algebraic operations.

[922] B.24–25 (with *sextarii* instead of *modii* and no equation set) are found only in MS $\mathcal{A}$. They might have been added later.

[923] Equation of the form $a_1 x + b_1 = a_2 x - b_2$.

LIBER MAHAMELETH 795

therefore a fourth of ten, which is two and a half. If there were two things, or three, then two and a half would be their price. This rule is valid only when the same number of things occurs in each price.[924]

(**b**) The rule applicable generally is the following. You know that the ratio of six to two corresponds to three times. Then ten minus a thing is the triple of a thing. So ten minus a thing is equal to three things.[925] Therefore ten, once completed, equals four things. Thus the price of the thing is two and a half. This is what you wanted to know.

(**B.28**) Someone asks: Four modii are sold for eight nummi minus a thing, and two modii are bought for a thing and one nummus. How much is this thing worth?

(**a**) You will do the following. Add two to four, thus making six. Next, add a thing and a nummus to eight minus a thing, thus making nine. Next, denominate two from six; this gives a third. The price of the thing is therefore a third of nine, which is three, minus one nummus [*which has been subtracted*], which leaves two. But if there were two things or more, two would be their price. This rule is valid only when the same number of things occurs in each price.[926]

(**b**) The rule applicable generally is the following. You know that the ratio of four modii to two modii corresponds to the double. Thus the ratio of the two prices is double. Therefore eight minus a thing is the double of a thing and one nummus. So two things and two nummi equal eight minus a thing.[927] Completing what is subtracted and subtracting what is added[928] will give six, equal to three things. Therefore the thing is worth two. This is what you wanted to know.

(**B.29**) Someone asks: Four modii are sold for twenty nummi minus two things, while one and a half modii are sold for two things minus three nummi. How much is this thing worth?

You will do here the same as in the previous problems, namely the following. Say to yourself: 'If one and a half modii are given for two things minus three nummi, what is the price of four modii?' You will find that it is five things and a third of a thing minus eight nummi. But this equals twenty nummi minus two things.[929] Restoring then anything that is subtracted[930] and putting as much on the other side will give five things and a third of a thing equivalent to thirty-two nummi.[931] Next, dividing the number of

---

[924] But with opposite signs.

[925] Equation of the form $b_1 - a_1 x = a_2 x$.

[926] Again, with opposite signs.

[927] Equation of the form $a_1 x + b_1 = b_2 - a_2 x$.

[928] That is, what is in excess (B.23 has: what is repeated).

[929] Equation of the form $a_1 x - b_1 = b_2 - a_2 x$.

[930] 'Restoring' is the usual expression for the elimination of subtracted terms (Arabic *jabara*) —thus same meaning as 'completing' in B.23*b*.

[931] *Sic*, instead of 'seven things and a third of a thing equivalent to twenty-eight nummi'.

796   PART TWO: TRANSLATION, GLOSSARY

nummi, which is thirty-two, by the number of things, which is five and a third, will give as the value of this thing four nummi and four elevenths.[932]

[*Knowledge of this kind of problem is mostly accessible only to those who are familiar with algebra or Euclid's treatise* [933]. *But what has been said about this should suffice as an introduction.*][934]

AGAIN. ANOTHER CHAPTER INVOLVING UNKNOWNS IN BUYING AND SELLING

(B.30) Someone asks thus: An unknown quantity of sextarii is given for ninety-three, and adding this unknown quantity of sextarii to the price of one of them makes thirty-four. How many sextarii are there?

You will do the following.[935]   Multiply half of thirty-four, which is seventeen, by itself, thus producing two hundred and eighty-nine. Subtract ninety-three from it; this leaves a hundred and ninety-six.   Add to its root, which is fourteen, seventeen, thus making thirty-one.   Subtract it from thirty-four, which leaves three.   Then, if the unknown quantity of sextarii[936] is greater than the amount of the price of each, say that the sextarii are thirty-one and the price of each, three. But if the amount of the price of each is greater than their quantity, say that the sextarii are three and the price of each, thirty-one.

The proof of this is the following.[937]   Let the unknown quantity of sextarii be AB and the price of each, BG; thus the whole AG is thirty-four. Now multiplying AB by BG produces ninety-three; for multiplying the price of each sextarius by their quantity produces ninety-three, which is the price of the sextarii.[938] Let AG be bisected; if the quantity of the sextarii is greater than the price of one, the section will be at the point D, but if the price of one of them is greater than their quantity, it will be at the point H.

Let first the quantity of sextarii be greater than the price of one, the section being at the point D. [*Now multiplying AB by BG produces ninety-three.*] Thus the products of AB into BG and DB into itself are equal to the product of DG into itself.[939] Now multiplying DG by itself produces two hundred and eighty-nine. Subtracting from this the product of AB into BG, which is ninety-three, will leave the product of DB into itself, namely a hundred and ninety-six. Thus DB is fourteen. Since AD is seventeen,

---

[932] The result of *this* division is 6.

[933] An alternative way was indeed the transformation of ratios.

[934] This looks like an interpolation, as is suggested by its place (at the end of the section) and the wording (*elgabre* instead of the usual *algebra* in MS $\mathcal{A}$ or *gebla* in MS $\mathcal{B}$).

[935] Solving $x^2 + c = bx$ ($b = 34$, $c = 93$).

[936] *numerus sextariorum ignotorum*; see notes 898, 900.

[937] The figures in the two manuscripts differ. Same situation in B.41a. See also notes 946 and 970.

[938] Sum and product of AB and BG thus known.

[939] *Elements* II.5 = Premiss PE_5.

AB is thirty-one, and this is the required quantity of sextarii. But DG being seventeen, and the part DB of it fourteen, this leaves as BG three, which is the amount of the price of each sextarius.

But if the price of each sextarius is greater than their quantity, the section will be at the point H, and the products of GB into BA and BH into itself will be equal to the product of AH into itself. Now the product of AH into itself is two hundred and eighty-nine. Subtracting from this the product of GB into BA, namely ninety-three, will leave as the product of BH into itself a hundred and ninety-six. Thus BH is fourteen. Since GH is seventeen, BG is thirty-one, and this is the price of each sextarius. But AH being seventeen and BH, fourteen, AB will be three, which is the required quantity of sextarii. This is what we wanted to demonstrate.

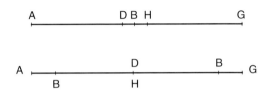

(B.30; figure above in MS $\mathcal{A}$, below in MS $\mathcal{B}$)

AGAIN.

(**B.31**) Someone asks: The price of an unknown quantity of sextarii is ninety-three, and subtracting this quantity of sextarii from the price of each leaves twenty-eight. How many sextarii are there?

You will do the following.[940] Multiply half of twenty-eight by itself, thus producing a hundred and ninety-six. Add ninety-three to this, thus making two hundred and eighty-nine. Adding to its root, which is seventeen, half of twenty-eight, which is fourteen, will give the price of each sextarius, namely thirty-one, while subtracting fourteen from the root will leave the quantity of the sextarii, which is three.[941]

The proof of this is the following. Let the price of each sextarius be AB and the unknown quantity of sextarii, GB; then AG will be twenty-eight. Now multiplying AB by BG produces ninety-three, as we said before.[942] Let AG be bisected at the point D. Then the products of AB into BG and DG into itself will be equal to the product of DB into itself. Since the product of AB into BG is ninety-three and the product of DG into itself, a hundred and ninety-six, the product of DB into itself is two hundred and eighty-nine. Thus DB is seventeen. Since DG is fourteen, this leaves as GB three, which is the quantity of the sextarii. But DB being seventeen

---

[940] Solving $x^2 = bx + c$ ($b = 28$, $c = 93$).
[941] These are not the two solutions of the same equation.
[942] Presumably B.30 rather than the formulation above. But here we know AB and GB by their *difference* and product, and will therefore apply *Elements* II.6 = PE_6.

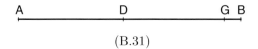
(B.31)

and AD, fourteen, the whole AB is thirty-one, and this is the amount of the price of each. This is what we wanted to demonstrate.

(B.31′) If it is said that the remainder after subtracting from the quantity of sextarii the price of each is twenty-eight, you will also operate in the same manner, and the resulting quantity of sextarii will be found to be thirty-one, and the price of each, three. The proof of this is the same as the preceding one.[943]

AGAIN.

(B.32) Someone asks: The price of an unknown quantity of sextarii is unknown and another unknown quantity of sextarii is given at a price also unknown, but the price set is the same as for the first sextarii, and multiplying the first sextarii by their price produces six while multiplying the second sextarii by their price produces twenty-four; moreover, adding the first and their price to the second and their price makes fifteen. How much is each quantity and what are their prices?

You will do the following. Divide twenty-four by six, which gives four. Always add to its root, which is two, one, which makes three. Dividing fifteen by this gives five. Such is the quantity of the first sextarii plus their price. Multiply half of this, which is two and a half, by itself; the result is six and a fourth. Subtracting six from this leaves a fourth. Subtract the root of this, which is a half, from two and a half, thus leaving two. Next, add this same root to two and a half; this makes three. If then the quantity of sextarii is greater than the price of all of them, say that the sextarii are three and their price, two. But if the sextarii are less than their price, say that the sextarii are two and their price, three.

You may want to know the quantity of the second sextarii and their price.[944] Divide six by twenty-four using denomination; this gives a fourth. Always add to its root, thus to a half, one, which makes one and a half. Dividing fifteen by this gives ten. This is the quantity of the second sextarii plus their price. Then multiply by itself half of ten, which is five, thus producing twenty-five. Subtracting twenty-four from this leaves one. Add the root of this, which is one, to five, which is half of the ten; the result is six. Next, subtract this same root from five; this leaves four. If then the first sextarii are three and their price, two, the second sextarii will necessarily be six and their price, four [*because it has been supposed that for each of the first and the second sextarii the price is the same*[945]]. But

---

[943] AB becomes the quantity of sextarii and GB the price of each.

[944] Computing their *sum*, as before for the first sextarii. As remarked at the end of the problem, this initial step can be directly inferred from the data and what has just been calculated.

[945] Literally: 'the price is one'; the meaning is clear but our author would

if the first sextarii are two and their price, three, the second sextarii will necessarily be four and their price, six [*for the price is for each of them the same*].

I shall (only) expound the proof for determining the quantity of the first sextarii and their price: the proof for determining the quantity of the second sextarii and their price will appear clearly from this proof. Let the first sextarii be AB and their price, BG. Let AB multiplied by BG produce D. Thus D is six. Let now the second sextarii be HZ and their price, ZK. Let HZ multiplied by ZK produce T. Thus T is twenty-four. It is known that the ratio of AB to HZ is the same as the ratio of BG to ZK. Therefore D and T are two similar figures, for their sides are proportional.[946] Then [947] their ratio to one another is the same as the ratio of one side of the first to the (corresponding) side of the second duplicated by repeating its name.[948] Thus the ratio of T to D is the same as the ratio of HZ to AB duplicated by repeating its name. Since T is the quadruple of D, HZ is the double of AB. Now it is clear that since the ratio of HZ to AB is the same as the ratio of ZK to BG, the ratio of HZ to AB will be the same as the ratio of the whole HK to the whole AG.[949] Thus the whole HK is the double of the whole AG. Therefore the whole HK and AG is the triple of AG. But HK and AG are (together) fifteen. Thus AG is five. Now AG is already divided at the point B into two parts the product of which is six.[950] So you will know that [*as said at the beginning of the problem*] if the quantity of sextarii is greater than their price, then AB will be greater than BG, so that, if AG is bisected, the section will occur on AB. And since [*it has been stated that the price of each of them is the same and*] the ratio of AB to HZ is the same as the ratio of BG to ZK,[951] then by alternation the ratio of AB to BG will be the same as the ratio of HZ to ZK. Since AB is greater than BG, HZ is greater than ZK. Therefore [952] AB will be three, and this is the quantity of the sextarii, and BG will be two, and this is their price; and HZ will be six, which is the quantity of sextarii, and ZK, four, which is their price.

You may want, having found the quantity and the price of the first

---

have avoided such ambiguity by writing, for instance, *nam sic positum fuit ut unusquisque primorum et secundorum esset unius pretii*. This occurs again in the subsequent interpolations.

[946] *Elements* VI.14. In MS $\mathcal{B}$, D and T are represented by rectangles.

[947] *Elements* VI.23.

[948] *comparatio geminata repetitione nominis*. This is how the multiplication of a ratio by itself is expressed.

[949] *Elements* V.12.

[950] Sum and product of AB and BG thus known. But the determination of AB and BG, known from B.30, is omitted.

[951] As stated shortly before.

[952] The values of the two required quantities are already known; still, their geometrical determination would be expected here.

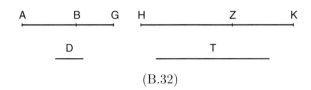

(B.32)

sextarii, to determine the quantity of the other sextarii and their price.[953] You will say that the quantity of the first and their price is five. Since it has been supposed that the first plus their price with the second plus their price is fifteen, subtract five from fifteen; this will leave ten, which is the quantity of the sextarii and their price. Continue with the remainder of the problem as we have explained before.[954]

(**B.33**) Someone asks: The price of an unknown quantity of sextarii is unknown, and another unknown quantity of sextarii is given for an unknown price, but the price set is the same as for the first, and multiplying the first by their price produces ten while multiplying the second by their price produces thirty; moreover, adding the first and their price to the second and their price makes twenty. How much is each quantity and what are their prices?

You will do what we have shown in the problem preceding this one, namely the following. If you want to know the sum of the quantity of the first and their price, you will divide thirty by ten, which gives three. Always add to its root, which is the root of three, one, thus making the root of three plus one. Divide twenty by this as we have taught in the chapter on roots[955]; the result will be the root of three hundred minus ten. This is the quantity of the first sextarii plus their price. You may now subtract the root of three hundred minus ten from twenty; this will leave the quantity of the second sextarii plus their price, which is thirty minus the root of three hundred.

You may also determine the sum of the quantity of the second modii[956] and their price using the rule seen before [*as you have done for determining the sum of the first and their price*][957]. That is, divide ten by thirty using denomination; this gives a third. Always add to its root one, which makes the root of a third plus one. Dividing twenty by this as we have shown in the chapter on roots[958] will give thirty minus the root of three hundred. This is how to find the sum of the quantity of the first modii and their price, and the sum of the quantity of the second and their price. But if

---

[953] That is, this quantity plus their price.
[954] Prior to the proof.
[955] A.298.
[956] Here and a few lines below instead of 'sextarii'. Although other problems in this chapter involve modii, it looks as if this paragraph had been added by the author later.
[957] See B.32 (rather than, as this early reader thought, just above).
[958] A.298 again.

LIBER MAHAMELETH

801

you want to know separately the quantity of modii and the amount of their price, you will do the following. First for the quantity of the first modii and, separately, their price.

Since it has been supposed that multiplying the first sextarii[959] by their price produces ten, the root of three hundred minus ten will be divided into two parts having ten as their product.[960] Therefore multiply half the root of three hundred minus ten, which is the root of seventy-five minus five, by itself, thus producing a hundred minus the root of seven thousand five hundred. Subtracting ten from this leaves ninety minus the root of seven thousand five hundred. Take the root of ninety minus the root of seven thousand five hundred; this is the root of ninety this ninety being diminished by the root of seven thousand five hundred, for it is a fourth binomial[961]. Subtract it from half the root of three hundred minus ten, thus from the root of seventy-five minus five; this leaves the root of seventy-five minus, I mean the root,[962] five and minus the root of ninety minus the root of seven thousand five hundred. And adding the binomial mentioned above to this half gives the root of seventy-five minus five, to which is added the root of ninety minus the root of seven thousand five hundred. Then if the first sextarii are greater than their price, say that the first sextarii are the root of seventy-five minus five to which is added the root of ninety minus the root of seven thousand five hundred, and their price will be the root of seventy-five minus five and minus the root of ninety minus the root of seven thousand five hundred. ⟨But if the first sextarii are less than their price, say that the first sextarii are the root of seventy-five minus five and minus the root of ninety minus the root of seven thousand five hundred, and their price will be the root of seventy-five minus five to which is added the root of ninety minus the root of seven thousand five hundred.⟩[963]

You will proceed in like manner to know separately the quantity of the second sextarii and their price, namely as follows. Multiply by itself half of thirty minus the root of three hundred, which is fifteen minus the root of seventy-five; the result will be three hundred minus the root of sixty-seven thousand five hundred. Subtract thirty from this, thus leaving two hundred and seventy minus the root of sixty-seven thousand five hundred. Subtract its root, namely two hundred and seventy minus the root of sixty-seven thousand five hundred the root of which is taken,[964] from half of thirty minus the root of three hundred,[965] and add to this the same root. If the

---

[959] Return to the 'sextarii' (note 956).

[960] Thus both sum and product of the two parts are known.

[961] Since $\sqrt{90 - \sqrt{7500}}$ is a fourth binomial, it is left in this form; see A.326.

[962] $\sqrt{75}$, not 75, is diminished by the quantity that follows.

[963] This is clearly a scribal omission; for, concerning this time the second sextarii, the two pairs of values are given below.

[964] Note the expression used for the root of a quantity with two terms, one of which is a root.

[965] Calculated before as $15 - \sqrt{75}$. Similar situation in B.36.

quantity of the first sextarii is the root of seventy-five minus five and minus the root of ninety minus the root of seven thousand five hundred, and their price is the root of seventy-five minus five, to which is added the root of ninety minus the root of seven thousand five hundred, then the quantity of the second sextarii will be fifteen minus the root of seventy-five, from which is subtracted two hundred and seventy minus the root of sixty-seven thousand five hundred, the root of which is taken[966], and their price will be fifteen minus the root of seventy-five, to which is added two hundred and seventy minus the root of sixty-seven thousand five hundred the root of which is taken. But if the quantity of the first sextarii is the root of seventy-five minus five, to which is added ninety minus the root of seven thousand five hundred the root of which is taken, and their price is the root of seventy-five minus five, from which is subtracted ninety minus the root of seven thousand five hundred the root of which is taken, then the second will necessarily be fifteen minus the root of seventy-five, to which is added two hundred and seventy minus the root of sixty-seven thousand five hundred the root of which is taken, and their price will be fifteen minus the root of seventy-five, from which is subtracted two hundred and seventy minus the root of sixty-seven thousand five hundred the root of which is taken.

The proof of all that has been said above is evident from what has been seen in the previous problem.

(B.34) Someone asks: An unknown quantity of sextarii is given for an unknown price, and the price of yet another unknown quantity is also unknown, but the price set is the same as for the first, and multiplying the first by their price produces six while multiplying the second by their price produces twenty-four; moreover, subtracting the quantity of the first and their price from the quantity of the second and their price leaves five. How much is each quantity and what are their prices?

You will do the following. Divide twenty-four by six; the result is four. Always subtract one from the root of this, which is two, thus leaving one. Dividing five by this gives five, which is the quantity of the first sextarii plus their price. Adding to this the other five makes ten, which is the quantity of the second sextarii plus their price. Continue then with the problem as we have taught in the preceding.[967] Those who have well understood what was above will easily understand what is here.

(B.35) Someone asks: The price of an unknown quantity of sextarii is unknown and the price of another unknown quantity is unknown, but the price set is the same as for the first, and multiplying the first by their price produces six while multiplying the second by their price produces twenty-four; moreover, multiplying the sum of the first and their price by the sum

---

[966] *excepta* instead of *accepta* in the MSS, here and five times below, obviously an alteration by a confused early reader.

[967] In fact the problem has become identical with B.32, so the required values are those seen there.

of the second and their price produces fifty. What are these two quantities and their prices?

You will do the following. Divide twenty-four by six; the result will be four. Divide by its root, which is two, fifty; the result is twenty-five. The quantity of the first sextarii plus its price is the root of this, thus five. Continue then with the problem as we have taught above[968]; the quantity of sextarii will be either three or two. You may want to know the quantity of the second sextarii plus its price. You know that multiplying the first quantity plus its price by the second plus its price produces fifty while the first and their price is five; so divide fifty by five, and this will give the second quantity and its price, namely ten. Continue then with the problem as we have taught above.[969] If the quantity of the first sextarii is two and its price, three, the quantity of the second will be four and its price, six; but if the quantity of the first is three and its price, two, the quantity of the second will be six and its price, four.

The proof of this is the following. Let the first quantity of sextarii be AB and its price, BG. Let AB multiplied by BG produce D, which is six. Let the second quantity be HZ and its price, ZK. Let HZ multiplied by ZK produce T, which is twenty-four.[970] Now the ratio of T to D is the same as the ratio of HZ to AB duplicated by repeating its name. But T is the quadruple of D. Thus HZ is the double of AB, and the whole HK will be the double of the whole AG according to what we have taught above[971]. Since the multiplication of AG by HK produces fifty, the multiplication of AG by its double produces fifty. Thus AG multiplied by itself will produce twenty-five. Therefore AG is the root of twenty-five, thus five, but HK is ten [*for the multiplication of AG by HK produces fifty*][972].

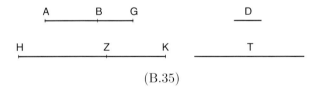

(B.35)

Continue with the rest of the problem as we have taught above.[973]

(**B.36**) Someone asks: There is an unknown quantity of sextarii, the price of which is unknown, and yet another unknown quantity, the price of which is unknown, but the price set is the same as for the first, and multiplying

---

[968] No need: the problem has become identical with B.32.
[969] Again, this is B.32.
[970] As in B.32, MS $\mathcal{B}$ represents D and T as rectangles.
[971] In the demonstration of B.32.
[972] This must be an addition: it has just been said that $HK = 2\,AG$.
[973] In B.30, B.32, B.33 where we learnt how to determine quantities and prices separately when sum and product are known.

804 PART TWO: TRANSLATION, GLOSSARY

the first by their price produces ten[974] while multiplying the second by their price produces twenty[975]; moreover multiplying the first plus their price by the second plus their price produces the root of five thousand seven hundred and sixty. How much are these two quantities and what are their prices?

You will do here the same as in the preceding problem, (for) the way of operating in this problem does not differ from the one above. Indeed, you will divide twenty by ten; this gives two. Divide by its root, which is the root of two, the root of five thousand seven hundred and sixty; this gives the root of two thousand eight hundred and eighty. The root of this, which is the root of the root of two thousand eight hundred and eighty, is the sum of the quantity of the first sextarii and its price. But it has been supposed that multiplying the first quantity of sextarii by its price produces ten. Therefore[976] take half the root of the root of two thousand eight hundred and eighty, which is the root of the root of a hundred and eighty, and multiply it by itself; the result will be the root of a hundred and eighty. Subtracting ten from this leaves the root of a hundred and eighty minus ten. Take its root as we have shown in the chapter on roots; it will be the root of the root of a hundred and twenty-five minus the root of the root of five.[977] Adding this to half the root of the root of two thousand eight hundred and eighty, which is the root of the root of a hundred and eighty, will give as a result the root of the root of a hundred and eighty and the root of the root of a hundred and twenty-five minus the root of the root of five. Or subtracting the same from half the root of the root of two thousand eight hundred and eighty[978] will leave the root of the root of a hundred and eighty and the root of the root of five minus the root of the root of a hundred and twenty-five. If the first quantity of sextarii is greater than its price, say that the first quantity is the root of the root of a hundred and eighty and the root of the root of a hundred and twenty-five minus the root of the root of five, and its price, the root of the root of a hundred and eighty and the root of the root of five minus the root of the root of a hundred and twenty-five. But if the first quantity is less than its price, say that it is the root of the root of a hundred and eighty and the root of the root of five minus the root of the root of a hundred and twenty-five, and its price, the root of the root of a hundred and eighty and the root of the root of a hundred and twenty-five minus the root of the root of five.

You may now want to know the second quantity and its price. Divide the root of five thousand seven hundred and sixty by the root of the root of two thousand eight hundred and eighty as we have shown in the chapter

---

[974] The MSS have 'twenty'.

[975] The MSS have 'ten', and thus invert the two given quantities.

[976] The determination of two quantities from their sum and product is considered known.

[977] A.325.

[978] That is, from $\sqrt[4]{180}$ (calculated before).

LIBER MAHAMELETH

on roots;[979] the result will be the root of the root of eleven thousand five hundred and twenty, which is the second quantity of sextarii plus its price. Now it has been supposed that multiplying the second quantity by its price produces twenty. Therefore[980] take half the root of the root of eleven thousand five hundred and twenty, which is the root of the root of seven hundred and twenty, and multiply this by itself; the result will be the root of seven hundred and twenty. Subtracting twenty from this leaves the root of seven hundred and twenty minus twenty. Take its root as we have taught above;[981] it will be the root of the root of five hundred[982] minus the root of the root of twenty. Add this to, and (also) subtract it from, the root of the root of seven hundred and twenty. If the first quantity of sextarii is the root of the root of a hundred and eighty and the root of the root of a hundred and twenty-five minus the root of the root of five, and its price, the root of the root of a hundred and eighty and the root of the root of five minus the root of the root of a hundred and twenty-five, then the second quantity will be the root of the root of seven hundred and twenty and the root of the root of five hundred minus the root of the root of twenty, and its price, the root of the root of seven hundred and twenty and the root of the root of twenty minus the root of the root of five hundred. But if the first quantity is the root of the root of a hundred and eighty and the root of the root of five minus the root of the root of a hundred and twenty-five, and its price, the root of the root of a hundred and eighty and the root of the root of a hundred and twenty-five minus the root of the root of five, then the second quantity will be the root of the root of seven hundred and twenty and the root of the root of twenty minus the root of the root of five hundred, and its price, the root of the root of seven hundred and twenty and the root of the root of five hundred minus the root of the root of twenty. [*For it has been supposed that the price of each sextarius is for both quantities the same.*][983]

The proof of all that has been said is the same as the preceding.[984]

(B.37) Someone asks:[985] There is an unknown quantity of sextarii such

---

[979] A.314.

[980] Determination of the two quantities from their sum and product.

[981] A.325.

[982] 'seven hundred and twenty' in the MSS.

[983] Similar interpolation in B.32.

[984] In B.35. In MS $\mathcal{B}$ we then find, by the same hand, the words 'after (this) leaf'; an early reader must have indicated where to find the next problem. (The disordered text contains cross-references, and the one found at the end of B.36 is repeated at the beginning of B.37.) This distance between these two problems also corresponds to the disorder in MS $\mathcal{B}$ itself, since B.36 ends on fol. 81va, 35 while B.37 begins on fol. 82vb, 5.

[985] In fact there is no question as such; the same in B.38, B.41, B.42, B.44, B.74, B.90″, B.211, B.213, B.224–226, B.231, B.237, B.239, B.346; see also B.98, B.103 ('someone says') and B.177–179, B.231–232, B.234–235.

that the root of its price is thrice the quantity of sextarii while subtracting the quantity of sextarii from its price leaves thirty-four.

You will do the following. Dividing the root as if it were one[986] by three —for he said 'thrice the quantity'— gives a third. Multiply[987] half of this, which is a sixth, by itself, thus producing a sixth of a sixth. Add it to thirty-four, thus making thirty-four and a sixth of a sixth. Add to its root, which is five and five sixths, a sixth; the result is six, which is the root of the price. Divide it by three; the result will be, for the quantity of sextarii, two, while the price will be thirty-six.

(B.37)

The proof of this is the following. Let the quantity of sextarii be AB and the root of its price, GD; thus the product of GD into itself is the price of the sextarii. Now we know that GD is thrice AB, so the product of GD into a third is AB, and (we also know) that subtracting AB from the square of GD leaves thirty-four; so subtracting from the product of GD into itself the product of GD into a third leaves thirty-four. I cut then from GD a third, say GH. So subtracting from the product of GD into itself the product of GD into GH leaves thirty-four. But the product of GD into itself, diminished by the product of GD into GH, is equal to the product of GD into DH. Therefore the product of GD into DH is thirty-four.[988] Let GH then be bisected at the point L. So[989] the products of GD into DH and LH into itself are equal to the product of LD into itself. Now the product of GD into DH is thirty-four, while the product of LH into itself is a sixth of a sixth. Therefore the product of LD into itself is thirty-four plus a sixth of a sixth, so LD is five and five sixths. Since GL is a sixth, GD is six, which is the root of the price, while AB is a third of the root. So AB is two and the price is thirty-six. This is what we wanted to demonstrate.

(B.38) Someone now asks: There is an unknown quantity of sextarii; the root of its price is twice this quantity, and adding quantity and price makes eighteen.

You will do the following. Dividing the root as if it were one by two —for he said 'is twice'— gives a half. Always multiply[990] half of this half,

---

[986] Better: 'Dividing one, which is the number of roots'. Same in B.38. In B.93, the author correctly speaks about the *numerus radicum*.
[987] Solving $x^2 = bx + c$ ($b = \frac{1}{3}$, $c = 34$).
[988] Difference and product of GD and HD thus known.
[989] *Elements* II.6 = PE_6.
[990] Solving $x^2 + bx = c$ ($b = \frac{1}{2}$, $c = 18$).

which is a fourth, by itself, thus producing half an eighth. Adding this to eighteen makes eighteen and half an eighth. Subtracting from its root, which is four and a fourth, a fourth leaves four, which is the root of the price of the sextarii. Thus the price of the sextarii is sixteen. Dividing four by two will give the quantity of sextarii.

The proof of this is the following. Let the quantity of sextarii be A and the root of its price, BG. Thus BG is twice A. Therefore multiplying BG by a half produces A. But we know that the product of BG into itself plus A is eighteen. Thus the products of BG into itself and into a half are eighteen. I draw the line BD, to be a half. Thus the products of BG into itself and into BD are eighteen. This is equal to the product of DG into BG. So multiplying DG by BG produces eighteen.[991] Then bisect BD, and continue with the rest of the problem as we have shown[992]. The result will be four for BG and two for A.

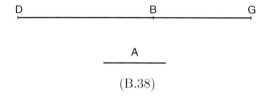

(B.38)

(**B.39**) Someone asks: Six modii are sold for four nummi and a thing, and two modii are bought[993] for three roots of this price. What is this thing worth?

You will do the following. Divide six by two, which gives three. Multiply this by three, which is the number of roots[994], thus producing nine. Multiplying this by itself produces eighty-one. Subtract four from this; this leaves seventy-seven, and such is the value of the thing. If there were more than one thing, seventy-seven would of course be their price.[995]

(**B.40**) Someone asks: Six modii are sold for a thing minus four nummi, and two modii are bought for three roots of this price. What is this thing worth?

You will do the following. Divide six by two, which gives three. Multiply this by three, namely the number of roots in the price, thus producing nine. Multiplying this by itself produces eighty-one. Add to this the four which was subtracted; this makes eighty-five, and so much is the thing

---

[991] Difference and product of DG and BG thus known.
[992] Preceding problem.
[993] As in B.23, at the selling price. Same in B.40–41.
[994] Not the previous result from the addition. Same distinction made in the next problem.
[995] Same remark (also in B.40) as in B.23, B.27, B.28, to which these problems are related.

worth. If there were more than one thing, eighty-five would of course be their price.

(B.41) Someone asks: Three sextarii are sold for two unequal things, the multiplication of which produces twenty-one, and one sextarius has been bought for the lesser thing and its ninth.

(a) You will do the following. Dividing the three sextarii by one sextarius gives three. Multiply it by one and a ninth, thus producing three and a third. Always subtract one from it; this leaves two and a third. Dividing twenty-one by this gives nine. Its root, which is three, is the lesser thing. Dividing twenty-one by this three gives seven, which is the greater thing.

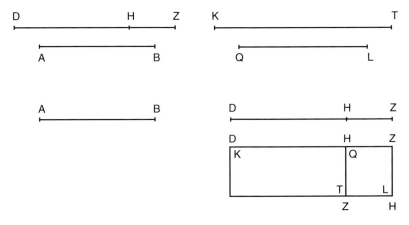

(B.41a, according to MS $\mathcal{A}$ above and to MS $\mathcal{B}$ below)

The proof of all these is the following.[996] Let the three sextarii be AB and the two unlike things, DH and HZ. We know by hypothesis that the multiplication of DH by HZ produces twenty-one; let then twenty-one be KT. It is then clear that the whole price of the three sextarii is DZ, which is the lesser and the greater things. Now one sextarius has been bought for the lesser thing, which is HZ, and its ninth. [*Therefore the ratio of one sextarius to three sextarii is the same as the ratio of HZ and its ninth to the whole DZ. Thus the result of dividing three sextarii by one is equal to the result of dividing the whole DZ by HZ and its ninth; for ratio is the same as division. But dividing three sextarii by one gives three. Therefore dividing DZ by HZ and its ninth gives three.*][997] So the product of HZ and its ninth [*which is one and its ninth*] into three is equal to DZ. But the result of multiplying the product of HZ into one and a ninth by three is equal to the result of multiplying the product of three into one and a ninth by HZ.[998] Now multiplying three by one and a ninth produces three

---
[996] As in B.30, different figures in MSS $\mathcal{A}$ and $\mathcal{B}$.
[997] This would seem to be superfluous.
[998] Premiss P_2.

LIBER MAHAMELETH

and a third. Thus the product of HZ into three and a third is equal to DZ. Therefore the whole DZ is the triple and a third of HZ. This leaves as HD the double of HZ and its third. Since multiplying HZ by DH produces twenty-one, multiplying HZ by its double and its third produces twenty-one. Let then HZ multiplied by itself produce QL.[999] Therefore multiplying HZ by itself produces QL while multiplying HZ by its double and its third produces twenty-one, which is KT. Thus the ratio of KT to QL is the same as the ratio of DH to HZ. Since DH is the double and a third of HZ, KT is the double and a third of QL. Therefore dividing KT by two and a third will give QL. ⟨So QL is nine.⟩ Since it is the square of HZ, HZ is three. This is what we wanted to demonstrate.

(b) Or otherwise, according to algebra,[1000] as follows. Put as the lesser thing a thing[1001] and as the greater thing, a dragma[1002]. So the price of the three sextarii will be one thing and one dragma. It has been supposed that one sextarius is bought for the lesser thing and its ninth, which is a thing and a ninth of a thing.[1003] This is equated to a third of the price of three sextarii, thus to a third of a thing and a third of a dragma. So you have a third of a thing and a third of a dragma equalling a thing and a ninth of a thing. Subtract a third of a thing from a thing and a ninth of a thing; this leaves seven ninths of a thing, which equal a third of one dragma. Therefore the whole dragma equals two things and a third of a thing. [*It is thus clear that a dragma is a thing and a third of a thing, taken twice.*[1004]] So multiply a thing by two things and a third of a thing, thus producing two squares and a third of a square; and this equals twenty-one.[1005] Therefore the square is nine, and the thing, three, which is the lesser thing. The dragma being the double and a third of a thing, it is seven, which is the greater thing. [*This is what we wanted to demonstrate.*]

(**B.42**) Someone asks: Five sextarii are given for two unequal things, the multiplication of which produces a hundred and forty-four, and one sextarius has been bought for a third of the lesser thing and two nummi.

(a) You will do the following. Divide five sextarii by one; this gives five.

---

[999] What follows is wordy. Since $HZ \cdot DH = KT$ while $HZ^2 = QL$, then (*Elements* VII.17) $DH : HZ = KT : QL = 2 + \frac{1}{3}$, so $KT = \left(2 + \frac{1}{3}\right) QL$, whence QL and HZ.

[1000] First allusion to algebra as a general way of treating problems.

[1001] Note the distinction between 'thing' = object and 'thing' = x.

[1002] Names of coins are commonly used in Arabic mathematical literature to designate various unknowns (see our *Introduction to the History of Algebra*, p. 76). A reader of our text was obviously not aware of this; see below.

[1003] See note 1001.

[1004] *Sic.* The same reader will intervene in the algebraic treatment of the next problem (B.42b).

[1005] Equation of the form $ax^2 = c$. First introduction of x^2 (*census*, Arabic *māl*).

Multiplying this by two nummi produces ten. Next, take a third of five —for he said 'a third of the lesser thing'— which is one and two thirds. Always subtract one from this[1006], thus leaving two thirds. Dividing a hundred and forty-four by this gives two hundred and sixteen. Dividing then the above ten[1007] by two thirds gives fifteen. Multiply[1008] its half, which is seven and a half, by itself; this produces fifty-six and a fourth. Add it to two hundred and sixteen, thus making two hundred and seventy-two and a fourth. Subtract from its root, which is sixteen and a half, half of the things[1009], which is seven and a half; this leaves nine, which is the lesser thing. Dividing a hundred and forty-four by this gives sixteen, which is the greater thing.

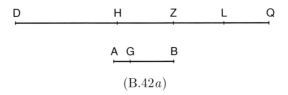

(B.42a)

The proof of this is the following. Let the five sextarii be AB, the two unequal things, DZ, the lesser being ZH and the greater, DH, and the sextarius bought, AG. We know that the ratio of the sextarius, which is AG, to the five sextarii, which is AB, is the same as the ratio of a third of HZ plus two nummi to the whole DZ. Since AG is a fifth of AB, a third of HZ plus two nummi is a fifth of the whole DZ. Thus the product of a third of HZ plus two nummi into five is equal to the whole DZ. But[1010] the product of a third of HZ plus two nummi into five is equal to the product of two nummi into five and the product of a third of HZ into five. Now[1011] the product of a third of HZ into five is equal to the product of a third of five into HZ, while the product of two nummi into five is ten; and a third of five is one and two thirds. Thus the product of one and two thirds into HZ plus ten is equal to the whole DZ. [*Therefore DZ less ten is equal to HZ and its two thirds. Thus DH less ten is two thirds of HZ.*][1012] It is therefore clear ⟨that two thirds of HZ plus ten nummi are DH. But multiplying DH by HZ produces a hundred and forty-four. It is therefore clear⟩ that the products of HZ into its two thirds and ten are a hundred and forty-four; thus HZ multiplied by itself and fifteen produces two hundred and sixteen [*as we have established in what precedes*[1013]]. Therefore I draw a line of fifteen, say ZQ. Thus the products of ZH into itself and into

---
[1006] *de qua*, referring to *tertia quinque*.
[1007] Just calculated.
[1008] Solving $x^2 + bx = c$ ($b = 15$, $c = 216$).
[1009] Rather: 'half of fifteen', for the equation has not been set.
[1010] *Elements* II.1 = PE_1 if need be.
[1011] Premiss P_2.
[1012] This seems to be a reader's attempt to fill the lacuna below.
[1013] In the computation before the proof.

LIBER MAHAMELETH   811

ZQ are two hundred and sixteen. Since the products of HZ into itself and into ZQ are equal to the product of HZ into HQ, the product of HZ into HQ is two hundred and sixteen.[1014] Let line ZQ be bisected at the point L. Continue the problem as we have taught before,[1015] and it will lead to what you wanted.

(*b*) Or otherwise, according to algebra, as follows. Let the lesser thing be a thing and the greater, a dragma. Thus the price of the five sextarii will be one thing and one dragma. It has been supposed that one sextarius is bought for a third of the lesser thing and two nummi, which is a third of a thing and two nummi. This equals a fifth of a thing and a fifth of one dragma. Subtracting a fifth of a thing from a third of a thing leaves two thirds of a fifth of a thing and two nummi, which equal a fifth of one dragma. Thus the whole dragma equals two thirds of a thing and ten nummi. [*It is thus clear that two thirds of the lesser thing plus ten nummi equal the greater thing. Having put a thing as the lesser, the greater will be two thirds of a thing and ten nummi.*] So the result of multiplying a thing by two thirds of a thing and ten nummi will be equal to a hundred and forty-four. Proceed with the rest as we have taught in algebra,[1016] and the result will be what you wanted.

ANOTHER CHAPTER, ON MODII AT DIFFERENT PRICES

(**B.43**) Someone asks thus: A modius of one corn is given for six nummi, of another for eight nummi, of yet another for nine nummi. With the thirty nummi which I had, I bought one and a fourth modii of the first and one and two thirds modii of the second. How much of the third corn shall I obtain with the remainder of the thirty nummi?

You will do the following. Seek the price of one and a fourth modii according to the selling price of six nummi to the modius; you will find seven nummi and a half. Next, seek the price of one and two thirds modii according to the selling price of eight nummi; you will find thirteen nummi and a third. Add to this the seven and a half, thus making twenty nummi and five sixths. It follows that he bought corn for twenty nummi and five sixths, which leaves of the thirty nummi nine and a sixth. Therefore consider how much he is to obtain of the corn sold at nine nummi to the modius for nine nummi and a sixth. You will find one modius and a sixth of one ninth, and this is what is obtained with the remainder of the thirty nummi.

In this way you will be able to find the answer when there are two, three or more corns.[1017]

(**B.44**) Someone asks: A modius of one corn is given for six nummi, of

---

[1014] Difference and product of HZ and HQ thus known.

[1015] Usual application of *Elements* II.6, as seen in B.31 & B.37–38.

[1016] Reducing to the form $x^2 + bx = c$, then solving as seen in B.38.

[1017] Sold at different prices, with, as here, just one quantity bought being unknown.

812   PART TWO: TRANSLATION, GLOSSARY

another for eight, and of yet another for ten. I want to buy equally of the three corns for eighteen nummi.

You will do the following. Add the individual prices of one modius; the sum will be twenty-four. Divide eighteen nummi by this, which gives three fourths. Therefore he bought of each corn three fourths of a modius, and this is the quantity you require.

(**B.45**) Someone says: A modius of one corn is given for six nummi and of another for eight. I want to buy three modii of the two corns for a same price. How much shall I receive of each?

(*a*) You will do the following. Seek a number divisible by six and by eight; such is twenty-four. Dividing it by six gives four; put this instead of six. Next, dividing twenty-four by eight gives three; put this instead of eight. Then say: 'Of two partners one has invested four nummi, the other three, and they have gained three modii; how will they divide them?'[1018]   The share of the investor of four is a modius and five sevenths, the share of the investor of three is a modius and two sevenths.   Therefore you are to receive,[1019] of the corn sold at six to the modius, a modius and five sevenths, the price of which is ten nummi and two sevenths; of the other corn, at eight to the modius, you are to receive a modius and two sevenths, the price of which is ten nummi and two sevenths. This is what you wanted to know.

(*b*) Or if you want to know how much he[1020] received of the corn at six nummi to the modius, divide six by itself and by eight, add the two quotients, [*which are one and three fourths*], and divide three by the sum; the result will be what you require.   Or if you want to know how much he received of the corn at eight nummi to the modius, divide eight by itself and by six, add the two quotients, and divide three by the sum; the result will be what you require.

If there are more than two corns, follow the two ways explained; the result will be what you require.

(**B.46**) Someone asks thus: A modius of one corn is given for six nummi and of another for eight nummi. I want to receive of both corns one modius for six and a half nummi. How much shall I receive of each?

Consider here whether the price at which he wants to buy, namely six and a half, lies between the two above prices, namely six and eight: if so, the problem will be solvable; but if this price is lower than the lesser or higher than the greater, the problem will not be solvable.[1021]   In the

---

[1018] See B.104.

[1019] The asker was 'I', and became 'you' for the computations. See p. 784, note 874.

[1020] Now, as usual, 'you' is the reader and 'he' the asker.

[1021] The case of the unknown price equal to one of the given prices set is not considered here ('lies between'), but it is in the next problem.

LIBER MAHAMELETH

present case six and a half lies between the two prices, and the problem will be solvable.

So you will do here the following. Take the difference between the two above prices, which is two; let it be your principal. If you want to know how much he will receive of the corn at six nummi to the modius: Take the difference between eight and the price at which you [1022] want to buy; this is one and a half nummi. Divide it by the principal; this gives three fourths. So much will you receive of the corn at six nummi, that is, three fourths of a modius. If you want to know how much he will receive of the corn at eight nummi: Take the difference between six and the price at which you want to buy; this is a half. Divide it by the principal; this gives a fourth. So much is he to receive of the corn at eight nummi, namely a fourth of one modius. The result is thus a modius of both corns for six and a half.

(B.47) Someone says: There were ten modii of barley and of wheat. I sold each modius of barley for six nummi and each modius of wheat for ten nummi, and I received altogether eighty-eight nummi. How many modii were there of barley, or how many of wheat?

You will do here the following. Take the difference between the two prices, which is four; let it be your principal. If you want to know how many modii of wheat there were: You must determine what the price of ten modii would be if they were all of barley, as follows. Multiply six, which is the price of one modius, by ten, which is the number of modii, thus producing sixty. Subtract it from eighty-eight; this leaves twenty-eight. Dividing this by the principal gives seven, and that many were the modii of wheat. If you want to know how many modii of barley there were: Consider what the price of ten modii would be if they were all of wheat; you will find a hundred. Subtracting eighty-eight from this leaves twelve. Dividing it by the principal gives three, and that many are the modii of barley. This is what you wanted to know.

Consider here likewise [1023] whether the price gathered from the whole sale, in this case eighty-eight, lies between the two prices which would result if all the modii consisted of barley or all of wheat. For if this amount is higher than the greater result or lower than the lesser result, the problem will not be solvable, as if here the asker had said that from the whole sale he obtained more than a hundred or less than sixty; for if he had said that he obtained a hundred nummi, all the modii would be of wheat, and if sixty, all would be of barley.

(B.48) Someone asks: A sextarius of one corn is given for three nummi, of another for four, and of yet another for five. Someone wants to receive for two nummi equally of each corn. How much will he receive of each?

You will do the following. Add three, four and five, thus making twelve. Denominate two from it; this is a sixth. He receives that much

---

[1022] This should also refer to the asker.
[1023] As in B.46.

of each sextarius. Thus he receives a sixth of a sextarius at three nummi for half a nummus, a sixth of a sextarius at four nummi for two thirds of a nummus, and a sixth of a sextarius at five nummi for five sixths of a nummus. Therefore he received of the three corns half a sextarius for two nummi.

The proof of this is the following. Let the quantity received from the corn at three nummi to the sextarius be A, and this is (also) what he receives from each of the other corns. Let its price [*that is, (of)* A] when the sextarius is given at three nummi be BG, its second price [*that is, (of)* A], when the sextarius is given at four, be GD, the third, when it is given at five, be DH. Therefore the whole BH is two nummi, according to what has been proposed. Now it is known [1024] that the ratio of A to one is the same as the ratio of BG to three; thus the product of A into three is equal to the product of one into BG; now the product of one into BG is BG, so the product of A into three is BG. The ratio of A to one being the same as the ratio of GD to four, the product of A into four is equal to the product of one into GD; since multiplying one by GD produces just GD, multiplying A by four produces GD. I shall likewise show that multiplying A by five produces DH. Thus multiplying A by three produces BG, by four, GD, and by five, DH. It is therefore clear [1025] that multiplying A by twelve produces the whole BH, which is two. So divide two by twelve; this gives a sixth, and so much does he receive from each of the three sextarii. This is what we wanted to demonstrate.

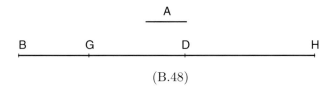

(B.48)

(**B.49**) A sextarius of one corn is given for three nummi, of another for four, and of yet another for five. Someone wants to receive of these three sextarii altogether one sextarius, but of each equally.

This problem is straightforward. For the sextarius is divided into three equal parts and he receives of each corn a third of a sextarius.

(**B.50**) Someone asks: A sextarius of one corn is given for three, of another for four, of yet another for five. A man buys of the three sextarii altogether one and a half sextarii and he receives of each equally. For how many nummi does he buy this?

You will do the following. Add three, four and five, thus making twelve. Next, denominate one and a half sextarii from three; this is a half, and the

---

[1024] From B.1 on (quantity : quantity = price : price). What follows is verbose, but usual (see p. 595, notes 64, 67).
[1025] *Elements* II.1 = PE_1.

LIBER MAHAMELETH

same fraction taken of twelve, which is six, is the number of nummi. If you want to know how much of each corn he has received: Denominate one and a half sextarii from three; this is a half, and so much does he receive of each sextarius.[1026]

(B.51) A sextarius of one corn is given for three nummi, of another for seven, and of yet another for twelve. Someone has received of all three one sextarius for ten nummi. How much of each sextarius did he receive?

This problem is indeterminate. You will do for it the following. Add three and seven, thus making ten. Next, double twelve, which gives twenty-four —you have doubled the price of the third corn because you have added the prices of the (other) two; if you had added three prices, you would triple the last.[1027] Next, subtract the ten of the addition[1028] from the double, which is twenty-four, thus leaving fourteen. Next, subtract ten, for which he has bought the sextarius,[1029] from twelve, which is the price of the third sextarius, thus leaving two. Denominate it from fourteen; this is a seventh. So much does he receive of the sextarius at three nummi, and so much also of the sextarius at seven nummi; he receives of the third sextarius, at twelve nummi, the remainder, namely five sevenths of a sextarius.

But if it were said in this problem that he bought one sextarius for twelve nummi or more, or for three or less, the problem would not be solvable.[1030]

(B.52) Someone asks: A man buys ten sextarii, the first for three and each of the subsequent prices exceeds the previous by four. What are the prices of the last and of all?

You will do the following. Always subtract one from the number of sextarii, which leaves in this case nine. Multiply it by the difference whereby they exceed one another, namely four, and add to the product twice the price of the first, which is six; the result is forty-two. Multiply this by half the number of sextarii, which is five, thus producing two hundred and ten. So much is the price of all sextarii. If you want to know the price of the last, which is in this case the tenth, multiply the difference by the number of sextarii minus one and add to the product the price of the first; the result will be the price of the last, namely thirty-nine.[1031]

(B.53) Someone buys twelve sextarii and a fourth, the first for three while all the others exceed one another by five. What is the price of all?

[1026] These two problems are, to say the least, particular.

[1027] This does not make the reason very clear; for a more satisfactory explanation the reader has to await B–XVIII.

[1028] Not that of the data.

[1029] Not the 'ten' of the addition.

[1030] Solution equal to zero again excluded (see note 1021).

[1031] MS $\mathcal{B}$ gives another reading of this same problem, at the end of which an early reader added an (inadequately expressed) reference to the problems in B–XII. Note that B–XII is found *before* this problem in the disordered text.

You will do the following. Always subtract one from the integral number of sextarii; this leaves (in this case) eleven. Multiply it by the common difference, add to the product twice the price of the first, and multiply the result by half of twelve; this gives three hundred and sixty-six. Next, multiply twelve by the (common) difference and add the price of the first sextarius to the product, which becomes sixty-three. Its fourth, which is fifteen and three fourths, is the price of a fourth of the last sextarius. Add this to three hundred and sixty-six; this makes three hundred and eighty-one and three fourths, and so much is the price of twelve sextarii and a fourth.

Chapter (B–II) on Profits

To this chapter belong five types, which follow.[1032]

— The first is when the capital is known and the profit unknown.

(B.54) For instance, someone asks thus: From what I have bought for five [*nummi*][1033] I have gained three. How much shall I gain from what I have bought for eighty?

You will do the following. Multiply eighty, which is the second capital, by three, which is the first profit, and divide the product by the first capital, which is five; the result will be what you require.

This rule is taken from a proportion. For in profits the ratio of the first profit to the first capital is the same as the ratio of the second profit to the second capital; you may also put the capital first, thus: the ratio of the first capital to the first profit is the same as the ratio of the second capital to the second profit. In the above problem the ratio of the first capital, namely five, to three, which is its profit, is the same as the ratio of the second capital, which is eighty, to its unknown profit. Thus the fourth is unknown. So[1034] it is necessary to multiply the second, which is three, by the third, which is eighty, and to divide the product by the first, which is five, and this will give the fourth, which is required; or to divide one of the factors by the divisor and multiply the result by the other, as has already been said,[1035] the result being what is required.

— The second type is the converse of the preceding, when the profit is known and the capital unknown.

(B.55) For instance, someone asks thus: In buying for five I have gained three and thereafter from another capital I have gained forty. How much was the capital from which I have gained forty?

We know[1036] that the ratio of the first capital to its profit is the same as the ratio of the required capital to forty. Therefore multiply the fourth, which is forty, by the first capital, which is five, and divide the product by the first profit, which is three; the result will be the third. [*For this is the appropriate way: always to multiply like by unlike, that is, neither profit by profit nor capital by capital, but profit by capital and conversely.*][1037] This

---

[1032] The five types are described in B.54–58, which form the first section.

[1033] These initial examples being general, no kind of unit needs to be mentioned.

[1034] Introduction to Book B, *i* (Rule of Three, p. 777).

[1035] Introduction to Book B, *ii* (banal 'other ways', p. 778).

[1036] B.54.

[1037] This gloss, commenting on the subsequent sentence, is found in MS C only (within the text).

818   PART TWO: TRANSLATION, GLOSSARY

rule is clear from the proportion with the numbers disposed as we have taught above.[1038]

CHAPTER ON UNKNOWN PROFITS.[1039]

— Such is the third type: each is unknown but their sum is given.

(B.56) For instance, someone asks: In buying for five I have gained three; I have gained from another capital, and adding profit and capital makes a hundred. How much of this hundred is profit and how much capital?

You will do the following. Add five to three, which is its profit, thus making eight; let it be your principal. If you want to know how much of the above hundred is capital, multiply the first capital, namely five, by a hundred and divide the product by the principal; the result will be the capital you require. If you want to know how much of the hundred is profit, multiply the first profit, namely three, by a hundred, and divide the product by the principal; the result will be the profit you require.

The proof of this is the following. Let five be AB, three, BG, the unknown capital, DH, and the profit, HZ; thus the whole DZ is a hundred. Now the ratio of AB to BG is the same as the ratio of DH to HZ. By composition, the ratio of AG to GB will be the same as the ratio of DZ to ZH. These are therefore four proportional numbers. Then multiplying the first by the fourth produces the same as multiplying the second by the third.[1040] So dividing the product of the second into the third by the first, which is eight, will give the fourth, which is the profit. This is what we wanted to demonstrate.

The proof for finding the capital takes the same course. For it is clear[1041] that the ratio of AG to AB is the same as the ratio of DZ to DH. So dividing the product of AB, the second, into DZ, the third, by AG, the first, will give DH, the fourth. This is what we wanted to demonstrate.

```
A        B    G    D                        H            Z
├─────────┴────┤    ├──────────────────────┴────────────┤
```

(B.56)

— The fourth type is when both are unknown but the remainder after subtraction is given.

(B.57) For instance, someone asks thus: In buying for six I have gained one and a half, and subtracting from another capital its profit leaves ninety. How much were profit and capital?

You will do the following. Subtracting from six its profit leaves four and a half, which is the principal. If you want to know the unknown

---

[1038] The two arrangements mentioned in B.54.

[1039] Problems with two unknowns.

[1040] *Elements* VII.19.

[1041] By composition again, but with the proportion inverted.

LIBER MAHAMELETH
819

capital, multiply the first capital, namely six, by ninety and divide the product by the principal; the result will be the capital. If you want to know the unknown profit, multiply the first profit, namely one and a half, by ninety and divide the product by the principal; the result will be the profit.

The proof of this is evident to those who think about it; for it is similar to the previous proof, except that this case involves converting the ratio instead of composing it. We have already explained these cases for the unknown modii and sextarii and we have established the proportion for them.[1042]

You may in this case divide six by the principal and multiply the quotient by ninety, and the result will be the capital; and dividing one and a half by the principal and multiplying the quotient by ninety will give the profit. Or else you may divide ninety by the principal; multiplying the quotient by six will produce the capital, and multiplying the quotient by one and a half will produce the profit. These two ways consist in dividing first and multiplying afterwards, as has already been said before about any two terms multiplying each other and a third dividing.[1043]

— The fifth type is when both quantities are unknown but their product is given.

(B.58) For instance, someone asks thus: In buying for five I have gained three and the multiplication of another capital by its profit produces sixty. How much was the profit and how much the capital?

We have already explained this case for the unknown modii,[1044] which is as follows. If you want to know the unknown capital, multiply five, which is the first capital, by sixty and divide the product by three, which is the first profit; the root of the result is what you require. Or, if you want to know the profit, multiply the first profit, namely three, by sixty and divide the product by five; the root of the result is what you require.

Or, having determined the capital first, divide sixty by it, and the result will be the profit. Or, having determined the profit first, divide sixty by it, and the result will be the capital.

The proof of what we said initially [1045] is the following. Let five, which is the first capital, be A, three, which is the first profit, B, the required capital, G, the second profit, D. Let G be multiplied by D and produce H; thus H is sixty. Let G be multiplied by itself and produce Z. And let D be multiplied by itself and produce K. It is then clear that Z, H and K form a continued proportion having the ratio of G to D.[1046] But the ratio of G to D is the same as the ratio of A to B. So Z, H and K form

---

[1042] See B.15–16 (involving modii).

[1043] Introduction to Book B, *ii* (p. 778).

[1044] B.17.

[1045] First way of solving, using the roots.

[1046] That is, $Z : H = H : K$ (and $H : K = G \cdot D : D^2 = G : D$).

a continued proportion having the ratio of A to B. Therefore the ratio of A to B is the same as the ratio of Z to H; so multiplying A by H and dividing the product by B will give Z; since this is the square of G, the root of Z is G. Likewise again, the ratio of A to B is the same as the ratio of H to K; so multiplying B by H and dividing the product by A will give K; its root is D. This is what we wanted to demonstrate.

(B.58)

CHAPTER ON PROFITS WHEREIN THAT WHICH IS SOLD OR BOUGHT IS NAMED [1047]

There are four cases in this chapter. We shall illustrate them by giving examples for modii;[1048] but these cases occur similarly for pounds or others involving prices.

(i) The first case concerns profit in nummi from nummi. For instance, someone buys a modius for five nummi which he then sells for eight; how many nummi will he gain from a hundred nummi?

(ii) The second (case) concerns profit in nummi from modii. For instance, someone sells for eight nummi a modius bought for five; how many nummi will he gain from a hundred modii?

(iii) The third case concerns profit in modii from nummi. For instance, someone sells for eight nummi a modius bought for five; how many modii will he gain from a hundred nummi?

(iv) The fourth case concerns profit in modii from modii. For instance, someone sells for eight nummi a modius bought for five; how many modii will he gain from a hundred modii?

In each of these four cases occur the two types spoken about a little before [1049], the first of which is to determine profit from capital and the second to determine capital from profit, and also the three types expounded above in the chapter on unknown profit [1050], the first of which is to determine capital and profit from their sum, the second to determine

---

[1047] That is, it is specified, by indicating the unit, whether the two quantities involved (profit and capital) are in cash or kind. B.59–82 illustrate the various cases.

[1048] In the subsequent problems, the capacity unit will be mostly the *caficius*, occasionally (B.59, B.65) the *modius*.

[1049] B.54–55 (one unknown).

[1050] B.56–58 ('chapter on unknown profit̲s'; two unknowns).

LIBER MAHAMELETH

each from their difference, the third to determine profit and capital from their product. There are therefore twenty types or cases in this chapter altogether [1051]. Two of the four cases arise from the problem 'what is the price?' and the other two from the problem 'how much shall I have?' [1052] I shall presently propose an example for each of the five types of the four cases.

(*i*) THE FIRST OF THE FOUR CASES CONCERNS PROFIT IN NUMMI FROM NUMMI.

(**B.59**) For instance, someone asks: I sell for seven and a half a modius bought for six; how many nummi shall I gain from a hundred nummi?

This case [1053], as we said before, arises from the problem 'what is the price?' In this case, you will do the following. You already know that here the profit from [*a modius and*] six nummi is one and a half nummi. [*So what is required here is the profit in nummi from nummi, and the capital is known.*] Then it is like saying: 'In buying for six nummi I have gained one and a half nummi; how much shall I gain from what I bought for a hundred?' Proceed here as has been said;[1054] the result will be twenty-five nummi. This is what you wanted. If the one hundred were solidi, the twenty-five would also be solidi.

Inversely, if the profit were indicated and the capital required, you would again proceed as shown before.

You will also adapt the three above mentioned types to this case, namely if he says: 'Adding the profit and the capital makes so many or so many nummi, or solidi, then how much is the profit or how much the capital?' Or if he says: 'Subtracting the profit in nummi from the capital in nummi', or 'the profit in solidi from the capital in solidi, leaves so much or so much, then how much is the profit or the capital?' Or if he says: 'Multiplying the profit (in nummi) by the capital (in nummi)', or 'multiplying the profit in solidi by the capital in solidi, produces so much or so much, then how much is the capital or the profit?'

The treatment of these types is evident from what we have taught above.[1055] For the purpose of greater clarity, though, I shall present an example of each of the five types of this first of the four cases.

[1051] We shall refer hereafter to the four *cases* and the five *types* occurring in each —a distinction which the text hardly makes, with its indiscriminate use of *modi* and *species*, or at most differentiating between the *four* main cases (*quatuor prime species*) and the *five* types (*quinque species*). The subheadings indicating the types 2 to 5 are probably early readers' additions, as are those found in the examples for the first type (B.60, B.66, B.72, B.78).

[1052] This refers to the profit, which is in nummi in the first two cases and in caficii in the last two.

[1053] Meaning the present problem, with profit in cash required. See B.78.

[1054] B.54. The references are, in this whole section, to the group B.54–58.

[1055] Again, B.54–58.

(B.60) For instance, someone asks: I have bought three caficii for ten nummi, and I have sold four for twenty. How many nummi shall I gain from a hundred nummi?

This case, as we said before [1056], arises from the problem 'what is the price?' Then you will say: 'When four caficii for twenty, what is the price of three?' It is fifteen. Then you will say: 'I have bought three caficii for ten and sold them for fifteen.' Therefore from ten nummi I gain five. Then you will say: 'If in buying for ten he gains five, how much will he gain in buying for a hundred?' Proceed as I have taught before; the result will be fifty, and so much does he gain in buying for a hundred. [*Such is the first of the five types into which each of the four cases is divided.*]

— The second of the five types is the following.

(B.61) A man buys three for ten and sells four for twenty.[1057] From how many nummi results his gain of fifty nummi?

You will do the following. You will say: 'When four for twenty, what is the price of three?' It is fifteen. He therefore bought three for ten and sold them for fifteen, thus gaining five from ten. Then you will say: 'When from ten he gains five, from how much will he gain his profit of fifty?' Proceed as I have taught above; the result will be a hundred, and this is the quantity of nummi from which he gains fifty.

— The third of the five types is the following.

(B.62) Someone asks: A man buys three caficii for ten and sells four for twenty, and adding his profit in nummi to his capital in nummi makes a hundred and fifty. What is the profit or the capital?

You will do the following. You will say: 'When four are given for twenty, what is the price of three?' It is fifteen. He therefore buys three caficii for ten and sells them for fifteen; thus from ten nummi he gains five. Then you will say: 'In buying for ten he gains five, and the sum of profit and capital is a hundred and fifty.' Proceed as I have taught above; the result will be for the capital a hundred and for the profit, fifty.

— The fourth of the five types is the following.

(B.63) Someone says: A man buys three caficii for ten nummi and sells four for twenty nummi, and subtracting the profit in nummi from the capital in nummi leaves fifty nummi.

You will do the following. You will say: 'When four for twenty, what is the price of three?' It is fifteen. He therefore bought three caficii for ten and sold them for fifteen; thus from ten nummi he gains five. Then you will say: 'In buying for ten he gains five, and subtracting profit from capital leaves fifty.' Proceed as I have taught above; the result will be for the capital a hundred and for the profit, fifty.

— The fifth of the above mentioned types is the following.

[1056] Beginning of B.59 and introduction to this section.

[1057] The unspecified capacity unit must be, as elsewhere, the caficius.

LIBER MAHAMELETH 823

(B.64) Someone asks: A man buys three caficii for ten and sells four for twenty, and multiplying the profit in nummi by the corresponding capital in nummi produces five thousand.

You will do the following. You will say: 'When four for twenty, what is the price of three?' It is fifteen. Thus it is known that from ten he has gained five. Then you will say: 'In buying for ten he gains five, and multiplying capital by profit produces five thousand.' Proceed as I have taught before; the result will be for the capital a hundred and for the profit, fifty.

(*ii*) THE SECOND OF THE FOUR CASES CONCERNS PROFIT IN NUMMI FROM MODII.

(B.65) Someone asks: I sold for seven and a half nummi a modius bought for six. How many nummi shall I gain from a hundred modii?

Here he wants to know from the profit in nummi, namely one and a half nummi gained from one modius, how much he will gain from a hundred modii. Then it is like asking: 'With one modius I have gained one and a half nummi; how much shall I gain from a hundred modii?' Say: a hundred and fifty nummi.

Assume the problem had been stated thus: 'With one caficius I have gained one and a half nummi, how much shall I gain with a hundred almodis?' Now with one caficius he has gained one and a half nummi; therefore he will gain with twelve caficii, which are one almodi, eighteen nummi. So multiply this profit from one almodi by a hundred; the result will be the profit he makes with a hundred almodis.

Assume he has indicated the profit [*which is one and a half nummi*][1058] and asked for the capital in caficii, stating the problem thus: 'I have gained with one caficius one and a half nummi and I have gained so many or so many nummi with another quantity of caficii; how many were these caficii?' Proceed as we have taught before; the quantity of caficii will be the result.

Assume he asked thus: 'I have gained a hundred solidi; with how many caficii did I gain it?' Consider first with how many caficii he gains a hundred nummi and multiply them by twelve; the product is what you require. We do this because a hundred solidi corresponds to twelve times as many (nummi).

You may likewise derive for this case the three types involving unknowns [1059], as if he says: 'Adding the profit in nummi to the corresponding capital in caficii makes so much or so much.' Or: 'Subtracting the capital (in caficii) from the profit (in nummi) leaves that much or that much' — for in this problem the quantity of the capital in caficii is smaller than the quantity of the profit in nummi. Or: 'Multiplying the capital in caficii by the profit in nummi produces so much or so much.'

[1058] The reader wanted to specify that the profit given is expressed in nummi.

[1059] *tres modos ignoti*. That is, with two unknowns instead of one.

824 PART TWO: TRANSLATION, GLOSSARY

The treatment in all these types is evident from what has been said before.[1060] For the sake of greater clarity, however, I shall present an example of each of the five types of this second of the four cases.

[Here the capital is known but the profit is unknown.][1061]

(B.66) For instance, someone says: A man buys three caficii for ten nummi and sells four for twenty. How many nummi will he gain with a hundred caficii?

You will do the following. You will say: 'When four are given for twenty nummi, what is the price of three?' It is fifteen. Thus he bought three caficii for ten and sold them for fifteen. Therefore with three caficii he gains five nummi. Then you will say: 'Someone with three caficii gains five; how much will he gain with a hundred?' Proceed as I have taught above; the result will be a hundred and sixty-six and two thirds, and so many nummi does he gain with a hundred caficii. *[Such is the first of the five types into which each of the four cases is divided.]*

— The second of the five is the following.

[Here the profit is known and the capital unknown.]

(B.67) A man buys three caficii for ten nummi and sells four for twenty. From how many caficii does he gain his profit of a hundred nummi?

You will do the following. You will say: 'When four caficii are given for twenty nummi, what is the price of three?' It is fifteen. Thus he bought these three caficii for ten nummi and sold them for fifteen. Therefore with three caficii he has gained five nummi. ⟨Then you will say: 'Someone with three caficii gains five nummi;⟩ with how many caficii does he gain his profit of a hundred [*nummi*]?' Proceed as has been explained before; the result will be sixty.

You can verify this as follows. You know that he has bought sixty caficii, three for nummi; thus he bought them for two hundred nummi. He sold them for three hundred, for he sold four for twenty. Therefore with sixty caficii he gains, according to the price set, a hundred nummi.

— The third type is the following.

[Here both are unknown, but their sum is known.]

(B.68) Someone buys three caficii for ten nummi and sells four for twenty, and adding the profit in nummi to the capital in caficii makes a hundred. What is the profit in nummi and the capital in caficii?

You will do the following. You will say: 'When four are given for twenty, what is the price of three?' It is fifteen. Subtracting ten from this leaves five. Then it is like saying: 'With three caficii he gains five nummi, and adding profit to capital makes a hundred.' Proceed as we have shown

---

[1060] Again, B.54–58, alluded to in each problem.

[1061] Early readers' additions found (from here to B.79) in the margin of MSS $\mathcal{A}$ and $\mathcal{B}$. ($\mathcal{C}$ has none of these problems.)

LIBER MAHAMELETH

above; there will result as the profit in nummi sixty-two and a half and as the capital in caficii, thirty-seven and a half.

— The fourth type is the following.

[*Here both are unknown but they are determined from the subtraction.*]

(B.69) Someone says: A man buys three caficii for ten and sells four for twenty, and subtracting the capital in caficii from its profit in nummi leaves a hundred.

If it were said that the profit in nummi is subtracted from the capital in caficii, this would not be solvable; for the profit in nummi is greater than the capital in caficii. Indeed, proceeding as we have taught before[1062], it will appear that he has gained five nummi with three caficii, thus that the profit in nummi is greater than the capital in caficii. So if (it is stated, as above, that) subtracting his capital in caficii from his profit in nummi leaves a hundred, the problem will be solvable. And his capital in caficii will be a hundred and fifty and the profit in nummi, two hundred and fifty.

Consider other similar cases like this one. But when the profit in nummi is greater than the capital in caficii and it is proposed to subtract the profit from the capital, the problem will not be solvable; and if the profit in nummi is less than the capital in caficii and it is proposed to subtract the capital in caficii from the profit in nummi, the problem will not be solvable either.

— The fifth type is the following.

[*Here both are unknown but they are determined from their product.*]

(B.70) A man bought three caficii for ten nummi and sold four for twenty nummi, and multiplying the profit in nummi by the corresponding capital in caficii produces a hundred.

You will do the following. You will say: 'When four are given for twenty, what is the price of three?' It is fifteen. Subtracting ten from this leaves five. Thus it is known that these five are gained from three caficii. Then you will say: 'Five are gained from three caficii, and multiplying the profit by the capital produces a hundred.' Proceed as has been shown above; the result will be, for the capital in caficii, the root of sixty; for the profit in nummi, the root of a hundred and sixty-six and two thirds.[1063]

(*iii*) THE THIRD OF THE FOUR CASES IS PROFIT IN CAFICII FROM NUMMI.

(B.71) For instance, someone asks thus: I have sold a caficius, bought for six nummi, for seven and a half. How many caficii shall I gain from a hundred nummi?

It is clear that, considering that a caficius is given for seven nummi and a half, four fifths of it will be given for six nummi; the remainder of the caficius [*worth one and a half nummi, that is*][1064] is the profit, which is

---

[1062] Beginning of B.68.

[1063] The values are interchanged in the MSS.

[1064] Probably the same interpolator as in B.65 (note 1058).

826   PART TWO: TRANSLATION, GLOSSARY

a fifth of a caficius. So with six nummi he gains a fifth of a caficius. Then it is like asking: 'With six nummi I have gained a fifth of a caficius; how much shall I gain with a hundred nummi?' Proceed as has been explained before;[1065] the result will be three and a third. Thus he has gained three and a third caficii with a hundred nummi.

But if this (capital) were a hundred solidi, you would multiply three and a third caficii by twelve, which is the number of nummi in a solidus [1066], and the result would be the amount of the profit from a hundred solidi. Or, since you already know that he has gained a fifth of a caficius with six nummi, and six is half a solidus, you will then say: 'When with a half he has gained a fifth, how much will he gain with a hundred?' Proceed as we have taught before; the result will be the quantity of caficii gained from a hundred solidi.

He may ask here: 'I have gained ten caficii; with how many nummi, or how many solidi, did I gain them?' You already know that a fifth of a caficius is his profit from six nummi, and this is half a solidus. Then you will say: 'When with six nummi he has gained a fifth of a caficius and with another capital he has gained ten, what is this capital?' Proceed as we have taught before; the result will be the amount of this capital in nummi. Next, you will convert it into solidi. But you may also say: 'With half a solidus he has gained a fifth; with what did he gain ten?' Proceed as has been explained before; the result will be the quantity of solidi.

You will likewise adapt the three (remaining) types to this case, when it is asked for profit in caficii and capital in nummi from their sum, or from the subtraction of profit in caficii from capital in nummi, or from their product. Proceed as has been explained before, and the result will be what you require.

For the sake of greater clarity, however, we shall present examples for each of the five types of this third case separately.

[Here the capital is known and the profit unknown.]

(B.72) For instance, someone asks: A man buys three caficii for ten nummi and sells four for twenty. How many caficii will he gain with a hundred nummi?

You will do the following. You will say: 'When four for twenty, how much shall I have for ten?' It is two. Thus with ten nummi he bought three caficii, from which he sold two for ten nummi. Therefore with ten nummi he has gained one caficius. Then you will say: 'When with ten he gains one, what will he gain with a hundred?' It is ten. *[This is the first of the five types found in the fourth of the four cases.]* The verification of this problem is evident.[1067]

— The second type is the following.

[1065] B.54 (and B.55–58 for the subsequent problems).
[1066] As stated in B.65.
[1067] The second quantities are ten times the first ones.

LIBER MAHAMELETH

[Here the capital is unknown and the profit known.]

(B.73) Someone says: A man buys three caficii for ten nummi and sells four for twenty. With how many nummi will he have a profit of a hundred caficii?

You will do the following. You will say: 'When four for twenty, how much shall I have for ten?' It is two. Subtracting this from three leaves one. Then you will say: 'When with ten the profit is one, from how many nummi will result his profit of a hundred?' Proceed as has been explained before, and the required quantity will be a thousand.

— The third type is the following.

[Here each is determined from their sum.]

(B.74) Someone asks: A man buys three caficii for ten and sells four for twenty, and adding the profit in caficii to the corresponding capital in nummi makes a hundred.

You will do the following. You will say: 'When four for twenty, how much shall I have for ten?' It is two. Subtracting this from three leaves one. Then it is like saying: 'Someone gains one with ten, and a hundred results from adding his profit to his capital.' Proceed as explained before, and the result will be for the profit nine and one eleventh and for the capital, ninety and ten elevenths.

— The fourth type is the following.

[Here each is determined from the subtraction.]

(B.75) Someone says: A man buys three caficii for ten and sells four for twenty, and subtracting his profit in caficii from his capital in nummi leaves a hundred.

You will do the following. You will say: 'When four are given for twenty, how much shall I have for ten?' It is two. Subtracting this from three leaves one. Then it is like saying: 'Someone gains one with ten, and subtracting his profit from his capital leaves a hundred.' Proceed as explained before, and the result will be for the profit eleven and a ninth and for the capital, a hundred and eleven and a ninth.

— The fifth type is the following.

[Here each is determined from their product.]

(B.76) Someone says: A man buys three caficii for ten and sells four for twenty, and multiplying his profit in caficii by his capital in nummi produces a hundred.

You will do the following. You will say: 'When four for twenty, how much shall I have for ten?' It is two. Subtracting this from three leaves one. Then you will say: 'With ten the gain is one, and multiplying profit by capital produces a hundred.' Proceed as explained before, and the result will be for his capital the root of a thousand and for his profit, the root of ten.

828   PART TWO: TRANSLATION, GLOSSARY

(*iv*) THE FOURTH OF THE FOUR CASES IS PROFIT IN CAFICII FROM CAFICII.

(B.77) For instance, someone asks: I have sold for seven nummi and a half a caficius bought for six nummi. How many caficii shall I gain from a hundred caficii?

You already know[1068] that the profit from one caficius is a fifth of a caficius. Then it is like saying: 'With one I have gained a fifth; how much shall I gain with a hundred?' Proceed as explained before;[1069] the result will be twenty caficii, which he has gained from a hundred caficii.   If the hundred were almodis, the twenty would also be almodis.

Suppose he had said here: 'The hundred caficii I have gained, from how many caficii did I gain them?' Or: 'The hundred almodis I have gained, from how many almodis did I gain them?' You will say: 'When he has gained a fifth from one, from how many did he have a profit of a hundred?' Proceed as explained before, and the result will be the capital in caficii or in almodis.

To this case too you will likewise adapt the three above-mentioned types of adding or subtracting or multiplying.

We shall here again present examples for each of the five (types) of this fourth case.

[*Here the capital is known but the profit unknown.*]

(B.78) For instance, someone asks: A man buys three caficii for ten nummi and sells four for twenty. How many caficii will he gain from a hundred caficii?

This case arises from the problem 'how much shall I have?'[1070] Then you will say: 'When four caficii are given for twenty nummi, how many shall I have for ten nummi?' It is two. So he buys three caficii for ten and sells two for ten, thus gaining one with three caficii. Say therefore: 'A man who gains one with three, how many will he gain with a hundred?' Proceed as I have taught above. The result will be thirty-three and a third, and this is what you wanted.   [*This is the first of the five types contained in each of the four cases.*]

— The second type is the following.

[*Here the capital is unknown but the profit is known.*]

(B.79) A man buys three caficii for ten nummi and sells four for twenty. From how many caficii did he have a gain of a hundred caficii?

Do as I have taught above, namely as follows. You will say: 'When four for twenty nummi, then how many caficii shall I have for ten?' It is two. Subtracting this from three leaves one. Then you will say: 'When with three he gains one, from how many caficii did he have a gain of a

---

[1068] B.71.

[1069] Again, B.54–58 for the five problems.

[1070] The profit is in kind.

LIBER MAHAMELETH   829

hundred?' Proceed as shown above. The result will be three hundred, and from that many caficii did he have a gain of a hundred caficii. The verification of this problem is evident [1071] from the previous one; there is therefore no need to repeat it in this case.[1072]

— The third type is the following.

(B.80) A man buys three caficii for ten and sells four for twenty, and adding his profit in caficii to his capital in caficii makes a hundred. How much is his profit or his capital?

You will do the following. You will say: 'When four caficii for twenty nummi, then how many shall I have for ten?' It is two. Subtracting this from three leaves one. Then you will say: 'One is gained with three, and adding the profit in caficii to the capital makes a hundred.' Proceed as I have taught before. The profit will be twenty-five and the capital, seventy-five.

— The fourth type is the following.

(B.81) Someone says: A man buys three caficii for ten and sells four for twenty, and subtracting his profit in caficii from his capital in caficii leaves a hundred.

Proceed as I have taught above, namely as follows. You will say: 'When four are given for twenty, then how many shall I have for ten?' It is two. Subtracting this from three leaves one. Then you will say: 'One is gained with three, and subtracting profit from capital leaves a hundred.' Do as I have taught above. The capital will be a hundred and fifty and the corresponding profit, fifty.

But if it were said here that the capital in caficii is subtracted from the profit in caficii [*and something has been left*], the problem would not be solvable. For the profit in caficii is less than the capital in caficii.

— The fifth type is the following.

(B.82) A man buys three caficii for ten and sells four for twenty, and multiplying his profit in caficii by his capital in caficii produces a hundred.

You will do the following. You will say: 'When four for twenty, then how many shall I have for ten?' It is two. Subtracting this from three leaves one. Then you will say: 'One is gained with three, and multiplying his profit by his capital produces a hundred.' Do as explained before. The profit will be the root of thirty-three and a third and the capital, the root of three hundred.

[1071] Without what follows: as in B.72, the second quantities differ from the first by a simple factor. With what follows: see next note.

[1072] If by 'previous one' is meant a *verification*, the reference must be to B.67, of the same type —indeed, the group B.66–70 immediately precedes that of B.78–82 in the disordered text. If it alludes to *problem* B.78, its verification is missing. (Note that the results found in the two problems merely differ by the factor 3.)

830   Part Two: Translation, Glossary

These are the twenty kinds which we have mentioned before.[1073] All coming chapters on profits are derived from them.[1074] Therefore those who have perfectly understood them and remember them well will find nothing obscure in the (following) problems on profit, each of which originates from them, is reduced to them, is proved by them.

Another chapter on profit [1075]

(B.83) Someone asks: I have sold for seven and a half nummi a caficius bought for six, and I have gained sixty nummi from another capital minus ten nummi. What is this capital?

You already know that his profit from six nummi is one and a half nummi. Therefore you will say: 'When from six nummi he has gained one and a half, then from what has he gained sixty?' Do as has been explained before.[1076] The result will be two hundred and forty. This is his capital minus ten nummi. Adding to it the ten subtracted will give his whole capital.

(B.83′) Suppose it were said here: 'From his capital plus ten nummi he has gained sixty; what is his capital?' Subtracting ten from the result, namely two hundred and forty, will leave his capital.

(B.84) Now someone asks: I sold for ten nummi a caficius bought for eight. I bought with an unknown capital an unknown quantity of caficii, from which I sold so many that I recovered three fourths of the unknown capital and forty caficii remained. How much was his[1077] capital in nummi?

You will do the following. Take three fourths of eight, for which he bought the caficius; this is six nummi. Sell for six nummi of a caficius according to its selling price of ten nummi; this will be three fifths of it. He therefore sold of the caficius bought for eight nummi its three fifths for three fourths of eight nummi, and two fifths remained. Then you will say: 'When the remainder of his capital of eight nummi is two fifths of a caficius, what will be the capital corresponding to a remainder of forty?' Multiply eight by forty and divide the product by two fifths; the result will be the capital in nummi, namely eight hundred. This is what you wanted to know.

Here the proportion is evident. For eight nummi are to two fifths of a caficius, which remained of the caficius, as the required capital is to forty caficii, which remained of it. Thus the third is unknown. So multiply the first, which is eight, by the fourth, which is forty, and divide the product by the second, which is two fifths; the result will be what you require.

[1073] Above, pp. 820–821.

[1074] B.83–83′ and B.85–88 are meant, as specified after B.88.

[1075] Should not include B.84, which belongs to the group B.89–90, not involving profit as such.

[1076] B.61.

[1077] Refers to 'someone'.

LIBER MAHAMELETH 831

The verification of this problem is the following. He buys for eight hundred nummi caficii according to a buying price of eight to the caficius and this results in a hundred caficii. Next, he sells of these caficii for three fourths of eight hundred nummi, which is six hundred, according to a selling price of ten nummi to the caficius; thus he sells sixty caficii. So his remainder from the hundred caficii will be forty caficii, as said above.

(B.84′) Now suppose the problem had been stated thus: 'From what I have sold I have recovered three fourths of my capital plus thirty nummi and forty caficii remained; what was the capital?' Look for the quantity of caficii corresponding to thirty nummi;[1078] you will find three. Adding them to the forty caficii will make forty-three. Therefore after recovering three fourths of his capital, he will be left with forty-three caficii. Working out the problem as explained before[1079] will lead to the required capital.

(B.84″) Suppose here he were said to have recovered three fourths of his capital minus thirty nummi and forty caficii remained to him; what was his capital in nummi?

You will do the following. Subtract three caficii, which correspond to thirty nummi,[1080] from the forty caficii; this will leave thirty-seven caficii. This is what remains after recovering three fourths of his capital. Then proceed as explained before, and the result will be his required capital in nummi.

You will again proceed likewise, according to the above rules, whenever a problem involves a loss, following all the ways seen with profit.[1081]

AGAIN ON PROFIT.

(B.85) Someone asks: Three caficii are bought for ten nummi and are later sold four for twenty. What was the capital in nummi of a man who, buying and selling[1082] with two thirds of his capital according to the prices set, gains ten caficii?

You will do the following. You will say: 'A man, having bought a quantity of caficii at three for ten nummi, sells four for twenty and gains ten caficii; with what capital did he gain them?' [*This is the second of the five types found in the fourth case.*][1083] Therefore do as I have taught above. The result will be that two thirds of his capital is a hundred nummi. Adding to it its half makes a hundred and fifty, and so much is the whole capital.

[1078] At the selling price since this sum has been obtained from the sale.

[1079] B.84. Same reference in the next problem.

[1080] Potentially (if they are sold).

[1081] That is, when only part of the initial capital is recovered and/or some of the merchandise remains unsold.

[1082] That is: trading.

[1083] It is the second type of the *third* case (B.73), but in the disordered text (MS *B*) the problems of the third case (B.72–76) follow those of the fourth (B.78–82).

(B.85′) Suppose, the problem remaining the same, he were said to have gained the ten caficii by trading with two thirds of his capital minus two nummi. Do as explained before. The result will be ⟨that⟩ two thirds of his capital minus two nummi ⟨is a hundred nummi⟩. Adding to it the two nummi will give two thirds of the capital.

(B.85″) Suppose now he were said to have gained ten[1084] caficii by trading with two thirds of the capital and two nummi. Do as explained before. The result will be ⟨that⟩ two thirds of the capital and two nummi ⟨is a hundred nummi⟩. Removing the two nummi will leave two thirds of the capital.

(B.86) Suppose it were said that someone sells caficii, bought three for ten, at four for twenty, and by trading according to the prices set with three fourths of his capital in caficii gains ninety-three nummi. Do as we have shown in the second of the five types found in the second of the four cases.[1085] The result will be three fourths of the capital in caficii. Adding to it its third will give the whole capital.

(B.86′) Suppose it were said that by trading with three fourths of his capital in caficii plus five caficii he gains so many or so many nummi. Do as has been shown.[1086] The result will be three fourths of his capital in caficii plus five further caficii. Removing the five additional caficii will leave three fourths of the capital.

To this chapter[1087] belong some unsolvable problems. We have deemed it appropriate to present one of them, whereby the others may be assessed. It is the following.

(B.87) Someone says: Three caficii are bought for ten nummi and four are sold for twenty. A man trading according to the prices set with three fourths of his capital in nummi plus ten nummi gains one caficius.

This is impossible. For trading according to the prices set with ten nummi only he will gain one caficius; if therefore three fourths of the capital are added, he will gain more. Thus when such a problem is proposed, you are to know that it is not solvable. It will not be solvable unless the quantity of caficii he gains is more than the quantity of caficii gained with the given quantity of nummi which is with a fourth, a third or whatever other proposed fraction of his capital. But if it is less than, or equal to, the quantity of caficii gained from the given quantity of nummi, the problem will not be solvable. Understand what I say and consider other such problems accordingly, and you will find that this is the way.

(B.88) Someone says: A man sells four caficii, bought three for ten nummi, for twenty. Having traded with two thirds of his capital in nummi plus ten

---

[1084] 'a hundred' in the MSS. Probably an early reader's marginal addition prompted by the lacunas in B.85′–85″, and then taken to be a correction.

[1085] B.67.

[1086] Previous problem.

[1087] B.83 & B.85–88.

LIBER MAHAMELETH

nummi, he gains a hundred nummi. What was the capital?

You will do the following. You will say: 'When three caficii are bought for ten nummi and are sold four for twenty, then from how many nummi did he gain his profit of a hundred nummi?' Do as has been shown in the second of the five types found in the first of the four cases.[1088] The result will be two hundred, which is two thirds of the capital and ten nummi. Thus two thirds of the capital is a hundred and ninety, and the whole capital is two hundred and eighty-five.

You can verify this problem as follows. Adding ten to two thirds of this capital, which is a hundred and ninety, makes two hundred. Buying, with these two hundred, caficii at the price set of three for ten, you will assuredly buy sixty; if you then sell them at the price set of four for twenty, you will assuredly sell them for three hundred nummi. Thus the profit from two thirds of his capital plus ten, which is two hundred nummi, will be a hundred nummi.

You could verify the preceding problems in the same way.[1089]

You must know that all problems which may belong to this chapter can be reduced to the twenty kinds mentioned before. We have already shown how these problems can be reduced to them; therefore there is no need to present all those which may arise.

Again on profits.[1090]

(B.89) A man bought an unknown quantity of sextarii, each for three nummi. He later sold so many, each for five nummi, that he recovered his capital and twenty sextarii remained to him. What was the unknown quantity of sextarii?

Multiply five by twenty and divide the product by the difference between three and five, namely two. This gives fifty, and so many were the sextarii which he bought.

(B.89′) A man bought an unknown quantity of sextarii, each for three nummi. He later sold so many, each for five nummi, that he recovered his capital plus ten nummi and twenty sextarii remained to him. How many sextarii were there?

Multiply five by twenty, add to the product the additional ten which he has gained, and divide the result by the difference between three and five, namely two. This gives fifty-five, and so many were the sextarii.

(B.89″) A man bought an unknown quantity of sextarii, each for three nummi. He later sold so many ⟨each for five nummi⟩ that he recovered his

---

[1088] B.61.

[1089] Allusion to B.83 and B.85–88 (which form 'this chapter' alluded to below). B.84 does not belong to this group and has its own verification.

[1090] This section belongs in B–II, but, like B.84, with the accent on recovery of capital.

834   Part Two: Translation, Glossary

capital minus ten nummi [*and he sold each for five nummi*][1091] and twenty
sextarii remained to him. How many sextarii were there?

Multiply five by twenty, subtract ten from the product, and divide the
remainder by the difference between five and three. This gives forty-five,
and so many were the sextarii.

(B.90) Someone asks: From a certain quantity of caficii, bought three for
ten, a man sells at four for twenty so many that he recovers three fourths of
his capital in nummi and twenty caficii remain for him to sell. How many
caficii were there and what is the capital in nummi?

You will do the following. Take three fourths of ten, which is seven
and a half. Buying with it corn at four for twenty, you will get one and
a half caficii.[1092] Subtracting these one and a half caficii from the three
caficii leaves one and a half. Then if you want to know the whole capital in
nummi, multiply ten by the twenty caficii which remained[1093] and divide
the product by the remaining one and a half;[1094] the result will be the
whole capital in nummi, namely a hundred and thirty-three and a third.
If you want to know the unknown quantity of caficii which he bought with
these nummi, multiply the three caficii by the twenty which remained and
divide the product by one and a half; the result will be forty, and such is
the unknown quantity of caficii he bought with the capital in question.

You can verify this as follows. With his capital in nummi, which was
a hundred and thirty-three and a third nummi, he bought forty caficii at
three for ten. Then he sold of these at four for twenty so many that he
recovered three fourths of his capital, thus a hundred nummi; that is, he
sold twenty caficii and another twenty remained, as stated above.

The proof of this rule is the following. Let the three caficii be line
AB, their price, which is ten, GD, the four caficii, HZ, the twenty nummi,
KT, the unknown quantity of caficii, QL, the unknown capital, MN, three
fourths of MN, ME, the twenty caficii left, CL. It is thus clear that the
ratio of AB to GD is the same as the ratio of QL to MN and that the
ratio of HZ to KT is the same as the ratio of QC to ME. The ratio of
AB to GD being the same as the ratio of QL to MN, the ratio of AB to
three fourths of GD will be the same as the ratio of QL to three fourths
of MN. But three fourths of MN is ME, and let three fourths of GD be
GF. So it is clear that the ratio of AB to GF is the same as the ratio of
QL to ME. [*Now the ratio of HZ to KT is the same as the ratio of QC
to ME.*][1095] Let the ratio of HZ to KT be the same as the ratio of AO to
GF. But the ratio of HZ to KT is the same as the ratio of QC to ME.
⟨So the ratio of AO to GF is the same as the ratio of QC to ME.⟩ But

---

[1091]   Early reader's gloss making up for the preceding lacuna.

[1092]   *you* buys from *him* (the 'man') at the selling price.

[1093]   'which remained' is a useless specification.

[1094]   'remaining', for 'one and a half' occurs both as subtrahend and remainder.

[1095]   Addition perhaps related to the lacuna below.

the ratio of AB to GF was already the same as the ratio of QL to ME. It follows[1096] that the ratio of the remainder OB to GF is the same as the ratio of the remainder CL to ME. Thus the ratio of OB to the whole GD, which is one and a third times GF, is the same as the ratio of CL to MN, which is one and a third times ME. Therefore multiplying GD, which is ten, by CL, which is the remaining twenty, and dividing the product by OB, which is one and a half, will give MN, which is his capital in nummi. Likewise again, the ratio of OB to AB will be the same as the ratio of CL, which is twenty, to QL, which is the whole unknown quantity of caficii. For the ratio of OB to GD is the same as the ratio of CL to MN, and the ratio of GD to AB is the same as the ratio of MN to QL; therefore, by equal proportionality,[1097] the ratio of OB to AB will be the same as the ratio of CL to QL. Thus multiplying AB, which is three, by CL, which is twenty, and dividing the product by OB, which is one and a half, will give QL, which is the unknown quantity of caficii, namely forty. This is what we wanted to demonstrate.

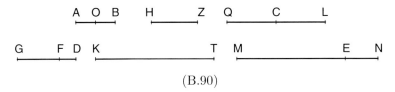

(B.90)

(B.90′) Now someone says: A man buys a certain quantity of caficii at three for ten nummi, then he sells part of it at four for twenty till he recovers three fourths of his capital in nummi plus ten nummi, and eighteen caficii remain to him. What is his capital and how many are the caficii?

You will do the following. We know that he sells two caficii for ten nummi. By hypothesis, after recovering, from the caficii sold, three fourths of his capital and ten nummi, he is left with eighteen caficii. So recovering just three fourths of the capital, he is left with twenty caficii.[1098] Then you will say: 'From a quantity of caficii, bought at three for ten, he sells some at four for twenty, recovering thereby three fourths of his capital and being left with twenty caficii to sell.' Do as I have taught above;[1099] it will lead to what you wanted.

AGAIN.[1100]

(B.90″) Someone asks: A man buys a certain quantity of caficii at three for ten nummi and sells them at four for twenty. After recovering three fourths of his capital minus twenty nummi, he is left with twenty-four caficii.

[1096] Premiss P_6.
[1097] *secundum equam proportionalitatem*. See note 60, p. 594.
[1098] As in B.84′: since the ten nummi result from the sale, they correspond to two caficii at the selling price.
[1099] No need, this is just B.90. Same remark for B.90″.
[1100] This heading should presumably follow B.90″.

836　　　　Part Two:　Translation, Glossary

We already know that he sells four caficii for twenty nummi and that after this sale, having recovered three fourths of the capital minus twenty nummi, he is left with twenty-four caficii. Thus from such a sale, after recovering just three fourths of his capital, he will be left with twenty caficii. Then you will say: 'A man buys three caficii for ten and later sells four for twenty, recovering thereby three fourths of the capital and being left with twenty caficii to sell.' Do as I have taught above; the result will be forty for the unknown quantity of caficii and a hundred and thirty-three and a third for the capital in nummi.

You can verify this problem as follows. We know that forty caficii were bought, of which sixteen were sold according to the price set, whereby he recovered three fourths of his capital in nummi minus twenty nummi. This is eighty, and he is left with twenty-four caficii.

Consider other such problems likewise, and you will find that this is the way.

There is another, foreign problem concerning profits which we deemed it appropriate to include in the present chapter, which is of the following kind.[1101]

(B.91) Someone says: A man gains three caficii from twelve caficii. How many nummi will he gain from a hundred nummi?

The meaning of this problem is here the following. A man has bought twelve caficii for an unknown quantity of nummi and then receives just as many nummi after selling nine of these caficii [*therefore from twelve caficii he has gained three caficii*], and he wants to know how many nummi he will gain if a hundred nummi are spent on caficii at the buying price set for the twelve, these caficii being in turn sold at the selling price for the nine.

You will do the following. Subtract the three caficii from the twelve, thus leaving nine. Next, multiply a hundred by three and divide the product by nine; the result will be what you wanted.

The proof of this is the following. Let the twelve caficii be AB and the price of each, GD. Let AB be multiplied by GD and the result be H; thus H is the quantity of nummi with which he bought the twelve caficii at the unit price GD. Now he has sold of these twelve caficii nine, say AQ, for H at a unit price GZ. Therefore the buying price of each caficius is GD and its selling price is GZ. Let the hundred nummi, with which he buys what he buys[1102] [*and at which he sells what he sells*] according to the price set, be K, and let what is sought, namely the profit he makes from them, be T. It is then evident that the ratio of DZ to GD is the same as the ratio of T to K. But we know that the product of GZ into AQ, which is nine, is equal to the product of GD into AB. Therefore the ratio of GZ to GD is the same as the ratio of AB to AQ. By separation,

---

[1101] Unlike before, one of the capitals is in cash and the other in kind, and the same holds for the profits.

[1102] We do not know how much he bought.

the ratio of DZ to GD will be the same as the ratio of QB to AQ. But the ratio of DZ to GD is the same as the ratio of T to K. So the ratio of QB, which is three, to AQ, which is nine, is the same as the ratio of the required quantity to a hundred. Multiplying therefore three by a hundred and dividing the product by nine will give what is required. This is what we wanted to demonstrate.

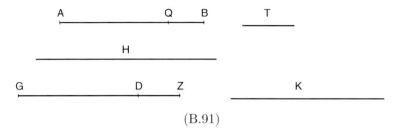

(B.91)

Again. Chapter on unknown profits[1103]

(B.92) For instance, someone asks: A man buys three caficii of wheat for ten nummi and an unknown number of caficii of barley for twelve nummi. He later sells each caficius of wheat at the price of each caficius of barley and each caficius of barley at the price of each caficius of wheat; he gains in the end four nummi. What is the unknown number of caficii of barley?

You will do the following. Add ten to twelve, thus making twenty-two. Add four to it; this makes twenty-six. Next, multiply ten by twelve; the result is a hundred and twenty. Next,[1104] multiply half of the twenty-six, which is thirteen, by itself; this produces a hundred and sixty-nine. Subtracting a hundred and twenty from it leaves forty-nine. You may add its root, which is seven, to thirteen; this makes twenty. Then you will say: 'When three caficii are given for ten, how many shall I have for twenty?' The result is six, and this is the unknown number of caficii of barley. You may also subtract seven from thirteen; this leaves six. Then you will say: 'When three for ten, how many shall I have for six?' The result is one and four fifths. Therefore you may either take as the unknown number of caficii of barley one and four fifths and as the price of each six and two thirds, or take as the number of caficii six and as the price of each two nummi.

The proof of this is the following. Let the three caficii be A, the ten nummi, B, the quantity of barley, D, the twelve nummi, H. Then you will say: 'When three caficii for ten nummi, what is the price of the unknown number of caficii of barley?' Let then also three caficii be Z, the unknown quantity of barley, G, the price of G, KT, the price of Z, TQ. Since the ratio of A to B is the same as the ratio of G to KT, then, by alternation, the ratio of A to G will be the same as the ratio of B to KT. But A is

---
[1103] This heading is misplaced. It applies to B.93–100 and B.102–103, with two unknowns, the profit and the capital.
[1104] Solving $x^2 + c = bx$ ($c = 120$, $b = 26$).

the same as Z and G the same as D; then the ratio of Z to D is the same as the ratio of B to KT. But we know [1105] that the ratio of TQ to H is the same as the ratio of Z to D. So the ratio of B to KT is the same as the ratio of TQ to H. Therefore the product of B into H is equal to the product of KT into TQ. But multiplying B by H produces a hundred and twenty. Thus the product of KT into TQ is a hundred and twenty. But it is known that KT and TQ is twenty-six.[1106] So let KQ be bisected, namely either at L or at M. If the point of section is at L, the number of caficii of barley will be six and their unit price, two nummi; if the section is at M, the number of caficii of barley will be one and four fifths and their unit price, six and two thirds. This is what we wanted to demonstrate.

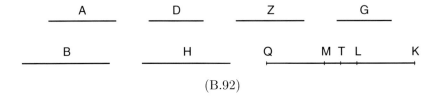

(B.92)

(B.93) A man gains three in buying for five. What is the capital from which he gains six roots of it?

You will do the following. Divide the number of roots by three, which he gains, multiply the quotient by five, and multiply the product by itself; the result will be a hundred, and so much was the capital.

(B.94) A man in buying for five gains three. What is the capital from which he gains six roots of its half?

Find the number by which a half when multiplied produces one; this is two. Multiplying it by three produces six. Keep it in mind. Next, multiply the number of roots by five. Divide the product by the six kept in mind,[1107] multiply the quotient by itself and double the product —because of the statement 'six roots of its half'. The result will be fifty, and this is what you wanted.

(B.95) Someone says: A man in buying for five gains three, and multiplying the root of a profit he makes according to the price set by the root of the corresponding capital produces ten. What is this profit of his or the capital?

You will do the following. It is evident that multiplying his profit by his capital produces a hundred, for multiplying a number by another and taking the root of the product is the same as multiplying the root of the multiplied number by the root of the multiplier; the proof of this rule has already been presented in the chapter on roots.[1108] This problem is then

---

[1105] As $A : B = G : TK$ (above), likewise we must have $D : H = Z : TQ$.
[1106] Sum and product of KT and TQ thus known.
[1107] Not that of the data.
[1108] A.284; but this explanation is unclear.

LIBER MAHAMELETH

839

like saying: 'For five the profit is three and multiplying another profit by another capital produces a hundred.' Do as we have shown above,[1109] and the result will be what you wanted.

(B.96) Someone says: A man buys three caficii for ten nummi and sells four for twenty, and multiplying the root of a profit he makes in nummi according to the prices set by the root of the corresponding capital in caficii produces twenty. What is the capital in caficii and the profit in nummi?

You will do the following. We know[1110] that multiplying his profit in nummi by his capital in caficii produces four hundred. So you will say: 'A man buys three for ten and sells four for twenty, and multiplying his profit in nummi by his capital in caficii produces four hundred.' Do as we have shown above in the fifth of the five types belonging to the second of the four cases.[1111]

(B.97) Someone says: In buying for nine a man gains four, and adding the root of his profit to the root of his capital makes twenty. What is the profit or its capital?

You will do the following. We know that the ratio of one capital to its profit is the same as the ratio of the other capital to its profit. Thus the ratio of the root of his capital to the root of his profit is the same as the ratio of the root of the unknown capital to the root of the unknown profit. By composition, the ratio of three to five will be the same as the ratio of the root of the unknown capital to twenty. Then proceed as I have taught above.[1112] The result will be, for the root of his unknown capital, twelve. Therefore his unknown capital is a hundred and forty-four. Inversely, the ratio of two to five will be the same as the ratio of the root of his profit to twenty. Proceed then as has been explained above. The root of his profit will be eight. Therefore his profit is sixty-four.

Consider other cases, similar to these, accordingly and you will find that this is the way.[1113]

(B.97′) And if he says that subtracting the root of his profit from the root of his capital leaves four, you will do as in the foregoing[1114], but this time using conversion.

(B.98) Someone says: A man buys three caficii for ten and sells four for twenty, and adding the root of his profit in caficii to the root of his capital in nummi makes fifty.[1115]

[1109] B.64.

[1110] Previous problem.

[1111] B.70.

[1112] Plain proportion, thus introduction to Book B (rule i, p. 777).

[1113] This would be expected after B.97′.

[1114] Thus B.97.

[1115] What is required does not really need to be specified, as in B.103. See note 985, p. 805.

840 PART TWO: TRANSLATION, GLOSSARY

Then you will say: 'When four for twenty, how much shall I have for ten?' It is two. So a man who buys three caficii for ten and sells two caficii for ten gains, with ten nummi, one caficius. Then you will say: 'With ten he gains one, and adding the root of his profit to the root of his capital makes fifty.' Proceed as has been explained before,[1116] namely as follows. Take the root of one and the root of ten, which are one and the root of ten. Then the ratio of one to the root of ten will be the same as the ratio of the root of his profit to the root of his capital. By composition, the ratio of one to one and the root of ten will be the same as the ratio of the root of his profit to fifty. So multiplying one by fifty and dividing the product by one and the root of ten will give the root of his profit. Likewise again, the ratio of the root of ten to one and the root of ten will be the same as the ratio of the root of his capital to fifty. So multiplying the root of ten by fifty and dividing the product by one and the root of ten will give the root of his capital.

(B.99) Now someone says: A man buys three for ten and sells four for twenty, and subtracting the root of his profit in caficii from the root of his capital in caficii leaves sixty. Do as has been explained above,[1117] and the result will be what you wanted.

(B.100) Someone says: A man sells caficii bought three for ten at four for twenty. He gains with an unknown quantity of nummi a quantity of caficii equal to the root of the nummi. What are his capital in nummi and his profit in caficii?

You will do the following. Say: 'When four for twenty, how many shall I have for ten?' It is two. Thus buying three for ten and selling two for ten, with ten nummi he gains one caficius. Next, say: 'When with ten he gains one and with an unknown quantity of nummi he gains as much as its root, what are the quantity of nummi and its root?' You know that the ratio of ten to one is the same as the ratio of the unknown quantity of nummi to its root. Since ten is the decuple of one, the quantity of nummi is the decuple of its root. Thus the quantity of nummi is a hundred, which is his capital in nummi; its root is ten, which is his profit in caficii.

(B.101) Someone says: A man buys an unknown quantity of caficii for ten nummi and sells four for twenty. With a hundred nummi he gains ten caficii. What is the unknown quantity of caficii?

You will do the following. That is, say: 'When four for twenty, how many shall I have for ten?' Proceed as has been explained above.[1118] The result will be two caficii; so he buys an unknown quantity of caficii for ten nummi and sells two of them for ten nummi. Now it is clear that the ratio of his profit in caficii from ten nummi to ten nummi is the same as the ratio

---

[1116] B.97.

[1117] B.(97–)97′.

[1118] Once again (notes 636, 1099), one wonders if the author does not sometimes write mechanically.

LIBER MAHAMELETH 841

of ten caficii to a hundred nummi. Proceed then as I have taught above.[1119] The result will be a profit of one caficius from ten nummi. Thus, when a man buys an unknown quantity of caficii for ten nummi and sells two of them for ten nummi, and one additional caficius remains, then the unknown quantity of caficii is three. [*This is what we wanted to demonstrate.*]

(**B.102**) Now someone says: A man buys an unknown quantity of caficii for the triple of its root, sells them for twenty nummi, and gains three hundred and fifty nummi with a hundred caficii. What is the quantity of caficii and what is three roots of it?

You will do the following. You already know that someone who gains with a hundred caficii three hundred and fifty nummi gains with one caficius three and a half. You also know that the ratio of three and a half to one is the same as the ratio of twenty minus three roots of the quantity of caficii to this quantity. Multiplying therefore the quantity of caficii by three and a half, the result will be equal to twenty minus three roots of this quantity. So we now have three and a half squares, which are equal to twenty minus three roots. Completing, you will have three and a half squares and three roots, which equal twenty nummi.[1120] After reduction of all the squares to one square, and of the three roots and the twenty in the same proportion, there will in the end be one square and six sevenths of a root, equal to five nummi and five sevenths. Proceed then as has been shown in algebra, which is as follows.[1121] Multiply half of the roots, which is three sevenths, by itself, thus producing a seventh and two sevenths of one seventh. Add it to the nummi; this will make five nummi and six sevenths and two sevenths of one seventh. Subtracting three sevenths from the root of the whole, which is two and three sevenths, leaves two. This is the root of the square. Thus the square is four, and such is the required quantity of caficii, while the price at which he bought them is six nummi. [*This is what we wanted to demonstrate.*]

(**B.103**) Someone says: A man buys a certain quantity of caficii for the triple of its root and sells them for the quintuple of its root. With a hundred caficii sold at this same price set he gains a hundred and fifty nummi.

You will do the following. It is known that when he buys a quantity of caficii for the triple of its root and sells them for the quintuple of its root, he gains with these caficii a quantity of nummi which is twice the root of the quantity of caficii. Now it was stated that he gains with a hundred caficii a hundred and fifty nummi. Therefore the ratio of a hundred and fifty to a hundred is the same as the ratio of two roots to the quantity of caficii. But a hundred and fifty to a hundred is one and a half times. Thus two roots are one and a half times the quantity of caficii. So one and a half

---

[1119] As in B.97, plain proportion.

[1120] Equation of the form $ax^2 + bx = c$. Thus 'nummi' here and below in the sense of 'units' (the constant term in the equation), like *dirham* in Arabic texts; see our *Introduction to the History of Algebra*, p. 158 n. 178.

[1121] Here clear description of the *algebraic* solution of $x^2 + bx = c$.

842 PART TWO: TRANSLATION, GLOSSARY

squares equal two roots; thus the root is one and a third, and the square, one and seven ninths. Such is the required quantity of caficii, that is, one and seven ninths. [*This is what we wanted to demonstrate.*]

You are to know that numerous other problems, indeed an innumerable quantity, can be formulated about unknown profits.[1122] We do not deal with them all here, only in so far as the present matter is concerned, for they cannot all be included.

Those who have understood what has been explained about these, and also the foregoing (problems) about profits according to the twenty kinds and the others,[1123] and our premises at the beginning of the book[1124], when they encounter any such problem which I have omitted, are to think about it and, by means of what has been explained above, will solve it with ease.

[1122] Alludes to B.95–100, with two unknowns.

[1123] B.59–82, B.83 & B.85–88 (see note 1089, p. 833).

[1124] Introduction to Book B (pp. 777–778).

Chapter (B-III) on Profit in partnership

(B.104) For instance, three partners, of whom the first has invested eight nummi, the second, ten, and the third, fourteen, have gained by trading twenty-two. You want to know what should be the share of each according to the amount of his invested capital.

You will do the following. Adding together all their capitals, the resulting sum will be thirty-two, with which they have gained twenty-two. Let the sum be your principal. Next, multiply the profit, which is twenty-two, by the capital of each and divide the product by the principal; the result is the profit of the investor of the capital in question. You may also consider what fraction eight is of thirty-two; it is a fourth, thus a fourth of twenty-two is the profit of the man who invested eight. Likewise, consider what fraction ten is of thirty-two; it is two eighths and half an eighth, thus this fraction of twenty-two is the profit of the man who contributed ten. You will do likewise for fourteen. You may also denominate twenty-two from thirty-two; it will be five eighths and half an eighth. Multiplying this by the capital of each produces each one's share of the profit.

This rule is taken from a proportion. For capital is to profit as capital to profit, or profit to capital as profit to capital.[1125] Thus in this case the ratio of the whole capital, which is thirty-two, to the whole profit, which is twenty-two, is the same as the ratio of eight to the corresponding share of the profit. Thus the fourth is unknown. So multiply the second by the third and divide the product by the first; the result will be what you require. Or divide one of the factors by the divisor and multiply the result by the other; the product will be what is required. Doing the same for ten and fourteen, you will have what you require.

You will proceed likewise for all partners, whether there are many or few. Always add all their capitals, and the ratio of the sum of their profits to the sum of their capitals will be the same as the ratio of each individual profit to the corresponding capital. Thus multiplying the individual capital by the sum of the profits and dividing the product by the sum of the capitals will give the individual share.

(B.105) Now three partners, of whom one has contributed eight, the second ten, the third fourteen, have gained by trading, and the profit of the investor of ten is four. You want to know what is due to the other two.

You will do the following. Let ten be your principal. Multiply the corresponding profit, which is four, by the capital of the man whose profit you want to know and divide the product by the principal; the result will be the profit of the investor of this capital. Or divide the profit corresponding to

---

[1125] As stated in B.54.

844 PART TWO: TRANSLATION, GLOSSARY

ten by ten and multiply the result by each one's capital; the product will be the profit corresponding to this capital. Or divide the capital corresponding to the profit you want to know by ten and multiply the result by four, the corresponding profit; the product is what you require.

This rule is taken from a proportion. For the ratio of ten to the corresponding profit, which is four, is the same as the ratio of the capital to the corresponding, required profit. Thus the fourth is unknown and the divisor will be the first, which is ten.

(B.106) Suppose the problem is stated thus: If together the contributor of eight and the contributor of ten make a profit of twelve, how much has each of the three gained?

Add eight to ten, thus making eighteen; let it be your principal. Multiply the corresponding profit, namely twelve, by the capital of each and divide the product by eighteen; the result will be what you require. Or divide one of the factors by eighteen, which is the divisor, and multiply the result by the other factor; the product will be what you require. This rule is evident from a proportion, as has already been explained.[1126]

(B.107) Someone asks thus: Three partners, one of whom contributed eight, the second ten, the third fourteen, have gained by trading and then divided the profit among themselves. Subtracting the share of the contributor of eight from the share of the contributor of fourteen leaves four. What is the share of each?

The treatment of this case is evident from what has been explained above, namely as follows.[1127] Subtract eight from fourteen, which leaves six. It is evident that the above four is the profit from this six. Then it is like asking: 'When from six is gained four, what will be gained from eight, from ten, and from fourteen?' Multiply four, which is the profit from six, by the capital of each, and divide the product by six; the result will be what you require.

(B.108) Suppose now he proposed the following, saying thus: Multiplying the share of the profit of the partner who invested eight by the share of the partner who contributed ten produces forty-five.

The treatment of this case is evident from what has been said before, in buying and selling and in profits, about similar problems involving unknowns [1128], namely the following. If you want to know the profit of the investor of eight, multiply eight by forty-five and divide the product by ten; the root of the result is the profit of the investor of eight. If you want to know the profit of the investor of ten, multiply ten by forty-five and divide the product by eight; the root of the result is what you require.

[1126] In the two preceding problems.

[1127] Subtraction replaces addition.

[1128] See B.17 on buying and selling and, on profits, B.58 (also B.64, B.70, B.76, B.82).

LIBER MAHAMELETH

(B.109) Someone asks thus: Two partners, one with a capital of ten and the other with twenty, have gained by trading. The share of the investor of ten is the root of the share of the investor of twenty. How much has each of them gained?[1129]

You will do the following. It is known that the ratio of ten to twenty is the same as the ratio of the profit of the investor of ten to the profit of the investor of twenty. Thus the profit of the investor of ten is half the profit of the investor of twenty. We then have half a square, which equals the root of this square. So the square is four and its root, two. Therefore the investor of ten gains two and the investor of twenty gains four. This is what you wanted to know.

AGAIN.

(B.110) There are three partners, one with a capital of ten, the second with thirty, the third with fifty. By trading they have gained an amount such that subtracting the profit of the first and the second from the profit of the third leaves three. What is each one's profit?

You will do the following. Add ten and thirty, thus making forty. Subtracting this from the capital of the third leaves ten. Denominate three from it; this gives three tenths. Each of them has gained so much, that is, three tenths, of his capital. Thus the first has gained three, the second, nine, the third, fifteen.

(B.111) There are three partners, one with a capital of twenty, the second with fifty, the third with thirty. By trading they have gained an amount such that subtracting from the profit of the first and the second the profit of the third leaves four.

You will do the following. Subtract the capital of the third from the sum of the capitals of the first and the second; this leaves forty. Denominating from it four gives a tenth. Therefore each of them gains a tenth of his capital. Thus the first gains two, the second, five, the third, three.

(B.112) There are three partners, one with a capital of ten, another with twenty, the third with forty. By trading they have gained an amount such that multiplying the profit of the first and the second by the profit of the third produces forty-eight.

You will do the following. Add the capital of the first to the capital of the second and multiply the sum by the capital of the third; the result will be one thousand two hundred. Denominate from this forty-eight; this gives a fifth of a fifth. The root of this is a fifth. Each one will gain this fraction of his capital, that is, a fifth. Thus the first gains two, the second, four, the third, eight.

(B.113) There are two partners, one with a capital of ten and the other with fifty. By trading they have gained an amount such that the profit of the first is half the root of the profit of the second.

[1129] This problem should be with B.113–114.

846   Part Two: Translation, Glossary

You will do the following. Divide fifty by ten; this gives five. Multiply it by a half, thus producing two and a half. Multiply this by itself. The result is six and a fourth, and so much does the second gain. The first gains half the root of this six and a fourth, which is one and a fourth. This is what you wanted.

(B.114) There are two partners, one with a capital of eight and the second with eighteen. By trading they have gained an amount such that multiplying the root of the profit of the first by the root of the profit of the second produces six.

(*a*) You will do the following. Multiply six by itself; this produces thirty-six. Next, multiply the capital of the first by the capital of the second; this produces a hundred and forty-four. Divide thirty-six by this using denomination; this gives a fourth. Multiply its root, which is a half, by eight; this produces four, and so much does the first gain. Next, multiply this half by eighteen; this produces nine, and so much does the second gain.

(*b*) Or otherwise. Divide eighteen by eight, which gives two and a fourth. Divide by its root, which is one and a half, six; this gives four, and so much does the first gain. Multiplying four by two and a fourth produces nine, and so much is the profit of the second.

(B.115) Someone asks: There are three partners, one with a capital of ten, the second with twenty, the third with a hundred. By trading they have gained, and then divided the profit among them. Multiplying the share of the investor of ten and the investor of twenty by the share of the investor of a hundred produces a hundred and twenty.

You will do the following. It is already known [1130] that the ratio of the sum of the profits of the investor of ten and the investor of twenty to thirty, which results from adding ten and twenty, is ⟨the same as the ratio of the profit of the investor of a hundred to a hundred. Thus the ratio of thirty to a hundred⟩ will be the same as the ratio of the sum of the profits of the investor of ten and the investor of twenty to the profit of the investor of a hundred. Then it is like saying: 'With a hundred he gains thirty, and multiplying his profit by his capital produces a hundred and twenty.' Proceed as I have taught before.[1131] His profit will be six. This is the profit from thirty. Then you will say: 'There are three partners, one investing ten, another twenty, the third thirty, and the partner who has invested thirty gains six; how much does each of the others gain?' You already know that the ratio of six to thirty is the same as the ratio of the profit of each of the others to his capital. Thus proceed as I have taught before;[1132] the profit of the investor of ten will be two, the profit of the investor of twenty will be four, and the profit of the one who has invested a hundred will be twenty.

[1130] B.112; or also the earlier problems of this chapter.

[1131] B.58 & B.64.

[1132] B.105.

LIBER MAHAMELETH

847

AGAIN ON THE SAME TOPIC.

(B.116) Someone asks thus: Two men had a hundred sheep, one sixty and the other forty. They admitted a third man into partnership for these sheep in such a manner that the hundred sheep became the property of the three of them equally; but this man gave the two others sixty nummi for his third-share. How will the two divide this amount between them?

You already know that the share of the owner of sixty dropped to thirty-three and a third and that he sold to the partner admitted the difference between these two numbers, which is twenty-six and two thirds. The second man's share, which was forty, dropped likewise to thirty-three and a third, and he sold to the partner admitted the difference between these two numbers, which is six and two thirds. Then it is like asking thus: 'Two partners, one of whom has contributed twenty-six and two thirds and the other six and two thirds, have gained sixty nummi; how will they divide this amount between them?' Proceed as we have taught before.[1133] The share of the contributor of twenty-six and two thirds will be forty-eight, and this is what the owner of sixty sheep is to receive. The share of the contributor of six and two thirds will be twelve, and this is what the owner of forty sheep is to receive.

You should know that there can be here many other problems of this kind.[1134] Although they are innumerable, those who have paid attention to what has been explained before will be able, whenever a problem about unknowns in partnership occurs, to solve it thereby.

[1133] B.106.

[1134] This remark, which recalls the one closing B–III (p. 842 & note 1122), means problems like B.110–112 & B.115.

Chapter (B–IV) on Division according to portions[1135]

(B.117) You want to divide thirty nummi between two men: for one of them a half and for the other a third.

The meaning of these words is that when one of them receives a half the other is to receive a third, (and so on) until nothing remains of the thirty.

You will do in this case the following. Multiplying the denominators of a half and a third produces six. Give half of this, which is three, to the recipient of a half, and give a third of it, which is two, to the recipient of a third. Then it is like asking thus: 'Two associates, one of whom has contributed three and the other two, have gained thirty; how will they divide this amount?' The result[1136] is for the owner of three eighteen, which is the share of the recipient of a half; the share of the owner of two is twelve, which belongs to the recipient of a third.

(B.118) You want to divide ten nummi between two[1137] men, for one of them a half and for the other a third[1138].

(*a*) The meaning of these words is that ten nummi are to be divided among three men and that to one of them is due a half, to the other, a third, and to the last, the remainder, which is a sixth of the amount, that is, of ten. But then the last partner gives his part, namely a sixth, of ten to be divided between the other two according to the ratio of their shares; thus it is known that the recipient of a half must receive half of this sixth and the recipient of a third, a third of this sixth. Next, from the remainder of this sixth, which is a sixth of a sixth, the recipient of a half must have a half and the other, a third; and so on until nothing remains. Now we know that when we proceed thus to infinity the only possibility is that the ratio of what the recipient of a half receives to what the recipient of a third receives is the same as the ratio of a half to a third. But the ratio of a half to a third is the same as the ratio of half of any number having a half and a third to a third of it. Then let this number be six, half of which is three, and a third, two. So the ratio of a half to a third is the same as the ratio of three to two. But the ratio of a half to a third is the same as the ratio

---

[1135] That is, prescribed fractions the sum of which does not make up the unit.

[1136] Found as in B.106.

[1137] The MS has 'three', from a correction prompted by the subsequent explanation.

[1138] Some of the text (in the margin) has become illegible.

LIBER MAHAMELETH

of what the recipient of a half receives of ten to what the recipient of a third receives of ten. So the ratio of what the recipient of a half receives of ten to what the recipient of a third receives of ten is the same as the ratio of three to two. By composition, the ratio of what the recipient of a half receives of ten to ten will be the same as the ratio of three to five. Therefore multiply three by ten and divide the product by five; the result will be six, and that much does the recipient of a half receive of ten. It is also evident that the ratio of two to five is the same as the ratio of what the recipient of a third receives of ten to ten. Therefore multiply two by ten and divide the product by five; the result will be four, and that much is the share of ten of the recipient of a third.

Because of this the treatment is as follows. We look for a number having a half and a third, such as six. Add its half, which is three, to its third, which is two; this makes five. Put it as the principal. If you want to know the share of the recipient of a half, or the share of the recipient of a third, multiply by ten half, or a third, of six and divide the product by the principal; the result will be what you wanted.

(*b*) This problem is solved in another way, which is the following. Add a third and a half; this makes five sixths. Always divide one by this; the result will be one and a fifth. Multiply this quantity by ten; the result is twelve. Therefore the recipient of a half must receive half of twelve, which is six, and the recipient of a third, a third of twelve, which is four.

The proof of this is evident. We must look for a number such that the sum of a third of it and half of it produces ten. Now we know that the ratio of one to five sixths is the same as the ratio of the required number to ten; since one is once five sixths and a fifth of it, the required number is once ten and a fifth of it, thus twelve. It is further evident that the ratio of a third to five sixths is the same as the ratio of a third of twelve to five sixths of it, which is ten; since the ratio of a third to five sixths is the same as the ratio of what the recipient of a third receives of ten to ten, the ratio of what the recipient of a third receives of ten to ten is the same as the ratio of a third of twelve to ten. Therefore a third of twelve is equal to what the recipient of a third receives of ten. It will likewise be shown that half of twelve is equal to what the recipient of a half receives of ten.

(*c*) You may solve it otherwise.[1139] Add half and a third of ten and put the sum as the principal. If you want to know the share of the recipient of a half, multiply half of ten by ten and divide the product by the principal; the result will be what you wanted. Likewise too, if you want to know the share of the recipient of a third, multiply a third of ten by ten and divide the product by the principal; the result will be what you wanted.

The proof of this way is similar to the proof of the former way,[1140] indeed they do not differ in anything. For it is clear that the ratio of a half to a third is the same as the ratio of half of ten to a third of it.

[1139] This should follow *a*.

[1140] Above, *a*.

850 PART TWO: TRANSLATION, GLOSSARY

(*d*) You will proceed likewise also if there are three or more men among whom a sum of money must be divided. That is, seek a number containing the fractions according to which the money is to be divided among the partners. Add together the fractions of this number and put the result as the principal. If you want to know the part of the whole sum due to the recipient of this or that fraction, take such a fraction of the number sought, multiply it by the sum and divide the product by the principal; the result will be what you wanted. The proof of this is evident from the foregoing.[1141]

(**B.119**) You want to divide thirty nummi among three men to one of whom is due two thirds of the amount, to the second as much as the whole, that is, thirty, and, to the third, one and a half times as much.

You will do the following. Form, from the denominators, the common number, which is six. Give two thirds of it, which is four, to the recipient of two thirds; to the man to whom as much as the whole is due, give six, and to the man to whom one and a half times are due, give as much as six and half of it, namely three, which makes nine. Then it is like asking thus: 'Three partners, one of whom has contributed four, another six, yet another nine, have gained thirty nummi; how will they divide this amount?' Proceed as has been explained before;[1142] the share of the investor of four is the share of the recipient of the two thirds, the share of the investor of six is the share of the recipient of the whole, and the share of the investor of nine is the share of the recipient of the one and a half times.

(**B.120**) You want to divide thirty nummi among three men to one of whom is due two thirds of thirty and three nummi, to the second as much as the whole, that is, thirty, and two nummi, to the third two and a fourth times thirty, and one nummus [*and so on until nothing remains of the thirty nummi*].

These (supplementary) nummi can be allotted in two ways.

(*i*) The first is if you choose to take them from the whole amount and give them (beforehand) to the men; the remainder will then be twenty-four nummi, which you will divide according to each one's due. Add then to the share of the recipient of two thirds the three nummi already given to him; to the share of the recipient of as much as the whole add the two nummi given before; and to the share of the recipient of two and a fourth times add the nummus mentioned above.

(*ii*) The second way is if you choose to allot these nummi with the parts. You will then do the following. Give two thirds of thirty, which is twenty, plus three nummi, to the recipient of two thirds and three nummi; next, give as much as the whole amount, plus two nummi, thus thirty-two nummi, to the recipient of as much as the whole and two nummi; next, give twice thirty and a fourth of it, plus one nummus, thus sixty-eight and a half, to

---

[1141] Also from B–III.
[1142] B.104.

LIBER MAHAMELETH 851

the recipient of twice and a fourth and one nummus. Then it is like asking: 'There were three merchants, one with a capital of twenty-three nummi, another with thirty-two, the third with sixty-eight and a half, who have gained by trading thirty nummi; how will they divide this amount?' The share [1143] of the owner of twenty-three will be the share of the first, the share of the owner of thirty-two will be due to the second, the share of the owner of sixty-eight and a half will be due to the third.

Treat other problems of this kind likewise.

[1143] Found as in B.104.

Chapter (B–V) on Masses

This is treated in all the same ways as the above subject of partnership. I shall present a single problem wherein will appear all the ways of treating such problems, which is the following.[1144]

(B.121) From a mass consisting of a mixture of ten ounces of gold, fourteen of silver and twenty of brass a part is taken which weighs twelve ounces. How much of each metal is there in this part?

This problem is like the one which states: 'Three partners, one of whom had a capital of ten nummi, the other of fourteen, the third of twenty, have gained twelve; how will they share it out?' Proceed as has been explained before.[1145] The share corresponding to the capital of ten is the amount of gold in the part, the share corresponding to the capital of fourteen is the silver, the share corresponding to the capital of twenty is the brass.[1146]

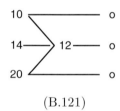

(B.121)

(B.122) There is a spherical mass with a diameter of ten palms. How many spherical masses can be made out of it having each a diameter of five palms?

You will do the following. Multiply ten by itself, thus producing a hundred; multiply this by ten; the result is a thousand. Next, multiply five by itself, thus producing twenty-five; multiply this by five; the result is a hundred and twenty-five. Dividing a thousand by it gives eight, which is the number of masses which can be made out of the original one.

Or otherwise. Divide ten by five, which gives two. Multiply it by itself, thus producing four; multiply it again by two; the result is eight, and this is the number of masses which can be made out of the first one.

[1144] This refers to B.121, which deals with the constituents of a piece taken from a larger mass of given composition.
[1145] B.104.
[1146] In the figure each of the three numbers on the left must be multiplied by 12 and the result divided by their sum; this will give the required quantities, symbolized by o (sign already seen in the figures pp. 781–783).

LIBER MAHAMELETH

(B.123) There is a spherical mass with a diameter of ten palms and a price of a hundred nummi. What is the price of another spherical mass of the same substance with a diameter of five palms?

You will do the following. Denominate five from ten; this is a half. Multiply this by itself, thus producing a fourth; multiply it by a half; the result will be an eighth. Thus the price of this smaller mass is an eighth of a hundred, which is twelve and a half.

Or otherwise. Multiply ten by itself, thus producing a hundred; multiply this by ten again, thus producing one thousand. Next, multiply five by itself, thus producing twenty-five; multiply this by five again, thus producing a hundred and twenty-five. Denominate this from one thousand; it is an eighth. So the price of this mass is an eighth of a hundred, which is twelve and a half.

Again, on something else

(B.124) There is a mass mixed from gold and silver weighing five hundred ounces and you want to know how much of each it contains without resorting to fire —that is, without melting it down— or taking a part from it or weighing it again.[1147]

(a) You will do the following. You must first know the weights of two bodies, one of gold and the other of silver, which are equal in volume. Then you will know the ratio of the weight of one of these bodies to the weight of the other; and this will enable you to determine what you require.

Now the method for ascertaining that they are equal in volume is the following. Weigh any piece of gold and, knowing thus its weight, put it in a vessel. Pour in enough water to cover the piece and mark the place reached by the water. Having removed the piece, weigh it again; if it weighs more than previously, add to the water an amount of water equal to the surplus in weight.[1148] Next, put in the water a certain piece of silver of a volume such that, when put in the water, the water level reaches the place marked. You know then that these two bodies, one of gold and the other of silver, are equal in volume. Weighing then the piece of silver, you will know the ratio of their weights. Suppose you find that the weight of the piece of silver is four fifths the weight of the piece of gold —I do not say that as a result of personal experimentation, but only as an example, in order to illustrate what I say.[1149]

(b) So find by the previous method some body of silver equal in volume to the given mass and weigh it; suppose you find a weight of four hundred and thirty-two ounces. Therefore, according to the above supposition, any body of gold equal in volume to this one will have a weight of five hundred and

---

[1147] Thus the mass must retain its integrity and we are not to use a hydrostatic balance.

[1148] This second weighing is not to determine the alloy but to compensate for any water remaining in the artefact.

[1149] This is certainly true since the actual ratio is slightly more than 2.7 : 5.

forty ounces. Next, take the difference in weight between these two bodies. This is a hundred and eight; keep it in mind. Take now the difference in weight between the silver body and the mixed mass, which is sixty-eight, and the difference in weight between the composite mass and the gold body, which is forty. These two differences divide the first difference, that kept in mind, according to the ratios of gold and silver in the composite mass.[1150] Now the ratio of the greater of these differences to the difference kept in mind is the same as the ratio of the part found in the mixed mass of the substance closest in weight, to the weight of this same substance.[1151] Since five hundred and forty is nearer to five hundred than four hundred and thirty-two, the ratio of the quantity of the gold which is in the mass to five hundred and forty is the same as the ratio of sixty-eight to a hundred and eight. Sixty-eight being five ninths and two thirds of a ninth of a hundred and eight, the quantity of gold which is in the mass is five ninths and two thirds of one ninth of five hundred and forty, which is three hundred and forty. Again, the ratio of forty to a hundred and eight is the same as the ratio of the quantity of silver which is in the mass to four hundred and thirty-two. Since the ratio of forty to a hundred and eight is three ninths and a third of a ninth, the quantity of silver which is in the mass is three ninths and a third of a ninth of four hundred and thirty-two, which is a hundred and sixty. This is what we wanted to demonstrate.

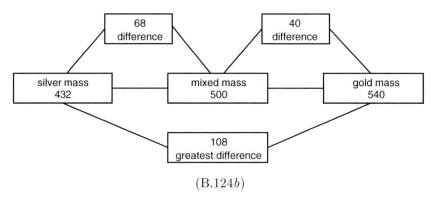

(B.124*b*)

[HOW TO DETERMINE IF GOLD OR SILVER IS PERFECTLY PURE AND, IF NOT, HOW MUCH OF ANOTHER SUBSTANCE HAS BEEN MIXED WITH IT.]

(**B.125**) Someone handed over to a craftsman one thousand ounces of gold to produce vessels. These vessels having been made and returned, their

---

[1150] The sum of these two differences (68, 40) is the difference 'kept in mind' (108), and from this we shall determine the alloy. Thus the heading below, which refers to that procedure, must be a former marginal gloss indicating the subject of this section.

[1151] The greater the difference with one metal, the more there is of the other in the mass; thus, here, 68 : 108 gives the proportion of gold found in the mixed mass.

LIBER MAHAMELETH

855

owner wishes to know whether the craftsman mixed anything with the gold, and how much.

The method for determining this is the same as the preceding. Suppose, then, you have found some piece of silver equal in volume to the piece of wrought gold, and found it to have a weight of eight hundred and sixty-four ounces [*there is no reason to object that it could weigh more or less, this is just given as an example*][1152]. It is thence known that any piece of pure gold equal in volume to this one will weigh one thousand and eighty ounces. Thus it is clear that the wrought gold is not pure; for if it were pure, it would weigh one thousand and eighty ounces while if it were pure silver it would weigh eight hundred and sixty-four ounces.

Having discovered this, we now want to know how much silver has been mixed in, or how much gold has been left. So take the difference in weight between the piece of pure gold and the piece of pure silver. This is two hundred and sixteen; keep it in mind. Next, take the difference in weight between the piece of pure silver and the vessels, which is a hundred and thirty-six, and the difference in weight between the vessels and the piece of pure gold, which is eighty. These two differences divide the difference kept in mind according to the ratios of gold and silver in the wrought vessels. Now the ratio of the greater of these differences to the difference kept in mind is the same as the ratio of the weight of the substance having a weight closer to the weight of the vessels, to the weight of this same substance. Since one thousand and eighty is closer to one thousand than eight hundred and sixty-four, the ratio of the weight of gold in the vessels to one thousand and eighty must be the same as the ratio of a hundred and thirty-six to two hundred and sixteen. The ratio of a hundred and thirty-six to two hundred and sixteen being five ninths and two thirds of a ninth, the (weight of) gold which is in the vessels is five ninths and two thirds of a ninth of one thousand and eighty, which is six hundred and eighty. Likewise also, the ratio of the (weight of the) silver in the vessels to eight hundred and sixty-four is the same as the ratio of eighty to two hundred and sixteen. The ratio of eighty to two hundred and sixteen being three ninths and a third of a ninth, the (weight of the) silver which is in the vessels is three ninths and a third of a ninth of eight hundred and sixty-four, which is three hundred and twenty. This is what we wanted to demonstrate.

[1152] This (which should come after the next sentence) reminds us of the assertion in B.124*a*.

Chapter (B–VI) on Drapery [1153]

(B.126) For instance, someone asks thus: From a cloth ten cubits long, eight cubits wide and fifty ounces in weight is cut a part six cubits long and four cubits wide. What is its weight?

You will do the following. Multiply the width of the cloth by its length; the result will be eighty. Do again the same for the part cut off; the result will be twenty-four. Then it is like asking: 'When eighty is given for fifty, what is the price of twenty-four, which is the size of the part?' [1154] Multiply twenty-four by fifty, which is the weight of the cloth and divide the product by eighty; the result will be fifteen, which is the weight of the part. You may also start by dividing one of the factors by the divisor and then multiply the result by the other; the product is what you require. [*You are to know that here by 'size' is meant, for cloths, the number of parts, one cubit long and wide, by which the whole is measured.*] [1155]

This rule is taken from a proportion. For the size of the cloth is to its weight as the size of the part is to its weight. Therefore eighty, which is the size of the cloth, is to fifty, which is its weight, as twenty-four, which is the size of the part, is to its weight, which is required. Thus the fourth is unknown, and the first, which is the divisor, is eighty.

(B.127) Someone says: From a cloth ten cubits long, eight wide and sixty ounces in weight is cut a part fifteen cubits long and nine wide. What is its weight? [1156]

This problem is impossible: the cloth is ten cubits long and therefore fifteen cannot be cut from it. But if the length of the part is less than the length of the cloth and the width of the part less than the width

---

[1153] *Cortine*, 'drapery', 'cloth-goods', the French *draps*: mediaeval treatises commonly distinguish the *commerce de draperie*, cloth-trade, dealing with woollen goods, from the *commerce de toilerie*, linen-cloth trade, dealing with linen goods, which is the subject of the next chapter; *cortina* will hereafter be referred to, for convenience, as 'cloth'. Both the whole cloth as its parts are supposed to be rectangular (sometimes square).

[1154] B.1.

[1155] Gloss explaining the meaning of *magnitudo* (which had the sense of 'volume' in the previous chapter). This is indeed our first encounter with measuring an area and its unit of measure. Note that the 'cubit' measures not only length but also area (our 'square cubit') and volume (our 'cubic cubit'; see B.272a, B.335a). This is not peculiar to our text (see, e.g., our *Introduction to the History of Algebra*, p. 21 n. 24, or p. 74 n. 64).

[1156] Only MS $\mathcal{A}$ has this problem. The treatment which follows what we are told about possibility may well be an addition. See note 1158.

LIBER MAHAMELETH

857

of the cloth[1157], then the problem will be solvable, and you will do the following.[1158] Multiply the length of the part by its width to produce its size. Multiply likewise the length of the cloth by its width to produce its size. Then the ratio of the size of the part to the size of the cloth will be the same as the ratio of the weight of the part, which is required, to the weight of the cloth. You may then either form the ratio of size to size and take the corresponding fraction of the weight or multiply the size of the part by the weight of the cloth and divide the product by the size of the cloth; the result will be the weight of the part, and this is what you wanted.

(B.128) Someone asks thus: From a cloth ten cubits long, eight cubits wide and weighing fifty ounces is cut a part which weighs fifteen ounces and is six cubits long. What is its whole size and its width?

You will do the following. You will say: 'If the whole size of the larger cloth, which is eighty, weighs fifty ounces, how many cubits[1159] are there for fifteen ounces?' There are twenty-four cubits, which is the size of the part. Divide this by the length, which is six; the result will be the width, which is four, and this is what you wanted to know. If, inversely, the width were known and the length unknown, you would divide the size of the part by its width, the result then being the length.

This rule, too[1160], is taken from a proportion. For the size of the larger cloth, which is eighty, is to its weight, which is fifty, as the size of the part, which is required, is to its weight, which is fifteen. Thus the third (term) is unknown, the divisor is the second, namely fifty, the first is one of the factors, namely the size of the larger cloth, which is eighty, and the other factor is fifteen, which is the weight of the part. Then you will proceed as said above.

(B.129) From a cloth of unknown length and width, but sixty ounces in weight, a part similar to the cloth is cut, five cubits long, four wide, and fifteen ounces in weight. What is the length or the width of the cloth?

You will do the following. Multiply the length of the part by its width; the product will be twenty. Next, divide the weight of the cloth by the weight of the part and multiply the result by twenty; the product will be eighty. Next, denominate the width of the part from its length and take this fraction, namely four fifths, of eighty; this is sixty-four. Its root, which is eight, is the width of the cloth. Then, if you want to know the length of the cloth, divide the length of the part by its width and multiply the result by the width of the cloth; the result will be ten, and this is the length of the cloth. Or, otherwise, divide the length of the part by its width and multiply the result by eighty; the root of the product is the length of the cloth.

[1157] 'less than', rather: less than or equal to.

[1158] This is superfluous, for it just repeats what we know from the previous problem.

[1159] That is, square cubits (note 1155).

[1160] See B.126 (also B.127).

(B.130) From a cloth of which the length is greater than the width by two cubits and sixty ounces in weight is cut a part five cubits long, four wide and fifteen ounces in weight. What is the length or the width of the cloth?

You will do the following. Multiply the length of the part by its width; the product will be twenty. Next, divide the weight of the cloth by the weight of the part and multiply the result by twenty; the product will be eighty. Next,[1161] multiply half of the two cubits by itself and add the product to eighty; the result will be eighty-one. Subtract from its root, which is nine, half of the two cubits; this will leave eight, which is the width of the cloth. Adding to it two cubits will make ten, which is its length.

(B.131) Someone asks thus: From a cloth ten cubits long, eight wide, and fifty ounces in weight is cut a square part weighing twenty-two and a half ounces. What are its size and its width or length?

Find the size of the part as we have taught before.[1162] It is thirty-six. Its root, which is six, is the length or width. If its size were a number having no root, you can still find an approximation as we have taught in the chapter on roots;[1163] this "root" [1164] will be the length.

(B.131′) Suppose he proposes thus: The part cut off has a weight of twenty ounces and a length twice the width. What is its length or width?

You will do the following. Find the size of the part as we have taught before.[1165] It is thirty-two. If you want to know the length of the part, find the root of twice thirty-two; this is eight, and such is the length. The root of half the size, or half the previous root, namely four, is the width. We do that because the length is twice the width, and (therefore) the width half the length. Then it is like saying: 'A number multiplied by its half produces thirty-two'; then multiplied by itself it will produce sixty-four, so this number is the root of sixty-four.

(B.131″) Likewise, if it were said that the length is thrice the width: you will find the root of thrice the size, and this will be the length, while the root of a third of the size will be the width.

(B.132) Now someone asks thus: From a cloth ten cubits long, eight wide, and sixty ounces in weight is cut a part similar to it and fifteen ounces in weight. What is its length or width?

[*'Similar' is used here to mean that the length of the part is to its width as the length of the larger (cloth) is to its width.*][1166]

[1161] Solving $x^2 + bx = c$.

[1162] B.128.

[1163] A.275–283.

[1164] *quasi radix.*

[1165] Again, B.128.

[1166] This looks like an interpolation, even if this is the first occurrence of *consimilis*. (B.129, where we have a part *similis* to the whole, does not (***a***) You will do the following. You should determine the size of the part as we have taught before.[1167] This is twenty. Then it is like asking: 'When ten caficii are given for eight nummi, what is the number of caficii which, when multiplied by the corresponding price, produces twenty?' Or like asking: 'Someone has gained eight from ten, and multiplying a profit of his by the corresponding capital produces twenty.' We have already taught above how to treat all that.[1168] Thus, if you want to know the length of the part, multiply the length of the cloth by the size of the part and divide the product by the width of the cloth; the root of the result is the length of the part. But if you want to know the width of the part, multiply the width of the cloth by the size of the part and divide the product by the length of the cloth; the root of the result is its width.

(***b***) You may proceed otherwise.[1169] ⟨Denominate the weight of the part from the weight of the cloth and multiply the root of the result by the length of the cloth; the product will be the length of the part. If you want to know its width, multiply the width of the cloth by the same root; the product will be the width of the part.

This is proved as follows.⟩ Let the cloth be the area $ABGD$ and the part cut off, the area $HZKT$. It is thus clear[1170] that the ratio of the area $HZKT$ to the area $ABGD$ is the same as the ⟨ratio of the weight of the first to the weight of the other, which is a fourth of it. But the ratio of the area $HZKT$ to the area $ABGD$ is the same as the⟩ duplicate ratio of one side of the first to the corresponding side of the other.[1171] Thus the ratio of ZK to BG ⟨is a half; so ZK⟩ is half ⟨of BG⟩. Likewise also, ZH is half of AB. Therefore HZ is four, which is the width, and ZK five, which is the length. This is what we wanted to demonstrate.

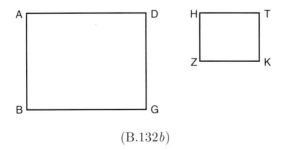

(B.132*b*)

AGAIN ON THE SAME TOPIC, BUT OTHERWISE[1172]

---

precede B.132 in the disordered text.) Other such interpolated passages in B.137 & B.146.
[1167] B.128.
[1168] B.17 and B.58, respectively.
[1169] This part *b*, found only in MS $\mathcal{A}$, suffers from several omissions.
[1170] B.126.
[1171] *Elements* VI.23. See also note 122, p. 608.
[1172] We are now dealing with mixed material.

(B.133) Someone asks thus: From a cloth ten cubits long and eight wide made of various materials, namely ten ounces of silk, fourteen of cotton and twenty of linen, is cut a part four cubits wide and six long. How much of each material does it contain and what is its weight?

You will do the following. Multiply the length of the part by its width; this will produce twenty-four. Do the same with the large cloth; this will produce eighty. Next, multiply the size of the part, which is twenty-four, by the amount of each of the materials and divide the product by the size of the larger cloth, which is eighty; the result will be how much the part contains of the material in question.

This rule is taken from a proportion. For as the size of the cloth is to the ounces of silk in it so is the size of the part to the ounces of silk in it; likewise for the cotton and the linen.

If you want to know the weight of the part, you will add together what it contains of each material; the sum is its weight. Or if you want to know this by means of the weight of the larger cloth, which is forty-four, proceed as has been explained before.[1173]

AGAIN ON THE SAME TOPIC.

(B.134) Someone asks thus: From a cloth of mixed material ten cubits long and just as wide, containing thirty ounces of silk, forty of linen and fifty of cotton, is cut a part fourteen ounces in weight. How much of each material does it contain and what is its size?

You will do the following. Determine the size of the cloth as explained above, which is as follows.[1174] Multiply ten by ten; this produces a hundred. Next, add together the ounces of the different materials; the result is a hundred and twenty, which is the principal. Next, multiply fourteen by a hundred; the result will be one thousand four hundred. Divide it by the principal; the result is the size of the part, which is eleven and two thirds. If you want to know how much silk, linen or cotton there is in the part, multiply fourteen by (the given weight of) any of these materials and divide the product by the principal; the result is how much it contains of the material in question.

(B.135) Someone asks thus: From a cloth of mixed material ten cubits long and eight wide, containing thirty ounces of silk, forty of cotton and fifty of linen, is cut a part thirty ounces in weight and five cubits in length. What is its width and how much of each material does it contain?

(a) You will do the following. Add all the ounces; this makes a hundred and twenty. Multiply this by five cubits, which is the length of the part; the result will be six hundred. Let it be your principal. Next, multiply the size of the cloth, which is eighty, by thirty, which is the weight of the part;[1175] the result will be two thousand four hundred. Divide it by the

---

[1173] B.126.

[1174] Again, B.126.

[1175] There is another 'thirty'. Same in B.136.

LIBER MAHAMELETH
861

principal; this gives four cubits, which is the width of the part. If you want to know how much it contains of each material, proceed as we have taught before,[1176] and the result will be what you wanted.

(*b*) Or otherwise. Denominate the weight of the part from the weight of the cloth; it is a fourth. Then divide a fourth of the size of the cloth, which is twenty, by five, which is the length of the part; the result will be four, which is the width of the part.

(**B.135′**) If he had proposed thus: 'The part being four cubits wide, how long is it?', you would do as we have taught before,[1177] namely as follows. Multiply the weight of the cloth by the width of the part; the product will be the principal. Next, multiply the weight of the part by the size of the cloth and divide the product by the principal; the result will be five, which is the length you require. Or, if you divide a fourth of the size of the cloth by the width of the part, the result will be the length you require.

(**B.136**) Someone asks thus: From a cloth of mixed material ten cubits long and just as wide, containing thirty ounces of silk, forty of cotton and fifty of linen, is cut a square part thirty ounces in weight. What is its length or width?

You will do the following. Multiply thirty, which is the weight of the part, by the size of the cloth, which is a hundred; this produces three thousand. Divide it by a hundred and twenty ounces, which is the weight of the cloth; the result will be twenty-five. Its root, namely five, is the width or length.

(**B.137**) Someone asks thus: From a cloth of mixed material ten cubits long and eight wide, containing ten ounces of silk, twenty of cotton and thirty of linen, is cut a part fifteen ounces in weight, similar to the cloth [*in the proportion of its length to its width*][1178]. What is the length or width of this part?

(*a*) You will do the following. Multiply the weight of the part by the size of the cloth, which is eighty; this will produce one thousand two hundred. Divide this by the weight of the cloth, which is sixty; the result will be twenty, which is the size of the part. If you want to know the length of the part, multiply the length of the cloth, which is ten, by twenty, which is the size of the part; this will produce two hundred. Divide this by eight, which is the width of the cloth; the result will be twenty-five. Its root, which is five, is the length of the part. If you want to know its width, multiply eight, which is the width of the cloth, by the size of the part, which is twenty; this will produce a hundred and sixty. Divide it by the length of the cloth; the result will be sixteen. Its root, which is four, is the width of the part.

[1176] B.133 (& B.134).

[1177] B.128 or, *mutatis mutandis*, B.135.

[1178] See note 1166, same kind of gloss.

(**b**) Or divide the size of the cloth by the size of the part; this will give four. Divide by its root, which is two, the length of the cloth; the result will be five, which is the length of the part. Divide by the same root the width of the cloth; the result will be four, which is the width of the part. Or divide the weight of the cloth by the weight of the part, and divide by the root of the quotient the length and the width of the cloth; the results will be the length and the width of the part. Or denominate the size of the part from the size of the cloth; you will find a fourth. Its root is a half. Thus half the length of the cloth is the length of the part and half the width of the cloth is the width of the part. Likewise also, if you denominate the weight of the part from the weight of the cloth then multiply by the root of the result as just said[1179], this will give what you wanted.

(**B.138**) Someone asks thus: From a cloth ten cubits long, eight wide, and sixty ounces in weight, is cut a part two cubits longer than it is wide and weighing eighteen. What is its length or width?

It is known[1180] that the ratio of eighteen to sixty is the same as the ratio of the size of the part to the size of the whole cloth, which is eighty. Thus proceed as I have taught above.[1181] The size of the part will be twenty-four. But we know that the size results from multiplying the length by the width. So the width is a number which when multiplied by itself plus two produces twenty-four.[1182] Let then the length be AB and the width, AG. Thus BG is two. On the other hand the multiplication of AB by AG produces twenty-four.[1183] Let BG be bisected at the point D. Then the products of AB into AG and GD into itself will be equal to the product of AD into itself. But the multiplication of AB by AG produces twenty-four, and the product of GD into itself is one. Therefore the product of AD into itself is twenty-five. Thus AD is five. GD being one, this leaves as AG four, which is the width. Likewise also: AD is five, DB is one, so AB will be six, which is the length. This is what we wanted to demonstrate.[1184]

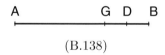

(B.138)

[1179] As for the half: by the length or the width of the cloth.
[1180] From B.126 on.
[1181] Rule of three (three terms known), thus rule i p. 777.
[1182] Equation of the form $x^2 + bx = c$, solved by *Elements* II.6.
[1183] Difference and product of AB and AG thus known.
[1184] Geometrical computation. See p. 724, note 584.

Chapter (B–VII) on Linens [1185]

(B.139) You want to know how many pieces [1186] four cubits long and three wide can be made out of a linen fifteen cubits long and eight wide.

You will do the following. Multiply the length of the linen by its width; the result will be its size, which is a hundred and twenty. Dividing this by the size of one piece, which is twelve, will give the number of pieces. Or divide the length of the linen by the length of one piece, which gives three and three fourths, and divide its width by the width of one piece, which gives two and two thirds. Next, multiply the two quotients; the result will be ten. You may also divide the length of the linen by the width of the piece, its width by the length of the piece, and multiplying the two quotients will produce what you require.

(B.140) From a linen ten cubits long and eight wide is cut a part six long and five wide. How much is cut from the linen?

You will do the following. Multiply the length of the linen by its width; the result will be eighty. Next, multiply the length of the part by its width; the result will be thirty. Denominating this from eighty gives three eighths [1187]. Such is the fraction cut from the linen, namely its three eighths [1188].

(B.141) Someone asks thus: From a linen fifteen cubits long and eight wide are cut ten pieces each four cubits long. How many cubits is each wide?

You will do the following. Divide the size of the linen by the number of pieces, which is ten; the result is twelve, which is the size of each piece. Divide it by its length, which is four; the result is three, which is the width of each piece.

Or divide the size of the linen by the product of the length of a piece into their number; the quotient is the width of each piece.

The proof of this is evident [1189]. We had to divide the size of the linen by the number of pieces, the result being the size of each piece, which we

---

[1185] *Lintei* (linen-cloths), which for convenience are referred to as 'linens'.

[1186] Hereafter for *gausapa*. *Gausape* is either a piece of linen used for household purposes ('Tuch', 'Tischtuch', 'Handtuch' according to the *Mittellateinisches Wörterbuch*) or —like the Arabic *thaub*— a garment; Alcuin also has a problem *de linteamine* (*Propositio* No 10) from which are taken pieces, each of which *sufficiat ad tunicam consuendam.*

[1187] 'three fifths' in the MS.

[1188] 'three fifths' in the MS.

[1189] And links the second way to the first.

864   Part Two: Translation, Glossary

then had to divide by the length of one piece, the result being the width of a piece. But this is the same as dividing the size of the linen by the product of the number of pieces into the length of each. For dividing one number by another and the result by yet another is the same as dividing the first dividend by the product of the two divisors, as said in the beginning.[1190]

(B.141′) If, the width being known, the length were unknown, you would divide the size of the piece, which is twelve, by its width, and the result would be its length.

(B.141″) He might propose the following: 'From this same linen ten square pieces are cut; what is the length or the width of each?' As explained in drapery [1191], find the root of the size of one piece; this will be its length or width.

(B.141‴) If he said that the length of the piece is twice or thrice its width, you would find its length or width by means of its size, as we have taught before in drapery.[1192]

(B.141^{iv}) Likewise, if he said that the piece is similar to the linen, you would find its length and its width by means of its size and by means of the length and width of the linen as we have taught a little before [1193].

AGAIN ON THE SAME TOPIC.[1194]

(B.142) Someone asks thus: From a linen fifteen cubits long, eight wide and with a price of twenty nummi, a piece is cut eight cubits long and four wide. What is its price?

You will do the following. Find the size of the linen, which is a hundred and twenty, and the size of the piece, which is thirty-two. It is then like asking: 'When a hundred and twenty, which is the size of the linen, is given for twenty, what is the price of thirty-two, which is the size of the piece?' You will proceed as we have taught before in the chapter 'what is its price?'.[1195]

This rule is taken from a proportion. For the size of the linen is to its price as the size of the piece is to its unknown price. The fourth is therefore unknown and the first, which is the divisor, is the size of the linen.

(B.143) Someone asks thus: A linen six cubits long and four wide costs ten. What is the price of a linen of the same sort fifteen cubits long and five wide?

You will do here the same as in the previous problem. For we already know that the ratio of the sizes of two linens is the same as the ratio of their prices. Thus you are to determine the size of the first linen, which is

---

[1190] Premiss P_4.

[1191] B.131.

[1192] B.131′–131″.

[1193] B.132. B.146 is a calculated example.

[1194] Section involving price.

[1195] B.1 (& B.3–6, B.11–12). Same reference for B.143.

LIBER MAHAMELETH

twenty-four; likewise the size of the other linen, which is seventy-five. Then it is like asking: 'When twenty-four is given for ten, what is the price of seventy-five?' Proceed as has already been explained.

(B.143′) You may want one linen only, or both, to be circular. Determine their respective sizes and proceed as has already been explained. Now the size of a circular (linen) is found by multiplying half its diameter by half the circumference, or by multiplying the diameter by itself and subtracting from the product a sixth and half a sixth of it;[1196] the remainder is the size of the circular linen.   The same would be done if the linens were triangular or of some other shape. The science of finding the size of such figures can be obtained from a book on taccir [*that is, on finding the size*].[1197]

Once the sizes of both linens are found, whatever their shape [⟨*or the size of one of them*⟩ *by means of the size of the other*][1198], the treatment will be the same as in a problem 'what is its price?',[1199] and the result will be what you wanted.

(B.144) Someone asks thus: A linen twenty cubits long and eight wide costs twenty nummi. How many cubits shall I have for twelve nummi of a linen of the same sort, and of unknown length and four cubits wide?

You will do the following. Determine the size of the first linen; this is a hundred and sixty. Then you will say: 'When a hundred and sixty is given for twenty nummi, how much shall I have for twelve of a linen four cubits wide?' Divide the size of the first by the width of the second, which gives forty. Divide this by twenty, which is the price of the first, and multiply the result by twelve, which is the price of the second; the product will be twenty-four, and this is the length in cubits of the linen four cubits wide.

Treat other problems of this kind as has been explained [*here*][1200].

(B.145) [1201] Someone asks thus: From a linen twenty cubits long and eight wide are cut six pieces each eight cubits long. How wide are they?

Multiply the number of pieces by their length, that is, six by eight; the result is forty-eight. Put it as the principal[1202]. Next, find the size of the linen by multiplying its length by its width; it will be a hundred and sixty.

[1196] *Sic*, instead of 'a seventh and half a seventh of it'. The same in B.152, but correct in B.341.

[1197] A practical geometry providing formulae for calculating areas and volumes of figures was called in Arabic a book on *taccir* (*taksīr* = breaking into parts) or on *misāḥa* (= measurement, surveying). The (presumably) interpolated explanation is found only in MS *A*.

[1198] This is irrelevant, but alludes to the group B.141.

[1199] B.1.

[1200] Same interpolation (but making sense) in B.242*a*.

[1201] The genuineness of this problem, similar to B.141 and thus in any event misplaced, is questionable.

[1202] Here in the sense of 'divisor'. Same in B.147, B.156–158. See note 1233, p. 872.

Divide it by the principal; this will give the width of each piece, namely three cubits and a third.

[*The proof of this is the following. You have to divide the size of the linen by the number of pieces in order to know the size of each, and dividing this by their length will give their width. But you are to know that dividing the size of the linen by the number of pieces and dividing the number of the result by the length of one piece is the same as multiplying the length of the piece by the number of them and dividing by the product the size of the linen, the result being the width of each piece. This is so because when one number is divided by another and the result divided by yet another, the result of the last division is the same as the result of dividing the first by the product of the two divisors.*[1203] *For instance, the number A is divided by the number B and the result is the number G, the number G is divided by the number D and the result is the number H, and the number B itself is multiplied by the number D and the result is the number T. Then I say that when the number A is divided by the number T the result will be H. This is proved as follows.*[1204] *Since the number D when multiplied by the number H produces the number G and, again, the number D when multiplied by the number B produces the number T, so the number D is multiplied by two different numbers, which are H and B. Then the ratio of the first product, which is G, to the second product, which is T, is the same as the ratio of the first multiplier, which is H, to the second multiplier, which is B. Then the ratio of G to T is the same as the ratio of H to B. So multiplying G by B produces the same result as multiplying H by T. But multiplying G by B produces A; thus dividing A by T will give H. This is what we wanted to prove. Thus it is known from this that when the size of the linen is divided by the product of the number of pieces into the length of a piece, the result is its width.*]

$$
\begin{array}{ccc}
\underline{H} & \underline{T} & \underline{A} \\
\underline{D} & \underline{G} & \underline{B}
\end{array}
$$

(B.145)

(**B.145′**) [1205] If he said that from the linen have been cut ten pieces the width of each of which is smaller than its length by three cubits, you would divide the size of the linen by the number of pieces, the result being the size of each piece. But this size results from multiplying the width by itself and by three. To determine this width, proceed as explained above in drapery,[1206] and you will find the answer.

AGAIN, ON THE SAME TOPIC.

[1203] This proposition, which is P_4, was already quoted in B.141.

[1204] Using *Elements* VII.18–19.

[1205] This should rather appear after B.141^{iv}.

[1206] B.138.

(B.146) Someone asks thus: From a linen ten cubits long and eight wide are cut four pieces similar to the linen [*that is, their width is to their length as the width of the linen is to its length*][1207]. What is their length or width?

(*a*) You will do the following. Divide the size of the linen, which is eighty, by the number of pieces; this gives twenty, which is the size of each piece. If you want to know the length of each, multiply the length of the linen by the size of the piece and divide the product by the width of the linen; the root of the result, namely five, is the length of the piece. If you want to know the width, multiply the width of the linen by the size of the piece, which is twenty, and divide the product by the length of the linen; the root of the result, namely four, is the width of the piece.

(*b*) Or, if you want to know the length, multiply the length of the linen by itself, the product by twenty, and divide the result by the size of the linen; the root of this is the length of the piece. If you want to know the width, multiply the width of the linen by itself, the product by twenty, and divide the result by the size of the linen; the root of this is the width of the piece. [*Or otherwise: multiply the width of the linen by the size of the piece and divide the product by the length of the linen; the root of the result is the width, namely four.*][1208]

AGAIN ON THE SAME TOPIC, BUT OTHERWISE [1209]

(B.147) For instance, someone asks thus: The price of a circular linen with a circumference of ten cubits is sixty nummi. What is the price of another circular linen with a circumference of five cubits?

You will do the following. Multiply ten by itself, thus producing a hundred. Put it as the principal. Next, multiply five by itself, thus producing twenty-five. Multiply this by sixty; the result will be one thousand five hundred. Divide it by the principal; this gives fifteen, which is the price you require. Or otherwise. Denominate the product of five into itself from the product of ten into itself; this is a fourth. Thus a fourth of sixty nummi, which is fifteen, is what you require.

AGAIN ON THE SAME TOPIC.

(B.148) Someone asks: Sixty nummi is the price of a circular linen with a circumference of ten cubits. What is the circumference of a linen with a price of fifteen nummi?

Multiply ten by itself, thus producing a hundred. Multiply this by fifteen; the result will be one thousand five hundred. Divide it by sixty nummi; this gives twenty-five. Its root, which is five, is the circumference of the linen with the price of fifteen nummi. Or otherwise. Denominate fifteen from sixty; it is a fourth. Multiply its root, which is a half, by ten; the result will be five, which is the circumference of the linen.

[1207] Just as one reader seemed to find *consimilis* in B.137 perplexing, so did he or another the *ad similitudinem* here.

[1208] Same as *a*.

[1209] Shape now circular.

868   PART TWO: TRANSLATION, GLOSSARY

These rules are generally applicable, whatever the value of the circumferences.[1210]

(B.149) Someone asks: From a circular linen ten cubits in diameter, how many circular linens, each with a diameter of two cubits, can be cut?

You will do the following. Divide the square of the diameter of the larger linen by the square of the diameter of the smaller; the result will be twenty-five, and this is the number of linens which can be cut.

The proof of this is clear. For the ratio of the area of any circle to the area of another is the same as the ratio of the square of the diameter of the first to the square of the diameter of the second.[1211] But the square of the diameter of the larger is twenty-five times the square of the diameter of the lesser. Thus the larger circle is twenty-five times the lesser circle. Therefore it is clear that from the larger piece twenty-five can be cut. This is what we intended to demonstrate.

You are to know, though, that such problems involving circles can never have any practical application.[1212]

AGAIN ON LINENS.

(B.150) Someone asks: From a circular linen with a diameter of ten cubits, how many pieces three cubits long and two wide can be cut?

This problem cannot actually be solved.[1213] For we cannot know the ratio of the area of a square to the area of a circle since the amount of the area of a circle cannot be determined exactly, and this is so because we cannot know the ratio of diameter to circumference since we cannot know the ratio (of the length) of a straight segment to (the length of) a curved one. Archimedes has found a ratio which is a very close approximation, namely that any circumference is thrice the corresponding diameter plus a seventh of it.[1214] Since it is known, but not with exactness, that any circumference is thrice the corresponding diameter plus a seventh (of it), while the value of the area of a circle cannot be determined except from its circumference[1215], this area cannot be determined with perfect accuracy. Thus dividing the approximate area of the circular (linen) by the size of one piece will give the approximate number of pieces which can be made from it. Since this can never give a perfectly accurate result, such problems cannot be actually solved.

(B.151) Someone asks: A circular linen with a diameter of eight cubits is sold for twenty nummi. What is the price of another circular linen with a diameter of two cubits?

[1210] For circular pieces, the price is exactly determinable.

[1211] *Elements* XII.2.

[1212] The number of pieces actually cut cannot equal the result of the computation.

[1213] That is, mathematically; its unfeasibility has already been stated.

[1214] In his *Dimensio circuli*, prop. 3 (approximation by excess).

[1215] As half the diameter by half the circumference.

LIBER MAHAMELETH
869

You already know [1216] that the ratio of the area of one circle to the area of another is the same as the ratio of one price to the other. Now the ratio of the area of one circle to the area of another is the same as the ratio of the square of one diameter to the square of the other. Thus the ratio of the square of one diameter to the square of the other is the same as the ratio of one price to the other. But the square of the greater diameter is sixty-four and the square of the lesser is four, while the price of the larger linen is twenty nummi. Therefore the ratio of sixty-four to four is the same as the ratio of twenty to the price of the smaller linen, which is required. Proceed then as has been explained above.[1217] The required result will be one and a fourth, and this is what you wanted.

(B.152) Someone asks: A circular linen with a diameter of ten cubits costs a hundred nummi. What is the price of a linen three cubits long and two wide?

You will do the following. You know that the ratio of the area of the circular (linen) to the area of the other linen is the same as the ratio of one price to the other; therefore the price will not be perfectly set.[1218] Now determine the size of the circular (linen). That is, multiply half the diameter by itself and the product by three and a seventh; or multiply the diameter by three and a seventh, which gives the circumference, then multiply half the diameter by half the circumference, which gives the area of the circular piece; or multiply the diameter by itself and subtract from the product a sixth and half a sixth of it [1219]. By whichever of these ways you proceed, the result will be, as the approximate area of the circular (linen), seventy-eight and four sevenths. Therefore the ratio of seventy-eight and four sevenths to six, which is the size of the other linen, is the same as the ratio of a hundred to what is required. Proceed as has been explained before [1220], and the result will be what is required, approximately.

(B.153) Someone says: A circular room has a floor with a diameter of six cubits. How many slabs, also circular but each three cubits in diameter, does it hold?

From what precedes [1221] it is known that, potentially, the floor holds four, but if we wanted to carry this out, we would not be able to.

This is shown as follows. Let the circumference of the floor be $ABGD$, the circumference of each slab, $HZKT$, and the diameter of the floor, AG. Thus AG is six. Now TZ is three, and let the centre of the circular floor be Q; so AQ is equal to TZ. Let one slab be $AMQL$, the second slab being $NQCG$. Indeed, it cannot contain more slabs: this leaves a figure,

---

[1216] B.142 & B.147–150.

[1217] Plain proportion, thus rule of three (p. 777).

[1218] See B.150.

[1219] *Sic*, for 'a seventh and half a seventh of it'. See note 1196, p. 865.

[1220] Plain proportion.

[1221] B.149.

encompassed by the arc ADG and the two curved portions AMQ and QNG, equal to the circular slab $HZKT$, and also a figure, encompassed by the arc ABG and the two curved portions ALQ and QCG, equal to the circular slab $HZKT$. If this slab could be superposed on either of these (two) figures and made to match it, our (above) statement [1222] would be applicable. But the slab can never be adjusted, that is, made to match any of these (two) figures. For when we draw the diameter BD, cut from it three cubits and describe on it a circle equal to the circle $HZKT$, say $DMQN$, it will cut a portion from the circles $QCGN$ and $AMQL$.[1223] Therefore this problem can never have any actual application, though it can be actually solved. Likewise, any problem of this kind can never be carried out in practice.

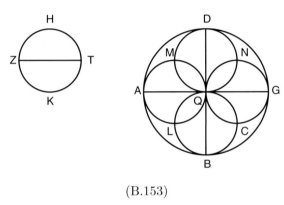

(B.153)

(**B.153′**) [1224] Likewise if he said: A linen is thirty cubits long and ten wide. How many circular linens can be cut from it having each a diameter of three cubits?

This problem again, like the previous one, can never be carried out in practice. This is clear from the foregoing.[1225]

Again, on something else

(**B.154**) Someone asks thus: A hall is twenty cubits long and eight wide. How many marble slabs each two cubits long and one and a half wide can it hold on its floor?

You will do the following. Determine the area of the floor by multiplying its length by its width; the result will be a hundred and sixty. Next, determine the area of one slab by multiplying its length by its width; this produces three. Divide the hundred and sixty by this; the result will be fifty-three and a third, which is the number of slabs. The treatment here is the same as for the linens.[1226] The same holds for all rectangular areas.

[1222] The above calculation.
[1223] If not superfluous, this last sentence is incomplete.
[1224] This should appear as B.150′.
[1225] Thus B.150.
[1226] B.139.

Chapter (B–VIII) on Grinding

This chapter comprises two cases.[1227] In the first, an agreement is made with the miller who specifies what fraction he is to receive from each caficius; in the second, the agreement with the miller is that he will receive for each, and not from each, caficius a certain fraction, also specified. Furthermore, in both cases there are four types. I shall present one problem for each.

THE FIRST TYPE OF THE FIRST CASE IS THE FOLLOWING.

(B.155) For the grinding of each caficius a fifth of it is handed over. Then for grinding a hundred caficii how much of it will be due, and how much taken away?

You will do the following. You already know that he wants the caficius to be divided into five parts and one of them be handed over for grinding the others. It is then evident that the ratio of one to five is the same as the ratio of what is due from a hundred to a hundred. Multiply therefore one by a hundred and divide the product by five; the result will be twenty, and this is what you wanted. Or take a fifth of a hundred, which is twenty. This is what is handed over for grinding a hundred, and eighty remain to be taken away. The same will hold if he asks that for each caficius a sixth, a ninth or whatever fraction of it be handed over.

[**(B.155′)**[1228] *For the grinding of a caficius a sixth of it is handed over. How much will be handed over for grinding a hundred and how much will be taken away?*

You will do the following. A sixth of a hundred, which is sixteen and two thirds, is what is handed over for grinding a hundred. This leaves eighty-three and a third to be taken away. The same holds if he asks that for each caficius a ninth or whatever fraction of it be handed over.]

THE SECOND TYPE OF THIS CASE IS THE FOLLOWING.

(B.156) For the grinding of each caficius two ninths of it are handed over. How much [*of another quantity of corn*] will be handed over for having a

---

[1227] This refers more specifically to B.155–162; the remaining B.163–174 involve other situations of grinding. Incidentally, the heading of this chapter is *de molare*, the word *molaris*, literally 'millstone', being used here in the general sense of 'what is *ad molendum pertinens*' (see the *Thesaurus linguae latinae*).

[1228] Obvious interpolation prompted by the last sentence; in MS $\mathcal{B}$ only, which omits B.155.

872   PART TWO: TRANSLATION, GLOSSARY

hundred caficii ground and also taken away?[1229]

You will do the following. Nine is the number from which a ninth is denominated. Now for grinding nine caficii two ninths, that is, two caficii, are handed over and seven remain to be taken away; therefore from nine two caficii are due for seven. Put this seven as the principal. Multiply two by a hundred and divide the product by seven; the result will be twenty-eight and four sevenths, and this is what is due for grinding a hundred. [*Adding then the caficii ground and those due for this work makes a hundred and twenty-eight caficii and four sevenths; this is what is brought for grinding, of which only a hundred is taken away.*][1230] You may also divide two by seven and multiply the result by a hundred, or divide a hundred by seven and multiply the result by two, according to what has been said before about first dividing and then multiplying.[1231]

THE THIRD TYPE OF THIS CASE IS THE FOLLOWING.

(B.157) For the grinding of a caficius two sevenths of it are handed over. How many did someone taking away eighty bring to be ground?

You will do the following. The number from which a seventh is denominated is seven. For grinding this seven two sevenths of it are handed over, that is, two caficii, and five remain. Let the five be your principal. Next, multiply seven by the eighty taken away and divide the product by the principal; the result will be a hundred and twelve, and this is what is brought. Or divide one of the factors by the divisor and multiply the result by the other, as we have taught above.[1232]

This rule is taken from a proportion. For the seven brought are to the five taken away as the unknown quantity brought is to the eighty taken away. Thus we do not know the third, and the second, namely five, is the divisor [1233].

THE FOURTH TYPE OF THIS CASE IS THE FOLLOWING.

(B.158) For the grinding of each caficius two ninths of it are handed over. A man gave the miller ten caficii. How many did he bring, and how many did he take away?

You will do the following. The number from which a ninth is denominated is nine. Now for grinding nine caficii two ninths are handed over, that is, two caficii, and seven remain. Let this two, due from nine for grinding seven, be your principal. If you want to know how much is brought to be

---

[1229] This time, the 'hundred ground' is taken away whereas before the 'hundred ground' meant the quantity brought. This may account for the two interpolations (the first of which is inappropriate). Such confusion is recurrent below.

[1230] This is correct, but not part of the question.

[1231] Beginning of Book B (rule *ii*, p. 778).

[1232] Again, beginning of Book B.

[1233] Called 'principal' before. Same in B.158. See note 1202, p. 865.

LIBER MAHAMELETH

873

ground,[1234] multiply nine by ten and divide the product by the principal; the result will be forty-five, and this is what is brought. If you want to know how much is taken away, multiply seven by ten and divide the product by the principal; the result will be thirty-five, and this is what is taken away. Or first divide and then multiply.

This rule is taken from a proportion. For the nine caficii, which are brought, are to the two, which are due, as the unknown quantity brought is to the ten, which are due. Likewise, as the seven, which are taken away, are to the two, which are due, so is the unknown quantity, which is taken away, to the ten, which are due for the work. In these two situations the third term is unknown and the divisor is the second, which is two caficii. Proceed as we have taught above.[1235]

THE FIRST TYPE OF THE SECOND CASE IS THE FOLLOWING.

(**B.159**) For the grinding of each caficius a quantity equal to a fifth of it is received from another caficius. How much is due, for grinding a hundred caficii, from other caficii?

You will do the following. As we know, when for each caficius ground a fifth of another is received, then for five parts ground a quantity equal to one part is received from another. Then it is evident that the ratio of one part, which is the remuneration given for grinding five parts, to the five parts is the same as the ratio of the remuneration given for grinding a hundred caficii to the hundred itself. Therefore the ratio of one to five is the same as the ratio of the required quantity to a hundred. Proceed then as explained above.[1236] The required quantity will be twenty. The treatment will be effected accordingly: You take the number from which a fifth is denominated, namely five, then multiply a fifth of it, which is one, by a hundred and divide the product by five; the result will be what you wanted.

THE SECOND TYPE OF THIS CASE IS THE FOLLOWING.

(**B.160**) For the grinding of each caficius a quantity equal to two sevenths of it is handed over from another caficius. Then when a hundred caficii are brought how many of them are due to the miller and how many taken away?

You will do the following. We know that seven parts are ground and a quantity equal to two such parts is received from another caficius. Thus the ratio of two to seven is the same as the ratio of what is due from the hundred caficii to what remains of them. By composition, the ratio of two to nine will be the same as the ratio of the remuneration to a hundred. Consequently the treatment is as follows. Take the number from which a seventh is denominated, namely seven, and add to it its two sevenths;

[1234] Including remuneration. In the lines above we find once 'for grinding nine' and once 'for grinding seven'.

[1235] Plain proportion (rule *i*, p. 777).

[1236] Plain proportion, or previous problems.

874　　　PART TWO: TRANSLATION, GLOSSARY

this makes nine, which you put as the principal. Next, multiply two by a hundred and divide the product by the principal; the result will be what you wanted.[1237]

THE THIRD TYPE OF THE SECOND CASE IS THE FOLLOWING.

(B.161) For the grinding of one caficius a quantity equal to two ninths of it is handed over. A man took away a hundred ground. How many did he bring to be ground?[1238]

You will do the following. The denominator is nine. Add to it two ninths of it, namely two; this makes eleven. Multiply it by a hundred and divide the product by nine; the result is what he brought. Or first divide one of the factors by the divisor and multiply the quotient by the other.

This rule, too, is taken from a proportion. For with nine to be ground he brought at the same time the corresponding remuneration, namely two.[1239] Now the sum of those which he took away and those which he gave is eleven. Then the eleven, which are brought, are to the nine, which are taken away, as the required number of caficii is to the hundred, which are taken away. Thus divide by nine, as indicated above.

THE FOURTH TYPE OF THE SECOND CASE IS THE FOLLOWING.

(B.162) For the grinding of one caficius a quantity equal to two ninths of it is handed over. Then if ten are due how much has been brought and how much taken away?

You will do the following. The denominator is nine. It is known that if he brings nine to be ground he will give as the remuneration two from another quantity. This added to nine makes eleven; these eleven he brought and nine he took away. Then it is like asking: 'From eleven two are due and nine are taken away; then when ten are due how much is brought and how much taken away?' Proceed as we have taught above.[1240] The result will be the quantity taken away, namely forty-five caficii; and the quantity brought is fifty-five.

Understand these problems, and treat others of the same kind in the same way.

(B.163) For the grinding of each caficius a quantity equal to a fifth of it is handed over. A man, after paying his due from an unknown number of caficii, took away a hundred. How many caficii did he bring?

We know that five parts are ground and a quantity equal to one of them is received. Thus if he brings six, he will take away five. Now the ratio of

---

[1237] The quantity taken away, also part of the question, is inferred from it. Here as in the previous problem, the solving method is recapitulated in more general terms. Same in B.163, B.198*b*, B.200*a*.

[1238] The quantity brought 'to be ground' thus includes the remuneration.

[1239] The quantity brought 'to be ground' is here the quantity taken away.

[1240] B.158.

LIBER MAHAMELETH

875

a quantity brought to the corresponding quantity taken away remains the same. Thus the ratio of six to five is the same as the ratio of the required quantity to a hundred. Proceed then as explained above.[1241] The result will be for the required quantity a hundred and twenty. Because of this you put as the principal the number from which a fifth is denominated, namely five. Adding to it its fifth makes six. This you multiply by a hundred, and you divide the product by the principal; the result is what you require. You may also add to a single caficius a quantity equal to a fifth of it and multiply the sum by a hundred; the result will be what you want.

(**B.164**) For the grinding of each caficius a seventh and half a seventh of it is handed over. A man, after paying his due from an unknown number of caficii, took a hundred away. How many did he hand over[1242] and how many did he bring?

You will do the following. We know that the man who brings a caficius to be ground divides it into seven parts, and, after giving to the miller one and a half of them, takes away five and a half ground. Thus it is known that he brings seven and takes away five and a half. Now the ratio of a quantity brought to the corresponding quantity taken away remains the same. Thus the ratio of seven to five and a half is the same as the ratio of the required quantity to a hundred. Because of this the treatment is as follows. You subtract from the number from which a seventh is denominated, namely seven, one and a half sevenths of it; this leaves five and a half, which you put as the principal. Next, you multiply seven by a hundred and divide the product by the principal; the result is a hundred and twenty-seven and three elevenths, which is what you require. You may also subtract one and a half sevenths from one[1243] and divide a hundred by the remainder; the result will be what you wanted.

(**B.165**)[1244] For the grinding of each caficius a fifth of it is handed over ⟨and in the process the quantity of corn increases by a tenth of itself⟩. A man brings an unknown number of caficii, pays his due from these, and takes away a hundred. How many caficii did he bring?

(**a**) We already know[1245] that for five parts brought one is the remuneration and four remain. These four, when increased by a tenth of themselves, become four and two fifths. Therefore if five are brought, four and two fifths will be taken away. Now the ratio of a quantity brought to the corresponding quantity taken away remains the same. Thus the ratio of five

---

[1241] Plain proportion.

[1242] Not answered.

[1243] Considering one caficius brought, as at the end of B.163.

[1244] This problem is misplaced and should appear with B.170–172. Its inclusion here may have to do with the fact that the characteristic part of the formulation has been omitted.

[1245] Same situation, *mutatis mutandis*, as in B.163.

876   PART TWO: TRANSLATION, GLOSSARY

to four and two fifths is the same as the ratio of the required quantity to a hundred. So multiply five by a hundred and divide the product by four and two fifths; the result will be what you wanted. Or otherwise. Reduce four to fifths; with the two this gives twenty-two. Likewise, reduce five to fifths; this gives twenty-five. Then the ratio of twenty-two to twenty-five will be the same as the ratio of a hundred to the required quantity. Proceed then as shown above.[1246] The required quantity will be a hundred and thirteen and seven elevenths, and this is what you wanted.

(**b**) You may also (do the following). You already know that when from each caficius a fifth of it is handed over, a fifth of the unknown number of caficii must be received, which will leave four fifths (of this number). This, when increased by a tenth of itself, becomes four fifths and two fifths of one fifth of this number. Thus look for the number of which four fifths and two fifths of one fifth is a hundred. You will find a hundred and thirteen and seven elevenths, and this is what you wanted.

(**B.166**)[1247] Someone asks thus: For the grinding of one caficius a fifth of a caficius is handed over, but not from the same, and multiplying the caficii brought to be ground[1248] by those of the remuneration produces one hundred and fifty. What is the unknown number of caficii?

(**a**) It has already been shown [*in the previous chapter on grinding, that is*][1249] that when a fifth of a caficius is given for grinding one caficius, then a sixth of the quantity of corn brought to be ground is the remuneration. Therefore multiplying the number of caficii brought by a sixth of it will produce a hundred and fifty. It follows that multiplying this number by itself will produce nine hundred. The root of this, namely thirty, is the required number of caficii brought to be ground.

(**b**) Or otherwise. Put as the unknown number of caficii a thing.[1250] Its sixth, which is a sixth of a thing, is the remuneration for a thing. Then multiply a sixth of a thing by a thing; the result will be a sixth of a square, which equals a hundred and fifty. Thus the square equals nine hundred, and the thing is thirty, which is the number of caficii brought to be ground.

(**c**) You may want to verify this. Multiply a sixth of thirty, which is five, by thirty; the result will be one hundred and fifty, as stated.

AGAIN ON THE SAME TOPIC.[1251]

[1246] Plain proportion, thus reference to the beginning of Book B (p. 777).
[1247] From now on we again encounter algebraic solutions.
[1248] Including remuneration.
[1249] B.163.
[1250] In MS $\mathcal{B}$, *res* appears to have been changed to *aliquid*, or at times deleted, by an early reader. Neither here nor in the next two problems do the algebraic solutions add anything significant.
[1251] As a matter of fact, B.166–168 form a coherent group.

LIBER MAHAMELETH

(B.167) Someone asks thus: For the grinding of one caficius a fifth of a caficius is given, and subtracting the caficii given in remuneration from those taken away leaves twenty. What is the unknown number of caficii?[1252]

(a) It is known[1253] that from the quantity of caficii brought to be ground a sixth of it is the remuneration. Handing over from an unknown quantity of corn a sixth of it leaves five sixths of it; such is therefore the quantity taken away ground. Subtracting from this the sixth given in remuneration leaves four sixths of it. The twenty remaining must then be four sixths of the quantity of corn brought to be ground. Thus look for the number of which four sixths are twenty; you will find that it is thirty, and this is the number of caficii brought to be ground.

(b) You may also put a thing as the number of caficii brought to be ground. After giving in remuneration from this thing a sixth of it, which is a sixth of a thing, five sixths of a thing are left, which is what is taken away ground. Subtracting from this the sixth of a thing given in remuneration leaves four sixths of a thing; this equals twenty. Thus the thing equals thirty caficii.

(c) You may want to verify this. Giving in remuneration from these thirty caficii a sixth of them, which is five caficii, leaves twenty-five, and this is the quantity taken away. What is subtracted from it is thus five, the remuneration; this leaves twenty, as has been proposed.

(B.168)[1254] Someone asks thus: For the grinding of a caficius a fifth of a caficius is handed over, and the number of caficii taken away when multiplied by the remuneration produces a hundred and twenty-five. How many caficii were brought to be ground[1255]?

(a) It is already known that a sixth of the unknown number of caficii[1256] is due, which leaves five sixths of this number as what is taken away ground. But multiplying the number of caficii taken away by the remuneration produces a hundred and twenty-five. ⟨It follows that multiplying the number of caficii taken away by itself produces six hundred and twenty-five. The root of this⟩[1257] [*Then a fifth of this*][1258] is the unknown number of caficii,[1259] namely twenty-five.

(b) You may also put as the number of caficii brought a thing.[1260] After

---

[1252] Namely the quantity brought.

[1253] Previous problem (or B.163).

[1254] The genuineness of this problem, with its errors and omissions, is questionable.

[1255] The quantity to be calculated is in fact the number of caficii taken away ground.

[1256] Here the 'unknown number' is the whole quantity brought.

[1257] Tentative reconstruction.

[1258] An early reader's attempt to fill in the lacuna.

[1259] The 'unknown number' is now the number of caficii taken away.

[1260] The 'thing' is thus again the whole quantity brought. Note, about the designation of the unknown, that in MS $\mathcal{B}$ not only do we sometimes find here *aliquid* added to *res* but also *habitus* to *census*.

878   PART TWO: TRANSLATION, GLOSSARY

paying a sixth of it, which is a sixth of a thing, five sixths of a thing is left, which is what is taken away ground. So multiply this by the remuneration, that is, by a sixth of a thing; the result will be five sixths of a sixth of a square; this equals a hundred and twenty-five. Thus a square[1261] is six hundred and twenty-five. The root of this, which is twenty-five, is the number of caficii brought to be ground.[1262]

AGAIN ON THE SAME TOPIC [1263]

(B.169) Someone asks thus: Working day and night, one mill grinds twenty caficii, another thirty, yet another forty. A man wants to grind ten caficii in these three working simultaneously.[1264] How much will he put in each one?

(**a**) This problem is in every respect like a problem on partnership and does not differ in anything from it.[1265] So add twenty, thirty and forty and put the sum as the principal. If you want to know how much is put in each mill, multiply the ten caficii by what is ground by the mill in question and divide the product by the principal; the result is what you require.

(**b**) Or otherwise. Take, as the number of caficii put in the mill grinding forty, something [1266]; what is put in the mill grinding thirty will be three fourths of something, and what is put in the mill grinding twenty will be half of something. Adding them all makes two somethings and a fourth; this equals ten. Then this something equals four and four ninths, and it is what is put in the mill grinding forty. Three fourths of it are put in the mill grinding thirty, and half of it in the mill grinding twenty. This is what you wanted to know.

AGAIN ON THE SAME TOPIC,
BUT ANOTHER CHAPTER INVOLVING INCREASE

(B.170) For the grinding of one caficius a sixth of it is given, and the corn while being ground increases by a third of itself. Someone brought a hundred caficii for grinding and gave from them his due. How many did he take away?

You will do the following. Multiplying the denominators of a sixth and a third produces eighteen. Therefore, for eighteen to be ground fifteen remain after the remuneration of a sixth, which is three; these fifteen increase in the process by their third and become twenty ground. Thus it is known that if eighteen are brought to be ground, twenty ground will be taken away. Then: If someone brings a hundred to be ground, how many will he take away ground? Multiply a hundred, which he brought, by

---

[1261] *Sic.*

[1262] The 'thing' is now again the quantity taken away.

[1263] This does not fit here: B.169 is of a different kind.

[1264] Thus working continuously for the same length of time.

[1265] See B.104.

[1266] Here both MSS have *aliquid* instead of *res.*

LIBER MAHAMELETH

879

twenty, which he took away, and divide the product by eighteen; the result will be what you require. [*This too is taken from a proportion.*][1267]

(B.170′) Suppose now the asker had said: Someone took away a hundred ground; how many did he bring to be ground?

It is already known that the ratio of eighteen, which he brought, to twenty, which he took away, is the same as the ratio of the unknown quantity, which he brought, to a hundred, which he took away. Then the divisor here is twenty. Proceed as we have taught above.[1268]

AGAIN ON THE SAME TOPIC.

(B.171) Someone asks thus: For the grinding of one caficius a quantity equal to a sixth of it is given and the corn while being ground increases by a third of itself. For thirty caficii taken away ground, how many were brought to be ground?

(a) You will do the following. The denominator of a sixth is six. Adding a third of it, which is two, makes eight; put it as the principal. Next, add to six a sixth of it, namely one; this makes seven. Multiply it by thirty and divide the product by the principal; the result will be twenty-six and a fourth, and this is what has been brought to be ground.

The reason for this is the following.[1269] We know that, of the corn brought to be ground, a seventh is due as the remuneration, since for each quantity ground an amount equal to a sixth of it is due. He is then left with six sevenths of the corn brought to be ground. These six sevenths, when ground, have increased by their third, thus becoming eight sevenths; these are the thirty caficii taken away. Therefore it is clear that the ratio of a quantity of corn brought, namely seven sevenths[1270], to the corresponding quantity of corn increased and taken away, namely eight sevenths, is the same as the ratio of the required quantity of corn to this quantity of corn increased to thirty. There are therefore four proportional numbers. Thus multiply seven by thirty and divide the product by eight, and the result will be the quantity brought.

(b) Or otherwise. Since the ⟨fraction remaining of the⟩ corn brought to be ground, namely six sevenths, when increased by its third becomes eight sevenths, eight sevenths must be like the quantity of corn brought to be ground, plus a seventh of it. Then consider what (fraction of eight) is to be subtracted from eight to leave seven; it is its eighth. Therefore subtract from the thirty caficii an eighth of them; this leaves twenty-six caficii and a fourth, which is the quantity of corn brought to be ground.

[1267] See B.161. B.170 follows B.161–162 in the disordered text.

[1268] Plain proportion.

[1269] Remember, for what is to come, that if the remuneration equals a sixth of the quantity actually ground, it is a seventh of the whole quantity brought (which includes the remuneration).

[1270] The MSS have 'six sevenths'. Whence, perhaps, the omission at the beginning of *b*.

880 PART TWO: TRANSLATION, GLOSSARY

(*c*) Or otherwise. Eight sevenths are six sevenths increased by a third of them. Then subtract from the thirty taken away a fourth of them; this will leave twenty-two and a half, which is the six sevenths ground. Adding its sixth makes twenty-six and a fourth, and this is what is brought.

(*d*) Or otherwise. Let the quantity of corn brought be a thing. After giving the seventh of it due, six sevenths of a thing will remain. Adding a third of them makes eight sevenths of a thing, which equal thirty. Thus the thing is twenty-six and a fourth, and this is what you wanted.

AGAIN.

(**B.172**) Someone asks thus: For the grinding of a caficius a sixth of a caficius is given, and the corn whilst being ground increases by a third of itself. Of thirty brought to be ground how many were taken away ground?

(*a*) You will do the following. The denominator of a sixth is six. Adding a sixth of it makes seven; put it as the principal. Adding again to six a third of it makes eight. Multiply this by thirty and divide the product by the principal; the result is the quantity of corn taken away, namely thirty-four caficii and two sevenths.

The reason for this is the following. It is known [1271] that if from a quantity of corn brought to be ground a seventh of it is the remuneration, six sevenths of it are left; these are therefore the caficii ground [1272], and since they have increased by their third, they have become eight sevenths. Then it is known that the ratio of the quantity of corn brought to be ground, namely seven sevenths, to the given quantity of corn brought, namely thirty caficii, is the same as the ratio of eight sevenths, found after the increase, to the unknown quantity of corn as increased. These are therefore four proportional numbers. Thus multiplying eight by thirty and dividing the product by seven will give the unknown quantity of corn, which is the fourth number.

(*b*) There is also another way, which is the following. Subtract from the thirty brought a seventh (of them); this will leave twenty-five and five sevenths. Adding to this its third will give what you wanted.

(**B.173**) Someone asks: A man buys one sextarius for three nummi, has it ground for half a nummus and baked for one nummus. What is the cost of ten sextarii bought, ground and baked?

You will do the following. You know that a sextarius is bought for three nummi, ground for a half, and baked for a nummus. Thus one sextarius bought, ground and baked costs four and a half nummi. Then you will say: 'When one sextarius is given for four and a half nummi, what is the price of ten sextarii?' Proceed as we have taught above [1273], and the result will be what you require.

[1271] See B.171*a*.

[1272] To be taken away.

[1273] Possibly B.3–4. But such a phrase often just means that the full computation is not performed. Same in the next problem.

LIBER MAHAMELETH
881

(**B.174**) Someone asks: A sextarius [1274] is bought for five, ground for three and baked for one and a fourth. How many sextarii bought, ground and baked will there be for a hundred nummi?

You already know that a sextarius bought, ground and baked costs nine nummi and a fourth. Then you will say: 'When one sextarius is given for nine nummi and a fourth, how many will be given for a hundred nummi?' Proceed as has been shown above, and the result will be what you require. If there are more than one sextarius, reduce them to one.

Treat other problems of this kind in the same manner, and you will find that this is the way.

[1274] 'caficius' here in the text.

Chapter (B–IX) on
Boiling must

(B.175) For instance, you want to boil ten measures of must until two thirds of them have evaporated, and thus a third remains. After two measures have evaporated, two overflow from the remainder. To what quantity does the rest have to be reduced[1275]?

(a) You will do the following. Subtract from the ten measures the two evaporated; this leaves eight, which you put as the principal. Subtract from the eight the two measures which have overflowed; this leaves six. Multiply it by a third of ten and divide the product by the principal; the result will be two and a half measures, which is to what the six have to be reduced by boiling.

The reason for this is the following. It is known that if the eight measures continued to boil without overflow, they would reach a third of ten measures, which is three and a third. Now it is known that the six remaining after the overflow of two have to be boiled as eight would be. Thus it is known that the amount of the eight which, should nothing overflow, evaporates in the process of reduction to a third of ten, that is, to three and a third, is the same as the amount of the six which evaporates in the process of reduction to the required number of measures. So the ratio of eight measures to three and a third is the same as the ratio of six measures to the quantity of measures to be attained. There are therefore four proportional numbers here. Since the first multiplied by the fourth produces as much as the second by the third, multiplying the third, which is six, by the second, which is three and a third, and dividing the product by the first, which is eight, will give the fourth, which is the number of measures to which the six have to be reduced. Or else[1276], since it is known that eight is to three and a third as six is to the unknown, required number of measures, then, by alternation, eight will be to six as three and a third is to the unknown, required number of measures. There are therefore four proportional numbers. Multiplying the second by the third and dividing the product by the first will give the fourth, which is what is required.

There is also another way. You will determine what fraction six is of eight, and this same fraction of three and a third is what you require.[1277]

[1275] 'have to boil', here and usually in the text.

[1276] Other way of dealing with the proportion set above. This adds nothing and may be a later addition.

[1277] One of the two 'other ways' taught at the beginning of Book B (rule *ii*, p. 778). The second appears in *d* below.

(b) Or look for the number which multiplying eight produces ten; you will find one and a fourth. Multiply this by six; the result will be seven and a half. A third of this, which is two and a half, is what you wanted to know.

The reason for this is that the number which increased by eight makes a sum having as its third three and a third is to eight as the number which increased by six makes a sum having as its third the required number is to six, while it is known that the number which increased by eight makes ten equals a fourth of eight. Thus add to six a fourth of it, and a third of the whole sum will be what you wanted.

(c) Or otherwise. Find the quantity which subtracted from eight leaves six; it is a fourth of it. Then subtract from three and a third a fourth of it; this will leave two and a half, which is what you wanted.

(d) Or otherwise. Consider what fraction three and a third is of eight; it is two and a half sixths. Thus two and a half sixths of six, which is two and a half, is what you wanted.

(e) Or otherwise. You already know that since ten measures are reduced by boiling to eight, the origin of these eight measures boiled down is ten measures. Now the origin of the two measures overflowed is two and a half measures not boiled. Thus it is as if these two and a half measures had been spilled from the ten before boiling down and the remainder, which is seven and a half, had to be reduced by boiling to its third, which is two and a half; and this is what you wanted.

(B.176) You want to boil ten measures of must until they are reduced to an unknown quantity. When they have boiled until two measures have evaporated, two have overflowed from the remainder and what is left has been reduced by boiling to two and a half measures. What is the unknown quantity to which he wanted [1278] to reduce the ten measures?

You will do the following. Subtract from the ten the two measures which have evaporated in the process, thus leaving eight. Subtract from this the two measures which have overflowed; this leaves six. Put it as the principal. Next, multiply the two and a half measures by eight; this produces twenty. Divide it by the principal; this will give three and a third, which is the unknown quantity to which you wanted to reduce the ten measures.

(B.177) You want to boil ten measures of must until they are reduced to their fourth. In the process, an unknown quantity has evaporated and from the remainder three measures have overflowed. What is left has been reduced to one measure and four eighths and half an eighth.[1279]

You will do the following. You already know [1280] that the ratio of ten minus the unknown quantity evaporated to two and a half is the same as the ratio of ten minus the unknown quantity and minus three measures to

---

[1278] 'he': see note 874, p. 784.

[1279] No question stated in B.177–179, but the context makes it clear.

[1280] B.175. Same proportion used in the next problems.

884   Part Two: Translation, Glossary

one and four and a half eighths. By alternation, the ratio of two and a half to one and four eighths and half an eighth will be the same as the ratio of ten minus the unknown quantity to ten minus the unknown quantity and minus three measures. But two and a half to one and four eighths and half an eighth is one and three fifths times. Thus ten minus the unknown quantity is one and three fifths times ten minus the unknown quantity and minus three measures.[1281] Let then ten be AB, the unknown quantity, BG, and the three measures, GD. Thus AG is equal to AD and three fifths of it. By subtraction[1282], DG will be three fifths of AD. Since DG is three, AD is five. Thus the whole AG is eight. Since AB is ten, this leaves two as GB. This is what we wanted to demonstrate.[1283]

A   D   G   B

(B.177)

(B.178) You want to boil ten measures of must until they are reduced to a fourth of them. In the process two measures have evaporated, and from the remainder an unknown quantity has overflowed. The rest has been reduced to one and a half and half an eighth.

You will do the following. You already know that the ratio of ten minus two measures, which is eight, to a fourth of ten, which is two and a half, is the same as the ratio of eight minus the unknown quantity overflowed to one and a half and half an eighth. Proceed as shown above.[1284] The result will be, for eight minus the unknown quantity, five. Thus the required quantity which has overflowed is three. [*This is what we wanted to demonstrate.*]

(B.179) You want to boil ten measures of must until they are reduced to their third. In the process, an unknown quantity has evaporated and as much has overflowed from the remainder. The rest has been reduced to two and a half measures.

You will do the following. You already know that the ratio of three and a third, which is a third of ten, to two and a half is the same as the ratio of ten minus the unknown quantity evaporated to ten minus the unknown quantity evaporated and the unknown quantity overflowed. Now the ratio of three and a third to two and a half is one and a third times. Therefore ten minus the unknown quantity evaporated is one and a third times ten minus the unknown quantities evaporated and overflowed. Thus[1285] ten minus the quantities evaporated and overflowed is six. So the number of measures evaporated and overflowed is four. Since both are equal, the

---

[1281] Equation of the form $c - x = a\,(c - x - b)$.

[1282] *dispergere* (like the Arabic *qalaba*) is used both for *converting* a ratio and *subtracting* a quantity.

[1283] Geometrical computation. See p. 724, note 584.

[1284] Plain proportion (thus i, p. 777).

[1285] Something is certainly missing here.

LIBER MAHAMELETH

number of measures evaporated is two and those overflowed, two. [*This is what we wanted to demonstrate*.]

(B.180) You want to boil sixty measures of must until they are reduced to their third. In the process, when ten measures have evaporated, five overflow from the remainder. The rest is then boiled down until nine measures have evaporated, and six overflowed from the remainder. To what does the last remainder have to be reduced?

This problem is composed of two problems. For it is as if you first wanted to boil the sixty measures of must until they are reduced to their third; in the process, when ten measures have evaporated, five overflow from the remainder; to what does the remainder have to be reduced? Proceed as we have taught before[1286]; the result will be eighteen. Next, it is as if you wanted to boil forty-five measures of must until they are reduced to eighteen; in the process, when nine have evaporated, six overflow from the remainder; to what does the remainder have to be reduced? Proceed also as we have taught above; the result will be fifteen, and this is what you wanted to know in the proposed problem[1287].

AGAIN ON THE SAME TOPIC.

(B.181) You want to boil ten measures of must until two thirds of them have evaporated. In the process once two measures have evaporated, two overflow from the remainder. The rest is then boiled until two measures have evaporated; and two overflow from this remainder. To what does the last remainder have to be reduced in order to be boiled down as desired?

Like the previous one, this problem is composed of two.[1288] Take then a third of ten, which is three and a third. Next, subtract from the ten measures the two evaporated; this leaves eight. Again, subtract the two which have overflowed; this leaves six. Subtract then also the two which have evaporated afterwards; this leaves four. Subtract then the two which have overflowed at the end; this leaves two. After that, multiply eight by four, thus producing thirty-two. Put it as the principal. Next, multiply the two measures remaining by six, thus producing twelve. Multiply this by three and a third; the result will be forty. Divide this by the principal; the result will be one and a fourth, and this is to what the last remainder is reduced.

(B.182) Now you want to boil a hundred measures of must until they are reduced to their fifth. In the process, ten measures have evaporated and nine have overflowed from the remainder. What remains is then boiled until eight measures have evaporated; and from the remainder seven overflow. What remains is then boiled until six measures have evaporated and

---

[1286] B.175.

[1287] That is, the one stated.

[1288] 'Like the previous one': this makes sense neither in MS $\mathcal{A}$ (where B.181 follows B.175) nor in the disordered text of MS $\mathcal{B}$ (where B.181 follows B.186).

886   PART TWO: TRANSLATION, GLOSSARY

from the remainder five measures have overflowed. To what does this last remainder have to be reduced?

This problem is composed of three problems; but the treatment is the same as in the previous case, and would remain the same if there were more.[1289] That is, you will work out, for each reduction and overflowing, a problem, until you reach the last problem; the final result is what you require.

To this chapter on boiling belong also many impossible problems. It is necessary to know about them beforehand so that when any ones similar to the problems explained in this and the previous chapter occur, you may realize that they are not solvable.[1290]

(B.183) For instance, you want to boil ten measures of must until they are reduced to their third. In the process, when seven have evaporated, from the remainder two overflow. To what does the remainder have to be reduced?

This problem is not solvable. For it has been supposed that (the ten measures) are reduced to their third, namely three measures and a third; that is, six and two thirds evaporate. Thus he has already gone beyond the end of the process since seven have evaporated during it [*which are more than the six and two thirds which he wanted to evaporate*]. The problem is therefore impossible.

An example of a problem similar to this one but not impossible is the following.[1291]

(B.184) For instance, you want to boil ten measures of must until they are reduced to their third, and two of these measures overflow. To what does the remainder have to be reduced?

In this case it is clear that if you want the eight to be boiled down as the ten were supposed to be, you will have to reduce these eight to their third, which is two and two thirds.

(B.185) You want to boil an unknown quantity of must until two thirds of it are evaporated. In the process two measures have evaporated and from the remainder two measures have overflowed. The rest has been reduced by boiling to two and a half measures. What is the unknown quantity of must?

(a) You will do the following. Multiply the number from which a third is denominated, namely three, by the two and a half measures, thus producing

---

[1289] 'as in the previous case': here, B.182 follows B.180 in MS $\mathcal{A}$ (neither problem found in MS $\mathcal{B}$).

[1290] The two 'chapters' (sections) in question are B.175–179, where a single overflow is involved, and B.180–182, where there are several.

[1291] 'similar': the data, except for one omission, are the same; 'not impossible': the usual proportion will lead to the right answer (and not to a meaningless one as in the previous problem).

LIBER MAHAMELETH

seven and a half. Adding to it the two measures evaporated and the two overflowed makes eleven and a half. Next, multiply seven and a half by the two measures evaporated; this produces fifteen. Next,[1292] multiply half of the eleven and a half, which is five and three fourths, by itself; the result will be thirty-three and half an eighth. Subtracting fifteen leaves eighteen and half an eighth. Add to its root, which is four and a fourth, the five and three fourths; the result is ten, and this is the unknown quantity of must.

The proof of this is the following. Let the unknown quantity of must be line AB; put as the remainder of must after evaporation and overflow line GB, the two measures evaporated in the process, line AD, the two overflowed, line DG, and the two and a half measures to which the must is reduced, line KB. We have already said in the foregoing, dealing with known quantities of must, that the ratio of the quantity added to the remainder of the must, after evaporation of two measures, so that a third of the resulting sum equals a third of the original quantity of must, to this remainder, is the same as the ratio of the quantity added to the second remainder, after evaporation of two and overflow of two more, so that a third of the resulting sum is equal to what is required, to this second remainder.[1293] Thus it is known that what is added to line BD to make it line AB is to line BD as what is added to line GB to make a third of the whole equal to two and a half, which is line KB, is to BG. We shall put, as the line added to line GB, line GH. Thus line KB is a third of line BH. Therefore HB is seven and a half. It is then clear[1294] that line AD is to line DB as line HG is to line GB. By composition, the ratio of line AD to line AB will be the same as the ratio of line HG to line HB. Thus the product of line AD into line HB is equal to the product of line AB into line HG. Now the product of line BH into line AD is fifteen, for line AD is two and line HB, seven and a half. Thus multiplying line HG by line AB produces fifteen. Next, I shall draw, from the point A on line AB, a line equal to line HG; this is line AT. Thus the product of line AT into line AB is fifteen. Now it is clear that line HT is four.[1295] Since line HB is seven and a half, line TB is eleven and a half.[1296] I shall now construct another line of eleven and a half, equal to line TB; this is line CQ. I shall cut from it a line equal to line TA; this is line CP. Thus the product of line CP into line PQ is fifteen.[1297] Next, I shall bisect line CQ at the point L. Thus the products of line CP into line PQ and line PL into itself are equal to the product of line CL into itself, as Euclid said in the second book.[1298] But the product of line CL into itself is thirty-three and half an eighth and the product of line CP into line PQ, fifteen. Therefore

---

[1292] Solving $x^2 + c = bx$ $(b = 11 + \frac{1}{2}, c = 15)$.

[1293] B.175b.

[1294] By separation.

[1295] Since $HT = AG$.

[1296] Sum and product of AB and AT thus known.

[1297] Sum and product of PQ and CP thus known.

[1298] *Elements* II.5 = PE_5.

the product of line PL into itself is eighteen and half an eighth. Thus line PL is four and a fourth. Line QL being five and three fourths, line PQ is ten, and this is the quantity of must which was required.[1299]

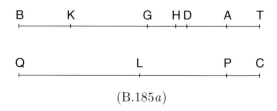

(B.185a)

(**b**) Or otherwise. Put as the unknown quantity of must a thing. Subtracting the two measures evaporated leaves a thing minus two. Subtracting the two measures overflowed leaves a thing minus four. It is therefore clear that the ratio of a thing minus two to a third of the original quantity of must,[1300] which is a third of a thing, is the same as the ratio of a thing minus four to two and a half. Thus the product of a thing minus two into two and a half is the same as the product of a third of a thing into a thing minus four. Then proceed as we have taught above in algebra.[1301] The result will be ten for the thing.

(**c**) Or otherwise. Let the unknown quantity of must be AB, the two measures, BG, the other two, GD, a third of the quantity of must, HZ, and the two and a half, HK. Thus the ratio of AG to HZ is the same as the ratio of AD to two and a half, which is HK. Now the ratio of AG to thrice HZ, which is AB, is the same as the ratio of AD to thrice HK, which is seven and a half. Thus the product of AG into seven and a half is equal to the product of AD into AB. Now[1302] the product of AB into AD is equal to the products of AD into itself and AD into DB which is four. Therefore the products of AD into itself and into four are equal to the product of AG into seven and a half. But the product of AG into seven and a half is equal to the products of AD into seven and a half and DG, which is two, into seven and a half. The product of DG into seven and a half being fifteen, the products of AD into itself and into four are equal to the product of AD into seven and a half, plus fifteen. Subtracting the product of AD into four from the product of the same into seven and a half will leave the product of AD into itself equal to the product of the same into three and a half, plus fifteen. So AD is greater than three and a half. Let thus three and a half be AT. Then the product of AD into itself will be equal to the product of AD into AT, plus fifteen. Now[1303]

[1299] No Q.E.D. here.
[1300] Here and in what follows, for our 'the original (or: initial) quantity of must' the text simply has 'the must' or 'all the must'. The proportion used is that known from B.175a.
[1301] To obtain an equation of the form $x^2 + c = bx$ (B.30).
[1302] PE_1; also below.
[1303] PE_2.

the product of AD into itself is equal to the products of AD into AT and AD into DT. Thus the products of AD into AT and AD into DT are equal to the product of AD into AT, plus fifteen. Removing the product of AD into AT, which is common, will leave, as the product of AD into DT, fifteen.[1304] Let now AT be bisected at the point Q. Then[1305] the products of AD into DT and TQ into itself will be equal to the product of QD into itself. But the product of AD into DT is fifteen while QT into itself is three and half an eighth. Thus the product of QD into itself is eighteen and half an eighth. Therefore QD is four and a fourth. AQ being one and three fourths, AD is six. But DB is four; therefore AB is ten. This is what we wanted to demonstrate.[1306]

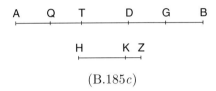

(B.185c)

(B.186) You want to boil an unknown quantity of must until two thirds of this quantity have evaporated. In the process, when a fifth of the quantity has evaporated, two measures overflow from the remainder; and the rest is reduced to two and a half measures. What is the unknown quantity of must?

(*a*) You will do the following. You already know that the ratio of four fifths of the original quantity of must to a third of this quantity is the same as the ratio of four fifths of this quantity minus two measures to two and a half measures. Proceed then as explained above;[1307] four fifths of the quantity of must minus two will be six, thus the quantity of must is ten. [*This is what we wanted to demonstrate.*]

(*b*) Or otherwise. Put as the unknown quantity of must a thing. Subtracting a fifth of it leaves four fifths of a thing. Subtracting two measures leaves four fifths of a thing minus two measures. It is then clear from the foregoing[1308] that four fifths of a thing are to a third of the quantity of must, which is a third of a thing, as four fifths of a thing minus two are to two and a half. These four numbers are therefore proportional. Thus the product of four fifths of a thing into two and a half, which is two things, is equal to the product of a third of a thing into four fifths of a thing minus two, which is a fifth of a square and a third of a fifth of a square minus two thirds of a thing. Adding two thirds of a thing to the two things makes two things and two thirds of a thing, which equal a fifth of a square and a

---

[1304] Difference and product of AD and DT thus known.
[1305] *Elements* II.6 = PE_6.
[1306] Geometrical computation. See p. 724, note 584.
[1307] Meaning: perform the computations.
[1308] Part *a* above.

890   PART TWO: TRANSLATION, GLOSSARY

third of a fifth of a square. Restore this [1309], that is, the fifth of a square and the third of a fifth of a square, to make it one whole square, namely by multiplying them by three and three fourths, and multiply this quantity also by two things and two thirds of a thing. The result will be ten things, which equal one square. Proceed then as shown before in algebra. [1310] The square will be a hundred and the thing, ten, which is the required quantity of must.

(B.187) You want to boil an unknown quantity of must until this quantity is reduced to its third. In the process, when two measures have evaporated, a fourth of the remainder overflows. The rest has been reduced to a fourth of the original quantity of must. [1311]

You will do the following. You already know that the ratio of a third of the initial quantity of must to a fourth of this initial quantity is the same as the ratio of the initial quantity minus two measures to three fourths of this. But a third of the initial quantity of must to a fourth of the same is one and a third times [1312]. Therefore the initial quantity of must minus two measures to three fourths of itself is one and a third times. Thus the present problem is indeterminate; for any number to three fourths of itself is one and a third times. [1313] But if it were said that the remainder had been reduced to a fifth of the initial quantity of must, this would not be solvable. For this quantity of must minus two measures to three fourths of itself would be one and two thirds times, which is impossible. Thus this problem cannot be stated [1314] unless it is said that the quantity of must has been reduced to a fourth of itself. Should this be so, and the other two fractions retained, the problem will be indeterminate.

Let thus the initial quantity of must be any number whatsoever, say thirty. Then this quantity of must minus two measures will be twenty-eight. Now the quantity overflowed is seven, which leaves twenty-one measures. Reduce them to a fourth of the initial quantity of must; this is seven and a half.

You will treat all problems of this kind in the same way. When there is agreement between the two fractions stated, the problem will be solvable, otherwise not. This is what we wanted to show.

(B.188) Someone wants to boil an unknown quantity of must until this quantity is reduced to its third. In the process, when a fifth of this quantity has evaporated, a fourth of the remainder overflows. The rest is reduced to two and a half measures. What is the unknown quantity of must?

[1309] Like *complere*, the verb *restaurare* applied to a term of the form ax^2 can mean changing it to x^2. See note 1371, p. 904.

[1310] Simple equation $ax^2 = bx$.

[1311] Required this initial quantity. Same in B.189.

[1312] 'times': expression of a ratio.

[1313] *questio interminata* means here that any number may be the solution.

[1314] *non potest fieri*.

LIBER MAHAMELETH 891

You already know that the ratio of four fifths of the initial quantity of must to three fifths of this initial quantity is the same as the ratio of a third of this initial quantity to two and a half. But four fifths of the initial quantity to three fifths of this same quantity is one and a third times. Thus a third of this quantity of must to two and a half is one and a third times. So a third of this quantity is three and a third; therefore the initial quantity of must is ten.

(B.189) You want to boil an unknown quantity of must until this quantity is reduced to its third. In the process, when two measures have evaporated, a fourth of the remainder overflows. The rest is reduced to two and a half (measures).

You already know that the ratio of the initial quantity of must minus two measures to three fourths of this (latter) quantity is the same as the ratio of a third of the initial quantity to two and a half measures. ⟨But the initial quantity minus two measures to three fourths of this (latter) quantity is one and a third times. Thus a third of the initial quantity to two and a half is one and a third times. So a third of the initial quantity is three and a third; therefore the initial quantity of must is ten.⟩

Chapter (B–X) on Borrowing

This chapter is like, and does not differ in any respect from, the chapter on buying and selling. For it is clear that the ratio of the number of sextarii returned to (their number to) the corresponding modius is the same as the ratio of the number of sextarii borrowed to (their number to) the corresponding modius. I shall present some problems and make clear here what has been said.[1315]

(B.190) For instance, someone has borrowed six sextarii of corn of which there are fourteen to the modius. He wants to discharge his debt in sextarii of which there are twenty to the modius. How many of them will he return?

This problem is like the one which states: 'When fourteen are given for six nummi, what is the price of twenty?'[1316]

(*a*) You will do the following. Multiply six by twenty and divide the product by fourteen; the result will be eight sextarii and four sevenths, and this is what you wanted to know. The reason for this is evident from what has been said above in buying and selling.

Or otherwise. Denominate the six sextarii from the fourteen; the same fraction of twenty will be what you wanted.

Or look for the number which multiplying fourteen produces twenty; multiplying it by six will produce what you require. The reason for this is the following. Since it is known[1317] that the ratio of fourteen sextarii to twenty is the same as the ratio of the six sextarii borrowed to the number of those returned, the amount by which fourteen becomes twenty is the same as the amount by which six becomes the number of sextarii returned.

(*b*) Or otherwise. Put as the unknown number of sextarii one thing. Now it is known that the ratio of six to fourteen is the same as that of the unknown number of sextarii, which is a thing, to twenty. Thus the product of the first into the fourth equals the product of the second into the third. Therefore the product of twenty into six equals the product of fourteen into a thing. So a hundred and twenty is equal to fourteen things. Thus the thing is eight and four sevenths, and this is what you wanted.

AGAIN ON THE SAME TOPIC.

(B.191) Someone who has borrowed six sextarii of corn, of which there are fourteen to the modius, discharges his debt with nine sextarii. How many sextarii are there to this other modius?

[1315] Indeed, for almost all these problems the equivalent form in terms of buying or selling will be explicitly given.

[1316] See B.1, B.3–6, B.11–12.

[1317] Evident, and stated in the introduction above. This whole justification could be an addition.

LIBER MAHAMELETH

Here, too[1318], the problem is like the problem which states: 'When fourteen modii are given for six nummi, how many shall I have for nine?' Now we have already explained how to treat this case and provided the proof in the chapter on buying and selling.[1319]

(*a*) The treatment consists in multiplying nine by fourteen and dividing the product by six; the result is what you wanted.

The reason for this is the following. Six sextarii are to fourteen as nine sextarii are to the unknown number of sextarii belonging to the other modius[1320]. There are therefore four proportional numbers. Thus multiplying the second by the third and dividing the result by the first will give the fourth.

Or otherwise. Look for the number which multiplying six produces fourteen, and multiply nine by it; the result will be what you require.

(*b*) Or put, as the number of sextarii to the other modius, a thing. Now it is known that the ratio of six to fourteen is the same as that of nine to the unknown number of sextarii, which is a thing. There are therefore four proportional numbers. So proceed as we have taught above in algebra[1321], and the result will be the value of the thing.

(**B.192**) Someone has borrowed six sextarii of corn, of which there are fourteen to the modius. He discharges his debt from another modius with a number of sextarii which, when multiplied by the number of sextarii to this modius, produces a hundred and eighty-nine. How many sextarii are there to this modius, or how many does he return?

You will do the following. If you want to know the number of sextarii to the modius used to discharge the debt[1322], multiply fourteen by a hundred and eighty-nine and divide the product by six; the root of the result will be the number of sextarii [*that is, how many there are to the modius*][1323], which is twenty-one. If you want to know how many of these he returns, denominate six from fourteen, and this fraction of twenty-one will be what you require. If you want to find that in another way,[1324] multiply six by a hundred and eighty-nine and divide the product by fourteen; the root of the result is what you require, namely nine.

(**B.193**) Someone has borrowed six sextarii of corn of which there are fourteen to the modius. He discharges his debt from another modius with a number of sextarii which, when added to the number of sextarii to this

---

[1318] As in the previous problem.

[1319] See B.2, B.7–10, B.13–14 (proof in B.13).

[1320] Literally (here and recurrently): 'to the sextarii of the unknown modius'.

[1321] Plain proportion, giving an equation of the form $ax = b$.

[1322] The MSS have: the number of sextarii returned.

[1323] Probably a gloss to rectify the above error (previous note).

[1324] That is, without knowing the previous result.

894 PART TWO: TRANSLATION, GLOSSARY

modius, makes thirty. How many sextarii ⟨are there to this second modius, or how many⟩ did he return?

This problem is like the problem which states: 'When fourteen modii are given for six nummi, what is the number of modii which, when added to their price, makes thirty, and what is this price?' We have already explained how to solve such a problem before.[1325]

(*a*) That is, you add fourteen to six, thus making twenty, which you put as the principal. If you want to know the number of sextarii to the modius used to discharge the debt, multiply fourteen by thirty and divide the product by the principal; the result will be twenty-one, which is the number of sextarii to this modius. If you want to know the number of sextarii returned, subtract twenty-one from thirty; this leaves nine, which is the number of sextarii returned. If you want to find this in another way, multiply six by thirty and divide the product by the principal; the result is what you wanted to know.

(*b*) There is also another way, as follows. You put, as the number of sextarii to the modius used to discharge the debt, a thing. Add three sevenths of it, which is three sevenths of a thing, to this thing; the result will be one thing and three sevenths of a thing; this is equal to thirty. Therefore the thing is twenty-one [*how many sextarii there are to this modius*][1326].

(*c*)[1327] Or denominate fourteen from twenty, and this fraction taken of thirty will give the number of sextarii to the second modius [*how many from it are used to discharge the debt*][1328]. Or also, denominate six from twenty, and this fraction taken of thirty will give the number of sextarii returned.

AGAIN ON THE SAME TOPIC.

(**B.194**) Someone has borrowed six sextarii of corn of which there are fourteen to the modius. He discharges his debt from another modius with a number of sextarii which, when subtracted from the number of sextarii to this modius, leaves twelve. How many sextarii of these are there to the modius, or how many does he return?

This problem is like the problem which states: 'When fourteen modii are given for six nummi, what is the number of modii which, when diminished by their price, leaves twelve?' We have already explained this in the chapter on buying and selling.[1329] Thus, solve ⟨this and⟩ others of the same kind accordingly.

[1325] B.15.

[1326] Early reader's marginal indication of what is computed. The same below. The lacuna in the formulation may account for these additions.

[1327] As one of the usual 'other ways', this rather belongs to *a*.

[1328] Refers to the next computation.

[1329] B.16.

Chapter (B–XI) on Hiring

This chapter does not differ from the chapter on selling and buying[1330]. For it is clear that the ratio of the number of days served to the number of days for which the person is hired is the same as the ratio of the wage for the days served to the wage for the days he is hired. But to this topic belong problems involving unknowns which are different from the problems involving unknowns in buying and selling, which we therefore have deemed appropriate to add here. But we shall first deal with the first, then with these others and with yet others related to them.[1331]

(**B.195**) For instance, a man hired at a monthly wage of ten nummi serves twelve days. What is his wage?

To this problem may be applied the same various treatments as the problem stated thus: 'When thirty modii are given for ten nummi, what is the price of twelve modii?'

(*a*) You will do the following. Multiply the twelve days served by the ten nummi and divide the product by the number of days in the month, namely thirty; this will give four, which is his wage. We have already indicated the reason for this in the chapter on buying and selling.[1332]

(*b*) There is also another way, as follows. Twelve is the same fraction of thirty as his wage is of ten. The reason for this is the following. It is known that twelve is to thirty as the wage due to him to ten.[1333]

(*c*) Or otherwise. Find the number which when subtracted from thirty leaves ten; it is two thirds of thirty. Therefore subtract from twelve its two thirds; this will leave four, which is his wage for the twelve days served.

[(*d*) *Or otherwise. Ten is the same fraction of thirty as his wage of twelve.*][1334]

(*e*) Or otherwise. Find the number which when subtracted from thirty leaves twelve; it is three fifths of thirty. Therefore subtracting from ten its three fifths will leave the wage, namely four.

AGAIN ON THE SAME TOPIC.

(**B.196**) A man is hired at a monthly wage of ten nummi. How many days must he serve to receive four nummi?

[1330] Usually: 'buying and selling'.

[1331] B.195–226 are of the known type (with the less simple B.218–219 perhaps misplaced since they appear between two similar sets of problems). They are followed by the three groups B.227–229, B.230–239, B.240–243.

[1332] In B.13, belonging to the group B.1, B.3–6.

[1333] In fact this justifies *a* as well.

[1334] This is found only in MS $\mathcal{B}$.

896   PART TWO: TRANSLATION, GLOSSARY

To this problem also may be applied the same various treatments as the problem stated thus: 'When thirty modii are given for ten nummi, how many shall I have for four nummi?'

(**a**) You will do here the following. Multiply four by thirty and divide the product by ten; this gives twelve, and so many days must he serve.

(**b**) Or otherwise. Four is the same fraction of ten, namely its two fifths, as the number of days he must serve is of thirty, that is, twelve. We have already indicated the reason for this in the chapter on buying and selling.[1335]

(**c**) Or otherwise. Find the number by which ten must be multiplied to produce thirty. This is three. Multiplying three by four produces likewise twelve, and this is what you wanted.

AGAIN ON THE SAME TOPIC.

(**B.197**) A man is hired for a month, but the first day at one nummus, the second at two, the third at three, and so on increasing the number of nummi with the number of days till the end of the month. How much will he receive for thirty days?

You will do the following. Add one to the number of days served; this makes thirty-one, which you multiply by half the number of days, namely fifteen. The result is four hundred and sixty-five, and so many nummi will he receive.

Likewise if he serves forty, or more, or less: add one to the number of days, multiply the resulting sum by half the number, and the product is what he is to receive.

CHAPTER INVOLVING UNKNOWNS IN HIRING FOR THINGS [1336]

(**B.198**) A man, hired at a monthly wage of ten nummi and a thing, receives for serving twelve days a thing; what is the value of this thing?

(**a**) You will do the following. Subtract twelve from thirty, thus leaving eighteen. Put this as the principal. Next, multiply twelve by ten and divide the product by the principal; the result is six nummi and two thirds. Such is the value of the thing.

The reason for this is the following. It is known that from the wage ten nummi remain for eighteen days. Then it is like saying: 'Since eighteen is given for ten, what is the price of twelve?' [1337] Multiply ten by twelve and divide the product by eighteen; the result will be what you require.

Another reason is the following. The ratio of days to days is the same as the ratio of wage to wage.[1338] Thus the ratio of thirty to twelve is the

---

[1335] This sentence might belong to part *a*. Compare with the previous problem.

[1336] In B.198–226 there are two unknowns: the wage and the value of the thing.

[1337] Beginning of B–I (rule of three, p. 777). We shall no longer refer to it.

[1338] As said in the introduction to this chapter.

LIBER MAHAMELETH

same as the ratio of ten and a thing to a thing. Now, by separation, the ratio of eighteen to twelve is the same as the ratio of ten to a thing. Thus the product of twelve into ten is equal to the product of eighteen into a thing. So multiply twelve by ten and divide the product by eighteen; this will give as the thing six and two thirds. Those who grasp well this proof will be able to apply it to all problems of this kind.[1339]

(*b*) Or otherwise. Divide the days in the month by the days served, which gives two and a half. Multiply this by the thing which he received, thus producing two and a half things; this is equal to ten and a thing.[1340] Subtracting then a thing from two and a half things leaves one and a half things, which equal ten. Thus the thing is six and two thirds.

The reason we divide the days in the month by the days served, multiply the result by a thing and equate the product to ten and a thing is the following. It is known that when the wage for twelve days is a thing, the wage received for thirty days must be two and a half things since thirty contains twelve two and a half times. It is also known that the wage for thirty days is ten nummi and one thing. Then two and a half things are equal to ten nummi and a thing. The reason for dividing thirty by twelve is to find out how many times twelve is contained in thirty, and this is two and a half. The reason for multiplying two and a half by a thing is to find out how many things are the wage for thirty days.

You will do likewise in what follows, which those who have understood what is here will easily grasp. Indeed, the general rule for working out such problems by algebra is the following. You must always divide the number of days he should serve, that is, the days in the whole month, by the days actually served, multiply the result by what he receives, and equate the product to the wage for the whole month; next, proceed as has been said before in algebra,[1341] and the result will be what you wanted.

(*c*) Or otherwise. It is known that twelve is to thirty as the wage for twelve is to the wage for thirty. Twelve being two fifths of thirty, the wage for twelve is two fifths of the wage for thirty. Therefore two fifths of ten and a thing, which is four and two fifths of a thing, is equal to a thing, which is what he received as a wage for the twelve days served. Do then as we have taught before in algebra[1342]: subtract two fifths of a thing from a thing, thus leaving three fifths of a thing; this equals four. Therefore the thing is six nummi and two thirds.

(*d*) Or otherwise. It is known that the ratio of thirty days to the corresponding wage, which is ten nummi and a thing, is the same as the ratio of twelve days to a thing, which is the corresponding wage. There are therefore four proportional numbers. Thus the product of the first into the

---

[1339] In B.202 composition is used instead.

[1340] Equation of the form $a_1 x + b_1 = a_2 x$.

[1341] Applying the two algebraic operations (restoration and reduction; here only the latter).

[1342] To reduce the equation $a_1 x + b_1 = a_2 x$.

898 PART TWO: TRANSLATION, GLOSSARY

fourth is equal to the product of the second into the third. Therefore the product of thirty into a thing is equal to the product of twelve into ten and a thing. Proceed then as we have taught before in algebra.[1343]

(*e*) You may want to verify this. You know the wage for twelve days; it is six and two thirds. Adding this to ten makes sixteen and two thirds, which is the wage for the whole month. Now it is known that he served two fifths of the month. Thus he must receive two fifths of sixteen and two thirds, which is six and two thirds, like the result found above for the value of the thing.

AGAIN ON THE SAME TOPIC.

(**B.199**) A man, hired at a monthly wage of ten nummi and a thing, serves twelve days and receives a thing and one nummus. What is this thing worth?

(*a*) You will do the following. Subtract twelve from thirty, which leaves eighteen; put this as the principal. Next, subtract from ten and a thing the nummus and the thing received, which leaves nine. Multiply this by twelve and divide the product by the principal; the result will be six. Subtracting from it the one (nummus) he received leaves five, and so much is the thing worth.

The reason we first subtracted the nummus and the thing is the following. Since a thing and a nummus are received for twelve days and this leaves nine, we know that nine is what is due for what remains of the month, namely eighteen days. Then it is like saying: 'When eighteen for nine, what is the price of twelve?' Multiply nine by twelve and divide the product by eighteen; the result is the wage for twelve days, namely six. Now it is known that the wage for twelve days is a thing and one nummus. Therefore this six must be a thing and one nummus. Removing a nummus leaves only the thing, which is therefore worth five nummi.

(*b*) Or otherwise. Divide the days in the month by the days served; the result is two and a half. Multiply this by a thing and a nummus; the result is two and a half things and two and a half nummi, which are equal to ten nummi and a thing.[1344] Taking away then two and a half nummi from ten leaves seven and a half. Next, taking away a thing from two and a half things leaves one and a half things. These equal seven nummi and a half. Thus the thing is worth five (nummi).

(*c*) Or otherwise. It is known that he served two fifths of the month. Thus he must receive two fifths of the wage, that is, four and two fifths of a thing, which is equal to a thing and a nummus. Taking away then a nummus from four nummi leaves three. Next, taking away two fifths of a thing from a thing leaves three fifths of a thing. These equal three nummi. The thing is therefore worth five nummi.

[1343] Equation of the form $a_1x + b_1 = a_2x$, as in *b*.

[1344] Equation of the form $a_1x + b_1 = a_2x + b_2$.

LIBER MAHAMELETH

(*d*) Or otherwise. The ratio of thirty days to the corresponding wage, which is ten and a thing, is the same as the ratio of twelve days to the corresponding wage, which is a thing and one nummus. There are therefore four proportional numbers. Thus the product of thirty into a thing and one nummus, which is thirty things and thirty nummi, is equal to the product of ten and a thing into twelve, which is a hundred and twenty and twelve things. Proceed then as shown before in algebra.[1345] The result will be the value of the thing, which is five nummi.

AGAIN ON THE SAME TOPIC.

(**B.200**) A man, hired at a monthly wage of ten nummi and one thing, served twelve days and received a thing minus one nummus. What is this thing worth?

(*a*) It is already known[1346] that if, for the twelve days served, he receives a thing minus one nummus, the wage for the eighteen remaining days must be eleven nummi [*since he has received a thing 'minus' one nummus*]. Then it is like saying: 'When eighteen are given for eleven, what is the price of twelve [*days*]?' Multiply twelve by eleven and divide the product by eighteen; the result will be seven and a third. This is the wage for twelve days. But since we know that the wage for twelve days is a thing minus one nummus, these seven nummi and a third must be a thing minus one nummus. Adding a nummus makes eight and a third, which is the value of the thing.

The treatment here is, generally speaking, the following. You must subtract twelve from thirty, which leaves eighteen, and put this as the principal. Next, add the nummus subtracted from the thing to ten, which makes eleven; multiply this by twelve, and divide the product by the principal. The result will be the value of the thing minus one nummus.

(*b*) Or otherwise. Divide the number of days in the month by the number of days served, and multiply the result by a thing minus a nummus; the result will be two and a half things minus two and a half nummi, which are equal to ten and a thing.[1347] Subtract then a thing from two and a half things, which leaves one and a half things. Next, add two and a half nummi to ten; this makes twelve and a half, which equals one and a half things. The thing is therefore worth eight and a third.

(*c*) Or otherwise. Since he served two fifths of the month, he must receive two fifths of the (monthly) wage. This is four nummi and two fifths of a thing; they are equal to a thing minus a nummus. Complete the thing by adding a nummus, and add a nummus to four nummi. Next, subtract two fifths of a thing from this thing, thus leaving three fifths of a thing; these are equal to five nummi. The thing is therefore worth eight nummi and a third.

[1345] Equation of the form $a_1x + b_1 = a_2x + b_2$, as in b and c.

[1346] Same situation as in B.199a.

[1347] Equation of the form $a_1x + b_1 = a_2x - b_2$.

(d) Or otherwise. It is known that the ratio of thirty days to the corresponding wage, which is ten nummi and a thing, is the same as the ratio of twelve days to a thing minus one nummus. There are therefore four proportional numbers. Thus the product of thirty into a thing minus a nummus is equal to the product of ten and a thing into twelve. Proceed then as shown before in algebra.[1348] The value of the thing will be eight nummi and a third.

AGAIN ON THE SAME TOPIC.

(B.201) A man, hired at a monthly wage of ten nummi and a thing, serves twelve days and receives ten nummi. How much is this thing worth?

(a) You will do the following. We know [1349] that if he receives ten nummi for twelve days served, the unknown thing must be due for the eighteen remaining days of the month. Then it is like saying: 'When he serves for twelve at ten, what is due to him for eighteen?' So multiply eighteen by ten and divide the product by twelve; the result will be fifteen, and so much is the thing worth.

(b) Or otherwise. Find the number which multiplying twelve produces eighteen; this is one and a half. Multiplying it by ten produces fifteen, and so much is the thing worth.

(c) Or otherwise. Divide the days in the month by the days served, which gives two and a half. Multiplying this by ten, which he received, produces twenty-five, which is the wage for the whole month. [*Then it is like saying: 'Hired at a monthly wage of twenty-five nummi, he serves twelve; what is his wage?' Do as we have taught above.*[1350] *Subtracting the result from twenty-five leaves the value of the thing.*

Or otherwise. Multiply two and a half by ten; this produces twenty-five.][1351] These are equal to ten nummi and a thing.[1352] Subtracting then ten nummi from twenty-five leaves fifteen, which is the value of the thing.

(d) Or otherwise. You know that since he serves two fifths of the month, he must receive two fifths of the wage, that is, four nummi and two fifths of a thing; this is equal to ten nummi. Subtracting then four from ten leaves six, which is equal to two fifths of the thing. Thus the thing is worth fifteen.

(e) Or otherwise. Thirty is to the corresponding wage, which is ten and a thing, as twelve days is to the corresponding wage, which is ten. There are therefore four proportional numbers. Thus the product of thirty into ten is equal to the product of ten and a thing into twelve. Proceed then as we

---

[1348] As before, equation of the form $a_1x + b_1 = a_2x - b_2$.

[1349] Previous problems, part a.

[1350] B.195. The words 'we have taught' are no proof of genuineness of authorship.

[1351] This may have been part of a verification. Otherwise, calculating the wage for twelve days is absurd since it is given.

[1352] Equation of the form $a_1x + b_1 = b_2$.

LIBER MAHAMELETH

901

have taught before in algebra.[1353] The result will be fifteen, which is the value of the thing.

(B.202) A man, hired at a monthly wage of ten nummi minus a thing, serves ten days and receives a thing and two nummi; what is this thing worth?

You will do the following. Add ten and thirty, which makes forty. Put this as the principal. Next, add two nummi to ten nummi, which makes twelve. Multiply this by ten and divide the product by the principal; the result will be three. This is a thing and two nummi. Removing the two nummi leaves one, which is the (value of the) thing.

The proof of this is clear, and is attained by composition. For we know that the ratio of ten minus a thing to a thing and two nummi is the same as the ratio of thirty to ten; then, by composition, the ratio of ten minus a thing plus a thing and two nummi, which is twelve nummi, to a thing and two nummi is the same as the ratio of thirty plus ten, which is forty, to ten. This is the reason for our adding ten to thirty and taking the sum as the principal, then adding two to ten, multiplying the sum by ten, and dividing the product by the principal; the result is what we require.

(B.202′) If, in the problem just stated, he served twenty days and received two things and four nummi, you would do the following. You already know that since he serves twenty days for two things and four nummi, he will serve ten days for a thing and two nummi. Then it is like saying: 'Hired at a monthly wage of ten nummi minus a thing, he serves ten days and receives a thing and two nummi.' Do as has been explained above [1354], and the result will be what you wanted.

(B.202″) Likewise, if he served five days and received half a thing and one nummus, you would proceed in the same way since you already know that for ten days he will receive a thing and two nummi.

The same holds for all the problems belonging to this case, that is, where he receives less or more than a thing [1355], as well as in all similar cases: no matter how (the coefficient of) the thing is set, you will always reduce to only one thing.

(B.203) Suppose, for instance, it were said in the first problem [1356]: Hired at a monthly wage of ten nummi and a thing, he serves three days and receives half a thing; what is this thing worth?

You already know that since for three days he receives half a thing, he will serve six days for a thing. Then it is like saying: 'Hired at a monthly wage of ten nummi and a thing, he serves six days and receives a thing;

[1353] Equation of the form $a_1 x + b_1 = b_2$, as in c & d.

[1354] Previous case.

[1355] That is, when the coefficient of the thing is other than 1 (B.202′, B.202″).

[1356] B.198.

902   PART TWO: TRANSLATION, GLOSSARY

what is this thing worth?' Do as has been shown before,[1357] and the result will be what you wanted.

AGAIN ON THE SAME TOPIC.

(B.204) A man, hired at a monthly wage of six nummi and a thing, served ten days and received six nummi minus half a thing.[1358]

You will do the following. Add the days served to half of the month; this gives twenty-five, which is the principal. Next, multiply six nummi by the days not served, that is, twenty; this produces a hundred and twenty. Divide this by the principal; the result is what the thing is worth, which is four nummi and four fifths.

(B.204′) Suppose he had received the nummi minus two thirds of a thing. Add two thirds of the month to the days served; this gives thirty, which is the principal. Next, multiply six by the days not served, namely twenty; this produces a hundred and twenty. Divide this by the principal; the result is what the thing is worth, namely four nummi.

Treat other problems of this kind in the same way.[1359]

(B.204″) If he had served ten days and received the nummi minus a thing, you would do the following. Add the days served to the days in the month; this gives forty, which is the principal. Next, multiply six by the days not served, namely twenty; this produces a hundred and twenty. Divide it by the principal; the result will be three, and this is what the thing is worth.

AGAIN ON THE SAME TOPIC.

(B.205) A man, hired at a monthly wage of three nummi and a thing, serves ten days and receives the product of three into the root of the wage. What is the value of the thing?

(a) You will do the following. It is known that since he receives for ten days, which is a third of the month, the product of three into the root of the wage, this product must be a third of the (monthly) wage. Therefore thrice three, which is nine, multiplied by the root of the wage is thrice a third of the wage. But thrice a third of the wage is the whole wage. Then it is clear that, as the product of nine into the root of the wage is the whole wage, nine must be its root; thus the wage is eighty-one. This is three nummi and a thing.[1360] Subtracting three nummi leaves seventy-eight, which is therefore the value of the thing.

(b) Or otherwise. It is known that the ratio of thirty days to the corresponding wage, which is three nummi and a thing, is the same as the ratio of ten days to the corresponding wage, which is the root of twenty-seven nummi and nine things.

[1357] B.198.

[1358] There are other statements with no question: see note 985, p. 805.

[1359] This sentence would be expected after B.204″.

[1360] Equation of the form $a_1 x + b_1 = b_2$.

LIBER MAHAMELETH

For it has been said that he receives for ten days the product of three nummi [1361] into the root of the wage, while it is known that the root of the wage is the root of three nummi and a thing. Now when you want to know the result of multiplying three nummi by the root of three and a thing, so as to determine the wage for ten days, you will do as Abū Kāmil has taught, saying that if you want to multiply one number by the root of another number, you are to multiply the former by itself and the result by the number of which the root was mentioned; the root of the product is what you wanted to know. [1362] Therefore multiply three by itself, thus producing nine. Multiplying this by three and a thing gives twenty-seven and nine things, and its root is the wage for ten days. This is the product of three into the root of the wage, namely the root of twenty-seven nummi and nine things.

There are therefore four proportional numbers. Thus the product of the first into the fourth is equal to the product of the second into the third. Multiply then thirty by the root of twenty-seven and nine things, as follows. Multiply thirty by itself, which produces nine hundred, and multiply this by twenty-seven and nine things as we have taught above; [1363] this gives twenty-four thousand three hundred and eight thousand one hundred things. Its root is the product of the first into the fourth. This is equal to the product of three nummi and a thing into ten, which is thirty nummi and ten things. Now it is known that, when thirty nummi and ten things are equal to the root of twenty-four thousand three hundred nummi and eight thousand one hundred things, the product of thirty nummi and ten things into itself must be equal to the product of the root of twenty-four thousand three hundred nummi and eight thousand one hundred things into itself, which is twenty-four thousand three hundred nummi and eight thousand one hundred things. But multiplying thirty nummi and ten things by itself produces a hundred squares and nine hundred nummi and six hundred things. These are equal to twenty-four thousand three hundred nummi and eight thousand one hundred things. Proceed as has been said above in algebra. [1364] That is, set out the two quantities side by side. Next, subtract the six hundred things from the eight thousand one hundred things, which leaves seven thousand five hundred things. Next, subtract the nine hundred nummi from the twenty-four thousand three hundred nummi, which leaves twenty-three thousand four hundred nummi. After all that, a hundred squares remain on one side, which equal the twenty-three thousand four hundred nummi and seven thousand five hundred things remaining on the other side. [1365] Reduce all the squares to one whole square [*and*

[1361] Here and once below instead of 'three'.

[1362] This whole passage might be an addition: reference should be made to Book A (A.285), as is the case in B.95 for a similar computation.

[1363] This vague reference may well be an interpolation.

[1364] Equation of the form $a_1 x^2 + b_1 x + c_1 = b_2 x + c_2$, reduced, as will be shown, like those of the first-degree.

[1365] Equation of the form $ax^2 = bx + c$.

904   PART TWO: TRANSLATION, GLOSSARY

all that which is with them in the same proportion], and reduce the others on the other side in the same proportion, that is, taking of them a tenth of a tenth. The result will be a square, which equals seventy-five things and two hundred and thirty-four nummi. Then halve the things,[1366] and multiply this half by itself, thus producing one thousand four hundred and six and a fourth. Add it to two hundred and thirty-four nummi; this makes one thousand six hundred and forty and a fourth. Adding its root, which is forty and a half, to half of the things, which is thirty-seven and a half, makes seventy-eight, which is the value of the thing.

(**c**) You may want to verify all this, that is, ascertain that for ten days' service is indeed received the product of three into the root of the wage. Add three nummi to the thing, which is seventy-eight; this gives eighty-one, which is the wage for the whole month. Since it is known that he served a third of the month, he must receive a third of the wage, which is twenty-seven nummi. Now this is equal to the result of multiplying three by the root of the wage for the whole month, which is nine [*that is, what he received for the ten days served*][1367].

[*All the previous problems can be proposed in the same form as this one.*[1368]]

AGAIN ON THE SAME TOPIC.

(**B.206**) A man, hired at a monthly wage of thirty nummi and a thing, serves ten days and receives a thing and the root of the (month's) wage. What is this thing worth?

(**a**) You will do the following. It is already known that if he serves a third of the month, he is to receive a third of the wage, which is ten nummi and a third of a thing. Now this equals a thing and the root of thirty nummi and a thing. Subtracting then a thing from ten nummi and a third of a thing leaves ten minus two thirds of a thing, which equal the root of thirty nummi and a thing. Multiplying ten minus two thirds of a thing by itself produces four ninths of a square and a hundred nummi minus thirteen and a third things, which equal thirty nummi and a thing. Complete[1369] the four ninths of a square and a hundred nummi by adding thirteen and a third things, and add the same to thirty nummi and a thing. Next, subtract thirty nummi from a hundred nummi, which leaves seventy. We are then left with four ninths of a square and seventy nummi, which equal fourteen and a third things.[1370] Complete[1371] four ninths of a square to make it only

---

[1366] Solving $x^2 = bx + c$. The expression 'halve the things' is commonly used in Arabic to introduce the solving of a quadratic equation; see our *Introduction to the History of Algebra*, p. 158 n. 175.

[1367] Formerly a gloss to the 'twenty-seven nummi' above.

[1368] This would more appropriately conclude B.203.

[1369] *Complere*, with the meaning of 'restore'.

[1370] Equation of the form $ax^2 + c = bx$.

[1371] *Complere*, this time with the meaning of changing the coefficient of x^2 to 1 (like *restaurare*, see note 1309, p. 890).

LIBER MAHAMELETH

905

one square; that is, multiply it by two and a fourth [*and it will be a square*], and then multiply two and a fourth by seventy, and by fourteen and a third things. This will give one square and a hundred and fifty-seven nummi and a half, which equal thirty-two and a fourth things.[1372] Halving the things gives sixteen and an eighth, and multiplying this by itself produces two hundred and sixty and an eighth of an eighth. Subtracting from this the nummi leaves a hundred and two and four eighths and an eighth of an eighth. Subtracting its root, which is ten and an eighth, from half of the things, that is, from sixteen and an eighth, leaves six. This is what the thing is worth.

(**b**) Or otherwise. The ratio of thirty days to the corresponding wage, which is thirty nummi and a thing, is the same as the ratio of ten days to the corresponding wage, which is a thing and the root of thirty nummi and a thing. There are therefore four proportional numbers. Thus the product of thirty into a thing and the root of thirty and a thing is equal to the product of thirty and a thing into ten. But the first product is thirty things and the root of twenty-seven thousand nummi and nine hundred things, and this equals the second product, which is three hundred nummi and ten things. Subtract thirty things from three hundred nummi and ten things; this leaves three hundred nummi minus twenty things, which equal the root of twenty-seven thousand nummi and nine hundred things. Thus the product of three hundred nummi minus twenty things into itself, which is four hundred squares and ninety thousand nummi minus twelve thousand things, is equal to the product of the root into itself, which is twenty-seven thousand nummi and nine hundred things. Complete four hundred squares and ninety thousand nummi by adding the lacking twelve thousand things, and add the same to twenty-seven thousand nummi and nine hundred things; next, subtract the nummi from the nummi, that is, twenty-seven thousand from ninety thousand. In the end we are left with four hundred squares and sixty-three thousand nummi, equal to twelve thousand nine hundred things.[1373] Reduce all the squares you have to only one square, reduce likewise in the same proportion the quantity with them, and likewise the quantity on the other side in the same proportion, that is, taking of all a fourth of a tenth of a tenth. We shall then have one square and a hundred and fifty-seven nummi and a half equal to thirty-two and a fourth things. Halve the things, and proceed then as I have taught above.[1374] The result will be the value of the thing, namely six. This is what you wanted.

(**c**) Or otherwise. Divide the days in the month by the days served and multiply the result by what was received (for the days served), which is a thing and the root of thirty and a thing; the product is then equal to thirty and a thing. Proceed then as I have taught above.[1375] This will give six

---

[1372] Equation of the form $x^2 + c = bx$.

[1373] Equation of the form $ax^2 + c = bx$.

[1374] This is just the equation seen in a.

[1375] Same equation as before (except for a multiplicative factor).

906　　PART TWO: TRANSLATION, GLOSSARY

as the value of the thing.

(**d**) You may want to ascertain that for ten days' service is indeed received a thing and the root of the wage. Add six, which is the thing, to thirty; this gives thirty-six, which is the wage for the whole month. [*Then it is like saying: 'Hired at a monthly wage of thirty-six nummi, he serves ten days; what is due to him?'*][1376] It is already known that he served a third of the month; thus he must receive a third of thirty-six, thus twelve, which is equal to a thing and the root of the wage.

AGAIN ON THE SAME TOPIC, BUT OTHERWISE.[1377]

(**B.207**) A man, hired at a monthly wage of ten nummi and a thing, serves twelve days and receives that by which the nummi exceed the thing. How much is this thing worth?

(**a**) You will do the following. Divide the days in the month by the days served; this gives two and a half. Multiply this by the excess of the nummi over the thing, which is ten minus a thing; for it is stated that the nummi exceed the thing. [*Now when you want to find out by how much they exceed the thing, subtract the thing from the nummi: the remainder will be this excess.*] The result of this multiplication is then twenty-five minus two and a half things, which is the wage for the month; and it is equal to ten nummi and a thing.[1378] Complete the twenty-five by adding two and a half things, and add the same to ten nummi and a thing; then subtract the ten nummi which are with the thing from twenty-five. The result will be fifteen nummi, which equal three and a half things. Thus the thing is worth four nummi and two sevenths.

Now if you want to ascertain that he indeed received for twelve days the excess of the nummi over the thing, you will do the following. Add four nummi and two sevenths, that is, the value of the thing, to ten nummi; this makes fourteen nummi and two sevenths, which is the month's wage. [*Then it is like saying: 'Hired at a monthly wage of fourteen nummi and two sevenths, he serves twelve days; what is due to him?'*] It is known that he served two fifths of the month; thus he must receive two fifths of the wage, namely five nummi and five sevenths. But this is the excess of the nummi over the thing: there are ten nummi, the thing is four and two sevenths, therefore this excess is five nummi and five sevenths.

(**b**) Or otherwise. The ratio of thirty days to the corresponding wage, which is ten nummi and a thing, is the same as the ratio of twelve days

---

[1376] This is superfluous; so also in B.207, B.209, B.216. Same glossator as in B.201*c*?

[1377] The coming problems involve an excess of the nummi over the thing, or conversely. This is just a formal change since nummi minus a thing, or conversely, has already occurred several times before. (The glossator will fail to grasp this equivalence: see B.207*a*, B.210*b*.) We also return to the first degree.

[1378] Equation of the form $a_1 x + b_1 = b_2 - a_2 x$.

LIBER MAHAMELETH **907**

to the corresponding wage, which is ten nummi minus a thing. There are therefore four proportional numbers. Thus the product of the first, which is thirty, into the fourth, which is ten minus a thing, is equal to the product of the second, which is ten and a thing, into the third, which is twelve. Proceed then as has been shown above in algebra.[1379] The value of the thing will be four and two sevenths.

(*c*) Or otherwise. Since he served two fifths of the month, two fifths of the wage are due to him, which is four nummi and two fifths of a thing. This is equal to ten nummi minus a thing. Complete ten nummi by adding one thing, and add the same to four nummi and two fifths of a thing. Next, subtract four nummi from ten. This will leave six, which is equal to a thing and two fifths of a thing. Thus the thing is worth four nummi and two sevenths.

AGAIN ON THE SAME TOPIC.

(**B.208**) A man, hired at a monthly wage of ten nummi and a thing, served twelve days and received the excess of the nummi over a fourth of a thing. How much is this thing worth?

(*a*) You will do the following. Divide the days in the month by the days served; this gives two and a half. Multiply this by ten minus a fourth of a thing; for it was stated that the nummi exceed a fourth of a thing. The product will be twenty-five minus five eighths of a thing, which equal ten nummi and a thing.[1380] Complete then the twenty-five by adding five eighths of a thing, and add the same to ten nummi and a thing; next, subtract ten from twenty-five. This leaves fifteen nummi, which equal a thing and five eighths of a thing. Thus a thing equals nine nummi and three thirteenths. [*The reason that a thing equals nine nummi and three thirteenths is that the ratio of a thing to a thing and five eighths of a thing is eight thirteenths, while eight thirteenths of fifteen is nine and three thirteenths.*][1381]

If you want to ascertain that he indeed receives as a wage for the twelve days served the excess of the nummi over a fourth of the thing, you will do the following. Add nine and three thirteenths to ten, making nineteen and three thirteenths. Since he served two fifths of the month, he will receive two fifths of the wage, which is nineteen nummi and three thirteenths, thus seven and nine thirteenths. This is the excess of the nummi over a fourth of the thing, since a fourth of the thing is two and four thirteenths.

(*b*) Or otherwise. Since he served two fifths of the month, he must receive two fifths of the wage, which is four nummi and two fifths of a thing. This equals ten nummi minus a fourth of a thing. Complete ten nummi by adding a fourth of a thing, and add this fourth to two fifths of a thing; next, subtract four from ten. This leaves six, which equals thirteen twentieths of a thing. Then find the number which when multiplying thirteen twentieths of

---

[1379] Equation of the form $a_1x + b_1 = b_2 - a_2x$, as in *a*.

[1380] Equation of the form $a_1x + b_1 = b_2 - a_2x$.

[1381] Early reader completing the computations.

908   PART TWO: TRANSLATION, GLOSSARY

a thing produces one thing. This is one and seven thirteenths. Multiplying it by six produces nine and three thirteenths, and this is what the thing is worth.

(**c**) Or otherwise. You already know that the ratio of thirty days to the corresponding wage, which is ten nummi and a thing, is the same as that of twelve days to the corresponding wage, which is ten nummi minus a fourth of a thing. There are then four proportional numbers. Thus the product of thirty into ten nummi minus a fourth of a thing is equal to the product of a thing and ten into twelve. Proceed then as explained above in algebra.[1382] The result will be nine and three thirteenths, and this is what the thing is worth.

AGAIN ON THE SAME TOPIC.

(**B.209**) A man, hired at a monthly wage of ten nummi and a thing, serves ten days and receives as much as the excess of the thing over a third of the nummi. How much is this thing worth?

(**a**) You will do the following. Divide the days in the month by the days served; this gives three. Multiply this by the excess of the thing over a third of the nummi, which is a thing minus three nummi and a third; for it has been stated that the thing exceeds a third of the nummi. The product is three things minus ten nummi, which is the wage for the month. This is equal to ten nummi and a thing.[1383] Complete the three things by adding ten nummi, and add the same to ten nummi and a thing; next, subtract a thing from three things. This leaves two things, which are worth twenty. Thus a thing is worth ten.

You may want to ascertain that the excess of a thing over a third of the nummi is indeed the wage for the ten days served. Add ten nummi, which the thing is worth, to ten nummi, making twenty, which is the month's wage. [*Then it is like saying: 'Hired at a monthly wage of twenty nummi, he serves ten days; what is due to him?'*] You know that he served a third of the month; thus to him is due a third of twenty nummi, which is six and two thirds. This is the excess of the thing over a third of the nummi; for the thing being ten nummi, and a third of the nummi, three and a third, the excess of a thing over a third of the nummi is six and two thirds.

(**b**) Or otherwise. You already know that the ratio of thirty days to the corresponding wage, which is ten and a thing, is the same as that of ten days to the corresponding wage, which is a thing minus three and a third. There are then four proportional numbers. Thus the product of thirty into a thing minus three and a third is equal to the product of ten and a thing into ten. Proceed then as we have taught above in algebra.[1384] The result will be ten nummi, and that much is the thing worth.

(**c**) Or otherwise. Since he served a third of the month, a third of the wage is due to him, namely three and a third and a third of a thing, which is

---

[1382] Equation of the form $a_1 x + b_1 = b_2 - a_2 x$, as in parts a & b.

[1383] Equation of the form $a_1 x + b_1 = a_2 x - b_2$.

[1384] Equation of the form $a_1 x + b_1 = a_2 x - b_2$, as in a.

LIBER MAHAMELETH

909

worth a thing minus three nummi and a third. Proceed then as we have explained before in algebra.[1385] The result will be ten, and that much is the thing worth.

AGAIN ON THE SAME TOPIC.

(B.210) A man, hired at a monthly wage of ten nummi and one thing, served twelve days and received the excess in value of the thing over the nummi. What is this thing worth?

(*a*) You will do the following. Subtract twelve from thirty; this leaves eighteen, which you put as the principal. Next, double ten; this gives twenty. Multiply it by twelve, thus producing two hundred and forty. Divide it by the principal; this gives thirteen and a third. Add it to the ten. The result is twenty-three and a third, and so much is the thing worth.

The reason for this is the following. It is known that the wage for the whole month is ten nummi and one thing, and that for the twelve (days) served he received the excess in value of the thing over the nummi, that is, a thing minus ten nummi. Now we know[1386] that if he were to receive for twelve days' service the whole thing, the wage for the remaining part of the month, which is eighteen (days), should be the remainder of the wage, which is ten nummi. But he receives a thing minus ten nummi; consequently, the wage for eighteen days must be those ten nummi, which are subtracted, and the ten with the thing, thus twenty nummi. Then it is like saying: 'When eighteen for twenty, then what is the price of twelve [*days*][1387]?' You will proceed as shown above. The wage for twelve ⟨days⟩ will be thirteen and a third. Now this is a thing minus ten nummi. Adding then ten nummi makes twenty-three and a third, and this is what the thing is worth.

(*b*) Or otherwise. Divide the days in the month by the days served, which gives two and a half. Multiply this by the excess in value of the thing over ten nummi, which is a thing minus ten. [*For when we take the excess in value of the thing over ten nummi, it is known that this thing must be greater than the nummi; thus when you want to know by how much the thing exceeds ten nummi, subtract ten from the thing; this leaves a thing minus ten nummi, which is the excess in value of the thing over the nummi. I now return to the problem to finish it. Namely:*][1388] Since the product is two and a half things minus twenty-five nummi, which is the wage for the month, this equals ten nummi and a thing, which is also the wage for the month.[1389] Complete then the two and a half things by adding twenty-five, and add the same to ten, which makes thirty-five; next, subtract a thing

---

[1385] Again, equation of the form $a_1 x + b_1 = a_2 x - b_2$.

[1386] B.198.

[1387] Misplaced addition.

[1388] A similar interpolation in the first problem on excesses (B.207). The present one is longer; in the disordered version now preserved by MS $\mathcal{B}$, this problem comes before B.207.

[1389] Equation of the form $a_1 x + b_1 = a_2 x - b_2$.

910   Part Two: Translation, Glossary

from two and a half things, which leaves one and a half things. This equals thirty-five nummi. Thus a thing is worth twenty-three and a third.

(*c*) You may want to verify that the excess of the thing over the nummi is indeed the wage for twelve days. Add the value of the thing, which is twenty-three and a third, to ten; this makes thirty-three and a third. This is the wage for the month. Since he served two fifths of the month, he will receive two fifths of the wage, which is thirteen and a third. This is the excess of the thing over the nummi:[1390] the value of the thing is twenty-three and a third, the nummi are ten, so the excess of the thing over the nummi is thirteen and a third.

(*d*) Or otherwise. Since he serves two fifths of the month, he must receive two fifths of the wage, which is four nummi and two fifths of a thing. This is equal to a thing minus ten nummi. Complete then the thing by adding ten nummi, and add the same to four nummi and two fifths of a thing; next, subtract two fifths of a thing from this same thing. This leaves three fifths of a thing, which equal fourteen nummi. Thus the thing is worth twenty-three nummi and a third.

(*e*) Or otherwise. It is known that the ratio of thirty days to the corresponding wage is the same as that of twelve to the corresponding wage. There are therefore four proportional numbers. Thus the product of thirty into a thing minus ten nummi is equal to the product of twelve into ten and a thing. Proceed then as we have taught before in algebra.[1391] The result will be the value of the thing, namely twenty-three and a third.

(**B.211**) Someone asks:[1392] Hired at a monthly wage of twenty nummi and a thing, he serves ten days and receives twelve nummi minus a thing.

This problem is solved by composition. The ratio of thirty to ten being the same as the ratio of twenty and a thing to twelve minus a thing, then, by composition, the ratio of forty to ten will be the same as the ratio of twenty and a thing plus twelve minus a thing, which is thirty-two, to twelve minus a thing. So proceed as has been shown above.[1393] Twelve minus a thing will be eight, thus the thing is four.

You will do the same in all problems of this kind. If there is more than one thing, reduce them to one as you did in the other instances.[1394]

Again on the same topic.

(**B.212**) A man, hired at a monthly wage of ten nummi minus one thing, serves twelve days and receives a thing minus two nummi. What is this thing worth?

[1390]   After this appears (in both MSS): 'Then add the value of the thing', which makes no sense. A similar expression is found, this time appropriately, four lines above.

[1391]   As in *b* & *d*, equation of the form $a_1x + b_1 = a_2x - b_2$.

[1392]   No question stated, see note 1358, p. 902.

[1393]   Plain proportion, leading to an equation of the form $b_1 - a_1x = b_2$.

[1394]   As stated in B.202″.

LIBER MAHAMELETH

(*a*) You will do the following. Divide the days in the month by the days served, which gives two and a half. Multiplying this by a thing minus two nummi produces two and a half things minus five nummi. This is equal to the wage for the month, which is ten nummi minus one thing.[1395] Complete the ten nummi by adding a thing, and add this same thing to the two and a half things on the other side; next, add the five subtracted to ten[1396]. The result is fifteen, which equals three and a half things. Thus the thing is worth four and two sevenths.

(*b*) Or otherwise. You know that since he served two fifths of the month, he is to receive two fifths of the wage, namely four nummi minus two fifths of a thing. This is equal to one thing minus two. Proceed then as we have taught before in algebra.[1397] The result will be, for what the thing is worth, four nummi and two sevenths.

(*c*) Or otherwise. Thirty days are to the corresponding wage, which is ten minus one thing, as twelve is to the corresponding wage, which is a thing minus two nummi. There are therefore four proportional numbers. Thus the product of twelve into ten minus a thing is equal to the product of a thing minus two into thirty. Proceed then as has been shown before in algebra.[1398] The result will be how much the thing is worth, namely four nummi and two sevenths.

(**B.213**) Someone asks: Hired at a monthly wage of ten nummi minus a thing, he serves six days and receives four nummi minus a thing.

This problem is solved by separation. The ratio of thirty to six being the same as the ratio of ten minus a thing to four minus a thing, then, by separation, the ratio of twenty-four to six will be the same as the ratio of ten minus a thing less four minus a thing to four minus a thing. Proceed as shown above.[1399] The result will be, for four minus a thing, one and a half; thus the thing is two and a half.

AGAIN ON THE SAME TOPIC, BUT OTHERWISE.[1400]

(**B.214**) A man, hired at a monthly wage of a thing, serves ten days and receives a thing minus ten nummi. What is this thing worth?

(*a*) You will do the following. Divide the days in the month by the days served, which gives three. Multiplying it by a thing minus ten nummi produces three things minus thirty nummi. This equals one thing.[1401] Complete then the three things by adding thirty nummi, and add the same to the thing; subtract this thing from three things. This leaves two things, which equal thirty nummi. Thus the thing is worth fifteen.

[1395] Equation of the form $a_1 x - b_1 = b_2 - a_2 x$.

[1396] *quinque demptos adde ad decem*: the absolute value is considered.

[1397] As in *a*, equation of the form $a_1 x - b_1 = b_2 - a_2 x$.

[1398] Again, equation of the form $a_1 x - b_1 = b_2 - a_2 x$.

[1399] Plain proportion, leading to an equation of the form $b_1 - a_1 x = b_2$.

[1400] This heading may well be misplaced.

[1401] Equation of the form $a_1 x - b_1 = a_2 x$.

912   PART TWO: TRANSLATION, GLOSSARY

(**b**) Or otherwise. Since he served a third of the month, he must receive a third of the wage, which is a third of a thing. This equals a thing minus ten nummi. Complete the thing by adding ten nummi, and add the same to a third of a thing; subtract this third of a thing from the thing. This leaves two thirds of a thing, which equal ten nummi. Thus the thing is worth fifteen.

(**c**) Or otherwise. Since the ratio of thirty days to the corresponding wage, which is a thing, is the same as the ratio of ten days to the corresponding wage, which is a thing minus ten nummi, the product of thirty into a thing minus ten nummi is equal to the product of a thing into ten [*nummi* [1402]]. Proceed then as has been shown in algebra.[1403] The result will be fifteen, and this is what the thing is worth.

AGAIN ON THE SAME TOPIC.

(**B.215**) A man, hired at a monthly wage of one thing, serves ten days and receives a fourth of a thing and two nummi. What is the thing worth?

(**a**) You will do the following. Divide the days in the month by the days served, which gives three. Multiply it by what he received, which is a fourth of a thing and two nummi; the result is three fourths of a thing and six nummi. This is the wage for the month, and it equals a thing.[1404] Subtract then three fourths of a thing from this same thing; this leaves a fourth of a thing, which equals six nummi. Thus the thing is worth twenty-four.

(**b**) Or otherwise. Since he served a third of the month, he should receive a third of the wage, which is a third of a thing. This equals a fourth of a thing and two nummi. Then subtract a fourth of a thing from a third of a thing; this leaves half a sixth of a thing, which equals two nummi. Thus the thing is worth twenty-four.

(**c**) Or otherwise. Since the ratio of thirty to the corresponding wage, which is a thing, is the same as the ratio of ten days to the corresponding wage, which is a fourth of a thing and two nummi, there are then four proportional numbers. Thus the product of the first, which is thirty, into the fourth, which is a fourth of a thing and two nummi, is equal to the product of the second, which is a thing, into the third, which is ten. Proceed according to what has been explained in algebra.[1405] The result will be twenty-four, which is the value of the thing.

AGAIN ON THE SAME TOPIC.

(**B.216**) A man, hired at a monthly wage of a thing minus ten [1406], serves twelve days and receives a fourth of a thing and two nummi. How much is this thing worth?

[1402] Nonsense.

[1403] Equation of the form $a_1 x - b_1 = a_2 x$, as in parts a and b.

[1404] Equation of the form $a_1 x + b_1 = a_2 x$.

[1405] Equation of the form $a_1 x + b_1 = a_2 x$, as in a & b.

[1406] *nummi* is omitted here, as elsewhere (B.221, B.222, B.224–226).

LIBER MAHAMELETH

(*a*) You will do the following. Divide the days in the month by those served, which gives two and a half. Multiplying this by a fourth of a thing and two nummi produces five eighths of a thing and five nummi. This equals a thing minus ten nummi.[1407] Complete the thing by adding ten nummi, and add the same to five nummi and five eighths of a thing; subtract this five eighths of a thing from this same thing. This leaves three eighths of a thing, which equal fifteen nummi. Thus the thing is worth forty nummi.

You may want to verify this. Subtract ten from forty, thus leaving thirty; for he was said to have been hired for a thing minus ten nummi, and you know that the thing is worth forty; therefore a thing minus ten nummi is worth thirty. [*Then it is like asking: 'Hired at a monthly wage of thirty nummi, he serves twelve days; what is his wage?'*] Since, as you know, he served two fifths of the month, he should receive two fifths of the wage, which are twelve nummi. This is like a fourth of a thing and two nummi: a fourth of a thing is ten nummi, and adding two to it makes twelve.

(*b*) Or otherwise. It is known that since he served two fifths of the month, he should receive two fifths of the wage, namely two fifths of a thing minus four nummi. This is equal to a fourth of a thing and two nummi. Complete two fifths of a thing by adding four nummi, and add another four to two nummi, which makes six nummi; next, subtract a fourth of a thing from two fifths of a thing. This leaves three fourths of a fifth of a thing, which equal six nummi. Consider then by what number three fourths of a fifth of a thing is to be multiplied to produce one thing;[1408] this is six and two thirds. Multiplying it by six produces forty, which is the value of the thing.

(*c*) Or otherwise. Since the ratio of thirty days to the corresponding wage, which is a thing minus ten nummi, is the same as the ratio of twelve days to the corresponding wage, which is a fourth of a thing and two nummi, there are therefore four proportional numbers. Thus the product of thirty into a fourth of a thing and two nummi is equal to the product of a thing minus ten nummi into twelve. Proceed then as before in algebra.[1409] The result will be the value of the thing, which is forty nummi.

AGAIN ON THE SAME TOPIC.

(**B.217**) A man, hired at a monthly wage of a thing minus ten nummi, served twelve days and received ten nummi. What is the value of the thing?

(*a*) You will do the following. Divide the days in the month by the days served, which gives two and a half. Multiplying this by ten produces twenty-five. This equals a thing minus ten nummi.[1410] Complete the thing by adding ten nummi, and add another ten to twenty-five. This makes thirty-five, which equals a thing.

[1407] Equation of the form $a_1x + b_1 = a_2x - b_2$.

[1408] A.218.

[1409] Equation of the form $a_1x + b_1 = a_2x - b_2$, as in parts *a* and *b*.

[1410] Equation of the form $a_1x - b_1 = b_2$.

914   PART TWO: TRANSLATION, GLOSSARY

(*b*) Or otherwise. It is known that the ratio of thirty days to the corresponding wage, which is a thing minus ten nummi, is the same as the ratio of twelve days to the corresponding wage, which is ten nummi. There are therefore four proportional numbers. Thus the product of thirty into ten is equal to the product of a thing minus ten nummi into twelve. Proceed then as we have taught in algebra.[1411] The result will be the value of the thing, which is thirty-five.

AGAIN ON THE SAME TOPIC.

(B.218) A man, hired at a monthly wage of a thing, serves a certain number of days of that month which, when multiplied by what is due to him as a wage, produces six, and the remainder of the month, when multiplied by the remainder of the thing, produces twenty-four. What is this thing worth?[1412]

You will do the following. Multiplying six by twenty-four produces a hundred and forty-four. Double its root, which is twelve; the result is twenty-four. Add it together with the sum of six and twenty-four, thus making fifty-four. Divide it by thirty, which gives one and four fifths. This is the value of the thing.   Now if you want to know the number of days he did not serve, add twelve to twenty-four and divide the sum by the value of the thing; the result is the number of days he did not serve.   If you want to know the number of days he served, add twelve to six and divide the sum by the value of the thing; the result is the number of days he served.

The proof of this is the following. Let thirty days be line AB, what he served of the month, line AG, what he did not serve, line GB, the thing, line DT, and what is due to him of the thing for his service, line DG.[1413] Now line AG when multiplied by line DG produces the area AD, which is six. Further, what he did not serve, which is line BG, when multiplied by the remainder of the thing, which is line GT, produces the area BT, which is twenty-four. I shall complete the figure, and it will give the area $ZKLM$. Now the ratio of what is served, which is line AG, to what is received of the thing, which is line DG, is the same as the ratio of what is not served, which is line GB, to the remainder of the thing, which is GT. There are therefore four proportional numbers. By alternation, the ratio of line AG to GB will be the same as the ratio of line DG to GT. Then the sides of the area AT and the sides of the area DB are mutequefia[1414]. Therefore[1415] the area AT is equal to the area DB.[1416] Now the ratio of AG to GB is

---

[1411] As in *a*, equation of the form $a_1 x - b_1 = b_2$.

[1412] Required also the days, worked or not worked.

[1413] DT is drawn perpendicular to AB.

[1414] That is, 'reciprocally proportional' (Greek ἀντιπεπόνθασιν —namely the πλευραί—, Arabic *mutakāfi'a*); in B.32, same situation, the sides are said to be 'proportional'). See B.230 & B.258. The text has 'equal', possibly a misplaced marginal gloss meant to be an emendation (see note 1416).

[1415] *Elements* VI.14.

[1416] For 'equal' the text has *talis qualis*.

the same as the ratio of the area AD to the area DB, and the ratio of DG to GT is the same as the ratio of the area DB to the area GM.[1417] But the ratio of AG to GB is the same as the ratio of DG to GT. Thus the ratio of the area AD to the area DB is the same as the ratio of the area DB to the area GM. There are therefore three proportional terms.[1418] Thus the product of the area AD, which is six, into the area BT, which is twenty-four, is equal to the product of the area BD into itself. But multiplying AD by BT produces a hundred and forty-four. Thus multiplying BD by itself produces as much. Therefore the area DB is twelve. Since it is equal to the area AT, the area AT is twelve. Therefore the whole area $ZKLM$ is fifty-four. Dividing it by line ZK, which is thirty, gives line KM, thus known; and this is the value of the thing since KM is equal to DT. Dividing now the area DM, which is thirty-six, by line DT gives line DK, which is what he did not serve, and it is equal to GB. Dividing now also the area ZT, which is eighteen, by line DT gives ZD, which is what he served, for it is equal to AG. The figure below illustrates this.[1419]

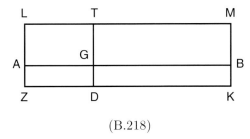

(B.218)

(B.219) Someone asks: Hired at a monthly wage of a thing, he serves a number of days of the month which when multiplied by the corresponding wage produces twenty-seven, while the multiplication of the days he did not serve by the corresponding wage produces a hundred and forty-seven. How much is the thing and how many days did he serve and how many did he not?

You will do the following. Multiply twenty-seven by a hundred and forty-seven and keep in mind the root of the product, which is sixty-three. If you want to know the number of days he served, add to this root twenty-seven and divide the sum by thirty; the result will be three, and such a part of the thing is due to him for what he served; dividing twenty-seven by these three gives nine, and so many days did he serve. If you want to know the number of days he did not serve, add to the above root a hundred and forty-seven and divide the sum by thirty; the result will be seven, and this is the wage for the days he did not serve; dividing a hundred and forty-seven by this gives twenty-one, and so many are the days he did not

---

[1417] *Elements* VI.1.
[1418] Three terms (*termini*) in continued proportion.
[1419] As in B.220*b*, instead of the usual closing words (Q.E.D.).

serve. If you want to know the (value of the) thing, add three and seven; this makes ten, which is the (value of the) thing.

The proof of all this is the following. Let the thing be AB, thirty days, DH, those served, DZ, the part of the thing due, AG, what he did not serve, ZH, the corresponding wage, GB. Let AG be multiplied by DZ and produce K; so K is twenty-seven. Let GB be multiplied by ZH and produce T; so T is a hundred and forty-seven. Then it is clear [1420] that the ratio of AG to DZ is the same as the ratio of GB to ZH; so the product of AG into ZH is equal to the product of GB into DZ. But it appears from the premisses [1421] that for any four numbers the result of multiplying the product of the first into the second by the product of the third into the fourth is equal to the result of multiplying the product of the first into the third by the product of the second into the fourth. Thus the result of multiplying the product of AG into DZ by the product of GB into ZH, which is the result of multiplying K by T, is equal to the result of multiplying the product of AG into ZH by the product of GB into DZ. Since the product of AG into HZ is equal to the product of GB into DZ, the root of the product of K into T is equal to the product of AG into ZH, and to (the product of) GB into DZ; this root is sixty-three. If you want to know the values of AG and DZ, you will do the following. You already know that the product of AG into ZH is sixty-three. By putting in common the product of AG into DZ, which is twenty-seven, the product of AG into the whole DH will be ninety. Dividing it by DH, which is thirty, gives as AG three. Dividing twenty-seven by this gives DZ, which is nine. If you want to know the values of ZH and GB, you will likewise do the following. You already know that the product of DZ into GB is sixty-three. Put in common the product of HZ into GB, which is a hundred and forty-seven. Next, continue as we have taught for the determination of AG and DZ; the result will be seven for GB and twenty-one for ZH. This is what we wanted to demonstrate.

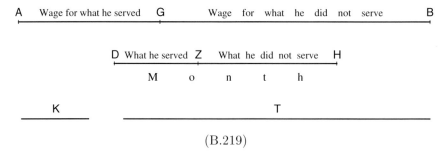

(B.219)

AGAIN ON THE SAME TOPIC.

(B.220) A man, hired for thirty days at a thing minus ten nummi, serves twenty days and receives eight nummi. How much is the thing worth?

[1420] Constancy of the ratio wage : days.
[1421] Premiss P_3 (irrespective of the order of the factors).

(*a*) You will do the following. Add eight nummi to a thing minus ten nummi, thus making a thing minus two nummi. Next, add twenty days to thirty days, thus making fifty days. Multiply this by eight and divide the product by twenty days; the result is the value of the thing minus two nummi. Adding two nummi to it will give the value of the thing. The reason for this is clear.[1422]

(*b*) Or otherwise. Multiply eight nummi by thirty days, divide the product by twenty days, and add to the result the ten nummi lacking from the thing; the result will be the value of the thing.

The proof of this is the following. Let thirty days be line AB, twenty days, line BG, the thing, line TQ. Subtracting from it ten nummi, say line DQ, will leave line DT as the wage corresponding to line AB. But the wage corresponding to line GB is eight; let line DL be eight. Now the ratio of line AB to line BG is the same as the ratio of line DT to line DL. There are therefore four proportional numbers. Thus dividing the product of the first, which is thirty, into the fourth, which is eight, by the second, which is twenty, will give DT; this is twelve. Adding to it DQ, which is ten, makes line TQ, namely twenty-two. This is what the thing is worth. The corresponding figure is appended.

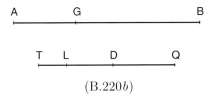

(B.220*b*)

(**B.221**) Someone asks: Hired at a monthly wage of a thing minus ten, he serves forty days and receives just one thing. How much is this thing worth?

This problem is solvable, for the days served are more than the days of the month.[1423] You will do in this case the following. You know that the ratio of thirty to forty is the same as the ratio of a thing minus ten to a thing. By separation, the ratio of thirty to ten will be the same as the ratio of a thing minus ten to ten. Proceed as has been shown above.[1424] The result will be, for a thing minus ten, thirty. Thus the thing is forty.

(**B.221′**) Now if the man hired had served twenty days and received half a thing, you would do the following.

Multiply the number from which a half is denominated, namely two, by twenty; this produces forty. Subtracting from it thirty days leaves ten;

[1422] Composition of the ratio.

[1423] This remark and the (incomplete) condition suggests that this problem should have followed B.226′.

[1424] Equation of the form $a_1 x - b_1 = b_2$ (immediately inferred since the denominators are the same).

put this as the principal. Next, multiply ten, which is lacking from a thing, by thirty and divide the product by the principal. Adding ten nummi to the result gives what the thing is worth.

The proof of this is the following. It is known that, since he serves twenty days for half a thing, he will serve forty for the whole thing. This is why we multiply twenty by two. Let then forty days be line GB and thirty, line AB; let the wage corresponding to line GB be line DT, which is a thing. But the wage for line AB is a thing minus ten nummi. I shall then cut off from line DT ten, say line QT, which will leave line DQ as the wage for line AB. Now the ratio of line GB to AB is the same as the ratio of DT to DQ. By separation, the ratio of line AG, which is ten, to AB, which is thirty, will be the same as the ratio of line QT, which is ten, to DQ. There are therefore four proportional numbers. Thus, dividing the product of line AB, which is thirty, into QT, which is ten, by AG will give DQ; this is thirty. Adding to it line QT makes DT, which is forty, that is, what the thing is worth. This is what we wanted to demonstrate.

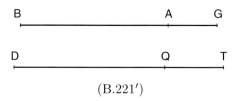

(B.221′)

(**B.221″**) If now he receives half a thing and a nummus, you will do the following.

Multiply the number from which a half is denominated, which is two, by twenty, which produces forty. Next, multiply half a thing and a nummus by two, thus producing a thing and two nummi. Next, subtract thirty days from forty; this leaves ten. Put this as the principal. Next, add two nummi to ten, which are lacking,[1425] making twelve. Multiply this by thirty, thus producing three hundred and sixty. Divide it by the principal, and add ten to the quotient. The result is the value of the thing, namely forty-six.

The proof of this is the following. It is known that serving twenty days for half a thing and one nummus, he will again[1426] serve forty days for the whole thing and two nummi. Let thus forty days be line BG, and let a thing and two nummi be line DL, the thing, DT, and the two nummi, line TL, and let thirty days be line AB. The wage for AB being a thing minus ten nummi, cut off from line DT ten, namely line QT, thus leaving line DQ, that is, the wage for line AB. Now the ratio of line BG to AB is the same as the ratio of line DL to QD. By separation, the ratio of line AG to AB will be the same as the ratio of line QL to QD. There are therefore four proportional numbers. Thus dividing the product of line AB, which

---

[1425] That is, 'lacking from a thing', as in B.221′.
[1426] As in B.221′.

is thirty, into QL, which is twelve, by AG will give DQ. Adding to it line QT, which is ten, will make line DT; this is forty-six, which is the value of the thing.[1427]

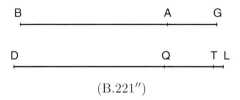

(B.221″)

(**B.222**) Someone asks: Hired at a monthly wage of a thing minus twenty, he serves fifty days and receives a thing and ten. What is the thing worth?

You already know that the ratio of thirty to fifty is the same as the ratio of a thing minus twenty to a thing and ten. By separation, the ratio of thirty to twenty will be the same as the ratio of a thing minus twenty to thirty. Proceed then as we have shown above.[1428] A thing minus twenty will be forty-five; thus the thing is sixty-five.

(**B.223**) Someone asks: Hired at a monthly wage of twenty squares, he serves six days and receives eight times the root of one square. What are the square and its root?

The treatment of this and similar problems using algebra or using multiplication is the same.[1429] I shall therefore (merely) explain the treatment using algebra, whereby the treatment using multiplication will appear. It is the following. Divide thirty by six; this gives five. Multiply it by eight things, thus producing forty things; this is equal to twenty squares.[1430] Since twenty squares are equal to forty things, one square is equal to two things ⟨and the thing is two⟩. So the square is four.

You will do the same in all similar problems.

(**B.224**) Someone asks: Hired at a monthly wage of a thing minus twenty, he serves five days and receives a thing minus forty-five.[1431]

You already know that the ratio of thirty to five is the same as the ratio of a thing minus twenty to a thing minus forty-five. By separation and alternation, the ratio of twenty-five to the twenty-five which is the difference between a thing minus forty-five and a thing minus twenty will be the same as the ratio of five to a thing minus forty-five. Thus a thing

---

[1427] No Q.E.D. here.
[1428] Plain proportion leading to an equation of the form $a_1 x - b_1 = b_2$ (B.217, B.221).
[1429] 'using multiplication': equating the products of the extreme and mean proportional terms.
[1430] Equation of the form $ax^2 = bx$.
[1431] As usual, the value of the thing is required.

920 PART TWO: TRANSLATION, GLOSSARY

minus forty-five is five;[1432] and so the thing is fifty. This is what you wanted.

(B.225) Someone asks: Hired at a monthly wage of a thing minus twenty, he serves three days and receives half a thing minus twenty.

You already know that if he served six days he would receive a thing minus forty. Then it is like saying: 'Hired at a monthly wage of a thing minus twenty, he serves six days and receives a thing minus forty.' Proceed as has been shown before,[1433] and the result will be what you wanted.

(B.226) Someone asks: Hired at a monthly wage of a thing minus twenty, he serves ten days and receives a thing minus five.

This problem is not solvable. Since a thing minus twenty is less than a thing minus five, he would have received more for a part of the month than for the whole month, and this is absurd[1434]. But if the days served were more than the days in the month, then the problem would be solvable.

(B.226′) Likewise if someone asks: Hired at a monthly wage of a thing minus ten, he serves five days and receives just a thing, or a thing and one nummus. This will not be solvable unless the days served are more than the days in the month.

AGAIN ON THE SAME TOPIC, BUT OTHERWISE[1435]

(B.227) Someone asks: A man is hired at a monthly wage of ten nummi and three sextarii of corn, each of which has a different value: the first exceeds the second by two nummi and the second the third also by two nummi. He serves ten days and receives half of the third, a third of the second, and a fourth of the first, but he must return five nummi to his hirer. How much is each sextarius worth?

You will do the following. Let a thing be the value of the third; the value of the second will be a thing and two nummi, and the value of the first, a thing and four nummi. So the value of the three sextarii is three things and six nummi. Adding to it ten nummi makes three things and sixteen nummi, which is the wage for the whole month. [*For it has been said that the month's wage is ten nummi and three sextarii, and as it is clear*[1436] *that the value of the three sextarii is three things and six nummi, so it is known that the month's wage is three things and sixteen nummi.*] Now after serving ten days he receives half of the third, a third of the second and a fourth of the first[1437]. But adding half of the third, a third

---

[1432] Equation of the form $a_1 x - b_1 = b_2$.

[1433] Previous problem.

[1434] *falsus*, also meaning (as above) 'not solvable'.

[1435] Part of the wage is now in kind.

[1436] Perhaps a copyist's error for 'it has been shown' (*manifestum est* instead of *monstratum est*).

[1437] This is what he receives in kind (thus not counting what he returns in cash).

LIBER MAHAMELETH

of the second and a fourth of the first makes one thing and half a sixth of a thing and one nummus and two thirds. Subtracting from this the five nummi he pays leaves a thing and half a sixth of a thing minus three nummi and a third, which is the wage for ten days. Therefore the wage for the whole month must be three things and a fourth of a thing minus ten nummi; this equals the wage for the month which is (also) three things and sixteen nummi. Proceed then as I have taught above in algebra.[1438] That is, complete the three things and a fourth of a thing by adding ten nummi and add the same amount to three things and sixteen nummi; next, subtract three things from three things and a fourth of a thing. This leaves a fourth of a thing, which equals twenty-six nummi. Therefore a thing is worth a hundred and four nummi. This is the value of the third sextarius, the value of the second being a hundred and six and the value of the first, a hundred and eight.

You may want to verify this, that is, ascertain that for ten days' service he does indeed receive half of the third, a third of the second, a fourth of the first but returns five nummi. Add the values of the three sextarii to ten nummi; this makes three hundred and twenty-eight nummi, which is the wage for the month. Now it is known that he served a third of the month. Thus to him is due a third of the wage, which is a hundred and nine and a third. But he receives for ten days' service half of the third, a third of the second and a fourth of the first[1439]. Add then together half of a hundred and four, a third of a hundred and six, and a fourth of a hundred and eight; this makes a hundred and fourteen and a third. Subtracting from it the five nummi he pays leaves a hundred and nine and a third, which equals what he received for ten days' service.

AGAIN ON THE SAME TOPIC.

(B.228) A man, hired at a monthly wage of ten nummi and three sextarii of unequal value, the first exceeding the second by three nummi and the second the third by three nummi also, serves ten days and receives half of the third, a third of the second, and a fourth of the first. What is the value of each sextarius?

You will do the following. Let the value of the third be one thing; thus the value of the second will be a thing and three nummi and the value of the first, a thing and six nummi. So the value of the three sextarii is three things and nine nummi. Adding to it ten nummi makes three things and nineteen nummi, which is the wage for the month. Since he receives for ten days' service half of the third, a third of the second and a fourth of the first, add half of the third, a third of the second and a fourth of the first; this makes one and half a sixth things and two and a half nummi. This is the wage for a third of the month; thus the wage for the whole month must be three things and a fourth of a thing and seven and a half

---

[1438] Reducing the equation of the form $a_1 x + b_1 = a_2 x - b_2$.

[1439] Wage received in kind.

nummi; this equals three things and nineteen nummi.[1440] Subtracting the things from the things and the nummi from the nummi leaves in the end a fourth of a thing equal to eleven and a half nummi. A thing therefore equals forty-six nummi. This is the value of the third sextarius; thus the value of the middle one is forty-nine and the value of the first, fifty-two.

You may want to verify this. Add the values of the three sextarii to ten nummi; this makes a hundred and fifty-seven, which is the wage for the month. But it is known that he served a third of the month; thus he must receive a third of the wage, which is fifty-two and a third. Since for ten days' service he receives half of the third (sextarius), a third of the second and a fourth of the first, add together half of forty-six[1441], a third of forty-nine and a fourth of fifty-two; this makes fifty-two and a third, which equals the wage for ten days.

AGAIN ON THE SAME TOPIC.

(B.229) A man, hired at a monthly wage of three sextarii of unequal value, the first exceeding the second by three nummi and the second the third by two nummi, serves ten days and receives half of the third, a third of the second and a fourth of the first. What is the value of each sextarius?

You will do the following. Let the value of the third[1442] be one thing; then the value of the second will be a thing and three nummi[1443] and the value of the first, a thing and five nummi. Adding the values of all these sextarii makes three things and eight nummi, which is the wage for the month. Since for ten days' service he receives half of the third, a third of the second and a fourth of the first, add half of the third, which is half a thing, a third of the second, which is a third of a thing and one nummus, and a fourth of the first, which is a fourth of a thing and one nummus and a fourth; this makes altogether one thing and half a sixth of a thing and two nummi and a fourth. This whole sum is the wage for a third of the month. Thrice that produces three things and a fourth of a thing and six nummi and three fourths, which is the wage for the month. This equals three things and eight nummi, which is also the wage for the month.[1444] Subtracting the things from the things and the nummi from the nummi leaves in the end a fourth of a thing equal to one and a fourth. Thus the thing is five. This is the value of the third sextarius; the value of the second is seven nummi[1445], and the value of the first, ten nummi.

AGAIN ON THE SAME TOPIC, BUT OTHERWISE[1446]

[1440] Equation of the form $a_1 x + b_1 = a_2 x + b_2$.

[1441] The MSS have 'sixty-four' (numerals inverted).

[1442] The MSS have: of the first (early reader's attempt to correct the faulty text).

[1443] According to the statement, it should be 'two nummi'.

[1444] Equation of the form $a_1 x + b_1 = a_2 x + b_2$.

[1445] In accordance with the initial statement.

[1446] The time is only partly served.

LIBER MAHAMELETH

(B.230) A man, hired at a monthly wage of ten nummi, receives it if he serves the whole month but if he does not serve at all, not only fails to receive it but must even pay two nummi to the hirer. After serving one part of the month and not the other, he leaves with neither receiving nor paying anything. How many days did he serve and how many did he miss?

(*a*) You will do the following. Add the two nummi, which he would pay, to the ten nummi; the result is twelve, at which someone else would be hired. Next, multiply two by the days in the month and divide the product by twelve; the result will be five days, and so many days did he serve [*while he did not serve any of the other days in the month*]. If you want to know how many days he did not serve,[1447] multiply ten by thirty, which produces three hundred, and divide this by twelve; the result will be twenty-five, and so many are the days not served.

The reason for this is the following. It is known[1448] that the ratio of the days served to the days in the month, which are thirty, is the same as the ratio of the wage for the days served to the wage for the whole month, which is ten. There are therefore four proportional numbers. Then the product of the first, which is what is served, into ten nummi is equal to the product of thirty into the wage for the days served. It is also known that the ratio of the days not served to thirty days is the same as the ratio of what he must pay of the two nummi to two nummi. There are therefore four proportional numbers. Then the product of the days not served into two nummi is equal to the product of thirty into what he must pay. But it is known that what he must receive for what he served is equal to what he must pay for what he did not serve; for it was said that he neither received nor paid anything. So it is clear that the product of thirty into what he must receive is equal to the product of thirty into what he must pay for the days not served. Now it has been shown[1449] that the product of thirty into what he must receive is equal to the product of the days served into the month's wage, which is ten nummi. It has also been shown[1450] that the product of thirty days into what he must pay is equal to the product of the days not served into two nummi. So the product of the days served into ten nummi must be equal to the product of the days not served into two nummi. There are therefore four proportional numbers. So the ratio of the days served to those not served is the same as the ratio of two nummi to ten nummi.[1451] By composition, the ratio of the days served to the sum of those served and those not is the same as the ratio of two nummi to the sum of two and ten. But it is known that those served and those not is the whole month. Thus the ratio of what he served to thirty days, which is the month itself, is the same as the ratio of two [*days*] to twelve, which is two

---

[1447] Direct formula, without knowing the previous result. The interpolator did not grasp that.

[1448] From the constancy of the ratio time : wage.

[1449] Above. (The MSS have *Manifestum est etiam*; see note 1436.)

[1450] Also above. (The MSS have *Manifestum est etiam*.)

[1451] This result will be used in part *d*.

and ten added together. There are therefore four proportional numbers. Then dividing the product of two nummi into thirty by twelve will give the number of days served. The origin of the determination of the days not served is analogous, but the proportion must be changed.

That this is right will appear from the following illustration; wherein let thirty be line AB, what he served of the month, line AG, and (thus) what he did not serve, line GB. Next, I shall draw from the point G a line of two, say GD.[1452] I shall multiply it by line GB to produce the area DB. Next, I shall draw from the point G a line of ten; this is line GH. I shall multiply it by line AG to produce the area AH. This area is then equal to the area DB, since it is known from the foregoing[1453] that the product of what he served into ten nummi is equal to the product of what he did not serve into two nummi. Now, as Euclid said in the sixth book, when two equal areas have two angles between parallel sides equal, the sides containing them will be mutequefia [*that is, coalternate*].[1454] Since the areas AH and BD are equal and their sides parallel, while the angle AGH is equal to the angle DGB, so the sides AG and GB, and DG and GH are mutequefia; that is, the ratio of AG to GB is the same as the ratio of DG to GH. By composition, the ratio of line AG to line AB is the same as the ratio of line DG to line DH. Now DG is a sixth of DH, thus AG is a sixth of AB. But line AB is thirty, thus line AG is five, and so much is what he served of the month. You may want to determine it by multiplication.[1455] In this case, the product of line AG into DH is equal to the product of line DG into AB; thus, dividing the product of line DG into AB by DH gives AG. This is what you wanted, and the figure below illustrates it.

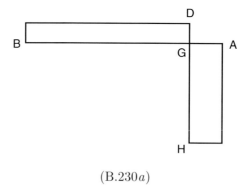

(B.230*a*)

(***b***) Or otherwise, namely dividing thirty into two parts such that multiplying one of them by ten and the other by two produces equal results. You will do the following. Let one part be a thing, namely the part of the

---

[1452] GD (and GH below) drawn perpendicular to AB.
[1453] Reasoning preceding the geometrical illustration.
[1454] *Elements* VI.14 (already applied in B.218).
[1455] To obtain a general form.

LIBER MAHAMELETH
925

month served, and (thus) the other part, thirty minus a thing, namely the part of the month not served. Multiply then a thing by ten, which produces ten things. Next, multiply thirty minus a thing by two, which produces sixty nummi[1456] minus two things; this equals ten things.[1457] Complete the sixty nummi by adding the two things which are lacking, and add two things to ten things; this makes twelve things, which equal sixty nummi. Thus a thing is five, and this is the part of the month served.

The reason for dividing thirty into two parts such that multiplying one of them by ten and the other by two produces equal results is because it is known (on the one hand) that the product of what was served into ten is equal to the product of what was not served into two and, on the other hand, the month being made up of the days served and those not served, that one part of the month was served and the other not.[1458]

(*c*) Or otherwise. Let a thing be what he served of the month. Since it is known that serving the whole month he would receive ten nummi, it is clear that the ratio of thirty days to the corresponding wage, which is ten, is the same as the ratio of the days served, which are a thing, to the corresponding wage. It is then like saying: 'When thirty for ten, what is the price of a thing?' Multiply a thing by ten and divide the product by thirty, as we have explained in the chapter on buying and selling;[1459] the result will be the price of the thing, which is a third of a thing, and so much is what he is to receive for what he served of the month. But it was said above that if he does not serve at all he pays two nummi. Now what he did not serve is thirty minus a thing: since he served a thing of the month, what he did not serve is thirty minus a thing. Thus it is clear that the ratio of the days in the month to the two nummi he pays is the same as the ratio of thirty minus a thing to the corresponding wage[1460]. Then it is like saying: 'When thirty is given for two nummi, what is the price of thirty minus a thing?' You will do the following. Multiply thirty minus a thing by two nummi and divide the product by thirty; the result is two nummi minus two thirds of a tenth of a thing. This is the price of thirty minus a thing, which are the days not served. They are equal to a third of a thing. The reason for their being equal to a third of a thing is because, according to the statement, what he was to receive for serving is equal to what he was to pay for not serving. Proceed then as we have taught in algebra.[1461] That is, complete the two nummi by adding two thirds of a tenth of a thing, and add this same amount to a third of a thing, thus making two fifths of a thing equal to two nummi. Thus a thing equals five, and this is what he served, what he did not serve being the remainder of the month.

[1456] In the sense of 'units' (see note 1120, p. 841).

[1457] Equation of the form $b_1 - a_1 x = a_2 x$.

[1458] These two arguments thus correspond to the two equations.

[1459] B.1 and preceding explanations (p. 780).

[1460] Not 'wage' but penalty.

[1461] Equation of the form $b_1 - a_1 x = a_2 x$, as in part *b*.

926   PART TWO: TRANSLATION, GLOSSARY

(*d*) Or otherwise. Put a thing as what he served of the month. What he did not serve will be five things. Indeed, we have established at the beginning of the problem [1462] that the ratio of what he served to what he did not serve is the same as the ratio of two, which he pays, to ten, which is the wage for the month; now, by inversion, the ratio of ten to two will be the same as the ratio of the days not served to the days served; since ten is obviously the quintuple of two, and we put a thing as what he served of the month, what he did not serve of the month must be five things. Adding then a thing to five things makes six things; this equals thirty. Therefore a thing is five, and this is what he served.

(*e*) If you want to verify that he left without receiving or paying anything, you will do the following. It is known that the wage for the month is ten nummi, while what he pays for the month is two nummi. On the other hand it is known that he served a sixth of the month; therefore he must receive a sixth of the wage, which is a nummus and two thirds. But what he did not serve is five sixths; therefore he must pay five sixths of two nummi, which is a nummus and two thirds. Thus he is to receive a nummus and two thirds and to pay a nummus and two thirds. These having paid for the others, he leaves exempted, without receiving or paying anything.

(**B.231**) Now someone asks:[1463] A man is hired for a month at ten nummi if he serves, and if he does not someone else is hired in his stead at eight nummi. He serves some days, misses the others, and leaves without receiving or paying anything.

This problem is not solvable.[1464] For if he served no part whatsoever of the month, he would be left with two nummi; and he must be left with more since he served some of the month. Such a problem will therefore not be solvable unless he is left with more than the difference between the two wages.

(**B.231′**) This is the case if someone asks: A man is hired for a month at ten nummi if he serves, otherwise someone else is hired in his stead at eight nummi. He serves one part and not the other and leaves with three nummi.

It appears that if he served nothing of the month he would leave with two nummi; since he leaves with three nummi, he served exactly what corresponds to one nummus according to his monthly wage of eight nummi. Thus the ratio of one to eight is the same as the ratio of the days served to thirty. Proceed as explained above.[1465] The result will be that he served three (days) and three fourths. It is further clear, since of his ten nummi he is left with only three, that someone else is hired in his stead who receives seven nummi according to the monthly wage of eight nummi. Thus the

---

[1462] In part *a* (note 1451).

[1463] In fact he does not ask anything. No question in the next two problems either.

[1464] *questio falsa*. Rather, as in B.226, absurd.

[1465] Plain proportion.

LIBER MAHAMELETH

927

ratio of seven to eight is the same as the ratio of the days not served to thirty. Thus the days not served are twenty-six and a fourth.

Consider other cases of this kind accordingly, and you will find that this is the way.

AGAIN ON THE SAME TOPIC.

(**B.232**) A man hired at ten nummi to serve the whole month, otherwise someone else is hired in his stead at twelve, serves some days, misses the others and receives one nummus.

(*a*) You will do the following. Subtract one nummus from the ten, which leaves nine. Next, multiply the difference between nine and twelve, namely three, by thirty, and divide the product by twelve; the result will be seven and a half, and these are the days served, those not served being twenty-two and a half.

The reason for this is that which we have explained in the foregoing, namely that the product of what he served of the month into ten nummi is equal to the product of what he did not serve of the month into two nummi; for the product of thirty into the wage for what he served is equal to the product of thirty into what he must pay, as said above, at the beginning of the problem preceding this one.[1466] Now it has been stated that the person hired left with one nummus. The meaning of these words is that the wage for the days served is greater by one nummus than what he must pay for the days not served. Then it is clear that the product of thirty into what he must receive of the ten nummi is greater than the product of thirty into what he must pay of the two nummi by as much as the product of one nummus into thirty. It is also clear that the product of thirty into what is due to him of the ten nummi is equal to the product of what he served of thirty into ten; thus the product of what he served into ten is greater than the product of thirty into what he must pay by the amount of the product of thirty into one nummus. But it is known[1467] that the product of what he did not serve into two nummi is equal to the product of thirty into what he must pay of the two nummi. Therefore it is clear that the product of what he did not serve of the month into two, plus the product of thirty into one, is equal to the product of what he served of the month into ten. I shall thus take a line of thirty, say line AB. Let what he served be line AG and (thus) what he did not serve, line GB. So the product of line AG into ten is equal to the product of line GB into two, plus the product of line AB into one. Now the product of AB into one is equal to the products of AG into one and GB into one, for the product of the whole line AB into any number is equal to the products of all its parts into this number.[1468] So the product of AG into ten is equal to the product of GB

[1466] The 'beginning of the problem preceding this one' is B.230, part *a* (B.231–231′ are particular cases, probably added later by the author). The difference with the present problem will now be explained.

[1467] B.230*a*.

[1468] *Elements* II.1 = Premiss PE_1.

into three plus the product of AG into one. Subtracting then the product of AG into one from the product of AG into ten leaves the product of AG into nine, which equals the product of GB into three. Thus it is clear that line AB, which is thirty, is divided into two unequal parts at the point G and that the product of AG into nine is equal to the product of GB into three.[1469] I shall draw from the point G a line of nine, which is line GD, and a line of three, which is line GH.[1470] Line AG when multiplied by line GD will produce the area AD; again, line GH when multiplied by line GB will produce the area HB. Thus the area AD is equal to the area HB. It is therefore clear that the ratio of line HG to GD is the same as the ratio of line AG to GB, as Euclid said in the sixth book[1471]. By composition, the ratio of line HG to line HD will be the same as the ratio of line AG to line AB. These four numbers are therefore proportional. Then dividing the product of line HG, which is the first, into AB, which is the fourth, by HD, which is the second, will give the third, which is AG. This is what he served of the month, what he did not serve being line GB. This is what we intended to demonstrate.

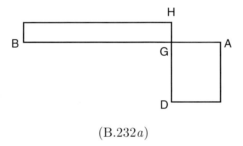

(B.232a)

(*b*) Or otherwise. Let what he served of the month be a thing; thus what he did not serve will be thirty minus a thing. Next, multiply a thing by nine, thus producing nine things. Then multiply thirty minus a thing by three, thus producing ninety minus three things. This equals nine things.[1472] Complete then ninety by adding three things, and add the same to nine things. This will make twelve things, which equal ninety. A thing therefore equals seven and a half. This is how much of the month he served, the remainder of the month being what he did not serve.

(*c*) Or otherwise. Denominate the difference between nine and twelve, namely three, from twelve; this is a fourth. Such a fraction of thirty, which is seven and a half, is what he served of the month.

(*d*) Or otherwise. The person hired served of the month a thing and three days; for he received for what he served one nummus, and this nummus, in accordance with the wage for a whole month's service, is due to him for

---

[1469] Sum and ratio of AG and GB thus known.
[1470] Both perpendicular to AB.
[1471] *Elements* VI.14 (the sides being 'mutequefia', see B.218 & B.230).
[1472] Equation of the form $b_1 - a_1 x = a_2 x$

LIBER MAHAMELETH   929

three days' service. Now what he did not serve is five things; for the wage for what he served is five times what he will pay for what he did not serve, as we have established [1473]. Thus, adding together a thing, five things and three days makes six things and three days; this equals thirty. Proceed as we have taught before [1474], and the thing will be four and a half. Now it is known that he served of the month a thing and three days. Thus he served of the month seven days and a half, and he missed twenty-two and a half.

(**e**) If you want to verify this, that is, ascertain that he indeed left with one nummus, you will do the following. It is already known that he served a fourth of the month; so he must receive a fourth of the wage, thus of ten, which is two and a half. Since he missed three fourths of the month, he must pay three fourths of two nummi, which is one nummus and a half. Thus he must pay one nummus and a half and receive two nummi and a half. Having paid his due, he is left with one nummus only.

AGAIN ON THE SAME TOPIC.

(**B.233**) A man is hired for a month at ten nummi and, if he does not serve, another is hired in his stead at twelve. He served only part and paid one nummus. How much did he serve and how much not?

(**a**) You will do the following. Add this nummus to the ten nummi, making eleven. Next, multiply the difference between eleven and twelve, which is one, by thirty and divide the product by twelve; the result will be two and a half. So much is what he served.

The reason for this is as explained in the foregoing,[1475] but you will add the nummus, which he paid, to the ten nummi; for what is due to him of the ten nummi is one nummus less than what he must pay of the two nummi; indeed, since by hypothesis the man hired pays one nummus, what he pays must be one nummus more than what he should receive. Proceed therefore as we have taught above [1476] and, having set the proportion, you will have a line of thirty divided into two unequal parts such that the product of one of them into eleven is equal to the product of the other into one.[1477] Then what he served of the month is two and a half, and what he failed to serve is twenty-seven and a half.

(**b**) The treatment of this problem by algebra is carried out in the same way as in the previous problem.[1478]

(**c**) The verification here is as follows. Since he served half a sixth of the month, he must receive half a sixth of ten nummi, which is five sixths of one nummus. But he missed five and a half sixths of the month, so he

---

[1473] B.230d.

[1474] Equation of the form $a_1 x + b_1 = b_2$.

[1475] B.232a.

[1476] B.232a.

[1477] Sum and ratio of the two parts thus known.

[1478] B.232b.

930 PART TWO: TRANSLATION, GLOSSARY

must pay five and a half sixths of two nummi, which is one nummus and five sixths. Thus he served and paid one nummus.

(**d**) Or otherwise. Subtract one from the difference between ten and twelve, thus leaving one. Denominate it from twelve; this gives half a sixth. So much is what he served of the month, that is, two days and a half, the remainder of the month being what he did not serve.

AGAIN ON THE SAME TOPIC.

(**B.234**) [1479] A man, hired for ten days, is to receive three nummi each day served and for each day not served he must pay five nummi. He serves only part of the ten days and leaves without receiving or owing anything.

(**a**) You will do the following. Add five to three, making eight, which you put as the principal. If you want to know how much he served, multiply five by ten and divide the product by the principal; the result is what he served. If you want to know how much he did not serve, multiply three by ten and divide the product by the principal; the result is what he did not serve.

(**b**) Or otherwise. It is clear that since he is hired for ten days at three nummi a day, he will earn for ten days' service thirty nummi; if, however, he does not serve at all he must pay fifty. Thus (on the one hand) to him are due thirty, (on the other) he himself owes fifty. Adding together thirty and fifty makes eighty. Then it is like asking: 'A man, hired for ten days, is to receive thirty nummi if he serves but if not another is hired in his stead at eighty; he himself serves in part only, and leaves without receiving or paying anything.' Do as we have taught above [1480] and the result will be what you wanted to know.

AGAIN ON THE SAME TOPIC. [1481]

(**B.235**) A man, hired for a month, is to receive a thing if he serves, but if not another is hired in his stead at a thing and two nummi. He serves in part only and leaves without receiving or paying anything, and the number of days he serves multiplied by his (month's) wage produces fifty. [1482]

You will do the following. Divide fifty by the two nummi; the result is what he did not serve of the month. Subtract it from the (number of days in the) month; this will leave what he served of the month. If now you want to know the value of the thing, denominate [1483] the days not served from those served, and this fraction of two nummi [1484] will be the

---

[1479] This is misplaced and should follow B.236.

[1480] B.230.

[1481] Perhaps 'with things' should have been added (*Item de eodem cum rebus*), as seen elsewhere (pp. 792, 896). In any event, B.235 should follow B.233.

[1482] Required what he served or did not serve and the value of the thing.

[1483] Here: divide.

[1484] Accordingly, it should be 'that many times two nummi'.

value of the thing. [*The proof of this is clear to those who reflect upon the problem.*][1485]

(B.235′) Suppose that, the problem remaining the same, it were said that the man hired left with one nummus.

You will do the following. Add the one nummus to the two, which makes three. Multiply this by thirty, thus producing ninety. Subtract from it fifty, which leaves forty. Divide this by the two nummi; the result is what he served, namely twenty days, and what he did not serve is ten days. If now you want to know the value of the thing, denominate what he did not serve from what he served; add to this fraction of three nummi, which is the sum of two and one[1486], thus one nummus and a half, one nummus; the result is two and a half, and so much is the thing worth.

The proof of this is the following. Let the thing be line DT and the two nummi, line TN; let what he served of the month be line AG, what he did not serve, line GB,[1487] the wage for what he served, line KZ, what he paid for what he did not serve, line ML, and let line ML be one nummus less than line KZ. Next, add one nummus to line TN, say line TH. Thus[1488] the product of line AG into DH is equal to the product of GB into HN. Let the product of AG into HN be (added in) common. Thus the product of HN into AB will be equal to the product of AG into DN. But the product of line AB into HN is ninety. Therefore the product of line AG into DN is ninety. Since the product of line AG into DT is fifty, the remainder is forty. But it results from multiplying line AG by TN which is two nummi; so dividing forty by two gives as line AG twenty. This is what he served; it leaves as line GB ten, which is what he did not serve. You may now want to know the value of the thing. It is already known[1489] that the ratio of line GB to AG is the same as the ratio of line DH to HN. But line GB is half of line AG. Thus line DH is one and a half; for it is half of line HN. Since line HT is one, line DT is two and a half, and such is the value of the thing. This is what we wanted to prove.

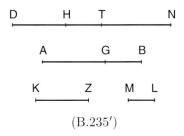

(B.235′)

[1485] The proof can be inferred from that in B.235′.
[1486] The sum calculated above.
[1487] Thus AB is the month.
[1488] B.232a.
[1489] Beginning of the proof.

(B.235″) If now it is said that the man hired pays one nummus and that the multiplication of what he serves by his (monthly) wage produces fifty nummi, the problem will not be solvable. [*The reason for this is clear to those who reflect upon this figure.*][1490]

Consider other (such) problems accordingly, except if it is said that the man hired serves a number of days which, when multiplied by his wage produces ⟨more than thirty, which is indeed not solvable. But if the result were less than thirty, the problem would be solvable.

(B.235‴) For instance, if it is said that the man hired pays one nummus and that the multiplication of what he serves by his wage produces⟩ twenty-five, then the problem will be solvable. [*The reason why is clear.*][1491] Here you will do the following. Multiply one by thirty and add twenty-five to the product; this makes fifty-five. Divide this by the two nummi; the result is twenty-seven and a half. This is what he did not serve of the month, and the remainder of the month, that is, two days and a half, is what he served. Now if you want to know the value of the thing, divide twenty-five by what he served of the month; the result is ten, and this is what the thing is worth.

AGAIN ON THE SAME TOPIC, BUT OTHERWISE.[1492]

(B.236) Someone asks: A man hired for twenty days is to receive three nummi each day if he serves and if not will pay the hirer two nummi. He served in part only and left without receiving or paying anything; how much did he serve and how much not?

The meaning of this problem is the following. For a day served he will receive three nummi but for a day not served he will not only fail to receive these three but even pay two. He served one part of the month and the other not and left without receiving or paying anything; that is, what he must pay for the days not served is equal to what he must receive for the days served.

It is clear, then, that twenty is divided into two parts such that the product of one of them, namely the days served, into three is equal to the product of the other, namely the days not served, into two.[1493] Let therefore the twenty days be AB, the days served, AG, the days not served, GB, the three nummi, DH, and the two nummi, HZ. Thus the product of DH into AG is equal to the product of HZ into GB. Therefore the ratio of DH to HZ is the same as the ratio of GB to AG. By composition, the ratio of the whole DZ to HZ will be the same as the ratio of the whole AB to AG. Thus the ratio of DZ, which is five, to HZ, which is two, is the same as the ratio of AB, which is twenty, to the days served, which is

---

[1490] Same interpolator as in B.235.

[1491] This is nonsense: the lacuna makes the text meaningless.

[1492] This heading (and no doubt B.236, where the meaning of this new kind of problem is explained) should appear before B.234.

[1493] Sum and ratio of the two parts thus known.

AG. Therefore the product of AG into DZ is equal to the product of HZ into AB. Dividing then by DZ the product of HZ into AB will give as AG eight, which is the days served. It is also clear that, by conversion, the ratio of DZ to DH is the same as the ratio of AB to GB [*which is the days not served*]. Thus multiplying DH by AB and dividing the product by DZ will give GB, which is the days not served.

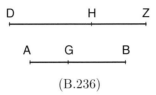

(B.236)

Because of this the treatment is the following. You add two and three, making five, which you put as the principal. If you want to know the days not served, multiply three, which is the wage for one day, by twenty and divide the product by the principal; the result will be what you wanted. If you want to know the days served, multiply two, which he pays for each day not served, by twenty, and divide the product by the principal. The results will be eight for the days served and twelve for the days not served. This is what we wanted to demonstrate.[1494]

(B.237) Someone asks thus: A man, hired for thirty days, is to receive for each day served three nummi but for each day not served he is to pay two. Having served in part only, he leaves with ten nummi.

You will do the following. Add two and three, making five, which you put as the principal. If you want to know the number of days served, multiply what he must pay for not serving, namely two, by thirty days; the result is sixty. Adding the ten nummi makes seventy. Divide this by the principal; the result is fourteen, and so many days did he serve. If you want to know the number of days not served, multiply what he receives for each day served, namely three, by thirty; the result is ninety. Subtracting the ten nummi leaves eighty. Divide this by the principal; the result is sixteen, and so many days did he not serve.

The proof of this is the following. Let the thirty days be AB, the days served, AG, the days not served, GB. It is known that since he earns ten nummi, the product of the days served into three will be greater by ten than the product of the days not served into two. If you want to know the days served, put in common the product of the days served into two. Then the products of AG, which is the days served, into three and into two will be greater by ten than the products of the days not served, which is GB, into two and AG into two. But we know that the products of AG into two and into three are equal to the product of AG into five. Thus the product of AG into five is greater by ten than the products of two into AG and into GB. But the products of two into AG and into GB are equal to

---

[1494] This should appear at the end of the demonstration.

the product of two into the whole AB. Thus the product of five into AG, which is the days served, is greater by ten than the product of two into AB, namely sixty. Therefore the product of AG into five is seventy. Thus AG is fourteen, and this is the days served. If you want to know the days not served, you will do the same, that is, the following. Since the product of three into AG is greater by ten than the product of two into GB, this time put in common the product of GB, which is what is required, into three. Then the products of AG into three and GB into three are greater by ten than the products of GB into two and into three. But the products of GB into two and into three are equal to the product of GB into five, while the products of AG and GB into three are equal to the product of the whole AB into three. Therefore the product of GB into five is less by ten than the product of AB into three, which is ninety. Subtracting ten from ninety leaves, as the product of five into BG, eighty. Thus GB is sixteen. This is what we wanted to demonstrate.

$$\text{A} \qquad \text{G} \qquad\qquad \text{B}$$
$$\vdash\!\!\!-\!\!\!-\!\!\!-\!\!\!-\!\!\!-\!\!\!-\!\!\!+\!\!\!-\!\!\!-\!\!\!-\!\!\!-\!\!\!-\!\!\!-\!\!\!-\!\!\!-\!\!\!\dashv$$

(B.237)

(**B.238**) Someone asks: Hired for thirty days, he is to receive for each day served three nummi and for each day not served he will pay two. He served in part only and paid ten nummi. How much did he serve and how much did he not?

You will do the following. Add the two nummi to the three, making five, which you put as the principal. If you want to know how much he served, multiply two, which he must pay for each day not served, by thirty, thus producing sixty. Subtract the ten nummi, which leaves fifty. Divide this by the principal; the result is ten, and so many days did he serve. If you want to know how many days he did not serve, multiply three, which he received for each day served, by thirty, thus producing ninety. Add ten, making a hundred. Divide this by the principal; the result is twenty, and this is the number of days not served.

The proof of this is clear. Indeed, let thirty days be AB, the days served, AG, those not, GB. Then the product of GB into two exceeds the product of AG into three by ten. If you want to know how much is AG, which is the number of days served, put in common the product of AG into two. Then the products of AG into three and into two are less by ten than the products of AG and GB into two. But the products of AG into three and into two are equal to the product of AG into five, while the products of AG and BG into two are equal to the product of AB into two. Therefore the product of AB into two, which is sixty, exceeds the product of AG into five by ten. Thus the product of AG into five is fifty; then AG is ten, and this is the number of days served. If you want to know how much he did not serve, put in common the product of what he did not serve, which is BG, into three. Continue with the rest as we have taught above, and the result will be what you wanted.

Treat other problems of the same kind like the present one, and you will find that this is the way.

$$A \qquad G \qquad\qquad B$$

(B.238)

You are to know that in order to relate the problems of this chapter to those of the preceding, you will have to do the following.[1495]

(**B.239**) For instance, someone asks: A man, hired for twenty days, is to receive for each day served two nummi and for each day not served is to pay one nummus and a half. He served in part only and did not receive or pay anything.

It is known that if the man hired at two nummi a day does not serve he will pay one nummus and a half. Thus it is as if he had received two nummi for each day, which he keeps if he serves, but if he does not another is hired in his stead at his expense for three nummi and a half: namely the two he had received and the one and a half which he has to pay. If then the other is hired in his stead for three and a half and he himself serves none of the twenty days, this other man will be hired in his stead at seventy nummi for the twenty days, whereas if he himself serves the whole twenty days he will have forty nummi in accordance with his being hired at two nummi a day. Then it is like saying: 'A man is hired at forty nummi for twenty days if he serves the whole time, but if he serves none another is hired in his stead at seventy nummi; he served in part only and did not receive or pay anything.' Do as has been shown above[1496], and the result will be what you wanted.

Similarly also if it is said that he leaves with a gain of one nummus or a loss of one nummus. Do as has been explained, and the result will be what you wanted.

Again on the same topic, but otherwise[1497]

(**B.240**) Three workers are hired, the first for thirty days at three nummi, the second for thirty days at five nummi, the third for thirty days at six nummi. They complete one month between them and receive equal wages. How much did each of them serve?

(*a*) You will do the following. Divide thirty days by three in order to know the number of days to be served for one nummus; this is ten. Next, divide likewise thirty days by six and by five in order to know the number of days to be served for one nummus; the result is five days for one of the six nummi and six days for one of the five nummi. Then it is like saying: 'Three

---

[1495] Thus relating (*adaptare*) the problems on daily wages (B.234, B.236–238) with those on monthly wages (B.230–233 & B.235).

[1496] B.230.

[1497] Case of three workers.

partners, one with a capital of ten, the second with six, the third with five, have gained by trading thirty; how will they share it among them according to the capital of each?' Do as we have taught before.[1498] For the capital of ten the result is fourteen and two sevenths, and this is the number of days served by the man hired for a month at three nummi. For the capital of five[1499] the result is seven and a seventh, and this is the number of days served by the man hired for a month at five. For the capital of six[1500] the result is eight and four sevenths, and this is the number of days served by the man hired for a month at six nummi.

You may want to verify that they received equal wages. You will say: 'A man, hired at a monthly wage of three nummi, serves fourteen days and two sevenths; what is his wage?' You will do as I have taught before;[1501] the result will be one nummus and three sevenths, and this is the wage of the man serving fourteen days and two sevenths. Next, you will say: 'Hired at a monthly wage of five nummi, he served eight days and four sevenths; what is due to him?' You will do as I have taught before; what is due to him will be found to be a nummus and three sevenths. Next, you will say: 'Hired at a monthly wage of six nummi, he served seven days and a seventh; what is due to him?' Do as I have taught above, and what is due to him will be found to be a nummus and three sevenths. Thus it is clear that they received equal amounts.

(**b**) Or otherwise. Look for a number divisible by three, five and six; you will find sixty[1502]. Dividing it by three gives twenty, by five, twelve, likewise by six, ten. You will then state (the problem) in this way: 'Three partners, one with a capital of twenty, another with twelve, and a third with ten, have between them gained thirty nummi by trading; how will they share this amount out according to the capital of each?' Do as I have taught above[1503], and the result will be what you wanted.

(**c**) Or otherwise. Let the number of days served by the man hired at three nummi be a thing, those served by the man hired at six, half a thing, since three is half of six, and those served by the man hired at five, three fifths of a thing. Next, add these things; this will make two things and a tenth of a thing, which are equal to thirty. Thus a thing equals fourteen and two sevenths, and so much does the man hired at three serve; the man hired at six serves half (of it), namely seven days and a seventh; the man hired at

---

[1498] B.104.

[1499] The MSS have 'six'. This confusion between the number of *days* served for one nummus and the number of *nummi* for the wage (recurring below) certainly originates with a reader since the subsequent verification has the correct results.

[1500] The MSS have 'five'.

[1501] B.195.

[1502] The author does not necessarily take the smallest (nor does he in B.319c; but see B.45a, B.118a, B.241, B.321c).

[1503] B.104 again.

LIBER MAHAMELETH

five serves three fifths of fourteen and two sevenths, namely eight and four sevenths. The wage of each of them for what he served is one nummus and four tenths and two sevenths of one tenth —which are three sevenths.

(*d*) Or otherwise. Since they had equal wages, put as the wage for what each man served one thing. Then consider which part of the month the man hired at three must serve for a thing. Then it is like saying: 'When thirty for three, how much shall I have for one thing?' You will find ten things. Next, consider which part of the month the man hired at six must serve for a thing. You will find five things. Consider also which part of the month the man hired at five must serve for a thing. You will find six things. So what they served between them all is, altogether, twenty-one things; this is equal to thirty. Thus a thing equals one and three sevenths, and this is the wage for what each man served of the month. You may want to know how much each man served of the month. You already know what the wage of each is. You will say: 'When thirty for three nummi, how much shall I have for one nummus and three sevenths?' Do as I have taught above, in the chapter on buying and selling [1504]; the result will be fourteen days and two sevenths, and so much of the month did the man hired at three nummi serve. Next, you will say: 'When thirty days for five nummi, how much shall I serve [1505] for a nummus and three sevenths?' Do as I have taught above; the result will be eight days and four sevenths, and so much of the month did the man hired at five serve. You will do similarly for six; the result will be what the man hired at six has served of the month, namely seven days and a seventh. This is what you wanted to know.

(**B.241**) Someone asks: Three workers are hired for a month, the first at four nummi, the second at six, the third at twelve. None of them completed the month himself but between them all they did, and they left with an equal wage. How much of the month did each one serve? [1506]

You will do the following. Look for a small number divisible by all these numbers; you will find twelve. Dividing it by twelve gives one, by six, two, by four, three. Add one, two and three; this makes six, which you put as the principal. If you want to know how many days the man hired at twelve served, multiply one, resulting from the division of twelve by itself, by thirty and divide the product by the principal; the result will be five, and that many days did the man hired at twelve serve. If you want to know how many days the man hired at six served, multiply two, resulting from the division of twelve by six, by thirty and divide the product by the principal; the result will be ten, and that many days did the man hired at six serve. If you want to know how many days the man hired at four served, multiply three [1507] by thirty and divide the product by the

---

[1504] B.2.

[1505] Not 'have' as before since he mentioned thirty *days*.

[1506] The equal wage is also required.

[1507] Its origin, unlike in the two previous cases, is not indicated.

principal; the result will be fifteen, and that many days did the man hired at four serve. If you want to know with what wage each one left, divide twelve by the principal; the result will be two, and so much did each one receive for his service.

The proof for all this is the following. Let thirty days be AB, four nummi, QL, what the man hired at four served, AG, six, LM, what the man hired at six served, GD, twelve, MN, what the man hired at twelve served, DB. It is then known that the ratio of AG to AB is the same as the ratio of each one's wage to QL, that the ratio of GD to AB is the same as the ratio of each one's actual wage to LM, likewise also that the ratio of DB to AB is the same as the ratio of each one's actual wage to MN. Since the ratio of AG to AB is the same as the ratio of each one's wage to QL, the product of QL into AG is equal to the product of the wage into AB; thus multiplying QL by AG and dividing the product by AB will give the wage. Likewise also, multiplying LM by GD and dividing the product by AB will give the same wage. Likewise also, multiplying MN by DB and dividing the product by AB will give the same wage. Consequently, the product of AG into QL must be equal to the product of GD into LM and to the product of DB into MN. This being so, the ratio of QL to LM is the same as the ratio of GD to AG, and the ratio of LM to MN is the same as the ratio of DB to GD. Next, I must look for three numbers such that the product of the first into QL is equal to the product of the second into LM and to the product of the third into MN. For this (purpose), take any number; divide it by QL to give HZ, by LM to give ZK, by MN to give KT. Therefore the product of HZ into QL will be this number, so also the product of ZK into LM, so again the product of MN into KT, whereby the product of HZ into QL is equal to the product of ZK into LM and to the product of KT into MN. Therefore the ratio of ZK to KT is the same as the ratio of MN to ML. But the ratio of MN to ML is the same as the ratio of GD to DB. Thus the ratio of GD to DB is the same as the ratio of ZK to KT. I shall likewise demonstrate that the ratio of HZ to ZK is the same as the ratio of AG to GD.

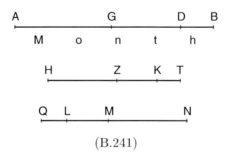

(B.241)

The reason we sought a small number divisible by four, six and twelve is to obtain from the division only integral numbers and no fractions, for it is thus easier; if we had found a number other than twelve, it would also

LIBER MAHAMELETH

be appropriate. Let us take then, say, sixty and let it be C[1508]. Then HZ will be fifteen, ZK, ten, KT, five, and the whole HT will be thirty.

We have already shown that the ratio of ZH to ZK is the same as the ratio of AG ⟨to GD; by composition, the ratio of ZH to HK will be the same as the ratio of AG to AD. Now we know that the ratio of ZK to KT is the same as the ratio of GD to DB and that the ratio of HZ to ZK is the same as the ratio of AG to GD; therefore, by equal proportionality, the ratio of HZ to KT will be the same as the ratio of AG to DB; by inversion, the ratio of TK to ZH will be the same as the ratio of BD to AG. But you know that the ratio of HK to HZ is the same as the ratio of DA to AG. So it is clear from the premisses[1509] that the ratio of the whole HT to ZH is the same as the ratio of the whole AB to AG. Then, by inversion, the ratio of HZ to the whole HT will be the same as the ratio of AG⟩ to the whole AB; thus multiplying HZ by AB and dividing the product by HT will give AG. It will likewise again be shown that the ratio of ZK to HT is the same as the ratio of GD to AB[1510]; thus, multiplying ZK by AB and dividing the product by HT will give GD. It will likewise again be shown that the ratio of KT to HT is the same as the ratio of DB to AB; thus multiplying KT by AB and dividing the product by HT will give DB. Therefore DB is five, GD, ten, and AG, fifteen. Taking a number other than sixty and proceeding as has been said would also lead to what you want.

I say also that dividing by the sum of the results of all the divisions the number found will give each one's wage. Let this number be sixty; then the sum of the results of the divisions will be thirty. Then I say that dividing sixty by thirty gives two, this being what each of them receives. The proof of this is the following. We know that the ratio of one of the numbers resulting from the division of sixty by one of the wages to the sum of all these results is the same as the ratio of each one's wage to the wage by which sixty is divided. This is proved as follows.[1511] Let one of these numbers be ZK. Then the ratio of ZK to HT will be the same as the ratio of GD to AB[1512]. But the ratio of GD to AB is the same as the ratio of the wage to LM. Therefore the ratio of the wage to LM is the same as the ratio of ZK to HT.[1513] Thus the product of ZK into LM is equal to the product of the wage into HT. But the product of ZK into LM is sixty. Then dividing sixty by HT will give the wage. This is what we intended to demonstrate.

AGAIN ON THE SAME TOPIC.

[1508] Not in the figure. The values calculated below correspond to this choice.

[1509] Premiss P'_7.

[1510] This will also be used below.

[1511] Proof of the previous assertion.

[1512] See note 1510.

[1513] Intermediate assertion thus proved.

(B.242) Someone asks: Three workers are hired for a month, one at six nummi, the other at five, the third at three, and they complete between them the month. The man hired at six nummi receives a wage greater by one nummus than the wage of the man hired at five, who himself receives a wage greater by one nummus than the wage of the man hired at three. How many days does each one of them serve?

(*a*) You will do the following. Divide the thirty days in the month by six nummi in order to determine how many days (the first) is to serve for one nummus; the result is five days. Doubling this gives ten, and so many days must he serve for two nummi. Next, divide thirty days by three and by five in order to determine how many days (each of the others) must serve for one nummus; dividing by five gives six days, by three, ten days. Now it was said above that the wage of the man hired at six is greater by one nummus than the wage of the man hired at five, which is itself greater by one nummus than the wage of the man hired at three; consequently, the wage of the man hired at six is greater by two nummi than the wage of the man hired at three. It was also said above that for these two nummi the (first's) service is ten days, while for the nummus the second man has more than the third his service is six days; so the man hired at six served ten days and the man hired at five, six days, and all three must between them serve the remainder of the month and receive an equal wage. Subtracting these sixteen days from thirty days leaves fourteen, and so many days did they serve between them so as to receive the same wage. Then it is like saying: 'Three workers are hired for a month, one at six nummi, another at five, a third at three; between them, they served fourteen days and received the same wage; how much did each of them serve?' Proceed as has been shown in what precedes [1514], thus as follows. Divide thirty by six, five and three, add the quotients and put the sum, namely twenty-one, as the principal. Next, multiply five, the result of the division by six, by fourteen and divide the product by the principal; the result will be three and a third, and so many days of the fourteen did the man hired at six serve. Add then the ten days found before to the three days and a third; this makes thirteen and a third, and so much of the thirty days did the man hired at six serve. You will then proceed [*here*] as I have taught above.[1515] The result will be that the man hired at five nummi served four of the fourteen days; adding this four to the six days found before makes ten, and so many days did he serve of the month. The man hired at three served, of the fourteen days, six and two thirds, and so many days did he also serve of the month.

(*b*) Or otherwise. Put a thing as what the man hired at six nummi served. Then what the man hired at five served will be a thing and a fifth of a thing minus six days. Indeed, if the wage for the service of the man hired for the month at six nummi were not greater by one nummus than the wage for the service of the man hired at five, the man hired at five would serve the same number of days as the man hired at six plus a fifth of it; thus

---

[1514] Last two problems.

[1515] In this same part *a*, as an early reader makes clearer.

LIBER MAHAMELETH 941

if we put a thing as the number of days served by the man hired at six, those served by the man hired at five are a thing and a fifth of a thing, and their wages are equal; but we know that the wage for the service of the man hired at six is greater by one nummus than the wage for the service of the man hired at five; consequently, the number of days the man hired at five served is a thing and a fifth of a thing minus six days: the nummus by which the wage of the first exceeds the wage of the second corresponds to six days' service by the man hired at five, the wages for what both served of the month being then equal.[1516] In the same manner it will also be shown that the number of days served by the man hired at three nummi is two things minus twenty days. Thus the entire service is four things and a fifth of a thing minus twenty-six days, and this is equal to thirty days. Proceed then as has been shown before in algebra[1517]. The thing will be thirteen and a third, and so many days did the man hired at six serve. The man hired at five serves as much and a fifth of it minus six days, which is ten days, and the third, twice as much minus twenty days, which is six days and two thirds. The man hired at six will have, as the wage for thirteen days and a third, two nummi and two thirds; the man hired at five will have, for his days' service, a nummus and two thirds; the man hired at three will have two thirds of a nummus.

(*c*) Or otherwise. Put, as the number of days served by the man hired at six, half a thing; the number of days served by the man hired at five will be three fifths of a thing minus six days and the number of days served by the man hired at three, a thing minus twenty days. Adding the various services makes two things and a tenth of a thing minus twenty-six days, which are equal to thirty days. Proceed then as has been shown before in algebra[1518]. The thing will be twenty-six and two thirds. The man hired at three nummi served this number minus twenty days, which is six days and two thirds; the man hired at five served three fifths of this same number minus six days, which is ten days; the man hired at six served half this same number, which is thirteen days and a third.

(*d*) Or otherwise. Put, as the wage due for the service of the man hired at three, a thing; the wage due for the service of the man hired at five will be a thing and one nummus, and the wage due for the service of the man hired at six, one thing and two nummi. Next, consider how many days the man hired at six must serve for a thing and two nummi; you will find that it is five things and ten days. Consider also how many days the man hired at five must serve for a thing and a nummus; you will find six things and six days. Next, consider how many days the man hired at three must serve for a thing; you will find ten things. So what the three workers served is altogether twenty-one things and sixteen days, which equal thirty days. Proceed then as has been shown before in algebra[1519]. The result will be

---

[1516] That is, if the second works six more days.

[1517] Equation of the form $a_1 x - b_1 = b_2$.

[1518] Again, equation of the form $a_1 x - b_1 = b_2$.

[1519] Equation of the form $a_1 x + b_1 = b_2$.

942 PART TWO: TRANSLATION, GLOSSARY

the wage for the service of the man hired at three nummi, namely two thirds of a nummus; the wage for the service of the man hired at five will be a nummus and two thirds, and the wage of the last, two nummi and two thirds. If you want to know how many days each of them served, you will say: 'A worker is hired at a monthly wage of six nummi; how many days will he serve for two nummi and two thirds?' You will do the following. Multiply two nummi and two thirds by thirty and divide the product by six; the result is thirteen and a third, and so many days did the man hired at six serve. Next, you will say: 'A man is hired at a monthly wage of five nummi; how many days will he serve for a nummus and two thirds?' Multiply a nummus and two thirds by thirty and divide the product by five; the result is ten, and so many days did the man hired at five serve. You will then say again: 'A man is hired at a monthly wage of three nummi; how many days will he serve for two thirds of a nummus?' Multiply two thirds by thirty and divide the product by three; the result is six days and two thirds, and so many days did the man hired at three nummi serve. This is what we wanted to find.

(B.243) Someone says: Three workers are hired for a month, one of them at a thing, the second at half a thing, the third at a third of a thing. None of them completed the month, but between them they did, and each of them left with two nummi. What is the thing worth and how much did each of them serve?

You will do the following. We know that dividing any number by a thing, half a thing, and a third of a thing, and multiplying the sum of the results by two will produce the dividend.[1520]   Let this number be a thing. Dividing a thing by a thing gives one, by half a thing, two, and by a third of a thing, three. Adding these quotients makes six. Multiplying this by two, which is the wage with which each of the workers left, produces twelve. This is the thing. Continue with the rest of the problem as we have taught above,[1521] and the result will be what you wanted.

[1520] See B.241. But the wording is unfortunate: this is not an identity, but the equation to solve.

[1521] Computation of the days worked as in B.242d.

Chapter (B-XII) on the Diversity of workers' wages

(**B.244**) Someone asks: Ten workers are hired, but one receives less than all others, namely three nummi, and another more than all others, namely twenty-one, and there is an equal difference between the wages.[1522] What is the common difference and how many nummi are there (altogether)?

To know the total quantity of nummi, you will do the following. Add the lowest wage[1523], which is three, to the highest wage, which is twenty-one; this makes twenty-four. Multiply this by half the number of men, namely five; this will produce a hundred and twenty, which is the total quantity of nummi. If you want to know the common difference between the wages, you will do the following. Subtract from the highest wage, which is twenty-one, the lowest wage, which is three; this leaves eighteen. Always divide it by the number of men less one, in this case by nine; the result is two, and this is the quantity by which the wages exceed one another.

This is proved as follows. It is clear[1524] that the sum of the wages of the first and last men is equal to the sum of the wages of the second and the ninth; that the sum of the wages of the second and the ninth is equal to the sum of the wages of the third and the eighth; that the sum of the wages of the third and the eighth is equal to the sum of the wages of the fourth and the seventh; and that the sum of these two wages is equal to the sum of the wages of the fifth and the sixth. In this way the sum of the wages of any two opposites is equal to the sum of the wages of two other opposites.[1525]

⟨Indeed, let the number of men be line AB, the first man, line AT, his wage, line TG, the second man, line TZ, his wage, line ZQ, the third man, line ZL, his wage, line LN, and so on to the tenth man, line FB, and his wage, line BK.⟩ I shall draw from the point G a line parallel to line AB, say line GI; from the point Q a line parallel to line GI, say line QR; and from the point N a line parallel to line QR, say line NE. Thus line TG is equal to line BI, as Euclid affirms in the first book, stating: 'When an area is formed by parallel lines, the opposite sides will be equal'.[1526] Since, as you know, line TG is the wage of the first, so line BI is the wage of the first. Again, line ZQ is equal to line BR, and line LN is equal to line BE. Then it is clear that the line (formed by) GT and BK is equal

---

[1522] *(operarii) superant se equali differentia* in the text. See also next note.

[1523] Literally: 'the wage of the smaller'. The same (*mutatis mutandis*) in what follows.

[1524] From Premiss P_1. This will, though, be proved below, but only in MS $\mathcal{B}$.

[1525] That is, opposite terms in the progression.

[1526] *Elements* I.34.

to the line (formed by) ZQ and FO; indeed, line FO is equal to line BC, and line TG is equal to line ZP, while line PQ is equal to line CK, for the difference by which they exceed one another is equal; therefore the line (formed by) BC and TG with, in addition, line CK, is equal to the line (formed by) FO and ZP with, in addition, line PQ; then it is clear that the line (formed by) TG and BK is equal to the line (formed by) ZQ and FO. In like manner it will also be shown that the line (formed by) FO and ZQ is equal to the line (formed by) DH and LN; indeed, line DH is equal to line FS, line ZQ is equal to line LX, and line XN is equal to line SO; therefore the line (formed by) FS and ZQ, with, in addition, line SO, is equal to the line (formed by) DH and LX, with, in addition, line XN. In like manner it will also be shown that the sum of the wages of any remaining (such) pair is equal to the sum of the wages of any other pair of opposites, which makes it thus clear that the sum of the wages of any two (opposites) is equal to the sum of the wages of any other pair of opposites. Therefore adding the wage of the first to the wage of the last will give the sum of the wages of a pair, and multiplying it by the number of men will produce twice the sum of the wages of all the men, so that multiplying it by half the number of men will produce the wages of (all) the men.

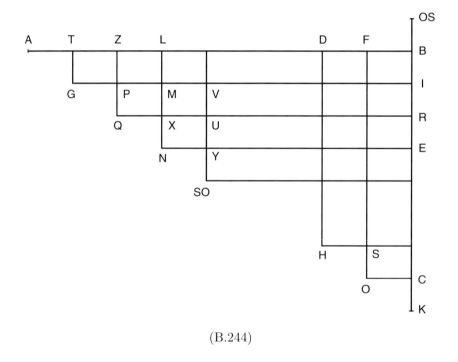

(B.244)

The proof of the rule for finding the difference is the following. You know that the wage of the last man is twenty-one and the wage of the first, three. The first man's exceeds nobody's but is exceeded by the second man's, exceeded by the third man's by twice the excess of the second man's,

LIBER MAHAMELETH

exceeded by the fourth thrice, by the fifth four times, by the sixth five times, by the seventh six times, by the eighth seven times, by the ninth eight times; so the tenth is as much as the first and nine times the difference. This is why you subtract the first wage from the last, so as to leave the sum of all the differences, and divide this remainder by the number of men minus one, the result being the common difference.

Indeed, you know that the first man is line AT and the corresponding wage, TG, which exceeds none. You know also that the excess of the second is line QP, while the excess of the third is line NM, which is twice the excess of the second: line MX is equal to line PQ, line XN also is equal to line PQ, so line NM is twice the difference by which they exceed one another. Line V,SO is thrice the same: line VU is equal to line PQ; line UY is equal to line XN, and line XN is equal to line QP, thus line QP is equal to line UY; and line Y,SO is equal to line QP; therefore line SO,V is thrice the difference. In the same way it will also be demonstrated that line IK is nine times the difference QP. Therefore, dividing line IK, which is nine times the difference, by nine will give the difference by which they exceed one another. This is what we wanted to demonstrate.

AGAIN ON THE SAME TOPIC.

(**B.245**) Someone asks: Ten workers are hired, the first at three nummi, and there is a difference of two between consecutive wages. How much is the last man's wage and what is the sum of all the wages?

(***a***) You will do the following. Always subtract one from the number of workers, leaving in this case nine. Multiply it by the difference, thus producing eighteen. Add to it the first man's wage, which is three; this makes twenty-one. This is the last man's wage. If you want to know the sum of all the wages, do as I have taught above; that is, add the first man's wage to the last man's and multiply the sum by half the number of workers, which is five, as we have shown in the foregoing (problem).

The reason for determining the last man's wage (thus) is the following. We have already shown in the previous problem that if you want to know by how much one worker's wage exceeds another's, you will subtract the first man's wage from the last man's and divide the remainder by the number of workers minus one; the result will be the difference between one wage and the next. Consequently, multiplying the difference, which is two, by the number of workers minus one will produce the highest wage less the lowest. Therefore adding to the result the lowest wage will give the highest.

(***b***) Or otherwise. Put as the highest wage a thing. Then subtract from it the lowest wage; this leaves a thing minus three. Since this is the highest wage less the lowest, dividing a thing minus three by the number of workers minus one will give the difference. Consequently, the result of multiplying the difference by the number of workers minus one is equal to a thing minus three nummi.[1527] Thus multiply the difference by nine; this gives eighteen,

---

[1527] *nummi* here rather in the sense of 'units' (see p. 841, note 1120). Same in the following problems.

946 PART TWO: TRANSLATION, GLOSSARY

which is equal to a thing minus three nummi. Proceed as I have taught before in algebra.[1528] The value of the thing will be twenty-one, which is the highest wage. The highest wage being known, you will, if you want to know the sum of all the workers' wages, do as I have taught before,[1529] and the result will be what you wanted.

AGAIN ON THE SAME TOPIC.

(B.246) Someone asks: Ten workers are hired at a hundred and twenty nummi, and there is a difference of two between consecutive wages. How much are the wages of the first and the last?

(a) You will do the following. Divide a hundred and twenty by half the number of workers; the result is twenty-four, which is the sum of the wages of the first and the last. Next, subtract one from the number of workers; this leaves nine. Multiply it by the difference; this produces eighteen. Subtract it from twenty-four; this leaves six, which is twice the first man's wage. Thus, half of it, namely three, is the wage of the first. If you want to know the last man's wage, add three to eighteen; this makes twenty-one, which is the wage of the last.

This is proved as follows. It is clear from the foregoing that multiplying the sum of the first and the last wages by half the number of men will produce the sum of all the wages.[1530] Thus dividing a hundred and twenty, which is the sum of all the wages, by half the number of workers will give twenty-four, which is the sum of the first and last men's wages. It is also clear that subtracting the wage of the first from the wage of the last and dividing the remainder by the number of men minus one will give the difference between consecutive wages.[1531] Thus multiplying the difference by the number of men minus one will produce the wage of the last minus the wage of the first. Then subtracting from twenty-four the product of the difference into the number of men minus one will leave twice the wage of the first. Half of this is therefore the wage of the first.

This will also be demonstrated by means of the previous figure, as follows. Draw from the point B a line equal to line TG, say line B,OS. But you know that line BI is equal to line TG. Thus it is clear that line OS,I is twice line TG. But line BK is the highest wage and line B,OS is the lowest. Thus line OS,K is the sum of the first and last men's wages, which is twenty-four. But we know that line IK is nine times the difference, that is, eighteen. Line OS,K being twenty-four, line OS,I is six. This is twice line TG; thus TG is three, which is the first man's wage. This is what we wanted to know.

(b) Or otherwise. Put as the first man's wage a thing. Next, always subtract one from the number of workers, because of what we said above;[1532]

[1528] Equation of the form $a_1 x - b_1 = b_2$ (e.g. B.242).

[1529] B.244.

[1530] B.244.

[1531] B.244.

[1532] This reference is either interpolated (as in B.251b) or misplaced.

this leaves nine. Multiply it by the difference, which is two; this produces eighteen. Add to it the wage of the first, which is a thing; this gives eighteen and a thing, which is the wage of the last. Add to it the wage of the first, which makes eighteen and two things. Multiply it by half the number of workers; the result is ninety nummi and ten things, which is equal to a hundred and twenty nummi.[1533] Subtracting ninety from the hundred and twenty leaves thirty, which equals ten things. Thus the thing equals three, which is the wage of the first.

AGAIN ON THE SAME TOPIC.

(**B.247**) Someone asks: Ten workers are hired, the last at twenty-one, and the difference between consecutive wages is two. How much is the first man's wage and what is the sum of all the wages?

You will do the following. Multiply the difference by the number of men minus one, that is, two by nine, thus producing eighteen. Subtract it from twenty-one; this leaves three, which is the first man's wage. If you want to know the sum of all the wages, add to the first man's wage that of the last and multiply the sum by half the number of workers; the result will be what you wanted to know.

AGAIN ON THE SAME TOPIC.

(**B.248**) Someone asks: Ten workers are hired at a hundred and twenty nummi, the first at three, and there is a common difference between consecutive wages. How much is the wage of the last and what is the common difference?

(**a**) You will do the following. Divide a hundred and twenty by half the number of workers, which gives twenty-four. Subtract from it the wage of the first, which is three; this leaves the wage of the last, which is twenty-one. If you want to know the difference between consecutive wages, subtract three from twenty-one and divide the remainder by the number of workers minus one.

The reason for this is the following. We know that the sum of all the wages, which is a hundred and twenty, is obtained by adding the wage of the first to the wage of the last and multiplying the sum by half the number of workers. Thus dividing the sum of all the wages by half the number of workers will give the sum of the wages of the first and the last. Subtracting from this the wage of the first will leave the wage of the last.

(**b**) Or otherwise. Put as the wage of the last a thing. Add to it the wage of the first, which is three nummi; this gives a thing and three nummi. Multiply it by half the number of workers, which is five; this produces five things and fifteen nummi, which is equal to a hundred and twenty nummi.[1534] Subtracting fifteen nummi from a hundred and twenty leaves a hundred and five, which equals five things. Thus a thing equals twenty-one, which is the wage of the last.

[1533] Equation of the form $a_1x + b_1 = b_2$.
[1534] Equation of the form $a_1x + b_1 = b_2$.

948   PART TWO: TRANSLATION, GLOSSARY

(**B.249**) Similarly also if it is said: The sum of all the wages is a hundred and twenty and the wage of the last, twenty-one. How much is the wage of the first?

Divide a hundred and twenty by half the number of workers, and subtract from the result the wage of the last; the remainder is the wage of the first.

AGAIN ON THE SAME TOPIC, BUT OTHERWISE [1535]

(**B.250**) Someone asks: An unknown number of workers are hired, the first at three and the last at twenty-one, and the difference between consecutive wages is two. How many workers are there?

(*a*) You will do the following. Subtract the wage of the first from the wage of the last, divide the remainder by the difference, which is two, and always add one to the result. This will give the number of workers.

The reason for this is the following. As we know [1536], if you want to know the difference for a given number of workers, you subtract the wage of the first from the wage of the last and divide the remainder by the number of workers minus one, the result being the difference. Consequently, the product of the number of workers minus one into the difference will be the wage of the last diminished by the wage of the first. Thus dividing the wage of the last minus the wage of the first by the difference will give the number of workers minus one. Adding one to this will give the number of workers.

(*b*) Or otherwise. Put as the number of workers a thing.[1537] Subtracting one from it leaves a thing minus one. Multiply this by the difference, which is two, thus producing two things minus two nummi. Adding the wage of the first, which is three, makes two things and a nummus. This is the wage of the last. It equals twenty-one, which is (also) the wage of the last. Proceed then as has been shown in algebra.[1538] The thing will be ten, and this is the number of workers.

AGAIN ON THE SAME TOPIC.

(**B.251**) Someone asks: An unknown number of workers are all hired at a hundred and twenty, the first at three, and the difference between consecutive wages is two. What is the wage of the last and how many workers are there?

(*a*) You will do the following, first in order to know the last man's wage. We know [1539] that if, after subtracting one from the number of men, you multiply the remainder by the common difference and add to the sum the

---

[1535] Number of workers now unknown.

[1536] B.244.

[1537] As usual, the text is more concise: *pone operarios rem.*

[1538] Equation of the form $a_1 x + b_1 = b_2$.

[1539] B.244. Same reference below.

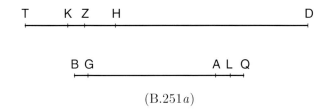

(B.251a)

first man's wage, which is the lowest, the result will be the last man's wage, which is the highest.

In order now to determine the number of workers, you will do the following. Let the number of workers be AB. We know that if, after subtracting one from the number of men, you multiply the remainder by the common difference and add to the sum the first man's wage, the result will be the last man's wage. Let then one be BG. Then multiplying two by AG and adding three to the product will give the last man's wage. Since multiplying AB by two gives twice AB, twice AB minus two is the product of the difference into the number of men minus one. Thus let twice AB be DZ, and two, HZ; therefore DH is the product of the difference into the number of men minus one. But we know[1540] that adding the first man's wage to the product of the difference into the number of men minus one gives the highest wage. Let the lowest wage be HK; thus the whole DK is the highest wage, and it is twice the number of men, plus one.[1541] Now we know[1542] that adding the lowest wage to the highest and multiplying the sum by half the number of men produces the total amount of the wages. Therefore add three, namely KT, to DK; then the whole DT will be twice the number of men, plus four. Thus it is clear that the product of DT into half of AB is a hundred and twenty. But[1543] the product of DT into half of AB is equal to the products of DZ into half of AB and ZT into half of AB. Now the product of DZ into half of AB is equal to the product of AB into itself, for DZ is twice AB, while the product of ZT into half of AB is equal to the product of half of ZT into AB. Therefore the products of AB into itself and into half of ZT, which is two,[1544] are a hundred and twenty. Let then two be AQ. So the products of AQ into AB and AB into itself are equal to the product of the whole QB into AB, while AQ is two.[1545] So let AQ be bisected at the point L. Complete the (treatment of the) problem according to what we have shown;[1546] the result for AB will be ten, which is the number of men.

[1540] B.244–245.
[1541] This is particular to this problem. Since $HK = 3$ and $HZ = 2$, so $KZ = 1$.
[1542] Again, B.244.
[1543] *Elements* II.1 = PE_1.
[1544] Since $HK = KT = 3$ and $HZ = 2$, so $KZ = 1$ and $ZT = 4$.
[1545] Difference and product of QB and AB thus known.
[1546] Using *Elements* II.6, as seen in B.38.

950 PART TWO: TRANSLATION, GLOSSARY

You are to know that this proof [1547] will suffice for you in all similar problems, whether the difference is equal to the lowest wage, or more, or less.

(**b**) Or otherwise. Put as the number of workers a thing. Subtract from it one [*which represents the worker whose wage does not exceed any other*][1548]; this leaves a thing minus one. Multiply it by the difference, thus producing two things minus two. Add the wage of the first, which is three; this gives two things and a nummus. Add the wage of the first; the result will be two things and four nummi, which is the sum of the wages of the first and the last. Multiplying it by half the number of workers, which is half a thing, will produce one square and two things, which are equal to the total amount of the workers' wages, namely a hundred and twenty. Proceed then as has been shown in algebra.[1549] The value of the thing will be ten, and this is the number of workers.

(**B.252**) Someone asks: An unknown number of men are hired. The wage of the last is twenty-one, the difference between consecutive wages is two, and the total amount of the wages is a hundred and twenty. What is the lowest wage and how many workers are there?

You will do the following. Let the number of men be AB. We know [1550] that multiplying what remains after subtracting one from AB by the difference and adding the lowest wage to the product will give twenty-one. Thus multiply two, which is the difference, by the number of men minus one, and let the product be GD. Therefore GD is twice AB minus two. Let two be DH. Thus GH is twice AB. Now we know [1551] that adding the lowest wage to GD gives twenty-one. Let the lowest wage be DZ [*which is greater than two since it cannot be equal or less*][1552]. Therefore GZ is twenty-one, namely the highest wage. Now we know [1553] that adding the wage of the first to the wage of the last and multiplying the sum by half the number of men produces the total amount of the wages. Let the lowest wage be ZK. Thus the product of GK into half of AB is a hundred and twenty. Therefore the product of GK into AB is two hundred and forty. But we know that GH is twice AB and GZ, twenty-one. Thus HZ is twenty-one minus twice AB. DH being two, DZ is twenty-three minus twice AB. This is the lowest wage. But ZK is that much also. Thus the whole DK is forty-six minus four times AB. Now we know that GD is twice AB minus two. Thus the whole GK is forty-four minus twice AB. But we know that the product of GK into AB is two hundred and forty. Thus the product

---

[1547] *probatio*. Rather: geometrical solution.

[1548] The formula has been used several times before and thus does not need to be justified.

[1549] Equation of the form $x^2 + bx = c$.

[1550] B.245.

[1551] As just stated.

[1552] Asserting that $DZ > DH$.

[1553] B.244.

of AB into forty-four minus twice AB is two hundred and forty. So the product of AB into twenty-two minus AB itself is a hundred and twenty. Let twenty-two be AL. Then the product of AB into BL is a hundred and twenty.[1554] Let AL be bisected at the point N [*for it cannot be otherwise, the proof of this is clear*][1555]. Then[1556] the products of AB into BL and NB into itself will be equal to the product of AN into itself. Now the product of AN into itself is a hundred and twenty-one, and the product of AB into BL is a hundred and twenty; this leaves, as the product of NB into itself, one. Thus NB is one. Since AN is eleven, AB is ten, which is the number of men. This is what we wanted to demonstrate.

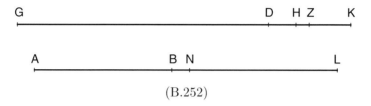

(B.252)

(**B.253**) Someone asks: An unknown number of workers are hired. The total amount of the wages is a hundred and twenty, the difference between consecutive wages is the same, the wage of the first is three and that of the last, twenty-one. How many workers are there?

(***a***) You will do the following. Add together the wages of the first and the last and divide a hundred and twenty by the sum; double the result. This gives the number of men.

The reason for this is the following. We know[1557] that adding the wages of the first and the last and multiplying the sum by half the number of men produces the total amount of the wages. Thus dividing the total amount of the wages by the sum of the wages of the first and the last will give half the number of men, which doubled will be their number.

(***b***) Or otherwise. Put as the number of workers a thing. Next, add the wages of the first and the last; the result is twenty-four. Multiply it by half the number of workers, which is half a thing; this gives twelve things, which are equal to a hundred and twenty. Thus a thing equals ten, which is the number of workers.

[1554] Sum and product of AB and BL thus known.
[1555] Asserting that $B \neq N$. We have encountered this same reader earlier in this problem and (presumably) in B.235″ & B.235‴.
[1556] *Elements* II.5 = PE_5.
[1557] Again, B.244.

Chapter (B-XIII) on Hiring carriers

(B.254) Someone, hired to carry twelve sextarii thirty miles at a wage of sixty nummi, takes three sextarii ten miles. How much is his wage?

(*a*) You will do the following. Multiply the twelve sextarii by the thirty miles; this produces three hundred and sixty, which you put as the principal. Next, multiply the three sextarii he actually carried by the ten miles and the product by the sixty nummi. Divide the result by the principal. This will give five (which is the number of) nummi, and that is what is due to him.

That is so because, as we know, his being hired to carry twelve sextarii thirty miles (for sixty nummi) is as if he were (hired) to carry one sextarius three hundred and sixty miles for sixty nummi. Now this is so because the ratio of one sextarius to twelve sextarii is the same as the ratio of the thirty miles he carries the twelve sextarii to the number of miles he carries one sextarius; now it is known [1558] that the ratio of one to twelve is half a sixth; it follows that thirty miles is half a sixth of the number of miles he carries one sextarius; therefore it is clear that these miles are three hundred and sixty, while the wage due to him is sixty nummi. It is further known that carrying three sextarii ten miles is the same as carrying one sextarius thirty miles. Then it is clear that the ratio of three hundred and sixty miles to the wage due, which is sixty nummi, is the same as the ratio of thirty miles to the corresponding wage. There are therefore four proportional numbers. Then the product of the first into the fourth is equal to the product of the second into the third. Therefore dividing the product of thirty into sixty by the first, which is three hundred and sixty, will give the five nummi.

(*b*) Or otherwise. Denominate the three sextarii from twelve; this gives a fourth. Keep in mind a fourth of sixty, namely fifteen. Next, denominate ten miles from thirty miles; this gives a third. Therefore a third of fifteen, which is five, is what you wanted to know.

The reason for this is the following. We know that if the carrier took three sextarii thirty miles at the same wage as for twelve, he would earn fifteen nummi: three sextarii being a fourth of twelve sextarii, he should receive a fourth of the wage. Then it is clear that, carrying three sextarii ten miles, which is a third of thirty miles, he must receive a third of fifteen nummi, which is five nummi.

(*c*) Or otherwise.[1559] If the carrier takes twelve sextarii thirty miles for sixty nummi, five nummi are due to him for carrying each sextarius thirty miles: since he carried half a sixth of the sextarii, to him is due half a sixth

---

[1558] A.217 (if need be).

[1559] Another 'reason' for the calculation seen in *b*.

LIBER MAHAMELETH 953

of the number of nummi, namely five nummi. Now it is known[1560] that carrying one sextarius thirty miles is the same as carrying three sextarii ten miles. [*This is shown as follows: Since he has carried three sextarii ten miles, then carrying each of the three sextarii ten miles is the same as carrying one sextarius thirty miles; now it is known that for carrying each of the twelve sextarii thirty miles he should receive five nummi.*][1561] Therefore a wage of five nummi is due to him for carrying three sextarii ten miles.

(*d*) Or otherwise. Denominate ten miles from thirty miles; this is a third. Thus take a third of sixty, which is twenty. Next, denominate three sextarii from twelve; this is a fourth. Then a fourth of twenty, which is five, is what you wanted to know.

Or multiply a fourth by a third; the result is half a sixth. Then half a sixth of sixty, which is five, is what you wanted to know.

The reason for this is the following. You were to multiply three sextarii by ten, and then what you wanted will be the same fraction of sixty as this product is of the product of twelve sextarii into thirty miles.[1562] But it is known[1563] that denominating the product of three into ten from the product of twelve into thirty ⟨and taking this fraction of sixty⟩ is the same as denominating three sextarii from twelve, which is a fourth, then denominating ten miles from thirty, which is a third, multiplying a third by a fourth, and taking from sixty the same fraction as the product gives, namely half a sixth. [*This is what we intended to demonstrate.*]

AGAIN, ANOTHER EXAMPLE.

(**B.255**) Someone asks: A man, hired to carry five sextarii ten miles at a hundred nummi, carries three sextarii four miles. What is due to him?

(*a*) We already know[1564] that the ratio of sextarii multiplied by miles to sextarii multiplied by miles is the same as the ratio of wage to wage. Thus multiply five by ten, which produces fifty. Next, multiply three by four, which produces twelve. Then the ratio of fifty to twelve is the same as the ratio of a hundred to what is required. Proceed as has been explained above.[1565] The required result will be twenty-four.

The proof of this is the following. Let the five sextarii be A, the ten miles, G, the three sextarii, B, the four miles, D, and the hundred nummi, H. If he were to carry five sextarii ten miles, he would receive a hundred nummi; carrying them four miles, which is two fifths of the whole distance, he should receive two fifths of the wage, namely forty. So let forty be Z.

[1560] Above, part *a*.

[1561] Gloss which should appear after the next sentence (and is superfluous anyway; see previous note).

[1562] That was the calculation in part *a*.

[1563] Premiss P_3'.

[1564] This has been used before, but will be demonstrated below.

[1565] Plain proportion.

Then the ratio of D to G is the same as the ratio of Z to H. It is then clear that for carrying these five sextarii four miles he would receive Z, which is forty. But he has carried only three sextarii, and should thus receive three fifths of the full wage. So what he should receive is three fifths of forty, which is twenty-four. Therefore let this be K. Then the ratio of K to Z is the same as the ratio of B to A. [*It would also be quite appropriate to proceed as follows: to form the ratio of miles to miles and take the same fraction of what we had kept in mind first; this would be what we wanted.*][1566] Thus the way of proceeding using multiplication is as we have said above.[1567] For the ratio of D to G is the same as the ratio of Z to H, while the ratio of B to A is the same as the ratio of K to Z. Now we know that for any three proportional numbers[1568] the ratio of the first to the third is the same as the ratio of the first to the second compounded with the ratio of the second to the third.[1569] Thus the ratio of K to H is the same as the ratio of K to Z compounded with the ratio of Z to H. But the ratio of K to Z is the same as the ratio of B to A, and the ratio of Z to H is the same as the ratio of D to G. Thus the ratio of K to H is the same as the ratio of D to G compounded with the ratio of B to A. Now the ratio of D to G compounded with the ratio of B to A is the same as the ratio of the product of D into B to the product of G into A.[1570] Thus the ratio of the product of G, which is the number of miles, into A, which is the number of sextarii, to the product of D, which is the number of miles actually covered, into B, which is the number of sextarii actually carried, is the same as the ratio of H, which is the wage, to K, which is the wage actually due. Therefore the ratio of the product of sextarii into miles to the product of sextarii into miles is the same as the ratio of wage to wage. This is what we intended to demonstrate.

(**b**) Or otherwise. Divide the sextarii actually carried by the sextarii he should have carried, divide the miles actually covered by the miles he should have covered, multiply the quotients one by the other, and multiply the result by the wage. In the present problem, divide[1571] three sextarii by five sextarii, which gives three fifths. Divide the miles by the miles, which gives two fifths. Multiply this by the three fifths, which produces a fifth and a fifth of a fifth. Multiply this by a hundred; the result will be twenty-four. The proof of this is clear from the foregoing.[1572]

[1566] In what precedes, we have first considered the fraction formed by the quantity of miles, then that formed by the quantity of sextarii. We could have proceeded in the inverse order, this early reader means (as will be done in *b* below).

[1567] Computations in part *a*, involving products in the dividend and the divisor.

[1568] *tres numeri proportionales.* That is, continuously proportional.

[1569] *Elements* V, def. 9. ('Compounded with': multiplied by.)

[1570] Premiss P_3'.

[1571] Rather 'denominate' (same further on).

[1572] Proof in part *a*.

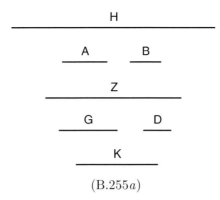

(B.255a)

AGAIN ON THE SAME TOPIC.

(**B.256**) A man is hired to carry twelve sextarii thirty miles at sixty nummi. How many miles will he carry three sextarii for five nummi?

(*a*) You will do the following. Multiply the three sextarii by the sixty nummi; the result is a hundred and eighty. Put it as the principal. Next, multiply twelve by thirty and then the product by five; the result is one thousand eight hundred. Divide it by the principal. The result is ten (which is the number of) miles, and this is what you wanted to know.

The reason for this is the same as that indicated in the previous problem, namely the following.[1573] The ratio of the twelve sextarii multiplied by the corresponding number of miles to the corresponding number of nummi is the same as the ratio of the three sextarii multiplied by the corresponding number of miles to the corresponding number of nummi. Now it is known[1574] that multiplying twelve sextarii by the corresponding number of miles produces three hundred and sixty. Thus it is clear that the ratio of this to sixty is the same as the ratio of the three sextarii multiplied by the corresponding number of miles to five. There are therefore four proportional numbers. Thus the product of three hundred and sixty into five nummi is equal to the product of sixty into three sextarii and then the result into the corresponding number of miles. Now multiplying sixty by three sextarii produces a hundred and eighty. Thus the product of a hundred and eighty into the unknown number of miles is equal to the product of three hundred and sixty into five, which is one thousand eight hundred. Dividing this by a hundred and eighty will give ten, and this is the required number of miles.

(*b*) Or otherwise. Put as the unknown number of miles a thing. Multiply it by three sextarii; the result is three things. Now it has been said[1575] that the ratio of the twelve sextarii multiplied by the corresponding number of miles to the corresponding number of nummi is the same as the ratio of the three sextarii multiplied by the corresponding unknown number of miles

---
[1573] The fundamental proportion seen in B.255a.
[1574] *constat*; expression also used for simple computations.
[1575] Part *a*.

956 PART TWO: TRANSLATION, GLOSSARY

to the corresponding number of nummi. Thus it is clear that the ratio of three hundred and sixty to sixty nummi is the same as the ratio of three things to five nummi. There are therefore four proportional numbers. Thus the product of five into three hundred and sixty is equal to the product of sixty into three things. Proceed then as I have taught before.[1576] In the end there will be a hundred and eighty things, equal to one thousand eight hundred. Thus the thing equals ten, and this is the required number of miles.

(**c**) Or otherwise. Denominate five from sixty; this is half a sixth. Multiply then half a sixth of the (given) number of miles, thus two and a half, by the number which multiplying three produces twelve, namely four; the result will be ten, and this is the required number of miles.

The reason for this is the following. We know that if, when a man is hired to transport twelve sextarii thirty miles at sixty nummi, he receives five nummi to carry twelve sextarii, he must carry them two and a half miles; for he receives, to carry the twelve sextarii, half a sixth of what he would have received to carry them thirty miles, that is, five nummi, and this is why he will carry these twelve sextarii half a sixth of thirty miles, which is two and a half miles. Now it is known that he received five nummi for carrying three sextarii only. So it is clear that carrying twelve sextarii two and a half miles is the same as carrying three sextarii ten miles. For the ratio of three sextarii to twelve is the same as the ratio of two and a half miles to the number of miles the three sextarii are carried; three being a fourth of twelve, two and a half miles are a fourth of the number of miles the three sextarii are carried, from which it follows that the number of miles the three sextarii are carried is ten.

(**d**) Or otherwise. Denominate three from twelve; it is a fourth. Take then a fourth of sixty, which is fifteen. Denominate five from this; it is a third. Thus a third of thirty miles, which is ten, is what you wanted.

The reason for this is the following. We know that, carrying three sextarii thirty miles, he is to receive a fourth of sixty nummi, which is fifteen nummi; for he carried a fourth of twelve sextarii. It is also known that for carrying three sextarii he received five nummi, which is a third of fifteen nummi. Therefore he must carry them a third of thirty miles, which is ten miles.

AGAIN ON THE SAME TOPIC.

(**B.257**) A man is hired to carry twelve sextarii thirty miles at sixty nummi. He carries three sextarii an unknown number of miles which when multiplied by the wage due produces fifty. What is this number of miles (and the corresponding wage) [1577]?

(**a**) You will do the following. Multiply the three sextarii by the sixty nummi; the result will be a hundred and eighty. Put this as the principal.

[1576] Plain proportion, thus rule of three (p. 777).

[1577] This is also part of the question.

LIBER MAHAMELETH

If you want to know the number of miles, multiply twelve by thirty and then the product by fifty, and divide the result by the principal; the root of the quotient is the number of miles, namely ten. If you want to know the wage due, divide fifty by this root, namely ten; the result is five, and this is the wage.

The reason for this is the following. We know[1578] that the ratio of the sextarii multiplied by the corresponding miles to the corresponding nummi is the same as the ratio of the three sextarii multiplied by the corresponding miles to the corresponding nummi. Thus it is clear that the ratio of three hundred and sixty to sixty nummi is the same as the ratio of the three sextarii multiplied by the corresponding number of miles to the corresponding number of nummi. There are therefore four proportional numbers. So the product of three hundred and sixty into the number of nummi corresponding to the three sextarii is equal to the product of sixty into three sextarii multiplied by the corresponding number of miles, that is, the number of miles corresponding to the three sextarii. Multiply in common by the number of miles of the three sextarii. Then the product of three hundred and sixty into the number of nummi corresponding to the miles of the three sextarii, multiplied by this number of miles, is equal to the product of sixty into three sextarii, multiplied by the number of miles by itself, that is, the number of miles of the three sextarii. But we know that multiplying three hundred and sixty by the number of nummi corresponding to the miles of the three sextarii, then the product by the number of miles of the three sextarii, is the same as multiplying the number of miles by the number of nummi and the result by three hundred and sixty [irrespective of which comes first][1579]. Now the multiplication of the miles of the three sextarii by the corresponding number of nummi produces fifty. Therefore the product of fifty into three hundred and sixty is equal to the product of sixty nummi into three sextarii, and the result by the product of the number of miles into itself. But fifty multiplied by three hundred and sixty produces eighteen thousand. Dividing this by the product of sixty into three sextarii, which is a hundred and eighty, will give a hundred. This is the product of the number of miles into itself; the required number of miles of the three sextarii is its root, which is ten.

You may want to know the wage.[1580] Multiply twelve by thirty; the result is three hundred and sixty, which you put as the principal. Next, multiply the sixty nummi by the three sextarii, the result by fifty, and divide the product by the principal; the root of the result, which is five, is the required wage. The reason for this will be known to those understanding the foregoing[1581], but you will (in this case) multiply in common by the

---

[1578] Fundamental proportion proved in B.255a.

[1579] Commutativity of the product, though known from *Elements* VII.16, always seems to bother some readers. See notes 39, 73, 90, 120.

[1580] By a direct computation.

[1581] Above reason.

958 PART TWO: TRANSLATION, GLOSSARY

number of nummi of the three sextarii, and continuing as I have taught above you will find what you require.

(**b**) Or otherwise. Put as the number of miles a thing. Then the number of nummi will be fifty divided by a thing. The reason for this is that, as was stated, the product of the number of miles into the corresponding number of nummi is fifty; therefore dividing fifty by the number of miles will give the number of nummi.[1582] Next, multiply three sextarii by the corresponding number of miles, which is a thing; the result will be three things. But we know[1583] that the ratio of the product of the sextarii into the corresponding miles to the corresponding nummi is the same as the ratio of the product of the sextarii into the corresponding miles to the corresponding nummi. It is thus clear that the ratio of three hundred and sixty to sixty nummi is the same as the ratio of three things to fifty divided by a thing. There are therefore four proportional numbers. Thus the product of three hundred and sixty into the quotient of fifty divided by a thing is equal to the product of sixty into three things. But we know that the product of three hundred and sixty into the quotient of fifty divided by a thing is equal to the result of dividing the product of fifty into three hundred and sixty by a thing; for when any number is divided by another and the quotient multiplied by a third number, the result is equal to the division of the product of the dividend into the multiplier by the divisor; this has been proved elsewhere.[1584] Now we know[1585] that the product of fifty into three hundred and sixty is eighteen thousand; thus the result is eighteen thousand divided by a thing; and this equals a hundred and eighty things. Multiplying then a hundred and eighty things by a thing gives a hundred and eighty squares, and this equals eighteen thousand.[1586] Reduce the squares to a single one and reduce eighteen thousand in the same proportion. This will leave in the end a square equal to a hundred. Its root, which is ten, is the value of the thing, and so much is the required number of miles. The number of nummi is the result of dividing fifty by ten, thus five.

(**c**) You may proceed otherwise to determine the number of miles. We know[1587] that the ratio of three hundred and sixty to the corresponding wage, which is sixty nummi, is the same as the ratio of three things to fifty divided by a thing. By inversion, the ratio of sixty nummi to three hundred and sixty will be the same as the ratio of fifty divided by a thing to three things. But the ratio of sixty to three hundred and sixty is a sixth. It follows that the ratio of the result of dividing fifty by a thing ⟨to three things is a sixth. It follows that the result of dividing fifty by a thing⟩ is a sixth of three things, which is half a thing. This being equal to the result

---

[1582] At this stage, such justification would seem to be superfluous.

[1583] Fundamental proportion.

[1584] Premiss P_5. The whole explanation may be an addition.

[1585] By computation.

[1586] Equation of the form $ax^2 = c$.

[1587] Above, *b*.

LIBER MAHAMELETH

959

of dividing fifty by a thing, multiply half a thing by a thing; the result will be half a square; this equals fifty. Thus a square equals a hundred, a thing equals ten, and so much is the number of miles. The number of nummi is the result of dividing fifty by ten, thus five. This is what you wanted.

AGAIN ON THE SAME TOPIC.

(B.258) A man, hired to carry twelve sextarii thirty miles at sixty nummi, carries three sextarii an unknown number of miles which when added to the corresponding wage makes fifteen. What is the number of miles and the number of nummi, that is, the wage?

(a) You will do the following. Multiply twelve by thirty; the result is three hundred and sixty. Next, multiply three sextarii by sixty nummi; the result is a hundred and eighty. Adding it to three hundred and sixty makes five hundred and forty; put it as the principal. If you want to know the number of miles, multiply three hundred and sixty by fifteen and divide the product by the principal; the number of miles is the result, namely ten. If you want to know the wage, multiply a hundred and eighty by fifteen and divide the product by the principal; the result will be the wage, namely five nummi.

The proof of this is the following. Let fifteen, which is the sum of the miles and the wage, be line AG; let the miles be part of this line, namely line BG, so the wage will be line AB. Now it is clear from the foregoing [1588] that the ratio of the product of the sextarii into the corresponding miles to the corresponding nummi is the same as the ratio of the product of the second sextarii into the corresponding miles to the corresponding nummi. Thus the product of three hundred and sixty into the wage for the miles of the three sextarii is equal to sixty multiplied by the three sextarii and then the product by the corresponding miles. Thus it is clear that the product of the part of fifteen which corresponds to the wage into three hundred and sixty is equal to the product of the other part, which corresponds to the miles, into the product of sixty into three sextarii, which is a hundred and eighty. Therefore the product of line AB into three hundred and sixty is equal to the product of line BG into a hundred and eighty. [1589] I shall now draw, from the point B, rectilinearly and on either side [1590], two lines: one, say line BH, of a hundred and eighty, and the other, say line BD, of three hundred and sixty. It is then clear that the product of line AB into BD, which is the area AD, is equal to the product of line BH into BG, which is the area HG. Since their sides are parallel and two of their angles, namely the angle ABD and the angle HBG, are equal, the sides of area AD and area HG are mutequefia [*that is, coalternate*], as Euclid said in the sixth book [1591]. Thus the ratio of line DB to line BH is the same as the ratio of line GB to line BA. By composition, the ratio of line DB to

---

[1588] Fundamental proportion.

[1589] Sum and ratio of AB and BG thus known.

[1590] And at right angles.

[1591] *Elements* VI.14.

line DH will be the same as the ratio of line GB to GA. Thus the product of line DB into GA is equal to the product of line DH into BG. Therefore dividing the product of line DB into GA by DH will give line BG, which is the number of miles. You will also be able to determine the wage in like manner. In that case, the ratio of line HB to BD is the same as the ratio of line AB to BG. By composition, the ratio of line HB to HD will be the same as the ratio of line AB to AG. Thus the product of line HB into AG is equal to the product of line HD into AB. Therefore multiplying HB by AG and dividing the product by HD will give AB, which is the wage. This is what we wanted to prove.

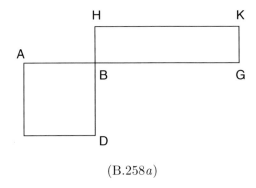

(B.258 a)

(**b**) You may want to know the number of miles and the wage in another way. Denominate a hundred and eighty from five hundred and forty; this gives a third. The wage is then a third of fifteen, namely five. If you want to know the number of miles, denominate three hundred and sixty from five hundred and forty; this gives two thirds. Multiply this by fifteen; the result will be ten, and this is the number of miles.

(**c**) Or otherwise. Put as the number of miles a thing. The number of nummi will be fifteen minus one thing. Next, multiply a thing by the three sextarii; the result will be three things. Now you already know[1592] that the ratio of three hundred and sixty, which is the product of the sextarii into the corresponding miles, to sixty nummi is the same as the ratio of three things to the corresponding nummi, which are fifteen minus a thing. There are therefore four proportional numbers. Thus the product of three hundred and sixty into fifteen minus a thing is equal to the product of sixty into three things. But the product of three hundred and sixty into fifteen minus a thing is five thousand four hundred nummi[1593] minus three hundred and sixty things, and this is equal to the product of sixty into three things, which is a hundred and eighty things.[1594] Complete then the five thousand four hundred nummi by adding three hundred and sixty

---

[1592] Again fundamental proportion.
[1593] Here and below (but not always: note 1595), in the sense of 'units'; see note 1120, p. 841.
[1594] Equation of the form $b_1 - a_1 x = a_2 x$.

LIBER MAHAMELETH

961

things and add as much to a hundred and eighty things; this will give five hundred and forty things; this equals five thousand four hundred nummi. Thus a thing equals ten, and this is the number of miles. The nummi[1595] are the remainder of fifteen, namely five.

(*d*) You may want to determine the wage according to algebra. Put as the wage a thing. So the number of miles will be fifteen minus a thing. Multiply this by the three sextarii; this will produce forty-five minus three things. Next, you will form the ratio between them as has been done before.[1596] Then the product of the first, which is three hundred and sixty, into the fourth, which is a thing, thus three hundred and sixty things, will be equal to the product of the second, which is sixty, into the third, which is forty-five minus three things, thus to two thousand seven hundred nummi[1597] minus a hundred and eighty things. Next, proceed as we have taught above in algebra.[1598] The thing will be five, which is the number of nummi.

AGAIN ON THE SAME TOPIC.

(**B.259**) A man, hired to carry ten sextarii fifty miles at a hundred nummi, carries three sextarii a number of miles which is unknown but which leaves four when the corresponding wage is subtracted from it. What is the number of miles, or the corresponding wage?

(*a*) You will do the following. Multiply the ten sextarii by the fifty miles; the result will be five hundred. Subtract the product of the three sextarii into a hundred; this will leave two hundred. Put this as the principal. If you want to know the number of miles, multiply four by five hundred and divide the result by the principal; in the case of the wage, multiply four by three hundred and divide the result by the principal. This will give what you wanted.

This will be demonstrated as follows. Let the unknown number of miles be line AB and the unknown wage, line GB. Thus it is clear that line AG is four. Now we have established in the foregoing that the ratio of the first sextarii multiplied by the corresponding miles to the corresponding nummi is the same as the ratio of the sextarii actually carried multiplied by the corresponding miles to the corresponding nummi.[1599] There are therefore four proportional numbers. Then the product of five hundred into the wage for the miles he carries the three sextarii, which is line GB, is equal to the product of a hundred into the three sextarii, multiplied by line AB.[1600] Next I shall draw line TN, which is five hundred. Let a line of three hundred be cut off from it, say line QT. It is then clear that the product of line NT into line GB is equal to the product of line QT into line AB. Thus the ratio of line NT to line TQ is the same as the ratio of

---

[1595] Actual monetary unit this time.

[1596] B.257*b* or *c* above.

[1597] Units.

[1598] Equation of the form $b_1 - a_1 x = a_2 x$, as in part *c*.

[1599] Again, fundamental proportion.

[1600] Difference and ratio of AB and GB thus known.

line AB to BG. By inversion, the ratio of line QT to TN will be the same as the ratio of line GB to AB. By separation, the ratio of line QT to QN will be the same as the ratio of line BG to GA. So the product of line QT into AG is equal to the product of line NQ into GB. Dividing then the product of line QT into AG by NQ will give line GB, which is the wage for the unknown number of miles.

The reason[1601] for the determination of the miles is the following. We know that if the ratio of line QT to QN is the same as the ratio of line BG to GA,[1602] then, by inversion, the ratio of line NQ to QT must be the same as the ratio of line AG to GB. By composition, the ratio of line NQ to NT will be the same as the ratio of line AG to AB. Thus the product of line NQ into AB is equal to the product of line NT into AG. Therefore dividing the product of line NT into AG by NQ will give AB, which is the number of miles. This is what you wanted to know.

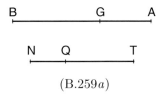

(B.259a)

(**b**) Or otherwise.[1603] Look for the number which multiplying two hundred produces five hundred; you will find two and a half. Multiply this by four; the result is ten, and this is the number of miles. If you want to know the wage, determine what fraction two hundred is of three hundred; it is two thirds. It follows that two thirds of the wage is four; thus the wage is six.

(**c**) Or otherwise. Put as the unknown number of miles a thing. The unknown wage will be a thing minus four nummi; for, as stated above[1604], subtracting the wage from the number of miles leaves four, which means that the sum of the wage and four is equal to the number of miles, thus the wage is a thing minus four. Next, multiply the number of miles, which is a thing, by three sextarii; the result is three things. Multiply it by a hundred, thus producing three hundred things. Next, multiply five hundred, which is the product of the ten sextarii into the corresponding miles, by a thing minus four; this will produce five hundred things minus two thousand nummi; this equals three hundred things.[1605] Complete the five hundred things by adding two thousand nummi, and add as many nummi to three hundred things; next, subtract three hundred things from five hundred things. This

---

[1601] *causa*; a demonstration (= *probatio*), in this case.
[1602] As just established.
[1603] 'Other way' of the computation in *a*; analogous to B.258*b*. Note the wording: 'look for the *number*' (> 1) and 'determine what *fraction*' (< 1).
[1604] Formulation of the problem. But this whole explanation might be an early reader's addition.
[1605] Equation of the form $a_1 x - b_1 = a_2 x$.

LIBER MAHAMELETH 963

leaves two hundred things, which equal two thousand nummi. Thus the thing is ten; this is the number of miles, and what you wanted to know. If you want to know the wage, subtract four from ten; this leaves six as the wage.

(*d*) You may put as the wage a thing; then the number of miles will be a thing and four. Next, multiply a thing by five hundred; this gives five hundred things. Next, multiply a hundred by the three sextarii and the product by a thing and four; the result will be three hundred things and one thousand two hundred nummi. This equals five hundred things.[1606] Subtract then three hundred things from five hundred things; this leaves two hundred things, which equal one thousand two hundred nummi. Thus the thing equals six nummi, and that much is the required wage.

AGAIN ON THE SAME TOPIC.

(**B.260**) Someone hires a man to carry twelve sextarii thirty miles at sixty nummi but gives him six nummi to carry, in proportion to this wage, as many sextarii as he covers miles. What is the number of sextarii or the number of miles?

(*a*) You will do the following. Multiply twelve by thirty, thus producing three hundred and sixty. Multiply this by six; the result is two thousand one hundred and sixty. Divide this by sixty, which gives thirty-six. Its root, namely six, is the number of sextarii, and that many are the miles.

The reason for this is the following. From what has been said in hiring carriers [1607] it is known that the ratio of the sextarii multiplied by the corresponding miles to the corresponding nummi is the same as the ratio of the sextarii multiplied by the corresponding miles to the corresponding nummi. Therefore the ratio of twelve sextarii multiplied by thirty miles to sixty nummi is the same as the ratio of the unknown number of sextarii multiplied by the corresponding miles to the corresponding nummi, namely six. So multiplying six by three hundred and sixty is the same as multiplying the unknown number of sextarii by the corresponding miles and the product by sixty. Thus dividing the product of six into three hundred and sixty by sixty and taking the root of the result will give the number of sextarii, and that many are the miles.

(*b*) Another way to determine the sextarii and the miles is the following.[1608] Multiply twelve by thirty, thus producing three hundred and sixty. Divide this by sixty, and multiply the result by six. The root of the product is the number of sextarii, and that many are the miles.

The reason for this is the following. We know that we had to multiply three hundred and sixty by six and to divide the product by sixty, the root of the result being what you wanted. But we know from what has been said

---

[1606] Equation of the form $a_1 x + b_1 = a_2 x$.

[1607] See B.255*a* or B.256*a*, both in the first section of this chapter.

[1608] The usual 'other way'.

964 PART TWO: TRANSLATION, GLOSSARY

before[1609] that multiplying three hundred and sixty by six, dividing the product by sixty, and taking the root of the result is the same as dividing three hundred and sixty by sixty, multiplying the result by six and taking the root of the product.

(*c*) Or otherwise. Find what fraction six is of sixty; it is a tenth. Multiply this same fraction of thirty, which is three, by twelve; the result is thirty-six. The number of sextarii or miles is the root of this, thus six.

The reason for this is the following. As we know, we had to denominate six nummi from sixty, take this fraction of three hundred and sixty, which results from multiplying twelve sextarii by thirty miles, and take the root of the result.[1610] Now you know[1611] that denominating six from sixty, taking this fraction of the product of twelve into thirty, and taking the root of the result is the same as denominating six from sixty, taking this fraction of thirty, multiplying the result by twelve and taking the root of the product.

You may also take this fraction of twelve, multiply the result by thirty and take the root of the product; this root will be what you wanted.

(*d*) Or otherwise. Put as the number of sextarii a thing. The number of miles will then also be a thing. Next, multiply twelve by thirty, thus producing three hundred and sixty, and multiply this by six nummi. The result is equal to sixty squares, which is the product of sixty into a thing and then into the second thing. Proceed then as has been shown in algebra.[1612] The value of the thing will be six, which is the number of sextarii or miles.

AGAIN ON THE SAME TOPIC.

(**B.261**) Someone, hiring a man to carry twelve sextarii thirty miles at a hundred nummi, gave him beforehand ten nummi to carry as many sextarii as three fourths of the number of miles that he had to carry them. What is the number of sextarii or of miles?

(*a*) You will do in this case the following. Multiply twelve by thirty miles, thus producing three hundred and sixty. Multiply this by ten nummi, which gives three thousand six hundred. Next, take three fourths of a hundred —because of the statement that he is to carry as many sextarii as 'three fourths' of the number of miles covered— which is seventy-five. Divide by this three thousand six hundred; the result is forty-eight. The approximate root of this, which is seven minus half a seventh, is the number of miles, and three fourths of it is the number of sextarii. The reason for this is the one already indicated in the previous problems on hiring carriers;[1613] those understanding them will find no difficulty in proving what is here.

[1609] Premiss P_5.

[1610] Recapitulating the previous computations.

[1611] Premiss P_2.

[1612] Equation of the form $ax^2 = c$.

[1613] In particular B.260, or also B.257, both of which rely on the fundamental relation seen in the first three problems B.254–256.

LIBER MAHAMELETH

965

(*b*) Or otherwise. Divide three thousand six hundred by a hundred; this gives thirty-six. Next, seek the number which multiplying three fourths produces one; this is one and a third. Multiply it by thirty-six, thus producing forty-eight. The root of this is the number of miles, and three fourths of it is the number of sextarii. The reason for this is clear from what has just been said.[1614]

(*c*) Or otherwise. Consider what fraction ten is of a hundred; it is a tenth. Multiply this same fraction of thirty, which is three, by twelve; the result is thirty-six. Next, seek the number which multiplying three fourths produces one; you will find[1615] one and a third. Multiply it by thirty-six; this gives forty-eight. The root of this is the number of miles, and three fourths of it is the number of sextarii.

You may also take a tenth of twelve, namely one and a fifth, and multiply it by thirty, thus producing thirty-six. Continue as shown above[1616], and you will find what you wanted.

(*d*) Or otherwise. Put as the number of miles a thing, and as the sextarii three fourths of a thing since, as stated, the man is to carry as many sextarii as three fourths of the number of miles covered. Next, multiply twelve by thirty, thus producing three hundred and sixty. Multiplying this by ten produces three thousand six hundred. Next, multiply a thing by three fourths of a thing; the result is three fourths of a square. Multiplying it by a hundred produces seventy-five squares; these are equal to three thousand six hundred.[1617] Then take of all these squares a single one, and take as much of the three thousand six hundred, (that is) according to the ratio that one (square) bears to seventy-five squares, which gives forty-eight. You have then one square equal to forty-eight nummi. Therefore the root of the square is seven minus half a seventh. This is the number of miles, and three fourths of it is the number of sextarii.

(**B.262**) A man, hired to carry fifteen sextarii forty miles at one square and its root, carries ten sextarii ten miles and receives the root. How much is the square?

You will do the following. Multiply fifteen by forty, thus producing six hundred. Next, multiply ten by ten, which produces a hundred. Divide six hundred by this, which gives six. Multiply it by one —because of the one root proposed: if there were two you would multiply by two— and subtract from the product the number of roots which are with the square; this will leave, in the present case, five, which is the root. Therefore the square is twenty-five.

(**B.263**) A man, hired to carry six sextarii ten miles at a cube and the corresponding square, carries four sextarii five miles and receives the square. How much is the square or the cube?

[1614] This is merely one 'other way', like that in *c* below.

[1615] Known from part *b*!

[1616] Multiplying by one and a third.

[1617] Equation of the form $ax^2 = c$.

You will do the following. Multiply six by ten, thus producing sixty. Next, multiply four by five, which gives twenty. Divide sixty by this; the result is three. Multiply it by one —because of the statement that one square is received— then subtract from the product one —because of his above assertion that one square is with the cube. The result will be the root of the square, which is two. Therefore the square is four, and the cube is the product of the square into its root, namely eight.

Chapter (B-XIV) on Hiring stone-cutters

(**B.264**) Someone hires three workers to cut four stones in thirty days at sixty nummi and two of them cut two stones in ten days. How much is due to them?

This problem lends itself to two interpretations, one being that with four stones two workers are as efficient as three; the other that two accomplish two thirds of what three would do.

(i) If you want to operate according to the first interpretation, you will do the following. Then it is like saying: 'Someone hires a worker to carry four sextarii thirty miles at sixty nummi, and he carries two sextarii ten miles; what is due to him?' You will follow for this problem all the ways we have taught above.[1618] The result will be ten, and this is what you wanted.

(ii) Taking now the second interpretation, you will do the following.

(**a**) Multiply three, the number of workers, by four, the number of stones, and the product by thirty days; the result is three hundred and sixty, which you put as the principal. Next, multiply the two workers by the two stones which they have cut, thus producing four. Multiply it by ten days; this produces forty. Multiplying this by sixty produces two thousand four hundred. Dividing it by the principal will give six and two thirds.

The reason for this is the following. We know[1619] that three workers cutting four stones in thirty days is the same as one worker cutting twelve stones in thirty days. But you know that one worker cutting twelve stones in thirty days is the same as one worker cutting one stone in three hundred and sixty days. Thus it is clear that one worker cuts one stone in three hundred and sixty days for sixty nummi. You also know that two workers cutting two stones in ten days is the same as one worker cutting one stone in forty days. Then it is like saying: 'When three hundred and sixty sextarii are given for sixty nummi, what is the price of forty sextarii?'[1620] So multiply forty by sixty and divide the product by three hundred and sixty. The result is what is due to the two workers for two stones.

Another reason is the following: the ratio of the product of three workers multiplied by four stones into thirty (days) to the product of two workers multiplied by two stones into ten days is the same as the ratio of sixty nummi to the required wage. The proof of this is the following. Let the three workers be A, the four stones, B, the thirty days, G, the two workers, D, the two stones, H, the ten days, Z, and the sixty nummi, K. If three workers cut four stones in thirty days, sixty nummi are due to

---

[1618] B.254–255.

[1619] The subsequent transformations have been known from B.254 on.

[1620] See B.1 *seqq.*

968   PART TWO: TRANSLATION, GLOSSARY

them; thus if two cut [*two thirds of*][1621] what three cut, they are to receive two thirds of sixty, which is forty. Let forty be T. Then the ratio of D to A is the same as the ratio of T to K.   Therefore, if two workers cut four stones in thirty days, they are to receive forty nummi; but since they have cut only half of the stones, half of forty, namely twenty, is due to them. [*Therefore, if two workers cut two stones in thirty days, they are to receive twenty nummi*][1622]. Now let the twenty nummi be Q. Therefore the ratio of H to B is the same as the ratio of Q to T.   Now they have worked only ten days; thus they should receive a third of twenty, which is six and two thirds. Let this six and two thirds be L. Then the ratio of L to Q is the same as the ratio of Z to G.   Thus it is that the ratio of D to A is the same as the ratio of T to K, that the ratio of H to B is the same as the ratio of Q to T, and that the ratio of Z to G is the same as the ratio of L to Q. Now we know that for any four proportional numbers the ratio of the first to the fourth is the same as the ratio compounded of the ratio which the first has to the second, the ratio which the second has to the third, and the ratio which the third has to the fourth.[1623] Thus the ratio of L to K is the same as the ratio compounded of the ratio which L has to Q, the ratio which Q has to T, and the ratio which T has to K. But the ratio of L to Q is the same as the ratio of Z to G, the ratio of Q to T is the same as the ratio of H to B, and the ratio of T to K is the same as the ratio of D to A. Therefore the ratio of L to K is the same as the ratio compounded of the ratio which Z has to G, the ratio which H has to B, and the ratio which D has to A. But the ratio compounded of the ratio which Z has to G, the ratio which H has to B, and the ratio which D has to A is the same as the ratio which Z multiplied by H and the product by D has to G multiplied by B and the product by A.[1624] Therefore the ratio of the number of workers multiplied by the number of stones and the product by the number of days to the second number of stones multiplied by the number of workers and the product by the number of days is the same as the ratio of the two wages. This is what we wanted to demonstrate.

(**b**) Or otherwise.[1625]   Find what fraction sixty is of three hundred and sixty; it is a sixth. The same fraction of forty, namely six and two thirds, is what you wanted.

(**c**) Or otherwise. Denominate the two stones cut from four; it is a half. Take the same fraction, that is, a half, of sixty; this is thirty. Next, denominate ten days from thirty; it is a third. Take the same fraction, that is, a third, of thirty; this is ten. Next, denominate two workers from three;

[1621] Common-sense addition by an early reader.

[1622] Same reader, summing up the reasoning.

[1623] The four numbers must be continuously proportional. See *Elements* V, def. 10. ('Compounded of': equal to the products of.)

[1624] P'_3, repeated

[1625] What follows is just the 'other ways' of dealing with the formula, now with four factors in the numerator.

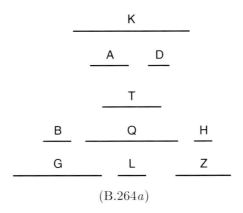

(B.264a)

this is two thirds. So two thirds of ten nummi, namely six and two thirds, is what you wanted.

The reason for this is evident. If three workers cut two stones in thirty days, half of the sixty nummi is due to them, namely thirty; if they do it in the ten days, a third of the thirty nummi is due to them, namely ten nummi. But we know that the two stones have been cut in ten days by two workers only, that is, by two thirds of the three workers. Therefore two thirds of ten nummi is due to them, namely six and two thirds.

(**d**) Or otherwise. Find what fraction two workers are of three; it is two thirds. Thus take two thirds of sixty nummi, which is forty. Next, find what fraction two stones are of four stones; it is a half. Thus take half of forty, which is twenty. Next, find what fraction ten is of thirty; it is a third. Thus take a third of twenty, which is six and two thirds. This is what you wanted.

The reason for this is evident. If two workers cut four stones in thirty days, forty nummi would be due to them since the work of two workers is two thirds of the work of three. You know too that if two workers cut two stones in thirty days, what they are owed would be half of forty nummi, which is twenty. But you know that, as a matter of fact, they have cut the two stones in a third of thirty days, which is ten days. Therefore a third of the twenty nummi is due to them, namely six and two thirds.

(**e**) Or otherwise. Denominate two workers from three; it is two thirds. Next, denominate two stones from four; it is a half. Next, denominate ten days from thirty; it is a third. Next, multiply two thirds by a half and then the product by a third; the final result will be two thirds of one sixth. Thus take two thirds of a sixth of sixty; this is six and two thirds. The reason for this is the same as that which I have taught in the chapter on carrying.[1626]

AGAIN ON THE SAME TOPIC.

(**B.265**) Someone, hiring three workers to cut four stones in thirty days at

---

[1626] B.254d.

970 PART TWO: TRANSLATION, GLOSSARY

sixty nummi, gave two workers six nummi and two thirds for cutting two stones. How many days will they be in his service?

(*a*) You will do the following. Multiply the two stones by the two workers, thus producing four. Multiply this by sixty; the result is two hundred and forty, which you put as the principal. Next, multiply the three workers by the four stones, thus producing twelve. Multiply it by thirty; the result is three hundred and sixty. Multiply this by six and two thirds, thus producing two thousand four hundred. Divide this by the principal; the result is ten, and this is the number of days.

The reason for this is the following. We know[1627] that the ratio of the three workers multiplied by the four stones and the product by the corresponding number of days to the corresponding number of nummi is the same as the ratio of the two stones multiplied by the two workers and the product by the corresponding number of days to the corresponding number of nummi. Thus it is clear that the ratio of three hundred and sixty to sixty nummi is the same as the ratio of four, resulting from the product of two workers into two stones, multiplied by the corresponding number of days to the corresponding number of nummi, which is six and two thirds. There are therefore four proportional numbers. Then the product of three hundred and sixty into six and two thirds is equal to the product of sixty into four, multiplied by the unknown number of days. Therefore it is clear that dividing the product of three hundred and sixty into six and two thirds by the product of sixty into four will give the unknown number of days.

(*b*) Or otherwise. Consider what fraction six and two thirds is of sixty; you will find that it is two thirds of a sixth. So take two thirds of a sixth of three hundred and sixty, which is forty, and divide this by four; the result will be ten, and this is the number of days.

The reason for this is the following. We have already shown in the previous chapter[1628] that three workers cutting four stones in thirty days for sixty nummi is the same as one worker cutting one stone in three hundred and sixty days for sixty nummi. ⟨But you know already that six nummi and two thirds have been given to two workers for cutting two stones.⟩[1629] You also know that two workers cutting two stones is the same as one worker cutting four stones.[1630] Thus it is clear that denominating six and two thirds from sixty and taking this fraction of three hundred and sixty, namely forty [*days*], is the number of days it takes one worker to cut four stones[1631] for six nummi and two thirds. [*But you know already that six nummi and two thirds have been given to two workers for cutting four*

[1627] B.264.*ii*.*a*.

[1628] B.264.*ii*.*a*.

[1629] This addition seems necessary.

[1630] Which is equivalent to one worker cutting one stone in four times as many days. This step is missing.

[1631] The MSS have 'one stone'. See subsequent interpolation.

LIBER MAHAMELETH

971

stones.] Thus divide forty by the four stones in order to determine the number of days needed for one stone, which is ten.

(*c*) Or otherwise. Multiply three workers by four stones, thus producing twelve. Next, multiply two workers by two stones, thus producing four. Next, consider what fraction six and two thirds is of sixty; you will find that it is two thirds of a sixth. Take then two thirds of a sixth of thirty days; this is three and a third. Next, find what number multiplying four gives twelve; this is three. Multiply it by three and a third; the result is ten, which is the number of days.

The reason for this is evident. We have already shown[1632] that three workers cutting four stones in thirty days is the same as one worker cutting twelve stones in thirty days and that two workers cutting two stones in an unknown number of days for six nummi and two thirds is the same as one worker cutting four stones in the unknown number of days for six nummi and two thirds. So it is clear that if sixty nummi are due to a man cutting twelve stones in thirty days, then, if he received six nummi and two thirds for cutting twelve stones, he will do this in three days and a third;[1633] for the ratio of six nummi and two thirds to sixty nummi is the same as the ratio of the three and a third days to thirty. But one worker cutting twelve stones in three days and a third is the same as one worker cutting four stones in thrice three days and a third, which is ten days,[1634] for twelve is thrice four. Therefore it is clear that the number which multiplying four stones produces twelve is the same as the number which multiplying three and a third produces the unknown number of days, and this is three. This is what we wanted to demonstrate.

(*d*) Or otherwise. Consider what fraction six and two thirds is of sixty nummi; you will find that it is two thirds of a sixth. Take then two thirds of a sixth of thirty days, which is three and a third. Next, consider what number multiplying two stones produces four; this is two. Multiply it by three and a third, thus producing six and two thirds. Consider then also what number multiplying two workers produces three; this is one and a half. Thus multiply one and a half by six and two thirds; the result is ten, and this is the number of days. The reason for this is evident from what has just been said.[1635] Those having understood what is above will understand what is here.

AGAIN ON THE SAME TOPIC, BUT OTHERWISE[1636]

(**B.266**) Someone hires five workers for thirty days at sixty nummi and two of them serve ten days. What is due to them?

[1632] See B.264.*ii.a* and part *b* above.

[1633] Which in practical terms is of course absurd.

[1634] Equally absurd (same wage).

[1635] Part *c*.

[1636] The kind of work is no longer specified.

972   PART TWO: TRANSLATION, GLOSSARY

You will proceed here using all the ways found in the first chapter on carrying.[1637]

(*a*) That is, multiply the five workers by the thirty days, thus producing a hundred and fifty, which you put as the principal. Next, multiply the two workers by ten, which produces twenty. Multiplying it by sixty gives one thousand two hundred. Divide this by the principal; the result is eight, and this is what is due to them.

The reason for this is evident from what has been said before,[1638] namely that five men working thirty days is the same as one man working a hundred and fifty days, and two working ten days is the same as one working twenty days. Then it is like saying: 'When a hundred and fifty sextarii are given for sixty nummi, what is the price of twenty sextarii?' So multiply twenty by sixty and divide the product by a hundred and fifty, as shown in the chapter on buying and selling.[1639]

(*b*) You may also denominate two workers from five; this gives two fifths. Then take two fifths of sixty nummi, which is twenty-four. Next, denominate ten days from thirty; this gives a third. This same fraction of twenty-four, namely eight, is what you wanted.

(*c*) Or otherwise. Denominate ten days from thirty; this gives a third. Then take a third of sixty, which is twenty. Next, denominate two workers from five; this gives two fifths. Then two fifths of twenty, which is eight, is what you wanted to know.

(*d*) Or otherwise. Denominate two workers from five, denominate ten days from thirty, and multiply one result by the other. Taking of sixty the fraction thus produced will give what you wanted.

All the ways of solving this problem are those shown in the chapter on carrying.[1640]

AGAIN ON THE SAME TOPIC.

(**B.267**) Someone, hiring five workers for thirty days at sixty nummi, gave two of them eight nummi. How many days do they have to serve him?

You will do here the same as in the second chapter on carrying.[1641]

(*a*) That is, multiply the two workers by the sixty nummi; the result is a hundred and twenty, which you put as the principal. Next, multiply the five workers by thirty, then the result by eight, and divide the product by the principal; the result will be ten, and this is the number of days you require.

[1637] B.254.

[1638] *Mutatis mutandis*, B.264–265.

[1639] See B.1 *seqq.*

[1640] Repetition (see beginning), thus possibly an addition.

[1641] See B.256. (B.254–255 form a single section, the latter being another example of the same type.)

LIBER MAHAMELETH

973

(*b*) Or otherwise. Denominate the two workers from five; this gives two fifths. Then take two fifths of sixty, which is twenty-four. Taking the same fraction of thirty as eight is of twenty-four will give what you wanted.

(*c*) Or otherwise. Denominate eight nummi from sixty nummi; this is a tenth and a third of a tenth. Take the same fraction, namely a tenth and a third of a tenth, of thirty, which gives four. Next, seek the number which multiplying two workers produces five; you will find two and a half. Multiply four by this; the result is ten, and this is what you wanted.

All the ways of solving this problem are those shown in the chapter on carrying.[1642]

AGAIN ON THE SAME TOPIC.[1643]

(**B.268**) Someone hires five workers for thirty days at sixty nummi, and two of them serve a number of days which when multiplied by the corresponding wage produces eighty. What are the number of days and the wage?

You will do the following. Multiply the two workers by sixty, thus producing a hundred and twenty; put it as the principal. Next, multiply the five workers by the thirty days, thus producing a hundred and fifty. Multiply this by eighty and divide the product by the principal; the root of the result is the number of days you require, namely ten.

The reason for this is the following. We know [1644] that the ratio of the five workers multiplied by the thirty days to the corresponding number of nummi is the same as the ratio of the two workers multiplied by the corresponding number of days to the corresponding number of nummi. There are therefore four proportional quantities. Then the product of a hundred and fifty into the unknown number of nummi is equal to the result of multiplying the product of sixty into two workers by the corresponding number of days. But the product of sixty into two workers is a hundred and twenty. Let the number of days of the two workers be a common factor.[1645] So the result of multiplying the product of a hundred and fifty into the number of nummi of the two workers by the corresponding number of days is equal to the result of multiplying a hundred and twenty by the number of days of the two workers multiplied by itself. Now the result of multiplying the product of a hundred and fifty into the number of nummi of the two workers by the corresponding number of days is equal to the result of multiplying the product of this number of days into the corresponding number of nummi by a hundred and fifty. But the product of the number of days into the corresponding number of nummi is eighty. Thus the result of multiplying eighty by a hundred and fifty is equal to the result of multiplying a hundred and twenty by the number of days of the two workers multiplied by itself. Therefore it is clear that multiplying

---

[1642] B.256 has in addition an algebraic treatment.

[1643] Case of two unknowns.

[1644] B.255*a* & B.257*a* (*mutatis mutandis*).

[1645] *Dies (. . .) sint commune in multiplicando.*

974 PART TWO: TRANSLATION, GLOSSARY

a hundred and fifty by eighty and dividing the result by a hundred and twenty will give the number of days multiplied by itself. Since this is a hundred, the number of days is its root, which is ten.

If you now want to know the wage, divide eighty by ten; the result is eight, and this is the wage. If you want to determine the wage in another way, multiply eighty by a hundred and twenty and divide the product by a hundred and fifty; the root of the result is what you wanted. The reason for this is the same as the one just given.[1646]

AGAIN ON THE SAME TOPIC.

(**B.269**) Someone hires five workers for thirty days at sixty nummi, and two of them have worked a number of days which when added to the wage due makes eighteen. What is this number of days and how much is the wage?

You will do here the same as in the chapter on carrying, and the enunciation of this problem is the same as the enunciation there.[1647] We have deemed it appropriate to repeat it here, in order that the reader does not think that it is different. It is the following:[1648] 'A man, hired to carry twelve sextarii thirty miles at sixty nummi, carried three sextarii an unknown number of miles which when added to the wage due to him makes fifteen.'

You will do here the same as for hiring.[1649]

(**a**) Multiply the five workers by the thirty days, thus producing a hundred and fifty. Next, multiply the sixty nummi by the two workers, thus producing a hundred and twenty. Adding this to a hundred and fifty makes two hundred and seventy; put it as the principal. If you want to know the number of days, multiply a hundred and fifty by eighteen and divide the product by the principal; the result will be ten, and this is the number of days. But if you want to know the wage, multiply a hundred and twenty by eighteen and divide the product by the principal; the result will be eight, and this is the wage. The reason for this is the same as the one in the chapter on carrying.[1650]

(**b**) The way of operating according to algebra is the following. Put as the number of days a thing; the wage will then be eighteen minus a thing. Next, multiply a thing by a hundred and twenty; this produces a hundred and twenty things. Next, multiply eighteen minus a thing by a hundred and fifty; the result is two thousand seven hundred minus a hundred and fifty

---

[1646] And the same as in B.257.

[1647] The MSS have twice *probatio*, proof, instead of *prolatio*, enunciation. The reference is to B.258.

[1648] The enunciation of B.258 is repeated almost word for word. There is some doubt about the genuineness of this whole passage.

[1649] Hiring carriers, thus B.258. Repetition, so either this or the previous passage must be an addition.

[1650] Again B.258.

LIBER MAHAMELETH

things; this is equal to a hundred and twenty things.[1651] Complete the two thousand seven hundred by adding a hundred and fifty things and add that many things to a hundred and twenty things; this will give two hundred and seventy things, which equal two thousand seven hundred. Thus a thing is worth ten, and this is the number of days.

(*c*) If you want to determine the wage, put as it a thing; the number of days will then be eighteen minus a thing. Next, multiply a thing by a hundred and fifty. The product equals the result of multiplying eighteen minus a thing by a hundred and twenty. Proceed then as we have taught in algebra.[1652] The result will be for the thing eight, and this is the wage.

AGAIN ON THE SAME TOPIC.

(**B.270**) Someone hires five workers for thirty days at sixty nummi, and two of them work a number of days which, when the wage due to them is subtracted from it, leaves two. What are the number of days and the corresponding wage?

You will proceed here by all the ways used in the chapter on carrying.[1653] Those who have understood how to operate there will understand how to operate here.

CHAPTER ON ANOTHER TOPIC [1654]

(**B.271**) Someone hires a shepherd to tend a hundred sheep for thirty days at ten nummi. He tends sixty of them for twenty days. What is due to him?

(*a*) You will do the following. Multiply a hundred by thirty; this produces three thousand, which you put as the principal. Next, multiply ten by sixty and the result by twenty, and divide the product by the principal; the result will be four, and this is what is due to him.

The reason for this is the following. We know [1655] that the ratio of a hundred sheep multiplied by the corresponding number of days, which is thirty, to the corresponding number of nummi is the same as the ratio of sixty sheep multiplied by the corresponding number of days to the corresponding number of nummi. There are therefore four proportional numbers. Then multiplying ten by sixty and the product by twenty and dividing the result by the product of a hundred into thirty will give four, and this is what you wanted.

(*b*) Or otherwise. Denominate sixty sheep from a hundred: this is three fifths. Then take three fifths of ten, which is six. Next, denominate twenty days from thirty days; this is two thirds. Thus two thirds of six, which is four, is what you wanted.

[1651] Equation of the form $b_1 - a_1 x = a_2 x$.

[1652] Equation of the form $b_1 - a_1 x = a_2 x$, as in *b*.

[1653] See B.259.

[1654] The kind of work is specified.

[1655] B.254 *seqq.*, B.266 *seqq.*

976 PART TWO: TRANSLATION, GLOSSARY

The reason for this is the following. If the shepherd tended sixty sheep for thirty days, three fifths of ten nummi, which is six nummi, would be due to him; for he tends three fifths of a hundred sheep. But you know that he tended them only for twenty days, which is two thirds of thirty days. Therefore two thirds of six, which is four, are due to him.

(*c*) Or otherwise. Denominate twenty days from thirty days; this is two thirds. So take two thirds of ten nummi, which is six and two thirds. Next, denominate sixty from a hundred; this is three fifths. Thus what is due to him is three fifths of six and two thirds, which is four.

(**B.271′**) If it is now said in the same problem: The shepherd receives four nummi for tending sixty sheep; how many days does he have to tend them?

(*a*) You will do the following. Multiply sixty by ten and put the product as the principal. Next, multiply four nummi by thirty days and the result by a hundred, and divide the product by the principal; the result will be twenty, and this is what you wanted.   The reason for this is clear to those who are acquainted with what has been said in the problems on carrying.[1656]

(*b*) Or otherwise. Denominate sixty from a hundred; this is three fifths. Then denominate four nummi from three fifths of ten nummi, thus from six; this gives two thirds. This same fraction of thirty, namely twenty, is what you wanted.

(*c*) You may also denominate four nummi from ten nummi; this is two fifths. Then take that much, that is, two fifths, of thirty, which is twelve. Next, consider what number multiplying sixty produces a hundred; this is one and two thirds. Therefore multiply twelve by one and two thirds; the result will be twenty, and this is the number of days.

CHAPTER ON ANOTHER TOPIC.[1657]

(**B.272**) Someone, hired to dig a hole ten cubits long, eight wide and six deep at eighty nummi, digs a hole four cubits long, three wide and five deep. What is due to him?

(*a*) You will do the following. Multiply ten by eight, thus producing eighty. Multiply this by six; the result is four hundred and eighty, which you put as the principal. Next, multiply three by four and the product by five; the result is sixty. Multiplying it by eighty produces four thousand eight hundred. Divide this by the principal; the result will be ten, and that much is due to him.

The reason for this is evident. We know that the volume of the larger hole, which is four hundred and eighty cubits, is to the corresponding wage, namely eighty nummi, as the volume of the smaller hole, which is sixty cubits, is to the corresponding wage.[1658] There are therefore four proportional

---

[1656] B.256.

[1657] Here too the kind of work done is specified, but now we have computation of areas and volumes.

[1658] Our 'volume' is variously expressed by: *spatium (totius) concavitatis*, or *tota concavitas*, or also *(tota) magnitudo*.

LIBER MAHAMELETH
977

numbers. Then dividing the product of sixty into eighty by four hundred and eighty will give the wage.

(**b**) Or otherwise. Find what fraction sixty is of four hundred and eighty; it is an eighth. This same fraction of eighty, namely ten, is what you wanted. The reason for this is the one we have explained,[1659] namely that the volume of the smaller hole is to the volume of the larger hole as the wage for the first is to the wage for the second. Or otherwise. Denominate eighty from the volume of the larger hole; it is a sixth. Thus a sixth of the volume of the smaller hole gives the corresponding wage.

(**B.272′**) If the holes were circular [1660], you would find both volumes as follows. Multiply half the circumference of each hole by half its diameter and multiply the product by the corresponding depth; the result is the measure of the volume of each hole. Next, denominate in the manner I have taught above for square [1661] holes, and the result will be what you require.

(**B.273**) Someone asks: A man, hired to dig a hole ten cubits long, six wide and five deep for a hundred nummi, receives ten nummi for digging a hole two wide and three long. What will its depth be?

We know [1662] that the ratio of the volume of one hole to the volume of the other is the same as the ratio of one wage to the other. Now the volume of the larger hole results from multiplying ten by six and the product by five, which gives three hundred. The volume of the smaller hole results likewise from multiplying two by three and the product by the unknown depth. Therefore the ratio of three hundred to the result of multiplying two by three and the product by the unknown depth is the same as the ratio of a hundred to ten. But multiplying two by three gives six. Thus the ratio of three hundred to the product of the depth into six is the same as the ratio of a hundred to ten. Therefore multiply three hundred by ten and divide the product by a hundred; the result will be thirty. This is the product of the depth into six; so the depth is five.

(**B.274**) Someone asks: A man, hired to dig a hole three cubits long, two wide and five deep for sixty nummi, receives ten nummi for digging a hole of equal length and width but one cubit and a fourth deep. What will its length and width be?

You will do the following. We know that the ratio of thirty, which is the volume of the first hole, to the volume of the other is the same as the ratio of sixty to ten. Proceed as we have taught above;[1663] the volume of the second hole will be found to be five. Now we know that the volume of the second hole results from multiplying its length by its width and the

---

[1659] Above, *a* (exchanging the terms of the proportion).

[1660] *rotundus*. This refers to the section.

[1661] *quadratus*, or right-angled. This refers to the section.

[1662] B.272.

[1663] Plain proportion, thus beginning of Book B.

978   PART TWO: TRANSLATION, GLOSSARY

product by its depth. So dividing five by one and a fourth will give four, which is the product of the length by the width. Since the length is equal to the width, the length is two, and the width likewise. [*This is what we wanted to demonstrate.*]

(B.275) Someone asks: A carpenter, hired to build a chest ten cubits long, five wide and eight high for a hundred and seventy nummi, built a chest two cubits long, three wide and four high. What is due to him?

Many consider this problem to be like those about holes, for they think that the ratio of the volume of the first chest to the volume of the second is the same as the ratio of the cost of the first to the cost of the second, which is certainly wrong: what is being sold is not the empty part [1664] of the chest but its sides.

The way to solve this problem is the following. We know that the (first) chest has six sides: the upper side, ten long and five wide, thus with an area of fifty, and the lower one opposite, equal (in area), so that both together make an area of a hundred; another side, eight long and five wide, thus with an area of forty, equal to the opposite side, so that both together make an area of eighty; yet another side, ten long and eight wide, equal to the opposite side, with an area of eighty, the area of both together being thus a hundred and sixty. Therefore the area of the first chest's sides is three hundred and forty altogether. We can determine in like manner the area of the smaller chest's sides. We know that it too has six sides: the upper one, three long and two wide, thus with an area of six, which is equal to the opposite side, the area of both being therefore twelve; another side, four long and two wide, thus with an area of eight, which is equal to the opposite side, the area of both being therefore sixteen; yet another side, four long and three wide, thus with an area of twelve, which is equal to the opposite side, the area of both being therefore twenty-four. So the area of the second chest's sides is fifty-two altogether. Therefore it is clear that the ratio of the first chest's whole area to the area of the second chest is the same as the ratio of the cost of the first to the cost of the second. Thus the ratio of three hundred and forty to fifty-two will be the same as the ratio of a hundred and seventy to the whole cost of the second chest's sides. Twenty-six nummi are therefore due to him. [*This is what we wanted to demonstrate.*]

[1664] *concavitas.*

Chapter (B-XV) on the Consumption of oil by lamps

Note that for all the problems in this chapter three quantities are involved: a number of lamps, a number of nights, a number of measures. This gives rise, therefore, to many types of problems according to whether the number of nights and the number of lamps are given and the number of measures is required, as in the first problem[1665]; or the number of nights and the number of measures are given and the number of lamps is required, as in the second problem[1666]; or the number of lamps and the number of measures are given and the number of nights is required, as in the fourth problem[1667]. Moreover, these three types vary according to whether the three quantities are one or larger integers, and according to the fractions[1668].

The first problem is the following.

(**B.276**) Someone asks: One lamp consumes in one night a fourth of an eighth of one arrova of oil. How many arrovae of oil will three hundred lamps consume in twenty nights?

(*a*) You will do the following. Multiply the numbers from which a fourth and an eighth are denominated; this gives thirty-two, which you take as the principal. Next, multiply three hundred by twenty, thus producing six thousand, and divide it by the principal. This gives a hundred and eighty-seven and a half, which is the quantity of arrovae.

The reason for this is the following. The consumption of one lamp was supposed to be one part of thirty-two parts of an arrova. Consequently, thirty-two lamps consume one arrova in one night. You further know that three hundred lamps multiplied by twenty nights produces (the same expense as) six thousand lamps; divide this by thirty-two lamps so as to attribute to each group of thirty-two lamps [*together*] an arrova. To them (all) correspond then a hundred and eighty-seven arrovae and a half.

(*b*) Or otherwise. One lamp was supposed to have a nightly oil consumption of one fourth of an eighth of one arrova. Consequently, three hundred lamps consume in one night three hundred fourths of an eighth. Now our intention is to know how many arrovae of oil three hundred lamps consume in twenty nights. Therefore multiply three hundred fourths of an eighth by twenty to produce six thousand fourths of an eighth of an arrova, and divide this by the product of the denominators of a fourth and an eighth, which

---

[1665] B.276.

[1666] B.277.

[1667] B.279. (B.278 is another example of the same kind as B.277.)

[1668] Which may themselves be aliquot (as in B.276, B.277, B.279, B.282) or non-aliquot (as in B.278, B.280, B.281, B.283–285).

980   PART TWO: TRANSLATION, GLOSSARY

is thirty-two. This will give a hundred and eighty-seven arrovae and a half. Or otherwise.[1669] You already know that three hundred lamps consume in one night three hundred fourths of an eighth, that is, nine arrovae and three eighths. Multiplying this by twenty nights produces a hundred and eighty-seven and a half, and this is what you wanted.

(*c*) Or otherwise. One lamp consumes in twenty nights twenty fourths of an eighth, that is, five eighths of an arrova. Therefore multiply this consumption of a single lamp by the three hundred lamps; the result will be a hundred and eighty-seven arrovae and a half, and this is what you wanted.

AGAIN ON THE SAME TOPIC.

(**B.277**) Someone asks thus: One lamp consumes in one night half an eighth of an arrova. How many lamps consume a hundred arrovae in thirty nights?

(*a*) You will do the following. Multiplying the numbers from which a half and an eighth are denominated, namely two and eight, produces sixteen. Multiply this by a hundred; this produces one thousand six hundred. Divide this by thirty; the result will be fifty-three lamps and a third.[1670]

The reason for this is the following. From the hypothesis we know that one lamp consumes in one night half an eighth (of an arrova). Consequently, sixteen lamps consume in one night one arrova, which is sixteen such parts. Multiplying these sixteen parts by a hundred arrovae produces one thousand six hundred parts, each of which is the nightly consumption of one lamp. Thus these thousand six hundred parts are the nightly consumption of one thousand six hundred lamps. You further know that these thousand six hundred parts, which make a hundred arrovae, are consumed in thirty nights. Thus dividing one thousand six hundred lamps by thirty nights will give the number of lamps consuming a hundred arrovae in that period, and this is what you wanted.

(*b*) Or otherwise. Divide a hundred arrovae by thirty nights and multiply the result by sixteen, which is the product of the numbers from which a half and an eighth are denominated. The result is fifty-three and a third, and such is the number of lamps.

The reason for this is the following. We know from the foregoing [1671] that the treatment consisted in multiplying sixteen by a hundred arrovae and dividing the product by thirty; which is the same thing as dividing a hundred by thirty and multiplying the result by sixteen.[1672] [*For when a number is multiplied by another number and the product divided by a third number, the result is the same as the result of dividing the second by the third and multiplying the quotient by the first* [*as stated in the sixth*

[1669] This just performs the division of 300 by 32 immediately. See B.286*a*.

[1670] Thus fractional lamp.

[1671] Part *a* above.

[1672] Premiss P_5.

LIBER MAHAMELETH   981

premiss [1673]. *The proof of this is the following. For instance, let the number G be multiplied by the number B to produce the number A, the number A be divided by D to give H, and also B be divided by D to give K, and K be multiplied by G to produce T. Then I say that T is equal to H. This is proved as follows. G multiplying B produces A. And A divided by D gives H; thus multiplying D by H will produce A. Therefore multiplying G by B is the same as multiplying D by H. Again, the division of line* [1674] *B by D gives K; thus multiplying K by D will produce B. But the multiplication of K by G produces T. Thus line K is multiplied by two numbers, namely D and G. Therefore the ratio of line B to T is the same as the ratio of line D to G. Thus the product of G into B is equal to the product of T into D. But the product of G into B is equal to the product of H into D. Then the product of T into D is equal to the product of H into D. Thus line T is equal to line H. This is what we wanted to demonstrate. From the aforesaid is thus clear that multiplying sixteen by a hundred and dividing the product by thirty amounts to the same thing as dividing a hundred by thirty and multiplying the result by sixteen.*]

(B.277*b*)

(*c*) Or otherwise. We know that one lamp consumes in thirty nights thirty halves of an eighth, which is an arrova and seven eighths. Thus divide by this a hundred arrovae; this gives fifty-three and a third, which is the number of lamps.

The reason for this is the following. As we know, one lamp consumes in thirty nights thirty halves of an eighth, which is one arrova and seven eighths. Now the ratio of one lamp to one arrova and seven eighths is the same as the ratio of the unknown quantity of lamps to the hundred arrovae they consume. There are therefore four proportional numbers. Then the product of one lamp into a hundred arrovae is equal to the product of one

---

[1673] Premiss P_5 (the sixth counting P_3'). If this reference had been inserted by the same interpolator, he would hardly have felt the need to add a proof.
[1674] Note here the change from 'number' to 'line'.

982 PART TWO: TRANSLATION, GLOSSARY

and seven eighths into the unknown quantity of lamps. Hence multiplying one lamp by a hundred and dividing the product by one and seven eighths will give the unknown quantity of lamps.

(B.278) Likewise if someone asks thus: One lamp consumes in one night two ninths of an eighth of one arrova of oil. How many lamps consume a hundred arrovae in thirty nights?

(*a*) You will do the following. Multiplying the denominators of a ninth and an eighth, which are nine and eight, produces seventy-two. Multiply this by a hundred arrovae and divide the product by thirty; half of the result is the number of lamps, which is a hundred and twenty.

The reason for our taking half of the result is the following.[1675] We know that if the nightly consumption of one lamp were a ninth of an eighth of one arrova, which is one part of seventy-two parts of an arrova, two hundred and forty lamps would consume a hundred arrovae in thirty nights. Since it was said above that the nightly consumption of one lamp is two ninths of an eighth, which is two parts of the seventy-two, a hundred and twenty lamps must consume a hundred arrovae in thirty nights: the consumption is doubled.

Likewise also if it were said: 'One lamp consumes in one night three fourths of an eighth; how many lamps consume a hundred arrovae in thirty nights?' You will do as I have taught before[1676] and take a third of the result, since 'three fourths of an eighth' was proposed. It will again be treated likewise for any proposed quantity of fractions.

(*b*) Or otherwise. Multiply half of seventy-two, which is thirty-six, by a hundred arrovae, and divide the product by thirty nights; the result will be a hundred and twenty, and this is what you wanted.

(*c*) Or otherwise. Double thirty —since he said 'two ninths of an eighth'— to give sixty nights. Therefore, one lamp consumes in each of these nights a ninth of an eighth of one arrova. Then it is like saying: 'One lamp consumes a ninth of an eighth in one night; how many lamps consume a hundred arrovae in sixty nights?' You will do the following. Multiply seventy-two by a hundred and divide the product by sixty; the result is a hundred and twenty, and this is what you wanted.

(*d*) Or otherwise. We know that since one lamp consumes nightly two ninths of an eighth, it must consume in thirty nights five sixths of an arrova; for it consumes sixty parts of the seventy-two. Then divide a hundred arrovae by five sixths as we have taught in the first Book [*chapter on division*][1677]; the result will be a hundred and twenty, which is the number

---

[1675] This is where the treatment differs from that seen in B.277*a*.

[1676] B.277 and just above.

[1677] This gloss adds nothing. In fact, the whole reference might be interpolated: in B.277*c* the divisor already contained a fraction and there was no allusion to A.246–248.

LIBER MAHAMELETH

983

of lamps. The reason for this is as we have established in the previous chapter.[1678]

AGAIN ON THE SAME TOPIC.

(B.279) Someone asks thus: One lamp consumes in one night a ninth of an eighth. In how many nights will forty lamps consume twenty arrovae?

(*a*) You will do the following. Multiply the denominators of a ninth and an eighth, thus producing seventy-two. Multiply this by twenty arrovae and divide the product by forty; the result will be thirty-six, which is the number of nights.

The proof[1679] of this is the following. We know that there are in an arrova seventy-two parts each of which one lamp consumes in one night. ⟨It follows then that seventy-two lamps will consume seventy-two ninths of an eighth, thus one arrova, in one night.⟩ Then multiplying the seventy-two parts[1680] by twenty arrovae produces one thousand four hundred and forty ninths of an eighth; this is therefore the nightly consumption of one thousand four hundred and forty lamps. You further know, from the statement of the problem, that this is the consumption of forty lamps during the unknown number of nights; dividing therefore one thousand four hundred and forty lamps by forty lamps will give the number of nights during which the forty lamps consume one thousand four hundred and forty parts, that is, twenty arrovae.

(*b*) Or otherwise. Forty lamps consume nightly forty ninths of an eighth, which is five ninths of an arrova. Thus divide twenty arrovae by this; the result will be thirty-six, which is the number of nights.

The reason for this is the following. We know that the nightly consumption of forty lamps is forty ninths of an eight, thus five ninths of an arrova. So it is clear that the ratio of one night to the five ninths which forty lamps consume in this night is the same as the ratio of the unknown number of nights during which forty lamps consume twenty arrovae to twenty arrovae. There are therefore four proportional numbers. So the product of one night into twenty arrovae, when divided by five ninths, will produce the unknown number of nights.

(*c*) Or otherwise. Put as the unknown number of nights a thing. You already know[1681] that the nightly consumption of forty lamps is five ninths of an arrova. Thus it is clear that the ratio of one night to five ninths is the same as the ratio of the unknown number of nights during which forty lamps consume twenty arrovae to twenty arrovae. There are therefore four proportional numbers. So the product of one night into twenty arrovae, which is twenty, is equal to the product of a thing into five ninths, which

---

[1678] B.277*c*.

[1679] *probatio*: a reasoning (*causa*), in this case as in B.282*a*. See note 1601.

[1680] Same reasoning as in B.277*a*.

[1681] Above, *b*.

984   Part Two: Translation, Glossary

is five ninths of a thing. Then the thing is thirty-six, which is the required number of nights.

AGAIN ON THE SAME TOPIC.

(**B.280**) One lamp consumes two ninths of an eighth in one night. In how many nights will forty lamps consume twenty arrovae?

(*a*) You will do the following. Multiply the numbers from which a ninth and an eighth are denominated, thus producing seventy-two. Multiplying this by twenty arrovae produces one thousand four hundred and forty. Divide this by forty [*and the result is thirty-six*], and half the result, which is eighteen [*because he said 'two ninths of an eighth'*][1682], is the required number of nights.

The reason for taking half the result of the division is the following.[1683] You know that an arrova is seventy-two ninths of an eighth and that one lamp consumes nightly two ninths of an eighth of these [*which are two parts of the seventy-two ninths of an eighth*][1684]. Multiplying these seventy-two ninths of an eighth by twenty arrovae will produce [*ninths of eighths, namely*] one thousand four hundred and forty ninths of eighths. This is the nightly consumption of seven hundred and twenty lamps; for the nightly consumption of each is two ninths of an eighth.[1685] You further know that these seven hundred and twenty double-parts are consumed by forty lamps. Dividing them by forty will produce the unknown number of nights in which forty lamps consume twenty arrovae.

(*b*) Or otherwise. Multiply half of seventy-two, which is thirty-six, by twenty and divide the product by forty; the result will be eighteen, which is the required number of nights.

(*c*) Or otherwise. Forty lamps consume nightly eighty ninths of an eighth, which is one arrova and a ninth. Dividing twenty by one and a ninth produces eighteen, which is the number of nights.

AGAIN ON THE SAME TOPIC [1686]

(**B.281**) Someone asks: Six lamps consume three eighths (of an arrova) in one night. How many arrovae do they consume in thirty nights?

You will do the following. It was supposed that the nightly consumption of six lamps is three eighths. Then multiply this by thirty nights; the result is eleven arrovae and a fourth, and this is what you wanted.

AGAIN ON THE SAME TOPIC.

[1682] Anticipating what will be explained below. However, what makes this suspect is the place rather than the repetition itself (another instance of which occurs in B.304′).

[1683] The explanation which follows is confused and surely cannot be what was in the original text.

[1684] 'of these' was considered unclear.

[1685] This is supposed to explain why we took half the result.

[1686] But from now on the first set of data applies to more than one lamp.

LIBER MAHAMELETH

(B.282) Someone asks: Three lamps consume an eighth of an arrova in one night. How many arrovae will ten lamps consume in thirty nights?

(*a*) You will do the following. Multiply the number from which an eighth is denominated, namely eight, by three lamps; this produces twenty-four, which you take as the principal. Next, multiply ten lamps by thirty nights and divide the product by the principal; the result will be twelve and a half, which is the required number of arrovae.

Such is the proof.[1687] If the nightly consumption of one lamp were an eighth, ten lamps should consume in thirty nights thirty-seven arrovae and a half according to our explanation in the first chapter on lamps[1688]; for ten lamps are multiplied by thirty nights and the product is divided by the number from which an eighth is denominated, namely eight. Since it has been supposed that three lamps consume an eighth in a night, you are to take a third of these thirty-seven arrovae and a half which resulted from the division; that is, you will divide them by three. But you know[1689] that multiplying ten by thirty, dividing the product by eight and then the quotient by three is the same as multiplying ten by thirty and dividing the result by the product of eight into three.

(*b*) Or otherwise. You know that since the nightly consumption of three lamps is an eighth, each lamp must consume nightly a third of an eighth. Then it is like saying: 'One lamp consumes in one night a third of an eighth; how many arrovae will ten lamps consume in thirty nights?' Do as explained above.[1690] That is, multiply ten by thirty and divide the product by twenty-four which results from (multiplying) the denominators of a third and an eighth. This will give what you wanted. You may then further treat this problem in all the various ways explained in the first chapter on lamps.[1691]

(*c*) Or otherwise. Since the nightly consumption of three lamps is an eighth, that of ten lamps must be three eighths and a third of an eighth; for ten lamps is three and a third times three lamps. Then multiplying three and a third eighths by thirty will give the result you require, which is twelve arrovae and a half.

AGAIN ON THE SAME TOPIC.

(B.283) Someone asks: Six lamps consume three eighths in one night. How many arrovae do ten lamps consume in thirty nights?

(*a*) You will do the following. Multiply the number from which an eighth is denominated, namely eight, by six; this produces forty-eight, which you take as the principal. Next, multiply three, which is the quantity of eighths, by ten lamps; this produces thirty. Multiplying it by thirty nights produces

---

[1687] Here too (as in B.279*a*), *probatio*.

[1688] B.276.

[1689] Premiss P_4, but with a product as the first term.

[1690] B.276.

[1691] In the various ways for one lamp, thus B.276*b–c*.

986   PART TWO: TRANSLATION, GLOSSARY

nine hundred. Divide it by the principal; this will give eighteen arrovae and three fourths of one arrova.

The reason for this is clear. If one lamp consumed three eighths nightly, ten lamps should consume a hundred and twelve arrovae and a half in thirty nights according to what was explained [*in the second chapter on candles* [1692]], namely: you multiply three eighths by ten, thus producing thirty eighths, which you multiply by thirty nights, thus producing nine hundred eighths, and you divide nine hundred by eight from which an eighth is denominated. And since it was stated that six lamps consume three eighths in one night, you are to take a sixth of one hundred and twelve and a half, that is, divide one hundred and twelve and a half by six. But you know [1693] that multiplying three by ten, multiplying the product by thirty, dividing the result by eight and dividing the quotient by six is the same as multiplying three by ten, the product by thirty, and dividing the result by the product of eight into six. This is what we wanted to demonstrate.

(**b**) Or otherwise. The nightly consumption of six lamps being three eighths, each lamp must consume half an eighth in a night. Then it is like saying: 'One lamp consumes half an eighth in one night; how many arrovae do ten lamps consume in thirty nights?' Solve it in all the ways shown above;[1694] the result will be eighteen and three fourths.

AGAIN ON THE SAME TOPIC.

(**B.284**) Someone asks: Six lamps consume three eighths in one night. How many lamps consume twenty arrovae in thirty nights?

(**a**) You will do the following. Multiply the number of eighths by thirty, thus producing ninety; take it as the principal. Next, multiply the number from which an eighth is denominated, which is eight, by six lamps, thus producing forty-eight. Multiply this by twenty, thus producing nine hundred and sixty. Divide this by the principal; the result will be ten lamps and two thirds of one lamp, which is what you wanted.[1695]

The reason for this is the following. Since the nightly consumption of six lamps is three eighths, they must consume in thirty nights ninety eighths. Now it has been supposed that the lamps in unknown number [1696] consume in thirty nights twenty arrovae, which is a hundred and sixty eighths. Therefore it is clear that the ratio of six lamps to their consumption in thirty nights, which is ninety eighths, is the same as the ratio of the unknown number of lamps to their consumption in thirty nights, which is a hundred and sixty eighths. There are therefore four proportional numbers.

[1692]   *Sic.* The author's allusion is to B.276, to which the present problem is reduced, and not to B.282, the second problem in the second section.

[1693]   Premiss P_4, with several factors as the first term. See note 1689.

[1694]   In B.276.

[1695]   Fractional lamp, as in B.277.

[1696]   *lampades ignote* or (below) *numerus lampadarum ignotarum.*

LIBER MAHAMELETH
987

Then multiplying six by a hundred and sixty and dividing the product by ninety will give the unknown number of lamps. Or otherwise.[1697] Consider which number multiplying ninety eighths produces a hundred and sixty eighths; you will find one and seven ninths. Then multiply one and seven ninths by six; the result will be ten and two thirds, which is the number of lamps.

(*b*) Or otherwise. You already know that six lamps consume in thirty nights ninety eighths, which is eleven arrovae and a fourth; take this as the principal. Next, multiply six by twenty and divide the product by the principal; the result will be ten and two thirds.

The reason for this is the following. We know that ninety eighths, which six lamps consume in thirty nights, correspond to eleven arrovae and a fourth. Thus it is clear that the ratio of six lamps to eleven arrovae and a fourth is the same as the ratio of the unknown number of lamps to the twenty arrovae they consume in thirty nights. There are therefore four proportional numbers. Then multiplying six by twenty and dividing the product by eleven and a fourth will give the unknown number of lamps.

(*c*) Or otherwise. You know that, since six lamps consume three eighths in a night, one lamp must consume half an eighth in a night. Then it is like saying: 'One lamp consumes half an eighth in one night; how many lamps consume twenty arrovae in thirty nights?' Do as has been shown above.[1698] The result will be ten and two thirds, which is the number of lamps.

AGAIN ON THE SAME TOPIC.

(**B.285**) Someone asks: Six lamps consume three eighths in one night. In how many nights will ten lamps consume twenty arrovae?

(*a*) You will do the following. Multiply the number of eighths, which is three, by ten, thus producing thirty; take it as the principal. Next, multiply the number from which an eighth is denominated, namely eight, by six, thus producing forty-eight. Multiply this by the twenty arrovae and divide the product by the principal; the result will be thirty-two, which is the required number of nights.[1699] The reason for this is clear from the foregoing.[1700]

(*b*) Or otherwise. Since the nightly consumption of six lamps is three eighths, each lamp must consume half an eighth in a night. Then it is like saying: 'One lamp consumes in a night half an eighth; in how many nights will ten lamps consume twenty arrovae?' Do as we have taught before;[1701] the result will be what you wanted.

[1697] This just computes the same keeping the fractional eighths.

[1698] B.277.

[1699] *numerus noctium ignotarum.*

[1700] See B.284*a*.

[1701] B.279.

Chapter (B-XVI) on the Consumption by animals

As far as the problems are concerned, this chapter does not differ from that on lamps, except for the later part, that is, the chapter on unknown numbers of animals.[1702]

(**B.286**) Someone asks: One animal eats a fourth of a caficius in one night. How many caficii will twenty animals eat in thirty nights?

(*a*) Take the number from which a fourth is denominated, namely four, as the principal. Next, multiply twenty by thirty, thus producing six hundred. Divide this by the principal; the result will be a hundred and fifty, which is the number of caficii.

The reason for this is the following. Since one animal eats a fourth of a caficius in one night, twenty animals must eat twenty fourths in one night. Multiplying this by thirty produces six hundred fourths of a caficius, which is the consumption of twenty animals in thirty nights. Divide this[1703] by the denominator of a fourth in order to know the number of caficii it contains. Or otherwise.[1704] You know that the twenty fourths eaten by twenty animals in one night are five caficii. Consequently, they will eat in thirty nights a hundred and fifty caficii; that is, multiplying five caficii by thirty nights will produce what you require.

(*b*) Or otherwise. You know that since one animal eats a fourth of a caficius in one night, it must eat in thirty nights thirty fourths of a caficius, which is seven caficii and a half. Multiply this by the number of animals; the result will be a hundred and fifty, and this is what you wanted to know.

(*c*)[1705] You are to know that the number of caficii eaten by one animal in a month is the same as the number of almodis it eats in a year. For when one animal eats seven caficii and a half in a month, it will perforce eat seven almodis and a half in a year: the number of almodis in a year will always be the same as the number of caficii in a month.

The reason for this is clear. When one animal eats seven caficii and a half in a month and you want to know the number of almodis it eats in a year, you will multiply seven caficii and a half by the number of months in the whole year, which is twelve, and divide the product by the number of caficii making one almodi, that is, twelve. But you know that multiplying seven and a half by twelve and dividing the product by twelve gives seven and a half. This is the cause of the above.

[1702] See B.294 *seqq.* It should be noted that B.286–293 are found only in MS $\mathcal{B}$, the text of which is not always reliable here (see notes below).

[1703] That is, six hundred.

[1704] This just performs the division in the first fraction. See B.274*a–b*.

[1705] This hardly seems to be the right place for such a conversion rule.

LIBER MAHAMELETH

989

(*d*) Or otherwise.[1706] You know that since one animal eats a fourth of a caficius in one night, it must eat a caficius in four nights. Then find the number which when multiplying four nights produces thirty; you will find seven and a half. Thus multiply a caficius by seven and a half and multiply the product by twenty; the result will be a hundred and fifty caficii, which is what you wanted.

AGAIN ON THE SAME TOPIC.

(**B.287**) Someone asks: One animal eats two fifths of a caficius in one night. How many caficii will twenty animals eat in thirty nights?

You will do the following. Put the number from which a fifth is denominated, thus five, as the principal. Next, multiply twenty by thirty, thus producing six hundred. Divide it by the principal; this gives a hundred and twenty. Double this; the result is two hundred and forty, which is what you wanted.

The reason for doubling the quotient is the following.[1707] You know that since one animal eats in one night two fifths of a caficius, it must in thirty nights eat sixty fifths of a caficius. Multiplying this[1708] by the number of animals and dividing the product by the denominator of a fifth will give two hundred and forty. Now this is the same as multiplying twenty by thirty, dividing the product by the denominator of a fifth and doubling the result, for the quantity of food eaten is doubled.[1709] The various ways of solving this problem are like those of the previous one.[1710]

AGAIN ON THE SAME TOPIC.

(**B.288**) Someone asks: One animal eats a third of a caficius in one night. How many animals will eat forty caficii in thirty nights?

(*a*) You will do the following. Multiply the number from which a third is denominated, which is three, by forty, thus producing a hundred and twenty. Divide this by thirty; the result is four, (which is the number of) animals.

The reason for this is the following. We know that when one animal eats a third of a caficius in one night, a hundred and twenty animals must eat forty caficii in one night. Thus divide a hundred and twenty by thirty nights so as to know how many animals correspond to each night.

(*b*) Or otherwise. Since one animal eats a third of a caficius in one night, it must eat ten caficii in thirty nights.[1711] Thus divide forty caficii by ten; this gives four, which is the number of animals.

[1706] Not significantly different from *b*.

[1707] The subsequent explanations are unclear.

[1708] That is, sixty.

[1709] With respect to the consumption of one fifth in one night.

[1710] B.286, *b* and *d*.

[1711] We would expect this in the justification below.

990   Part Two: Translation, Glossary

The reason for this is the following. We know that the ratio of one animal to its monthly [1712] consumption, which is ten caficii, is the same as the ratio of the unknown number of animals to their monthly consumption, which is forty caficii. There are therefore four proportional numbers. Then the product of one animal into forty caficii when divided by ten will give the unknown number of animals, and this is what you wanted.

AGAIN ON THE SAME TOPIC.

(B.289) Someone asks: One animal eats a fourth of a caficius in one night. In how many nights do fifty animals eat a hundred caficii?

(a) You will do the following. Multiply the number from which a fourth is denominated by a hundred, thus producing four hundred. Divide it by fifty; this gives eight, which is the number of nights.

The reason for this is the following. Since one animal eats a fourth of a caficius in a night, it must eat in four hundred nights a hundred caficii. Now you know that fifty animals eat a hundred caficii. Thus dividing four hundred by fifty will give the number of nights.

(b) Or otherwise. We know that since one animal eats a fourth of a caficius in a night, fifty animals must eat twelve caficii and a half in one night. Thus divide a hundred caficii by twelve and a half; the result is eight, which is the required number of nights. The reason for this is clear to those understanding what has gone before.[1713]

AGAIN ON THE SAME TOPIC.

(B.290) Someone asks: One animal eats three eighths of a caficius in one night. In how many nights will twenty animals eat a hundred caficii?

You will do the following. Multiply the number from which an eighth is denominated by a hundred and divide the product by twenty; this will give forty. A third of it —since it was said that 'three eighths' are eaten in a night—, which is thirteen and a third, is what you wanted.

The reason for our taking a third of the quotient is the following. We know that if one animal were to consume in a night an eighth of a caficius, it would consume a hundred caficii in eight hundred nights. But since it has been supposed that one animal consumes in a night three eighths of a caficius, it must consume a hundred caficii in two hundred and sixty-six nights and two thirds. Dividing this by the number of animals, which is twenty, will give the number of nights. Now you know that multiplying eight by a hundred and dividing a third of the product by twenty is the same as multiplying eight by a hundred, dividing the product by twenty and taking a third of the result. For this problem you will again follow the various ways we have taught before.[1714]

AGAIN ON THE SAME TOPIC.

[1712] The thirty nights mentioned before.

[1713] Setting the proportion as, e.g., in B.279b.

[1714] 'again': as for B.287; 'various ways': see B.279–280.

LIBER MAHAMELETH   991

(B.291) Someone asks: Five animals eat eight caficii in six nights. How many caficii will twenty animals eat in thirty nights?

(**a**) You will do the following. Multiply five animals by six nights, thus producing thirty; take it as the principal. Next, multiply eight caficii by twenty animals, multiply the product by thirty nights and divide the result by the principal; the quotient will be a hundred and sixty caficii, and this is what you wanted.

The reason for this is the following. We know that the ratio of five animals to the eight caficii they eat in six nights is the same as the ratio of twenty animals to their consumption in six nights. There are therefore four proportional numbers. Then dividing the product of eight into twenty by five will give the number of caficii consumed by twenty animals in six nights, which is thirty-two. It is also clear that the ratio of six nights to thirty nights is the same as the ratio of the number of caficii for six nights, which is thirty-two, to the number of caficii for thirty nights. There are therefore four proportional numbers. Thus dividing the product of thirty into thirty-two by six will give the number of caficii eaten by twenty animals in thirty nights. Therefore it is clear that multiplying eight by twenty, dividing the product by five, multiplying the result by thirty and dividing the product by six will give the number of caficii eaten by twenty animals in thirty nights. But we have established before[1715] that multiplying eight by twenty, dividing the product by five, multiplying the result by thirty and dividing the product by six is the same as multiplying eight by twenty, the result by thirty and dividing this product by the product of six into five. This is what we wanted to demonstrate.

(**b**) Or otherwise. We know that since five animals eat eight caficii in six nights, they must eat forty caficii in thirty nights. Then it is like asking: 'Five animals eat forty caficii in thirty nights; how many caficii will twenty animals eat in thirty nights also?' There are therefore four proportional numbers. Then multiplying twenty by forty and dividing the product by five will give a hundred and sixty, which is the number of caficii consumed by twenty animals. Or otherwise. Consider what number when multiplying five animals produces forty, which is the quantity of their caficii; you will find eight. Multiply it by twenty; this produces a hundred and sixty, which is what you wanted. [*The reason for this is clear to those who reflect upon what has been said before.*][1716]

(**c**) Or otherwise. You know that when five animals eat eight caficii in six nights, one of them must eat in one night a fifth of a caficius and a third of a fifth of a caficius. Then it is like asking: 'One animal eats a fifth of a caficius and a third of a fifth of a caficius in one night; how many caficii will twenty animals eat in thirty nights?' Do as we have taught above,[1717] which is as follows. Take the number from which a fifth is denominated

---

[1715] Extension of Premiss P_4. See notes 1689, 1693.

[1716] Analogous interpolations in B.235–235‴.

[1717] See B.286–287; but here with a compound fraction.

992 PART TWO: TRANSLATION, GLOSSARY

as the principal.[1718] Next, multiply twenty by thirty, thus producing six hundred. Dividing this by the principal gives a hundred and twenty. Next, consider what number when multiplying a fifth makes a fifth and a third of a fifth; you will find one and a third.[1719] Multiplying this by a hundred and twenty will produce a hundred and sixty, which is what you wanted. Or otherwise.[1720] You know that when one animal consumes in a night a fifth and a third of a fifth, twenty animals must consume in one night five caficii and a third. Thus multiply five caficii and a third by thirty nights; the result will be a hundred and sixty caficii.

AGAIN ON THE SAME TOPIC.

(B.292) Someone asks: Five animals eat six caficii in eight nights. How many animals will eat fifty caficii in thirty nights?

(a) You will do the following. Multiply thirty by six caficii; the result is a hundred and eighty, which you take as the principal. Next, multiply five animals by eight nights, multiply the result by fifty and divide the product by the principal; the result will be eleven animals and a ninth of the consumption of one animal.[1721]

The reason for this is the following. We know that the animals in unknown number consume fifty caficii in thirty nights. But we know that five animals consume six caficii in eight nights. Then it is clear that the ratio of five animals to their consumption in eight nights, which is six caficii, is the same as the ratio of an unknown number of animals[1722] to their consumption which, in eight nights too, is fifty caficii. There are therefore four proportional numbers. Then dividing the product of five animals into fifty by six caficii will give the number of animals eating fifty caficii in eight nights[1723], which is forty-one and two thirds[1724]. It is also clear that the ratio of thirty nights to eight nights is the same as the ratio of the number of animals for eight nights, which is forty-one and two thirds, to the number of animals for thirty nights; the number of animals eating fifty caficii in eight nights is actually greater than the number of animals eating fifty caficii in thirty nights, which is why the proportion is formed in this way. Thus multiplying eight by forty-one and two thirds and dividing the product by thirty will give the number of animals eating fifty caficii in thirty nights. But you know[1725] that multiplying five by fifty, dividing the product by six, multiplying the quotient by eight and dividing the product by thirty is the same as multiplying five by fifty, multiplying the product by eight and dividing the result by the product of six into thirty.

[1718] For $\frac{1}{5} + \frac{1}{3}\frac{1}{5} = \frac{1}{5}\left(1 + \frac{1}{3}\right)$.

[1719] Namely the numerator of the compound fraction (note 116, p. 606).

[1720] Direct, and more appropriate, computation.

[1721] Note here the expression of the fractional result.

[1722] *animalia ignota*, but not the required number of the problem.

[1723] This is specified for it is not the required number.

[1724] Fractional animal.

[1725] Extension of Premiss P_4. See note 1715.

LIBER MAHAMELETH 993

(**b**) Or otherwise. We know that when five animals eat six caficii in eight nights, they must eat three fourths of a caficius in one night. Then it is like saying: 'Five animals eat three fourths of a caficius in one night; how many animals eat fifty caficii in thirty nights?' Do as has been shown [*in the first chapter*], namely as follows.[1726] Multiply the quantity of fourths, which is three, by thirty nights; the result is ninety, which you take as the principal. Next, multiply the number from which a fourth is denominated by five animals, thus producing twenty. Multiplying this by fifty caficii gives a thousand. Divide it by the principal; the result will be eleven and a ninth.

(**c**) Or otherwise. You know that when five animals eat three fourths of a caficius in one night, one of them must eat a fifth of three fourths of a caficius (in one night). Then it is like saying: 'One animal eats in a night three fourths of a fifth[1727] of a caficius; how many animals eat fifty caficii in thirty nights?' You will do as I have taught above[1728], which is as follows. Multiply the product of the denominators of a fourth and a fifth, which is twenty, by fifty; this produces a thousand. Next, multiply the quantity of fourths, namely three, by thirty; this gives ninety. Divide a thousand by it; the result is eleven and a ninth, which is what you wanted to know. Or otherwise. Divide a thousand by thirty; this gives thirty-three and a third. Take a third of it, which is eleven and a ninth —for it was said above: 'three fourths of a fifth'; we have already established the reason for this.[1729]

AGAIN ON THE SAME TOPIC.

(**B.293**) Someone asks: Six animals eat eight caficii in ten nights. In how many nights will forty animals eat a hundred caficii?

(**a**) You will do the following. Multiply eight by forty, thus producing three hundred and twenty; take it as the principal. Next, multiply six animals by ten nights, multiply the product by a hundred caficii and divide the result by the principal; this will give eighteen and three fourths, which is the number of nights. The reason for this is clear to those understanding what has gone before.[1730]

(**b**) Or otherwise. Since six animals eat eight caficii in ten nights, they must eat eight tenths of a caficius in one night. Then it is like saying: 'Six animals eat eight tenths of a caficius in one night; in how many nights will forty animals eat a hundred caficii?' Do as has been explained above, which is as follows.[1731] Multiply eight by forty animals, thus producing three hundred and twenty; take it as the principal. Next, multiply the

---

[1726] The reference is to the computations seen in part *a*.

[1727] With the terms inverted: usual form of a fraction of a fraction.

[1728] B.288.

[1729] B.290.

[1730] Vague reference, possibly an early reader's addition. As a matter of fact, the justification follows (B.292*b*).

[1731] It repeats the computations seen in *a*.

994 PART TWO: TRANSLATION, GLOSSARY

number from which a tenth is denominated, which is ten, by six, multiply the product by a hundred and divide the result by the principal; this will give eighteen and three fourths.

(**c**) Or otherwise. Consider what fraction one animal is of six animals; it is a sixth. Thus take a sixth of eight tenths; this is a tenth and a third of a tenth. Then it is like saying: 'One animal eats a tenth of a caficius and a third of a tenth in one night; in how many nights will forty animals eat a hundred caficii?' Do as has been shown before,[1732] which is as follows. Multiply the denominators of a third and a tenth; the result is thirty. Take a tenth and a third of a tenth of it; this is four. Multiply it by forty, thus producing a hundred and sixty; take it as the principal. Next, multiply thirty by a hundred; this produces three thousand. Divide it by the principal; the result will be eighteen and three fourths, and this is what you wanted.

⟨CHAPTER ON UNKNOWN NUMBERS OF ANIMALS⟩[1733]

(**B.294**) For instance, someone asks: An unknown number of animals eat in a month the quintuple of that number and in ten nights six of them eat a quantity equal to a fourth of the unknown number of animals. What is the unknown number of animals?

(**a**) You will do the following. You know that since the unknown number of animals eat in a month the quintuple of that number, six of them must eat (in a month) the quintuple of six, namely thirty caficii. You also know that since six animals eat in a month thirty caficii, they must eat in ten days[1734] a third of thirty caficii, which is ten caficii; for ten days is a third of thirty days, and this means a third of the consumption. Now it was said above that six animals eat in ten days a quantity equal to a fourth of the unknown number of animals. Therefore ten, the number of caficii, must be a fourth of the number of animals; thus the unknown number of animals is forty, and this is what you wanted.

(**b**) Or otherwise. Put as the unknown number of animals a thing. Then their monthly consumption will be five things. Again, the consumption of six animals in ten days will be a fourth of a thing, for it has been supposed above that the consumption of six animals in ten days is a quantity equal to a fourth of the unknown number of animals. But you know that when six animals consume in ten days a fourth of a thing, they will perforce consume in a month three fourths of a thing. Thus it is clear that the ratio of the unknown number of animals, which is a thing, to their monthly consumption, which is five things, is the same as the ratio of six animals to their monthly consumption, which is three fourths of a thing. There are therefore four proportional numbers. Then the product of a thing into three fourths of a thing is equal to the product of five things into six. Therefore

---

[1732] B.289.

[1733] Mentioned at the beginning of this chapter (p. 988).

[1734] Day-time eating from now on.

LIBER MAHAMELETH   995

three fourths of a square equal thirty things. Thus a square equals forty things. So the thing is forty, which is the required number of animals.

AGAIN ON THE SAME TOPIC.

(B.295) Someone asks: An unknown number of animals eat in a month sixty caficii and six of them eat in five days a quantity equal to three fifths of that unknown number. How many animals are there?

(*a*) You will do the following. You already know that since the unknown number of animals consume in a month sixty caficii, they must consume in five days ten caficii. You also know that six animals consume in five days a quantity equal to three fifths of the unknown number of animals. Thus it is clear that the ratio of the unknown number of animals to their consumption in five days, which is ten caficii, is the same as the ratio of six animals to their consumption in five days, which is three fifths of the unknown number of animals. There are therefore four proportional numbers. Then the product of the unknown number of animals into three fifths of itself, which is the consumption of six animals, is equal to the product of six animals into ten caficii. Since the product of six into ten is sixty, the product of the unknown number of animals into three fifths of itself is sixty. Since the product of this number into three fifths of itself is sixty, its product into itself is a hundred. Then the unknown number of animals is the root of a hundred, which is ten.

A short solution of this problem is as follows.[1735] Denominating five days from the whole month gives a sixth. Then multiply a sixth of sixty caficii, which is ten, by six animals, which produces sixty, and divide this by three fifths. The root of the result is the number of animals.

(*b*) Or otherwise. Put as the unknown number of animals a thing. It is known that they eat in a month sixty caficii, while six animals eat in five days (a number of caficii equal to) three fifths of the unknown number of animals, which is three fifths of a thing, and so they must eat in a month three things and three fifths of a thing. Therefore it is clear that the ratio of the unknown number of animals, which is a thing, to sixty caficii, which they eat in a month, is the same as the ratio of six animals to their monthly consumption, which is three things and three fifths of a thing. Thus the product of a thing into three things and three fifths of a thing, namely three squares[1736] and three fifths of one square, is equal to the product of sixty caficii into six animals, which is three hundred and sixty.[1737] Reduce then all squares to one square, and reduce in proportion what they are equal to. This will in the end leave a square equal to a hundred nummi.[1738] Thus the root of the square is ten. So much is the thing worth and so much the required number of animals. [*This is what we wanted to demonstrate.*]

[1735] Summing up the above reasoning.

[1736] MS $\mathcal{B}$ has *census* each time followed by the adjective *habitus*.

[1737] Equation of the form $ax^2 = c$.

[1738] Here as in B.296*b*, in the sense of 'units' (see p. 841, note 1120).

996   PART TWO: TRANSLATION, GLOSSARY

AGAIN.

(B.296) Someone asks: An unknown number of animals consume in a month the decuple of that number and five of them consume in six days a quantity equal to the root of this unknown number of animals. How many animals are there?

(a) You will do the following. It is perfectly clear that since the animals in unknown number consume in a month the decuple of that number, five of them must consume in a month the decuple of their number, namely fifty. They will thus perforce consume in six days a fifth of fifty, which is ten caficii. But it has been supposed that they consume in six days as much as the root of the unknown number of animals. Ten must therefore be the root of the number of animals and this number, a hundred.

(b) Or otherwise. Put as the unknown number of animals one square. Then their monthly consumption will be ten squares and the monthly consumption of five animals, five things. It is therefore clear that the ratio of a square, which is the unknown number of animals, to ten squares, which is their monthly consumption, is the same as the ratio of five animals to the five things they consume in a month. There are therefore four proportional numbers. Then the product of a square into five things is equal to the product of ten squares into five animals. The final result will then be five cubes equal to fifty squares.[1739] Dividing the whole by one square gives five things equal to fifty nummi. Thus a thing equals ten. This is the root of the number of animals, and there are therefore a hundred animals. This is what you wanted to know.

Such problems about unknown numbers of animals can be stated for lamps as well.[1740]

(B.297) Thirty geese consume in a month six sextarii. The first day of the month, after receiving its food, one is killed, the second day likewise another one, and so on each day till the end of the month. What remains in the end of the six sextarii?

You will do the following. Divide thirty by six, which gives five. Multiply this by thirty, which produces a hundred and fifty. Take it as the principal. Next, always add one to thirty; the result is thirty-one. Multiply it by half of thirty; this gives four hundred and sixty-five. Divide it by the principal; the result is three sextarii and a tenth, and so much is consumed in the month according to the conditions. The remainder of the six sextarii is what is left, namely three sextarii minus a tenth.

We have divided thirty by six because when thirty geese consume in a month six sextarii, five must consume in a month one sextarius. Each of them eats in a day a one-hundred-and-fiftieth of a sextarius: for it eats a

---

[1739] Equation of the form $ax^3 = bx^2$. Second (after B.263) and last occurrence of the cubic power of the unknown.

[1740] That is, with an unknown number of lamps and another quantity depending upon it.

LIBER MAHAMELETH

fifth of a third of one tenth of one sextarius, which is this fraction. Then it is like saying: 'There are thirty workers; the first worker's wage is a one-hundred-and-fiftieth and the other wages exceed one another by this same difference'. Do as has been explained in the chapter on workers,[1741] namely as follows. Multiply the difference, which is a one-hundred-and-fiftieth, by the number of workers minus one and add to the result the lowest wage, that of the first worker, which is a one-hundred-and-fiftieth; this produces thirty one-hundred-and-fiftieths, which is the highest wage, that of the last worker. Add to it the wage of the first, which is a one-hundred-and-fiftieth; this makes thirty-one (such parts), which is the sum of the first and last workers' wages. Multiply it by half the number of workers, thus producing four hundred and sixty-five (which are) one-hundred-and-fiftieths of a sextarius. Dividing this by the number from which they are denominated, namely a hundred and fifty, will give three sextarii and a tenth. This is what the geese eat, and the remainder of the six sextarii is what is left.

Chapter on another topic

(**B.298**) For instance, one Segovian modius and a fourth are equal to one caficius and two thirds in Toledo. How many modii equal twenty caficii?

This problem is like saying: 'A modius and a fourth is given for a nummus and two thirds, how much shall I have for twenty nummi?' You will solve this problem in all the ways I have taught in the chapter on buying and selling.[1742]

(*a*) That is, multiply one modius and a fourth by twenty; this produces twenty-five. Divide it by one caficius and two thirds; this gives fifteen. So many modii equal twenty caficii.

The proof of this is the following.[1743] We know that the ratio of one modius and a fourth to one caficius and two thirds is the same as the ratio of the unknown number of modii to twenty caficii. There are therefore four proportional numbers. Then dividing the product of one modius and a fourth into twenty by one caficius and two thirds will give the required number of modii.

(*b*) Or otherwise. Multiply the denominators of a third and a fourth, thus producing twelve. Multiply this by one caficius and two thirds; the result is twenty, which you take as the principal[1744]. Next, multiply twelve by one modius and a fourth; this produces fifteen. Then it is like saying: 'When fifteen for twenty, how much shall I have for twenty nummi?' Proceed then

---

[1741] B.245.

[1742] B.9–10 & B.13–14.

[1743] The mention of the capacity units *modius* and *caficius* in what follows might be confusing to the reader since the quantities indicated are in fact equal.

[1744] The point here seems to be to distinguish this twenty from the one in the data, since the division is not performed.

using all the ways I have taught above;[1745] the result will be fifteen, and this is what you wanted.

AGAIN ON THE SAME TOPIC.

(B.299) One modius and a fourth in Segovia are equal to one caficius and two thirds in Toledo. How many Toledan caficii equal fifteen modii?

Solve this problem in all the ways I have taught in the chapter on buying and selling.

We have introduced these problems merely to make you aware that they are similar to those concerning buying and selling.

[1745] But here two corresponding terms are simply identical.

Chapter (B-XVII) on the Consumption of bread by men

Note that in like manner [1746] there are four quantities involved here: a number of arrovae, the number of loaves made from them, the number of men eating bread, the number of days in which they eat. This gives rise to four types of problem, according to whether the numbers of loaves, men and days are given and the number of arrovae is required, or the numbers of arrovae, men and days are given and the number of loaves is required, or the numbers of arrovae, loaves and men are given and the number of days is required, or the numbers of arrovae, loaves and days are given and the number of men is required.[1747] These types vary according to the (given amounts of these) quantities, which may be (integers) larger than one, one, fractions.

(B.300) For instance, someone asks thus: From one arrova twenty loaves are made and one man eats one of them each day. How many arrovae will forty men consume in thirty days?

(**a**) You will do the following. Multiply thirty by forty and divide the product by twenty; the result will be sixty, and this is the number of arrovae.

The reason for this is the following. We know [1748] that from an arrova twenty loaves are made and that forty men eat forty loaves in a day. Thus multiplying forty loaves by thirty days will produce the number of loaves [*corresponding to thirty days*][1749] eaten by forty men in thirty days, namely one thousand two hundred loaves. Divide it by the number of loaves to the arrova, which is twenty, so as to know how many arrovae it corresponds to.

(**b**) Or otherwise. You know that forty men eat forty loaves in a day. This is two arrovae since twenty loaves are made from an arrova. Thus multiply two arrovae by thirty days; the result is sixty arrovae, and this is what you wanted to know.

(**c**) Or otherwise. One man eats thirty loaves in thirty days. Now this is one arrova and a half. Thus multiply one arrova and a half by forty men; the result is sixty, and this is what you wanted.

AGAIN ON THE SAME TOPIC.

[1746] See introduction to the chapter on lamps.

[1747] In actual fact, since the number of loaves and that of bushels (*caficii* or *arrovae*) differ merely by a constant, there are three types of problem: bushels or loaves required (B.300–302); days required (B.303–304); men required (B.305).

[1748] Data and immediate inference.

[1749] Misplaced gloss, perhaps referring to 'arrovae' below.

1000   PART TWO: TRANSLATION, GLOSSARY

(B.301) From one arrova forty loaves are made and one man eats two of them each day. How many arrovae do twenty men eat in thirty days?

(a) You will do the following. Multiply twenty by thirty; this gives six hundred. Dividing it by forty gives fifteen. Double it; the result is thirty, which is what you wanted to determine.

The reason for doubling is the following. You know that twenty men consume in a day forty loaves, for each man eats two. Thus multiplying forty by thirty and dividing the product by the number of loaves to the arrova will give what you wanted. But you know [1750] that multiplying two loaves by twenty, then the result by thirty and dividing the product by forty is the same as multiplying twenty by thirty, then dividing the product by forty and doubling the result.

(b) Or otherwise. It has been supposed that one man eats two loaves (in a day); so twenty men must consume in a day forty loaves, which is one arrova. Thus multiply one arrova by thirty days; the result is thirty arrovae, and this is what you wanted.

(c) Or otherwise. It has been supposed that two loaves are eaten in a day by one man; thus he must consume in thirty days sixty loaves, which is one arrova and a half. Multiply this by the number of men, which is twenty; the result will be thirty arrovae, and this is what you wanted.

(B.302) Someone asks: Two men eat in three nights ten loaves, forty of which are made from one caficius. How many (caficii) will twenty men eat in forty-five nights? [1751]

(a) From what has been explained in the (chapter on) hiring carriers it appears that the ratio of one product of men into nights to another product of men into nights is the same as the ratio of the corresponding numbers of loaves; for what we said, namely 'Two men eat in three nights ten loaves; how many will twenty men eat in forty-five nights?', is as if we would say: 'A man is hired to carry two sextarii three miles at ten nummi, but he carried twenty sextarii forty-five miles; what is due to him?' [1752] Thus we must treat the present problem like the other, namely as follows. Multiply two men by three nights and take the product as the principal. Next, multiply twenty men by forty-five nights and the product by ten; the result is nine thousand, which, divided by the principal, will give the number of loaves. Now we want to know the number of caficii they correspond to. We know that from one caficius forty loaves are made. Thus divide the number of loaves by forty. But all this is the same as dividing nine thousand by the product of six into forty. This is why [1753] we multiply the (first) number

---

[1750] P_5 extended.

[1751] In B.302–303 *caficius* replaces *arrova* and we have night-time eating (see B.294). These two problems, found only in MS $\mathcal{A}$, may well have been added by the author later.

[1752] See B.254.

[1753] Explanation of the general formula.

LIBER MAHAMELETH 1001

of men by the (first) number of nights, the result by the number of loaves to the caficius, and take the product as the principal. Next, we multiply the second number of men by the second number of nights and the result by the number of loaves (they eat). Dividing the product by the principal will give what you wanted.

(**b**) Or otherwise. You already know that when two men eat ten loaves in three nights, one man will eat five loaves in three nights. This being so, in one night he will eat one loaf and two thirds. Therefore twenty men consume in one night thirty-three loaves and a third, and thus one thousand five hundred loaves in forty-five nights. Dividing then one thousand five hundred loaves by forty will give the number of caficii, namely thirty-seven and a half, and this is what you wanted to know.

(**B.303**) Someone asks: Two men eat in one night four loaves, thirty of which are made from a caficius. In how many nights will forty men consume fifty caficii?

(**a**) We already know [1754] that the ratio of one product of men into nights to another product of men into nights is the same as the ratio of loaves to loaves. Therefore convert fifty caficii into loaves, that is, multiply the caficii by the number of loaves made from one caficius; the result is one thousand five hundred. Then the ratio of the product of two men into one night, which is two, to the product of forty men into the unknown number of nights is the same as the ratio of four to one thousand five hundred. Multiplying two by one thousand five hundred and dividing the result by four will therefore give the product of forty men into the unknown number of nights. Dividing this product by forty will then give the unknown number of nights. Now all this is the same as dividing the product of two into one thousand five hundred by the product of four into forty. This is why [1755] we multiply thirty, which is the number of loaves to the caficius, by fifty caficii, then the result by two men, then the product by one night, and divide this last result by the product of the number of loaves into the second number of men. This gives what we want.

(**b**) Or otherwise. We know that when two men eat four loaves in one night, one man eats two in one night. Then forty men eat in one night eighty loaves, thus two caficii and two thirds. We further know that forty men consume fifty caficii. Therefore dividing fifty caficii by two and two thirds will give the unknown number of nights in which forty men consume fifty caficii, which is eighteen and three fourths.

AGAIN ON THE SAME TOPIC.

(**B.304**) Someone asks: From one arrova forty loaves are made and one man eats one of them in a day. In how many days will twenty men consume sixty arrovae?

[1754] B.302*a*.

[1755] General formula thus deduced.

1002 PART TWO: TRANSLATION, GLOSSARY

(*a*) You will do the following. Multiply forty loaves by sixty arrovae and divide the product by twenty men; the result will be a hundred and twenty, and this is what you wanted.

The reason for this is evident. Since forty loaves are made from an arrova and one man eats one loaf in a day, multiplying forty loaves, which are made from an arrova, by sixty arrovae will produce two thousand four hundred loaves, which is the number of loaves ⟨consumed by one man in two thousand four hundred days. But you know that this is the number of loaves⟩ consumed by twenty men in the unknown number of days. Then dividing this result by twenty men will give the unknown number of days [*since twenty men consume each day twenty loaves, you must know how many times twenty is found in two thousand four hundred loaves, and the result is the number of days*][1756].

(*b*) Or otherwise. By hypothesis, twenty men consume each day twenty loaves, which is half an arrova. Divide then sixty arrovae by half an arrova; this gives a hundred and twenty, and this is what you wanted.

The reason for this is that which we have indicated in the chapter on lamps, for the problem is the same.[1757]

(**B.304′**) If he had said that one man ate two loaves:

(*a*) You will proceed as above, and half the result will be what you wanted, because he said 'two loaves'. If he had said that one man ate three or four loaves, or two and a half, what you wanted would be a third, a fourth, or two fifths of the result: that is, the same fraction as that arising from denominating one from the number of loaves, irrespective of the initial choice of this number.[1758]

The reason for taking half the result is the following. You know that forty loaves are made from one arrova; therefore two thousand four hundred loaves will be made from sixty arrovae. So if twenty men consumed twenty loaves in a day, they would consume two thousand four hundred in a hundred and twenty days, as established in the previous problem;[1759] but since the consumption is doubled, for they consume forty loaves each day, they must consume two thousand four hundred loaves in sixty days. This is the reason for taking the half. Consider other, similar cases using the same reasoning.

(*b*) Or otherwise. We know that twenty men consume in a day forty loaves, which is an arrova. Consequently, they will consume in sixty days sixty arrovae.

AGAIN ON THE SAME TOPIC.

[1756] An early reader's attempt to fill in the lacuna.

[1757] B.279*b* and B.280*c* have similar treatments. But this remark is probably misplaced and must apply to B.304′ (reduced to B.304 by assuming a unitary consumption).

[1758] B.278*a*, on lamps, contains a similar remark. See previous note.

[1759] Result found in B.304.

LIBER MAHAMELETH 1003

(**B.305**) From one arrova twenty loaves are made and one man eats one loaf in a day. How many men will consume forty arrovae in thirty days?

(**a**) You will do the following. Multiply twenty by forty, which produces eight hundred. Divide it by thirty, which gives twenty-six and two thirds, and this is the number of men.[1760]

The reason for this is the following. It has been supposed that twenty loaves are made from each of the forty arrovae. Thus, eight hundred loaves will be made from the forty arrovae. It has also been supposed that one man eats one loaf in a day. Thus, eight hundred men will consume eight hundred loaves in a day. But you know that these eight hundred loaves are consumed in thirty days. So dividing eight hundred men by thirty days will give the number of men who eat each day, which is twenty-six and two thirds of one man's consumption.

(**b**) Or otherwise. You know that when one man eats one loaf in a day, he must eat in thirty days thirty loaves, that is, one arrova and a half. Therefore divide forty arrovae by one arrova and a half; the result will be twenty-six and two thirds, and this is what you wanted.

(**B.305′**) If in this problem it had been said that one man eats two loaves:

(**a**) You will do as I have taught above [1761] and take half of the result, because of the 'two loaves' proposed; if there were three, you would take a third of the result.

The reason for this is evident. Eight hundred loaves correspond to forty arrovae. Thus if one man eats one loaf (in a day), these eight hundred loaves will be eaten in thirty days by twenty-six men and two thirds, as we have shown in the foregoing [1762]. But it has been supposed that one man eats two loaves in one day. Consequently, the eight hundred loaves must be eaten in thirty days by thirteen men and a third [1763], since the consumption is doubled. This is why the half is taken.

(**b**) Or otherwise. It has been supposed that one man eats in thirty days sixty loaves, which is three arrovae. Therefore divide forty by three; the result will be thirteen and a third, and this is what you wanted.

Again on the same topic, but otherwise: wherein are involved measures of different regions

(**B.306**) For instance, someone asks: One arrova and a half in Toledo is equal to one emina and a fourth in Segovia, and from one arrova are made twenty loaves, one of which is eaten by one man each day. How many eminae will forty men eat in thirty days?

[1760] Expressed otherwise at the end of the subsequent justification. See also p. 992, notes 1721, 1724.

[1761] That is, in B.305.

[1762] B.305a.

[1763] Fractional man.

(a) You will do the following. You are to know the number of arrovae forty men eat in thirty days as we have taught above and convert the result into Segovian eminae. The way to determine the number of arrovae forty men eat in thirty days is the following. You will say: 'From one arrova twenty loaves are made, one of which is eaten by one man (in one day); how many arrovae will forty men consume in thirty days?' You will proceed as I have taught above,[1764] multiplying thirty by forty and dividing the product by twenty. The result will be sixty arrovae, and this is the number of Toledan arrovae forty men eat in thirty days. Then convert them into eminae, saying thus: 'One Toledan arrova and a half are equal to one Segovian emina and a fourth; how many eminae correspond to sixty arrovae?' You will proceed here as I have taught above,[1765] multiplying sixty by one emina and a fourth and dividing the product by one arrova and a half. The result will be the Segovian eminae, namely fifty, and this is the consumption of forty men in thirty days.

(b) Or otherwise. It has been supposed that one emina and a fourth is equal to one arrova and a half. Consequently, one arrova must be equal to five sixths of an emina. Since from an arrova twenty loaves are made, twenty-four loaves will be made from a whole emina. Thus you will say: 'From an emina twenty-four loaves are made and one man eats one loaf each day; how many eminae will forty men eat in thirty days?' Proceed as I have taught before,[1766] multiplying thirty by forty and dividing the product by twenty-four. The result will be fifty, (which is the number of) eminae, and this is what you wanted.

(c) Or otherwise. The ratio of one emina and a fourth to one arrova and a half [1767] is the same as the ratio of one loaf made from an emina to one loaf made from an arrova. Since one emina and a fourth is five sixths of one arrova and a half,[1768] five sixths of one loaf made from an emina must be equal to one loaf made from an arrova. Then it is like saying: 'One man eats five sixths of one loaf, which shall be one twentieth of an emina [1769]; how many eminae will forty men eat in thirty nights [1770]?' Proceed as I have taught above:[1771] multiply thirty by forty, divide the product by twenty, and take five sixths of the result, because of the 'five sixths of one loaf, which is one twentieth of an emina' proposed. The result will be what you wanted, namely fifty eminae.

[1764] B.300.

[1765] Presumably B.298a.

[1766] Again B.300.

[1767] Here again (see B.298), mentioning the capacity units is unfortunate since the amounts are then equivalent.

[1768] This (with the words 'emina' and 'arrova') is not just unfortunate but wrong.

[1769] 'arrova' in the text, probably a correction by an early reader who misunderstood the (unclear) text.

[1770] Nightly consumption here.

[1771] B.301a.

(*d*) Or otherwise. You already know that since a man eats five sixths of one loaf (in a day), he must eat in thirty days twenty-five loaves, which is one emina and a fourth. Thus multiply one emina and a fourth by the number of men, namely forty. The result will be fifty, and this is what you wanted.

You will do the same [*here*] in all problems involving arrovae of different regions [1772] [*as in the foregoing problems where the arrova was of just one region*].

[1772] Rather (as in the title) 'involving measures of different regions' (whence the early reader's intervention).

Chapter (B-XVIII) on the Exchange of morabitini

(B.307) For instance, someone asks: Thirty nummi are given for a morabitinus. What is the equivalent of ten such nummi in nummi at forty to the morabitinus?

You will do the following. It is known that someone having ten nummi has a third of the morabitinus exchanged for thirty. But thirteen and a third is a third of a morabitinus exchanged for forty. Thus ten of the former are equivalent to thirteen and a third of the latter.

Indeed, it is clear that the ratio of ten to thirty is the same as the ratio of the required quantity to forty. Proceed then as explained for four proportional numbers[1773], and the result will be what you want.

(B.308) Someone asks: A man exchanges ten nummi, of which there are twenty-five to the morabitinus, for fifteen nummi of another money. How many of these nummi are given for a morabitinus?

It is evident from what has been said[1774] that the ratio of ten to twenty-five is the same as the ratio of fifteen to the required quantity. Proceed then as explained above and the result will be what you are looking for, namely thirty-seven[1775] and a half.

Treat other, similar problems in the same way and you will succeed.

AGAIN.

(B.309) Someone asks: A hundred morabitini are exchanged each for fourteen nummi. How many solidi will there be for a hundred morabitini?

(*a*) You will do the following. Convert the hundred morabitini into nummi thus. Multiply a hundred by the nummi in one morabitinus, namely fourteen; this produces one thousand four hundred. Dividing this by the nummi in one solidus, namely twelve, will give the number of solidi in a hundred morabitini, namely a hundred and sixteen and two thirds. This is what you wanted.

(*b*) Or otherwise. You are to know that taking as many solidi as the quantity of nummi in one morabitinus —whether this quantity is an integer only or an integer and a fraction— will always give twelve morabitini. When, for instance, a morabitinus is exchanged for fourteen nummi, that many solidi will give twelve morabitini and, inversely, twelve morabitini will give fourteen solidi; or, when a morabitinus is exchanged for five solidi, which correspond to sixty nummi, sixty solidi will give twelve morabitini;

[1773] Beginning of Book B (p. 777, *i*).

[1774] B.307.

[1775] The text (MS $\mathcal{A}$) has 'thirty-six'.

LIBER MAHAMELETH

1007

if it is exchanged for sixty nummi and a half, sixty solidi and a half will give twelve morabitini. [*Keep this small rule in mind.*]

The reason for this is evident. For suppose someone says: 'Twelve morabitini are exchanged each for fourteen nummi; how many solidi will be given for twelve morabitini?' You are to solve this as we have taught,[1776] that is, multiplying twelve morabitini by the nummi in a single morabitinus, namely fourteen, in order to convert them into nummi, and then dividing the product by twelve, which is the number of nummi in a single solidus, in order to convert them into solidi. But you know that dividing the product of twelve into fourteen by twelve will always give fourteen.

This being now clear,[1777] divide a hundred morabitini by twelve in order to determine how many times it contains twelve, and multiply the quotient by fourteen [*solidi*]; the result will be what you wanted. Now dividing a hundred by twelve gives eight and a third, and multiplying eight and a third by fourteen produces one hundred and sixteen solidi and two thirds. You will always proceed likewise in all the problems of this kind[1778], that is, dividing the morabitini by twelve, always, and multiplying the quotient by the number of nummi in one morabitinus; the result will be the number of solidi you wanted.

(*c*) Or otherwise. Find the number which multiplying twelve produces the quantity of nummi in one morabitinus, and multiply it by the given quantity of morabitini; the result will be the number of solidi you wanted. Thus, in the present problem, find the number which multiplying twelve produces fourteen —for such is the quantity of nummi in one morabitinus. You will find one and a sixth. Then multiply one and a sixth by a hundred; this produces one hundred and sixteen and two thirds, and this is what you wanted.

The reason for this is the following. As we have already established,[1779] fourteen solidi correspond to twelve morabitini. It is therefore clear that the ratio of twelve to fourteen is the same as the ratio of a hundred morabitini to the corresponding unknown quantity of solidi.[1780] Now you know that the number which when multiplying twelve produces fourteen is the same as the number which when multiplying a hundred produces the unknown quantity of solidi. This is why we have proceeded in this way.

(*d*) Or otherwise. You know that if each morabitinus were exchanged for two solidi, one hundred morabitini would give two hundred solidi.[1781] But it has been supposed that a morabitinus is exchanged for fourteen

---

[1776] Above, part *a*.

[1777] Return to the proposed problem.

[1778] 'of this kind' (*huius capituli*); that is, problems involving nummi and solidi like the present one.

[1779] Above, *b*.

[1780] What follows seems superfluous.

[1781] And thus 2400 nummi.

1008   PART TWO: TRANSLATION, GLOSSARY

nummi.[1782] So you see that there is for two hundred solidi a surplus of a thousand nummi: ten nummi are added to the value of each of these one hundred morabitini. Thus you should determine how many solidi there are in a thousand nummi by dividing them by twelve, thus finding eighty-three and a third; subtracting this from two hundred will leave a hundred and sixteen solidi and two thirds.

You will proceed likewise if the number of nummi comprises a fraction.[1783]

AGAIN ON THE SAME TOPIC.

(B.310) For instance, someone asks: A morabitinus is exchanged for fifteen nummi. How many morabitini will there be for five hundred solidi?

(a) You will do the following. Convert all the solidi into nummi by multiplying them by twelve and divide the result by fifteen; this will give what you wanted.

(b) Or always denominate one from the quantity, whatever it is,[1784] of the nummi in one morabitinus; this fraction taken of the total quantity of nummi, resulting from all the solidi, will give the number of morabitini you wanted. Since, as you know, one is a third of a fifth of fifteen, a third of a fifth of the total quantity of nummi will be what you wanted.

(c) Or otherwise. Always divide the solidi by the quantity of nummi in one morabitinus and multiply the result by twelve; this will give what you require. So divide five hundred solidi by fifteen, which gives thirty-three and a third; multiplying it by twelve produces four hundred, and this is what you wanted.

The reason for this is evident. We have already established[1785] that taking as many solidi as there are nummi in one morabitinus always gives twelve morabitini. Since it has been supposed that a morabitinus is exchanged for fifteen nummi, fifteen solidi will give twelve morabitini [*and, inversely, twelve morabitini fifteen solidi, as said above*]. This is why we divided five hundred solidi by fifteen solidi, in order to know how many times fifteen solidi are in five hundred solidi, and multiplied the result by twelve morabitini, in order to count twelve morabitini for every group of fifteen solidi. We found that fifteen occurs thirty-three times and a third in five hundred; counting for each time twelve morabitini makes four hundred morabitini.

Or otherwise.[1786] As you know,[1787] the treatment consisted in converting five hundred solidi into nummi by multiplying them by twelve and then

---

[1782] And thus 100 morabitini for 1400 nummi.

[1783] This refers to the various treatments we have just seen. See also B.310′.

[1784] Integer only or integer with fraction.

[1785] B.309*b*.

[1786] This is another 'reason' (applying P_5 to the computation in *a*).

[1787] Computation in part *a*.

LIBER MAHAMELETH

1009

dividing the quantity of nummi by the number of nummi in one morabitinus, namely fifteen. But it is clear that multiplying five hundred by twelve and dividing the result by fifteen is the same as dividing five hundred by fifteen and multiplying the result by twelve, as we have established in what precedes.[1788]

(*d*) Or otherwise. You are to determine how many morabitini correspond to ten solidi according to the rate of exchange given;[1789] this is eight. Multiply it by a tenth of five hundred, namely fifty, thus producing four hundred. This is what you wanted. The same would hold if you had determined how many morabitini correspond to twelve solidi, namely nine morabitini and three fifths: multiplying this by a twelfth, or half a sixth, of five hundred solidi will produce what you wanted.

(*e*) Or otherwise. Always denominate the nummi in one solidus from the nummi in one morabitinus;[1790] this fraction taken of the given quantity of solidi will be what you require. Thus, in the above problem, denominate twelve from fifteen; this is four fifths. Such a fraction of five hundred, namely four hundred, is what you wanted to know.

The reason for this is the following. We have already established[1791] that fifteen solidi correspond to twelve morabitini. It is therefore clear that the ratio of twelve morabitini to fifteen solidi is the same as the ratio of the unknown quantity of morabitini to five hundred solidi. Twelve being four fifths of fifteen, the unknown quantity of morabitini is four fifths of five hundred; that is, since four hundred is four fifths of five hundred, four hundred. This is what you wanted.

(*f*) Or otherwise. You know that when a morabitinus is exchanged for two solidi, five hundred solidi give two hundred and fifty morabitini. But it has been supposed that a morabitinus is exchanged for fifteen nummi.[1792] So there is for each one of the two hundred and fifty morabitini a surplus of nine nummi, which makes two thousand two hundred and fifty altogether. You are therefore to determine to how many morabitini they correspond at an exchange rate of fifteen nummi. To do so, divide the above quantity of nummi by fifteen; this gives a hundred and fifty, which you add to two hundred and fifty; this makes four hundred, and this is what you wanted.

(*g*) Or always denominate the surplus in each morabitinus, in this case nine, from those in one morabitinus, namely fifteen; this is three fifths. Add this fraction of two hundred and fifty, which is a hundred and fifty, to two hundred and fifty; this makes four hundred, and this is what you wanted.

[1788] Premiss P_5 or beginning of Book B (*ii*, p. 778).

[1789] Computation as in *a*.

[1790] In actual fact we may have to 'divide' since there can be less than twelve nummi to one morabitinus, as in the problems below (B.310 *seqq.*). See note 1999, p. 1065.

[1791] B.309*b*.

[1792] Whereas a morabitinus at two solidi corresponds to 24 nummi.

(B.310′) The quantities of nummi may include fractions, as is the case if someone asks: A morabitinus is exchanged for fourteen nummi and three fourths; how many morabitini are there for five hundred solidi? Use all the ways found in the previous problem, where integers only occurred, and the result will be what you require.

Again on the same topic

(B.311) For instance, someone asks: A morabitinus, exchanged for ten nummi of one money or for twenty of a second, is exchanged for nummi of the two moneys. A man receives two nummi from the money at ten to the morabitinus. How much is due to him from the other money in order to make up the value of the morabitinus?

(*a*) You will do the following. Subtract two from ten, thus leaving eight. Multiply it by twenty; this produces a hundred and sixty. Divide it by ten; the result is sixteen, and such is the quantity of nummi due to him from the other money.

The reason for this is the following. It has been supposed that he receives two nummi from the money at ten nummi to the morabitinus. Thus he has two tenths of it, and eight tenths remain from the total value of the morabitinus. Therefore he is to receive these eight tenths from the twenty of the other money. [*Thus the ratio of eight to ten is the same as the ratio of the required quantity to twenty. There are therefore four proportional numbers.*] Thus if you multiply eight by twenty and divide the product by ten, the result will be what you require. [⟨*For multiplying a fraction by*⟩ *some other number is the same as multiplying the number of parts by this number and dividing the product by the number of which they are parts, as we have taught above* [1793]. *Thus it is clear that multiplying eight parts of ten by twenty is the same as multiplying eight by twenty and dividing the product by ten.*][1794]

(*b*) Or otherwise. You know that when two nummi are received from the money at ten to the morabitinus, a fifth of the morabitinus is received and four fifths of it remain to be received. Therefore four fifths of twenty, which is sixteen, is what you wanted.

Again on the same topic.

(B.312) For instance [1795], someone asks: A morabitinus is exchanged for ten nummi of one money, for twenty of another, or for thirty of yet another. A man changing a morabitinus for nummi of these three moneys receives two nummi from the money at ten to the morabitinus and four from the money at twenty to the morabitinus. In order to make up the total value of the morabitinus, what is he still to receive from the third money?

[1793] A.90*b* or P_5. See note 1350, p. 900.

[1794] This gloss, which links the last two steps of the demonstration, does not appear in MS $\mathcal{A}$.

[1795] We would have expected to be told that now three currencies are involved.

(a) You will do the following. Multiply two by thirty and divide the product by ten; this will give six. Next, multiply four by thirty and divide the product by twenty; this will give six as well. Add it to the other six, which makes twelve. Subtract it from thirty; this leaves eighteen, and such is the quantity of nummi received from the money at thirty nummi to the morabitinus.

The reason for this is the following. We know that, receiving two from ten, he receives a fifth of this quantity and that, receiving four from twenty, he receives its fifth. He is therefore still to receive three fifths of the total value of the morabitinus, but from the money at thirty nummi to the morabitinus. But you know that taking three fifths of thirty is like subtracting from it its two fifths. You know also that taking a fifth of thirty is like multiplying two by thirty and dividing the product by ten, or like multiplying four by thirty and dividing the product by twenty. [*For we know that the ratio of two to ten is the same as the ratio of a fifth of thirty to thirty; there are therefore four proportional numbers; thus if the first, which is two, is multiplied by the fourth, which is thirty, and the product is divided by the second, which is ten, this will give the third, which is a fifth of thirty. Likewise, the ratio of four to twenty is the same as the ratio of a fifth of thirty to thirty; there are therefore four proportional numbers; thus if the product of four into thirty is divided by twenty, the result will be a fifth of thirty.*][1796] This is why we have multiplied two by thirty and divided the product by ten, likewise again multiplied four by thirty and divided the product by twenty, added the two quotients and subtracted the sum from thirty, thus leaving its three fifths, which is what he must receive from thirty, and this is what you wanted.

(b) Or otherwise. You know that when two are received from ten, a fifth of this quantity is received, and when four are received from twenty, its fifth is received. Three fifths of the morabitinus still remain to be received, thus three fifths of thirty, which is eighteen. This is what you wanted.

(c) Or otherwise. Convert the nummi received from one money into nummi of the other, say in this case the two nummi received from the money at ten to the morabitinus into nummi of the money at twenty to the morabitinus;[1797] this will give four. Add it to the other four, received from the money at twenty to the morabitinus; this makes eight. Then it is like asking: 'From one money twenty are given for a morabitinus and from another thirty, and a man exchanging a morabitinus for nummi of the two moneys receives eight nummi from the money at twenty to the morabitinus; how many nummi is he to receive from the other money?' Do as has been explained above[1798], and the result will be what you wanted.

But if the sum of the nummi after conversion happens to be larger than the quantity of nummi obtained for one morabitinus from the money

---

[1796] Same kind of interpolation found twice in B.311a.

[1797] A.307.

[1798] B.311.

1012 PART TWO: TRANSLATION, GLOSSARY

into which the conversion was made, the problem will not be solvable: the man wanted to receive the value of the morabitinus from three moneys, but he actually receives more than this value from two moneys only; now this cannot be[1799]. And if this sum is equal to the quantity of nummi, he will receive nothing from the third money since the value of the morabitinus is made up by the nummi of these two moneys.

You are to know that when a morabitinus is exchanged for nummi of three moneys and only what is received from one of them is specified, the problem is indeterminate unless some other additional condition renders it determinate. Likewise, in the case of more than three moneys, the problem will not be determinate unless it is specified how much he receives from each except one, or unless the problem is made determinate in some other way.

Treat other, similar cases accordingly and you will find that this is the way.[1800]

AGAIN ON THE SAME TOPIC.

(B.313) Someone asks: Twenty nummi of one money are given for a morabitinus, or thirty of another, and a man changing a morabitinus for nummi of the two moneys receives twenty-four of them. How many does he receive from each?

(*a*) The treatment here must be as follows. We subtract the smaller number of nummi given for a morabitinus from the larger one; this leaves ten, which you must keep in mind. When you want to know what he receives from those at thirty[1801], subtract twenty from twenty-four, leaving four; denominate it from ten, which gives two fifths; two fifths of thirty, which is twelve, is what he received from those at thirty. When you want to know what he received from those at twenty, subtract twenty-four from thirty, which leaves six; denominate it from ten, which gives three fifths; three fifths of twenty, which is twelve, is what he receives from those at twenty.

The proof of this is the following. Let the morabitinus be AB, its part received from the nummi at twenty to the morabitinus, AG, its part received from the nummi at thirty to the morabitinus, BG. So the product of twenty into AG added to the product of thirty into GB is twenty-four.

If this number[1802] were greater than thirty, the problem would not be solvable: we know that the products of AG into thirty and GB into thirty are thirty, whereas the products of GB into thirty and AG into less than thirty would be larger than thirty, which is impossible. It would likewise be impossible if it were less than twenty: we know that the products of AG into twenty and GB into twenty make twenty, whereas the products of AG into twenty and GB into more than twenty would be less than

---

[1799] *hoc autem esse non potest.* See note 1314, p. 890.

[1800] This should appear before the concluding remarks.

[1801] 'from thirty', here and recurrently in the text.

[1802] The total number of coins received.

twenty, which is impossible. It is also clear that if what he received were twenty ⟨or thirty⟩ [*or less than twenty*], the problem would not be solvable. Therefore it cannot be solvable unless what he receives from both moneys is less than the larger number and greater than the smaller one, as is the case here for twenty-four.

Thus the products of AG into twenty and GB into thirty are twenty-four. But the product of GB into thirty is equal to the products of GB into ten and GB into twenty. Thus the products of AG into twenty, GB into twenty, and GB into ten are twenty-four. But the product of AG and GB into twenty is twenty, for AB is one. This leaves, as the product of ten into GB, four. Therefore GB is two fifths of the morabitinus. Now the ratio of GB to AB is the same as the ratio of the value of GB[1803] to thirty. GB being two fifths of AB, this value is two fifths of thirty, which is twelve.

When you want to know the value of AG, you will find it by means of the following proof.[1804] We know that the products of AG into twenty and GB into thirty are twenty-four. Let the product of AG into thirty be (added in) common. Then the products of AG into twenty, GB into thirty, and AG into thirty will be twenty-four increased by the product of AG into thirty. Now the products of AG into thirty and GB into thirty are thirty. Thus the product of twenty into AG increased by thirty will be equal to the product of thirty into AG increased by twenty-four. Subtract the product of twenty into AG from the product of thirty into AG; this leaves the product of ten into AG. Subtract also twenty-four from thirty; this leaves six. Thus the product of ten into AG is six, so AG is three fifths. But the ratio of AG to AB is the same as the ratio of the value of AG to the value of the morabitinus. Since AG is three fifths of AB, he is to receive three fifths of twenty, which is twelve.

Treat other problems of this kind in the same way and you will succeed.[1805]

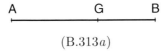

(B.313a)

(**b**) Or otherwise. Put, as what he receives from the money at twenty nummi to the morabitinus, a thing. Thus what he is to receive from the money at thirty nummi to the morabitinus is a morabitinus[1806] minus a thing. Now you know that the amount received for the morabitinus, which is twenty-four nummi, consists of two parts: one taken from the money at twenty nummi to the morabitinus and the other from the money at thirty

---

[1803] That is, its amount in nummi (*pretium*).

[1804] *probatio*, but also calculation using a geometrical figure (see note 584, p. 724).

[1805] This seems out of place.

[1806] Here and below, instead of 'one'.

1014 PART TWO: TRANSLATION, GLOSSARY

nummi to the morabitinus. Thus it is clear that adding the product of one of these two parts [*of the value*] of the morabitinus into twenty and the product of the other into thirty will give twenty-four. So multiply a thing by twenty, which produces twenty things. Next, multiply the morabitinus minus a thing by thirty, which produces thirty minus thirty things. Add it to the twenty things; this will give thirty minus ten things, which is equal to twenty-four.[1807] Complete then the thirty by adding ten things, and add as much to twenty-four; after that, subtract twenty-four from thirty, thus leaving six, which equals ten things. Therefore a thing is equal to three fifths, and that much does he receive from the money at twenty nummi to the morabitinus, namely twelve. This leaves to be received two fifths from the money at thirty nummi, namely twelve.

(B.314) Someone asks: A morabitinus is exchanged for ten nummi of one money or twenty of another. A man changing a morabitinus for nummi of the two moneys receives for the total value of the morabitinus just fifteen nummi from both. How much does he receive from each?

You will do the following. Take as the principal the difference between ten and twenty, namely ten. Next, subtract from twenty the fifteen nummi received from both moneys; this leaves five. Denominate it from ten, which gives a half. He receives that much from the money at ten nummi to the morabitinus, and he also receives a half from the money at twenty nummi to the morabitinus.

This will be demonstrated in the following way. Let the morabitinus be line AB, its first part [*which is received from the money at twenty nummi to the morabitinus*][1808], line AG, the other part, line GB. Next, let a line of ten be drawn[1809] from point G, say line GD. Let it be multiplied by line AG to produce the area $AGQD$.[1810] Then let line GK be twenty. Let it be multiplied by line GB to produce the area GH. Let the (whole) area be completed. It is then clear that the areas AD and GH are fifteen. But we know that line AB is one, for it is one morabitinus, while line BH is twenty; so the area AH is twenty. The two areas AD and GH being fifteen, the area QK is five. Now line DK is ten; thus line QD is a half. Since it is equal to line AG, line AG is half the morabitinus, and this is what he receives from the money at ten nummi to the morabitinus, while line GB is the other half [*that is, what he receives from the money at twenty nummi to the morabitinus*]. This is what we wanted to demonstrate.

AGAIN ON THE SAME TOPIC.

(B.315) Someone asks: There are two different moneys, one at ten nummi to the morabitinus and the other at thirty, and a man exchanges a morabitinus for nummi of the two moneys in such a way that subtracting those he

---

[1807] Equation of the form $b_1 - a_1 x = b_2$.

[1808] Gloss referring to GB.

[1809] Perpendicularly.

[1810] Area designated by four letters at its first occurrence. Same in B.316*c*.

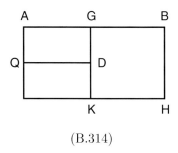

(B.314)

receives from the money at ten nummi to the morabitinus from the others leaves twenty. How many nummi does he receive from each money?[1811]

(*a*) You will do the following. Add ten and thirty, which makes forty; take it as the principal. Next, subtract twenty from thirty, which leaves ten. Divide it by the principal; the result is a fourth, (thus a fourth) of the morabitinus, and this is what he receives from the money at ten nummi to the morabitinus, while he receives the remaining three fourths of the morabitinus in nummi of the second money.

That will be demonstrated by a figure like this. Let the morabitinus be line AB, the part he receives from the money at ten nummi to the morabitinus, line AD, the remaining part, line DB. Let a line of ten be drawn[1812] from point D, say line DG. Let it be multiplied by line DA to produce the area AG. Next, let a line of thirty be drawn[1813] from point D, say line DK. Let it be multiplied by line DB to produce the area KB. Next, let from the surface KB an area equal to the area AG be cut off, say the area DH [*for he said that subtracting what is received in nummi at ten to the morabitinus from the number of nummi of the other money leaves twenty*][1814]. Therefore the area KBH[1815], left from the larger area after the area DH has been cut off, is twenty. Let the (whole) area IB be completed. Now we know that line AB is one, for it is the morabitinus, and that line AI is thirty. Therefore the area IB is thirty. But it has been demonstrated[1816] that the area KBH is twenty. This leaves, as the two areas DH and DI, ten. But the area DT is equal to the area DH. Thus the area IG is ten, while line GK is forty. Then let the area IG, which is ten, be divided by GK; line TG will thus be a fourth, and it is equal to line AD. This is what he receives from the money at ten nummi to the morabitinus. This leaves for DB three fourths of the morabitinus,

---

[1811] In fact, here as in B.316–318, it is the *fractions* of the morabitinus which are required.
[1812] Perpendicularly.
[1813] Again perpendicularly, but in the opposite direction.
[1814] Gloss to the next sentence.
[1815] Not being a rectangle, it is designated by three letters.
[1816] The MSS have 'it is clear', doubtless a copyist's error (*manifestum est* instead of *monstratum est*). Same error elsewhere (p. 920, note 1436; p. 923, notes 1449 & 1450), but not in B.316c.

to be received in nummi of the other money. This is what we wanted to demonstrate.

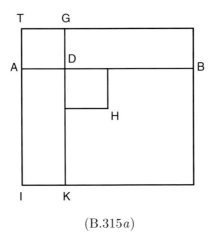

(B.315a)

(**b**) Or otherwise. Let what he receives from the money at ten nummi to the morabitinus be a thing. Therefore what he receives from the money at thirty nummi to the morabitinus is a morabitinus[1817] minus a thing. Then multiply a thing by ten, thus producing ten things. Next, multiply a morabitinus minus a thing by thirty, thus producing thirty minus thirty things. Subtract ten things from it; this leaves thirty minus forty things, which are equal to twenty.[1818] Then complete thirty by adding the lacking forty things, and add as much to twenty; this gives thirty, which equals twenty and forty things. Then subtract twenty from thirty; this leaves ten, which equals forty things. Thus the thing is a fourth. This is what he receives from the money at ten nummi to the morabitinus, while he receives the remainder [*which is left*] in nummi of the other money. You may also[1819] add ten things to twenty; this makes twenty and ten things, which are equal to thirty minus thirty things. Proceed then as I have taught above in algebra.[1820] The thing will be a fourth. This is what he receives from the money at ten nummi to the morabitinus, while he receives the remainder from the other.

Again on the same topic

(**B.316**) Someone asks: Ten nummi of one money are given for a morabitinus, twenty of another, or thirty of yet another. A man exchanges a morabitinus for nummi of the three moneys and receives twenty-five of them. How many does he receive from each one?

In such problems, where there are more than two moneys, the problem will be indeterminate unless there is some additional condition to make it

---

[1817] As in B.313 (note 1806), instead of 'one'.
[1818] Equation of the form $b_1 - a_1 x = b_2$.
[1819] Rearranging the terms.
[1820] Equation of the form $a_1 x + b_1 = b_2 - a_2 x$.

determinate. For instance, it may be stated (in this case) that he has received as much [1821] from the nummi at ten to the morabitinus as from those at twenty to the morabitinus, or that he has received from one of the moneys such or such fraction of the morabitinus; or something else of that kind may be stated [1822].

(*a*) Thus, if in this problem it were said that he received as much from the nummi at ten to the morabitinus as from those at twenty, you would do the following. Add ten to twenty, which makes thirty. ⟨Now the nummi of the third money are thirty.⟩ Double it, which gives sixty. Next, double twenty-five, which gives fifty. Then it is like saying: 'Thirty nummi of one money are given for a morabitinus and sixty of another, and a man exchanging a morabitinus for nummi of the two moneys receives fifty; how many does he receive from each?' The original problem is solvable, for this one is. Then proceed as we have taught above.[1823] This will give, for what he receives from the nummi at thirty, a third of the morabitinus, which is equal to what he receives (altogether) from the nummi at ten to the morabitinus and from the nummi at twenty to the morabitinus. Therefore he receives from the nummi at ten to the morabitinus a sixth of the morabitinus, and from those at twenty a sixth of the morabitinus likewise; from those at thirty he thus receives the remainder of the morabitinus, namely two thirds.

The proof of this is the following. Let the morabitinus be AB, the part he receives in nummi at ten to the morabitinus, AD, what he receives from those at twenty, DG, and from those at thirty, GB. Thus AD is as much as GD. It is then clear from what has been said before [1824] that the products of AD into ten, DG into twenty, and GB into thirty are twenty-five. But the products of AD into ten and DG into twenty are equal to the product of AD into thirty, for AD is as much as DG. So the products of AD into thirty and GB into thirty are twenty-five. Thus the products of twice AD into thirty and twice GB into thirty are twice twenty-five, which is fifty. But the product of twice GB into thirty is equal to the product of GB into twice thirty, which is sixty. Therefore the products of twice AD [1825] into thirty and GB into sixty are fifty. Then proceed as has been shown above.[1826] The result will be that AG is a third of AB, half of which is AD, and so much is DG, while this leaves as GB two thirds. This is what we wanted to demonstrate.

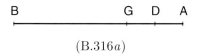

(B.316*a*)

[1821] That is, the same fraction of the morabitinus.
[1822] According to the first approach (thus: equating the fractions of two other currencies).
[1823] B.314.
[1824] In the data.
[1825] Which is AG.
[1826] Again, B.314.

1018 PART TWO: TRANSLATION, GLOSSARY

(b) Or otherwise. Put a thing as what he receives from the money at ten nummi to the morabitinus. Likewise, put a thing as what he receives from the money at twenty nummi to the morabitinus. What he receives from the third money is then a morabitinus minus two things. Next, you will say: 'One morabitinus is divided into three parts in such a way that the sum of the products of the first into ten, the second into twenty and the third into thirty is twenty-five.' Then multiply a thing by ten, which produces ten things, next a thing by twenty, which produces twenty things, next a morabitinus minus two things by thirty, which produces thirty minus sixty things, next add all these products. The result is thirty nummi minus thirty things, which are equal to twenty-five nummi.[1827] Complete the thirty nummi by adding the lacking thirty things, and add as many things to twenty-five; this gives thirty nummi, which equal twenty-five nummi and thirty things. Subtract then twenty-five from thirty; this leaves five, which equal thirty things. Thus a thing is equal to a sixth, and this is what he receives from the money at ten nummi to the morabitinus, and he receives as much also from the money at twenty nummi to the morabitinus; he will receive the remainder of the morabitinus, namely its two thirds, from the third money.

(c) Or otherwise. Add ten and twenty, which makes thirty. Then always multiply the quantity of nummi to the morabitinus of the last money by the number of other moneys, in this case by two, thus producing sixty; but if there were besides the last money three, four or more moneys, you would always multiply the quantity of nummi to the morabitinus of the last money by the number of other moneys, namely three, four or more. Next, you will subtract from the product the sum of the nummi of all the other moneys; in this case, you will subtract from sixty thirty, the sum of ten and twenty, thus leaving thirty, which you take as the principal. Next, subtract twenty-five from the nummi of the last money, namely thirty, thus leaving five. Divide[1828] this by the principal; the result is a sixth. So much does he receive from the money at ten nummi to the morabitinus, so much also from the money at twenty nummi to the morabitinus; and he will receive the remainder of the morabitinus from the last money.

This is demonstrated by means of the following figure. Let the morabitinus be line AB, that which is received from the first money, at ten nummi to the morabitinus, line AG, that from the second, line GD, equal to line AG; then let what remains to be received from the last money be line DB. Let next from the point G be drawn[1829] a line of ten, say line GK; let it be multiplied by line AG to produce the area $AGQK$. Then I shall extend line GK to the point Z so as to make it twenty. Let then line GZ be multiplied by line GD to produce the area GT. This being done, I shall extend line DT to the point I so as to make it thirty. Let then line DI be multiplied by line DB to produce the area BI. Then it

---

[1827] Equation of the form $b_1 - a_1 x = b_2$ ('nummi': see note 1120, p. 841).

[1828] Denominate.

[1829] Perpendicularly.

is clear[1830] that the areas AK, GT, BI are (together) twenty-five. ⟨Let the area AN be completed.⟩ Now you know that line AB is one, for it is the morabitinus, while line BN is thirty; so the area AN is thirty. Since it has been established that the areas AK, GT, BI are twenty-five, the areas QZ and HI are five. But you know that the amount of the areas QZ and HI results from multiplying line KZ by ZH and IT by TH. [*You also know that line IT is equal to line KZ and that line HZ is equal to line ZT. Therefore the multiplication of line IT by TH and by ZH gives the two areas QZ and HI.*][1831] You also know that the product of line IT into TH is equal to the products of line IT into TZ and into ZH [*whether joined or separate, the result is the same*][1832]; but the product of line IT into TZ is equal to the product of line IT into ZH; so the product of line IT into TH is equal to twice the product of IT into ZH. Now you know that line IT is equal to line ZK. Thus multiplying line ZK by thrice ZH produces the two areas which are five. But the product of line ZK into thrice ZH is equal to the product of line ZH into thrice line KZ, which is thirty. Therefore the multiplication of line HZ by thirty produces five. Thus if five is divided[1833] by thirty, the result is a sixth, which is line ZH. Since it is equal to line AG, line AG is a sixth of the morabitinus. Line GD is also a sixth. Thus line DB is two thirds. This is what we wanted to demonstrate.

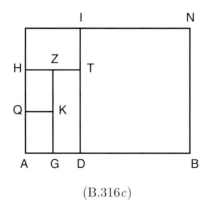

(B.316c)

(***d***) If it were said that he received as much from the nummi at ten to the morabitinus as from those at thirty, the problem would not be solvable. Indeed, to reduce it to two moneys, we add ten to thirty, thus making forty, double twenty, which gives forty, and double twenty-five, which gives fifty. So you will say: 'Forty of one money is given for a morabitinus, and likewise forty of another, and a man exchanging a morabitinus for nummi

---

[1830] By the data.

[1831] This seems unnecessary.

[1832] There is no need for a justification of this common operation, which is just Premiss PE_1.

[1833] Denominated.

1020   PART TWO: TRANSLATION, GLOSSARY

of the two moneys receives fifty of them.' This problem not being solvable, neither is the original one.

(B.316′) [1834] If it were said: Ten nummi of one money are given for a morabitinus, twenty-four of another and thirty of a third, and someone exchanges a morabitinus for nummi of the three moneys and receives twenty-three.

This problem is solvable with each of the two additional conditions.[1835]

For if he receives as much from those at ten as from those at twenty-four, the problem will be solvable. To reduce it to two moneys we add ten to twenty-four, thus making thirty-four, double thirty, which gives sixty, and double twenty-three, which gives forty-six. This is then like saying: 'Thirty-four nummi of one money are given for a morabitinus and sixty of another, and someone exchanges a morabitinus for nummi of the two moneys and receives forty-six of them.' This problem being solvable, so is the original one.

Likewise also, if it were said that he received as much from the nummi at ten to the morabitinus as from those at thirty, the problem would be solvable. Indeed, adding ten to thirty makes forty, doubling twenty-four gives forty-eight, and doubling twenty-three gives forty-six; you will then say: 'Forty nummi of one money are given for a morabitinus, or forty-eight of another, and someone exchanges a morabitinus for nummi of the two moneys and receives forty-six of them.' This problem being solvable, so is the original one.

It would still be solvable by adopting the third determination, according to which he received as much from those at twenty-four as from those at thirty. For adding twenty-four and thirty makes fifty-four, doubling ten gives twenty, and doubling twenty-three gives forty-six. Then it is like saying: 'Twenty of one money are given for a morabitinus and fifty-four of another, and a man exchanging a morabitinus for nummi of the two moneys receives forty-six of them.' This problem being solvable, so is the original one.

(e) [1836] If it were said that he received from one of the three moneys such or such fraction of the morabitinus and the problem resulting from the reduction to two moneys were solvable, so would the original problem be. This is the case if it is said that 'Ten nummi of a money are given for one morabitinus, or twenty of another, or thirty of a third, and a man exchanging a morabitinus for nummi of the three moneys receives twenty-five of them', and that he is to receive some arbitrary fraction of the morabitinus from one of the three moneys provided the remaining problem is solvable. For instance if it were said that he received from the nummi at ten to the morabitinus a fifth of the morabitinus, which is two nummi. This leaves four fifths of the morabitinus and twenty-three nummi. Then you will

---

[1834] We shall return to the previous problem after treating this example.
[1835] Those seen in *a* and *d*.
[1836] We now return to B.316.

LIBER MAHAMELETH 1021

say: 'Twenty of a money are given for a morabitinus, or thirty of another, and a man receives twenty-three nummi for four fifths of a morabitinus.' Take four fifths of twenty, which is sixteen, and four fifths of thirty, which is twenty-four. Then it is like saying: 'Sixteen nummi of one money are given for a morabitinus, or twenty-four of another, and a man exchanging a morabitinus for nummi of the two moneys receives twenty-three of them.' Proceed as I have taught above.[1837] What he receives from the nummi at sixteen will be found to be an eighth of a morabitinus, that is, an eighth of four fifths, which is a tenth, of the original morabitinus, and so much is he to receive from the money at twenty nummi to the morabitinus. What he receives from those at twenty-four will be seven eighths of a morabitinus, that is, seven eighths of four fifths, which is seven tenths, of the original morabitinus, and so much is he to receive from the money at thirty nummi to the morabitinus. Thus he has received from the nummi at ten, for a fifth of the morabitinus, two nummi; from those at twenty, for a tenth of the morabitinus, two nummi likewise; and from those at thirty, for seven tenths of the morabitinus, twenty-one nummi. The (whole) morabitinus is thus made up, so also is the number of nummi he received. The proof of all this is evident from the proof seen above[1838].

(*f*) If it were said that he had received from those at thirty to the morabitinus a fifth of the morabitinus, namely six nummi, the problem would not be solvable. For this would leave four fifths of the morabitinus, and also nineteen nummi; then you would say: 'From one money are given ten to the morabitinus, from another twenty, and a man receives nineteen nummi for four fifths of a morabitinus.' This problem not being solvable, neither is the original one. Indeed, taking four fifths of ten, which is eight, and four fifths of twenty, which is sixteen, you will say: 'Eight nummi of one money are given for a morabitinus, and sixteen of another, and someone receives for a morabitinus nineteen of them.' This problem being not solvable, so too is this determination inappropriate.[1839]

(*g*) If it were said that he had received from the nummi at twenty to the morabitinus a fourth of the morabitinus, the problem would be solvable.

Although there are numerous determinations for this kind of problem, the principles whereby all these problems can be worked out are those we have mentioned.[1840] [*If there are more than three moneys, the problem will be indeterminate unless some additional condition renders it determinate.*][1841] That is, it must either be stated that he shall receive from each and every money except one whatever fraction of the morabitinus

---

[1837] B.314.

[1838] B.314 as well.

[1839] 'too': our 'not solvable' and 'inappropriate' are both expressed by *falsus*.

[1840] B.316, beginning.

[1841] This sentence is, at best, misplaced.

he wishes [1842] and the remainder of the morabitinus from the money left, provided the problem thus reduced to two moneys is solvable; or he is to receive whatever fractions of the morabitinus he wishes,[1843] whether equal or not, from each and every money except two, and the remainder of the morabitinus from the (two) moneys left, provided the problem thus reduced is solvable. I shall propose (two) problems concerning all of this, which are to illustrate the above assertions.[1844]

(B.317) Someone asks: Ten nummi of one money are given for a morabitinus, twenty of another, thirty of yet another, forty of a fourth. A man exchanging a morabitinus for nummi from all these moneys receives twenty-five of them.

(a) If in the present case he receives as much from each of the first, second and third moneys, and the remainder from the fourth, the problem will be solvable. Indeed, adding ten, twenty and thirty makes sixty, tripling forty gives a hundred and twenty, and tripling twenty-five gives seventy-five. Then you will say: 'Sixty nummi of one money are given for a morabitinus and a hundred and twenty of another, and a man exchanging a morabitinus for nummi of the two moneys receives seventy-five of them.' This problem is solvable. Then proceed as I have taught above.[1845] The amount he receives from the money at sixty —which sixty is the sum of ten, twenty and thirty— will correspond to three fourths of the morabitinus. Thus he receives a fourth of the morabitinus from each of these three moneys, while he receives from the money at forty the remainder of the morabitinus.

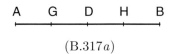

(B.317a)

The proof of all this is the following. Let the morabitinus be AB, what he receives from the first money, AG, from the second, GD, from the third, DH, and from the fourth, HB. Thus the products of ten into AG, twenty into GD, thirty into DH and forty into HB are twenty-five. But the products of ten into AG, twenty into GD and thirty into DH are equal to the products of ten, twenty and thirty into AG, since AG, GD and DH are equal. Thus the products of AG into sixty and HB into forty are twenty-five. Therefore the products of thrice AG into sixty and HB into thrice forty are thrice twenty-five, which is seventy-five. Thus the products of AH into sixty and HB into a hundred and twenty are seventy-five. Then proceed as I have taught above.[1846] The result will be that AH is three

---

[1842] Better: he is to receive in each case the same (unknown) fraction of the morabitinus.
[1843] This time they are set numerically.
[1844] They are B.317 & B.318.
[1845] B.314.
[1846] Again, B.314.

fourths; AG being a third of AH, AG is a fourth, and so is each of GD and DH, and this leaves for HB a fourth also. This is what we wanted to demonstrate.

(**b**) If it were said that he had received as much from each of the first, second and fourth moneys, and the remainder from the third, it would also be solvable.

(**c**) If it were said that he had received as much from each of the second, third and fourth moneys, and the remainder from the first, it would also be solvable.

(**d**) If it were said that he had received as much from the first as from the second and as much from the third as from the fourth, it will also be solvable. Adding ten and twenty makes thirty, adding thirty and forty makes seventy, and doubling twenty-five gives fifty. Then it is like saying: 'Thirty of one money are given for a morabitinus, and seventy of another, and someone exchanging a morabitinus for nummi of both moneys receives fifty; how many does he receive from each?'[1847] Proceed as has been explained above.[1848] What he receives from those at thirty will correspond to half the morabitinus, so that he receives from each of those at ten and at twenty the equivalent of a fourth of the morabitinus. He receives also from those at seventy the equivalent of half the morabitinus; therefore he receives from those at thirty the equivalent of a fourth of the morabitinus and from those at forty the equivalent of a fourth of the morabitinus.

The proof of this is the following. Let the morabitinus be AB, what he receives from the money at ten be AG, from the money at twenty, GD, from the money at thirty, DH, and from the money at forty, HB. Thus the products of AG into ten, GD into twenty, DH into thirty and HB into forty are twenty-five. But the products of DH into thirty and HB into forty are equal to the product of HB into seventy, and the products of AG into ten and GD into twenty are equal to the product of AG into thirty. Thus the products of AG into thirty and HB into seventy are twenty-five. Therefore the products of twice AG, which is AD, into thirty and twice HB, which is DB, into seventy are twice twenty-five, that is, fifty. Thus the products of AD into thirty and DB into seventy are fifty. Continue then with the rest of the problem as we have taught.[1849] The result will be that AD is half the morabitinus; thus AG will be a fourth, and so much is GD also. Likewise again DB will be a half; thus DH will be a fourth, and HB a fourth also. This is what we wanted to demonstrate.

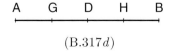

(B.317d)

[1847] But here too (see p. 1015, note 1811) it is fractions of the morabitinus which are required.
[1848] Also B.314.
[1849] B.314.

1024 PART TWO: TRANSLATION, GLOSSARY

(e) If now it were said that he had received some fraction of the morabitinus from any two of the moneys and the remainder from the remaining two; for instance, if it is said that he has received from the money at ten and from the money at twenty some fraction of the morabitinus, say from the money at ten three tenths of the morabitinus, which is three nummi, and from that at twenty a tenth, which is two nummi; this leaves twenty nummi and three fifths of the morabitinus. Then it is like saying: 'Thirty of one money are given for a morabitinus and forty of another, and a man exchanging a morabitinus for nummi of the two moneys receives twenty nummi for three fifths of the morabitinus; how much does he receive from each[1850] money?' Proceed as I have taught above:[1851] that is, take three fifths of thirty, which is eighteen, and three fifths of forty, which is twenty-four. Then it is like saying: 'From one money are given eighteen nummi for a morabitinus and from another twenty-four, and a man exchanging a morabitinus for nummi of the two moneys receives twenty of them; how many does he receive from each one?' Then proceed as I have taught above.[1852] What he receives from the money at twenty-four will be for a third of the morabitinus, that is, for a third of the three fifths, which is a fifth; and he receives that much from the nummi at forty, namely eight, which is a fifth of forty. From the money at eighteen he will receive for two thirds of the morabitinus, which is two thirds of the three fifths, thus two fifths of the original morabitinus; and so much also, namely twelve, does he receive from the nummi at thirty, namely two fifths of it.

Treat in the same manner all similar problems and you will find that this is the way.

You are to know that when there are more than two moneys such problems can admit of innumerable determinations. But more essential are the two following principles. The first is to state that the person shall receive from each money except one a certain arbitrary fraction[1853] of the morabitinus, the remainder of the morabitinus being received from the money omitted provided that the problem thus reduced to two moneys is solvable. The second is to state that he shall receive from each money except two, and individually, some arbitrary fraction[1854] of the morabitinus provided that the problem left with two moneys is solvable. [*That is, he is to receive some fraction of the morabitinus from each of the moneys except two, and, since this leaves to be received nummi and some fraction of the morabitinus, which fraction he is to receive from the two moneys singled out, the remaining nummi will be received from the two moneys.*][1855]

[1850] *unaqueque* is sometimes found instead of *utraque*.

[1851] B.316e.

[1852] B.314.

[1853] As above (note 1842), not set numerically.

[1854] Which is numerically set.

[1855] Early reader's gloss to make clear which two currencies are left and what the new problem is.

LIBER MAHAMELETH

1025

These determinations form the foundation of this chapter. We therefore considered it appropriate to bring in one problem in which such determinations will be applied; and, with what has been said above, this should suffice for all other instances.

(**B.318**) Someone asks: From one money eight nummi are given for a morabitinus, from another twelve, from a third fifteen, from a fourth eighteen and from a fifth twenty, and a man exchanging a morabitinus for nummi of all the moneys receives sixteen of them. How many does he receive from each one?

(**a**) If the determination in this problem were that he should receive as much from each of the first, second, fourth and fifth moneys, and the remainder of the morabitinus from the third money, it would not be solvable. For reducing to two moneys will mean adding eight, twelve, eighteen and twenty, which makes fifty-eight, quadrupling fifteen, which gives sixty, and quadrupling sixteen, which gives sixty-four. Then it is like saying: 'From one money fifty-eight nummi are given for a morabitinus, from another, sixty, and a man exchanging a morabitinus for nummi of the two moneys receives sixty-four of them; how many does he receive from each?' This problem not being solvable, the above determination is inappropriate.

(**b**) You have then to introduce determinations which will render the problem solvable. If, for instance, it is said that he receives equally [1856] from the first four moneys and the remainder from the fifth, namely the money at twenty nummi to the morabitinus, it will be solvable. Then adding eight, twelve, fifteen and eighteen makes fifty-three, while quadrupling twenty gives eighty and quadrupling sixteen, sixty-four. Then it is like saying: 'For a morabitinus are given from one money fifty-three nummi and from another eighty, and a man exchanging a morabitinus for nummi of these two moneys receives sixty-four of them; how many does he receive from each?' Do as I have taught above.[1857] It will be found that he receives from those at fifty-three for five ninths and a third of a ninth of the morabitinus; a fourth of this, which is a ninth and a third of a ninth, is what he receives from those at eight, namely one nummus and a ninth and two thirds of a ninth. So much does he receive from those at twelve, namely a ninth and a third of a ninth of the morabitinus, which is one nummus and seven ninths of a nummus; so much also from those at fifteen, namely a ninth and a third of a ninth of the morabitinus, which is two nummi and two ninths; and so much from those at eighteen, namely a ninth and a third of a ninth of the morabitinus, which is two nummi and two thirds. This leaves to be received from those at twenty three ninths and two thirds of a ninth, which is eight nummi and a ninth and a third of a ninth. Thus are made up the sixteen nummi for the morabitinus.

(**c**) If it were said that he had received equally from the last four moneys and the remainder from the first, namely the money at eight to the mora-

---

[1856] That is, an equal fraction of the morabitinus. So also below, *c*.

[1857] B.314.

1026   PART TWO: TRANSLATION, GLOSSARY

bitinus, this problem would also be solvable. Do then as I have taught above, and the result will be what you wanted.

What has been said about all these ways should suffice.

(*d*) The second way for this case is that he receives from each of three moneys a part of the morabitinus which is arbitrary provided that the problem thus left is solvable; the reason for our (choosing to) receive from three moneys is to leave two, if there were six we would receive from four so as always to leave two. For instance, let him receive from the first, namely the money at eight, an eighth of the morabitinus, which is one nummus; from the money at fifteen a fourth of the morabitinus, which is three nummi and three fourths; and from the money at eighteen an eighth of the morabitinus, which is two nummi and a fourth. Thus he receives altogether from these three moneys half the morabitinus and seven nummi; this leaves nine nummi and half the morabitinus, and two moneys, at twelve and at twenty. Then it is like saying: 'From one money are given twelve nummi for a morabitinus and from another twenty, and a man exchanging half a morabitinus for nummi of the two moneys receives nine of them; how many does he receive from each?' Do then as I have taught above.[1858] That is, take half of twenty, namely ten, and half of twelve, which is six. Then it is like saying: 'From one money are given ten nummi for a morabitinus and from another, six, and a man exchanging a morabitinus for nummi of the two moneys receives nine of them; how many does he receive from each?' Do then as I have taught above.[1859] What he receives from the money at six will be for a fourth of the morabitinus, that is, a fourth of the above half, which is an eighth of the (original) morabitinus; so much does he receive from the money at twelve, namely one nummus and a half; from the money at ten he receives for three fourths, which is three fourths of the above half, namely three eighths of the morabitinus; so much does he receive from the money at twenty, namely three eighths, which is seven nummi and a half. Thus he receives from the money at eight one nummus, corresponding to an eighth of the morabitinus; from the money at twelve one nummus and a half, corresponding to an eighth of the morabitinus; from the money at fifteen three nummi and three fourths, corresponding to a fourth of the morabitinus; from the money at eighteen two nummi and a fourth, corresponding to an eighth of the morabitinus; and from the money at twenty seven nummi and a half, corresponding to three eighths of the morabitinus. The morabitinus is thus made up, so also the sixteen nummi.

If it were said that he had received from a different choice of moneys these or other fractions, and the problem thus reduced to two moneys were solvable, so would the original problem be.

Treat all other similar cases in like manner, and you will find that this is the way.

[1858] B.316*e*.
[1859] B.314.

LIBER MAHAMELETH

(B.319) Someone asks: There are two moneys, one at ten nummi to the morabitinus and the other at twenty, and a man exchanging a morabitinus for nummi of the two moneys receives as many nummi from one money as from the other.[1860] How many nummi does he receive from each?

(*a*) You will do the following. Divide twenty by ten, which gives two. Next, divide twenty by itself, which gives one. Adding it to two makes three, by which you divide twenty; this gives six nummi and two thirds, and so much did he receive from the nummi at ten to the morabitinus, and so much likewise from the nummi at twenty to the morabitinus.

The reason for this is the following. We know that twenty is twice ten; therefore the fraction received from the nummi at twenty to the morabitinus is half the fraction received from the nummi at ten to the morabitinus, whereby the quantities of the nummi will be equal. Now we know that taking from those at ten two parts and from those at twenty one part makes up the value of the morabitinus. Thus it is clear that the ratio of the two parts to the morabitinus, which consists of three parts, is the same as the ratio of the quantity received from the nummi at ten to the morabitinus to ten itself. There are therefore four proportional numbers. Thus dividing the product of the two parts into ten by the three parts will give the quantity received of nummi at ten to the morabitinus. It is also clear that the ratio of one part to the three parts is the same as the ratio of the quantity received of nummi at twenty to the morabitinus to twenty itself. Thus dividing the product of one part into twenty by the three parts will give the quantity received of nummi at twenty to the morabitinus.

(*b*) Or otherwise. Denominate from the three parts the two (parts) received from the nummi at ten to the morabitinus; taking such a fraction of ten will give what you wanted. Next, denominate one part from the three; taking such a fraction of twenty will be what you wanted.

(*c*) Or otherwise. Look for a number which when divided by ten and by twenty gives two results without any fraction; such is forty. Divide it by ten, which gives four, and by twenty, which gives two. Adding them makes six, by which you divide forty; this gives six and two thirds, and so much does he receive from the nummi at twenty to the morabitinus, and so much also from the nummi at ten to the morabitinus.

(*d*) Or otherwise. You know that the morabitinus consists of two parts such that the product of one of them into ten is equal to the product of the other into twenty [*and that there are as many nummi received from one money as from the other, and the morabitinus is made up*][1861].

AGAIN ON THE SAME TOPIC.

(B.320) Someone asks: From one money are given twenty nummi for a morabitinus and from another, thirty, and a man exchanging a morabitinus for nummi of the two moneys receives equally from each —that is, in

---

[1860] The equality is now in the *quantity* of nummi received.

[1861] Mere repetition. The rest of the solution is missing (only MS B has preserved B.319).

quantities of nummi.[1862] What fraction of the morabitinus does he receive from each?

(*a*) You will do the following. Take any number, divide it by twenty and by thirty and add the results. Divide, for instance, thirty by thirty, which gives one, and by twenty, which gives one and a half; add this to one, found before, thus making two and a half. Divide by it thirty; this gives twelve, and so much does he receive from each money. Thus he receives from those at twenty twelve for three fifths of the morabitinus, and from those at thirty twelve for two fifths of the morabitinus, thus receiving equally from each money.

The proof of this is the following. Let the morabitinus be AB, what is received from those at twenty [*for a part of the morabitinus*][1863], AG, and what is received from those at thirty, GB. Thus it is clear that the product of AG into twenty is equal to the product of GB into thirty. Therefore the ratio of twenty to thirty is the same as the ratio of GB to AG. I shall now look for two numbers bearing to one another the same ratio as twenty to thirty, that is, such that the product of twenty into one of them is equal to the product of thirty into the other. This is done as follows. Take any number[1864], say thirty, and divide it by thirty, which gives one, and by twenty, which gives one and a half. The product of one into thirty is thus equal to the product of one and a half into twenty; therefore the ratio of one to one and a half is the same as the ratio of twenty to thirty. The ratio of twenty to thirty being the same as the ratio of GB to AG, the ratio of GB to AG is the same as the ratio of one to one and a half. By composition, the ratio of GB to AB will be the same as the ratio of one to two and a half. But the ratio of GB to AB is the same as the ratio of the nummi at thirty received for GB to thirty. Thus the ratio of one to two and a half is the same as the ratio of the nummi at thirty received for GB to thirty. Thus the product of one into thirty is equal to the product of the nummi (at thirty) received into two and a half. But the product of one into thirty is the number taken above[1865] and divided by twenty and thirty, namely thirty.[1866] Thus the product of what he receives into two and a half is thirty. Therefore divide thirty by two and a half; this gives twelve, and so much does he receive from those at thirty. He receives as much from those at twenty, since it has been supposed that he would receive equally from each of the two moneys. This is what we wanted to demonstrate.

You are to know that when there are more than two moneys, the

---

[1862] Remember that before B.319 'receiving equally' meant the same fraction of the morabitinus (thus, with the different rates, unequal quantities of nummi).

[1863] To make clear 'what is received'.

[1864] Literally: 'Seek any number' (*quere* instead of *accipe*).

[1865] Literally: 'the number sought above'. See previous note.

[1866] Thirty occurs twice here: once in the data and once as an arbitrarily chosen number.

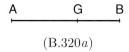

(B.320a)

treatment and the proof remain the same as here without differing in anything.[1867]

(**b**) Or otherwise. Let what he receives from the nummi at twenty to the morabitinus be a thing.[1868] Then what he receives from the nummi at thirty to the morabitinus is a morabitinus[1869] minus a thing. So let a thing be multiplied by twenty, which produces twenty things, and then a morabitinus minus a thing by thirty, which produces thirty minus thirty things. This equals the twenty things, since it was said that he received as many from one money as from the other. Next, proceed as I have taught in algebra.[1870] As a result, the value of the thing will be three fifths. That much does he receive from the nummi at twenty to the morabitinus, namely twelve, and the remainder of the morabitinus, that is, two fifths, from the other money.

(**c**) Or otherwise. Put, as what he receives from each money, a thing.[1871] Next, denominate the first thing from twenty, which is half a tenth of a thing, and then the second thing from thirty, which is a third of a tenth of a thing. Adding them makes five sixths of a tenth of a thing. Thus he receives from both moneys five sixths of a tenth of a thing, which equals a morabitinus. Then consider what number when multiplying five sixths of a tenth of a thing produces a thing; you will find twelve. Therefore multiply one by twelve. This produces twelve, and so many nummi at twenty to the morabitinus does he receive, and so many nummi at thirty to the morabitinus does he receive likewise.

AGAIN ON THE SAME TOPIC.

(**B.321**) Someone asks: There are three moneys from one of which ten nummi are given for a morabitinus, from another twenty, from a third thirty, and a man exchanging a morabitinus for nummi from all of them receives from each an equal quantity of nummi. How many does he receive from each?

(**a**) You will do the following. Divide thirty by ten, which gives three. Next divide thirty by twenty, which gives one and a half. Add it to three, which makes four and a half. Next, divide thirty by itself, which gives one. Add it to four and a half, which makes five and a half. Divide thirty by it; this gives five and five elevenths, and so much does he receive from each money. The reason for this is that which we have indicated in the

---

[1867] See the next problem.
[1868] The thing is here a fraction of the morabitinus.
[1869] Again (notes 1806, 1817) instead of 'one'.
[1870] Equation of the form $b_1 - a_1 x = a_2 x$.
[1871] The 'thing' is now a number of nummi.

1030 PART TWO: TRANSLATION, GLOSSARY

previous chapter.[1872]

(**b**) Or otherwise. Denominate the three parts received from the nummi at ten to the morabitinus from five and a half parts; taking such a fraction of ten gives five and five elevenths, and such is the number of nummi received at ten to the morabitinus. Next, denominate one and a half parts from the five and a half; taking such a fraction of twenty gives the number of nummi received at twenty to the morabitinus. Next, denominate one part from the five and a half; taking such a fraction of thirty gives the number of nummi received at thirty to the morabitinus.

(**c**) Or otherwise. Find a number which, when divided by ten, twenty and thirty, gives each time a result without a fraction. Such is sixty.[1873] Dividing it by ten gives six, again by twenty gives three, again by thirty gives two. Adding all the results makes eleven, by which you divide sixty; this gives five and five elevenths, and so much does he receive from each and every money. The reason for this is that which we have indicated in the chapter on dividing according to shares.[1874]

(**d**) Or otherwise. Let what he receives from each money be a thing. Next, denominate a thing from ten, which is a tenth of a thing, then also a thing from twenty, which is half a tenth of a thing, then also a thing from thirty, which is a third of a tenth of one thing. Thus he receives from all the moneys a tenth of a thing and five sixths of one tenth of one thing, which equal a morabitinus. Therefore consider what number multiplying a tenth of a thing and five sixths of a tenth of a thing will produce the whole thing; you will find five and five elevenths. Multiplying it by one[1875] gives five and five elevenths, and so much does he receive from each money.

(**e**) You may want to verify this problem. You already know that he has received from the money at ten nummi to the morabitinus six elevenths of it, from the money at twenty three elevenths of it, and from the money at thirty two elevenths of it. This makes up the value of the morabitinus, and the nummi are in equal quantity.

(**B.322**) Someone asks: From one money ten nummi are given for a morabitinus, from another twenty and from a third thirty, and a man exchanges a morabitinus for nummi of the three moneys in such a way that he receives from the second money twice the quantity of nummi received from the first, and from the third thrice the quantity of nummi received from the second.[1876]

You will do the following. Take half of twenty, namely ten; for it was said that he receives from the money at twenty twice what he receives

---

[1872] B.319*a*.

[1873] The author does not necessarily choose the smallest possible, as already remarked (note 1502, p. 936).

[1874] Presumably B.118*d*.

[1875] The morabitinus.

[1876] Required the fractions of the morabitinus.

LIBER MAHAMELETH 1031

from the money at ten. Since, as we know, he receives from the money at twenty twice the quantity received from the money at ten, while the quantity received from the money at thirty is thrice the quantity received from the money at twenty, the quantity received from the money at thirty is the sextuple of the quantity received from the money at ten; so take a sixth of thirty, which is five. Then it is like saying: 'From one money is given ten nummi for a morabitinus, from another ten and from another five, and a man exchanging a morabitinus for nummi of all moneys [1877] receives an equal quantity of each [1878].' Do as has been shown above.[1879] The quantity received from the money at ten will be two and a half; he receives twice this from the money at twenty, thus five, and thrice the latter from the money at thirty, thus fifteen. Therefore he receives from the money at ten for a fourth of the morabitinus, from the money at twenty for a fourth likewise, and from the money at thirty for half the morabitinus. Thus the morabitinus is completed, and the stipulation is fulfilled [1880].

The proof of this is the following. Let the morabitinus be AB, the part received from the money at ten, AG, from the money at twenty, GD, and from the money at thirty, DB. Thus the product of twenty into GD is twice the product of ten into AG; therefore the product of half of twenty into GD is equal to the product of ten into AG; thus take half of twenty, namely ten. Likewise again, the product of thirty into DB is the sextuple of the product of ten into AG; thus the product of a sixth of thirty into DB is equal to the product of ten into AG, and a sixth of thirty is five. Therefore the product of ten into AG is equal to the product of ten into GD, and (to the product of) five into DB. Complete then (the treatment) as we have taught [1881], and the result will be what you wanted.

Do the same in all similar cases.

$$A \quad G \quad D \quad\quad B$$

(B.322)

(**B.323**) Someone asks: From one money ten nummi are given for a morabitinus, from another twenty and from a third thirty, and a man exchanges a morabitinus for nummi of all the moneys in such a way that he receives two nummi from the money at ten, and as many from the money at thirty as from the money at twenty. How many does he receive from these two?

[1877] 'of each of the two moneys' (*utriusque* instead of *uniuscuiusque* in the text). Probably a change by an early reader since the exchange rate for two of the three currencies is the same.

[1878] 'of each of the two' (*de utraque* instead of *de unaquaque*). See p. 1024, note 1850.

[1879] B.321.

[1880] That is, that two quantities of nummi are given multiples of one other.

[1881] B.321 or here above.

1032 PART TWO: TRANSLATION, GLOSSARY

You will do the following. We already know that when two nummi are received from the money at ten, a fifth of the morabitinus is received and four fifths of the morabitinus remain to be received in equal quantities from the money at twenty and the money at thirty. Then it is like saying: 'From one money twenty nummi are given for a morabitinus, and from another thirty, and a man exchanging a morabitinus for nummi of the two moneys receives equal quantities of them for four fifths of the morabitinus.' Do then as I have taught above [1882], and the result will be what you wanted.

(B.324) Someone asks: From one money fifteen nummi are given for a morabitinus and from another sixty, and a man exchanging a morabitinus for nummi of the two moneys receives from those at fifteen the root of what he receives from those at sixty. How many does he receive from each?

You will do the following.[1883] Let the morabitinus be AB. It is then known that the morabitinus is divided into two parts such that the product of one of them into fifteen, when multiplied by itself, is equal to the product of the other into sixty. Let then one part be AG and the other, GB. Thus the product of fifteen into AG, when multiplied by itself, is equal to the product of sixty into GB. But the product of fifteen into AG, when multiplied by itself, is equal to the product of the square of fifteen into the square of AG. The square of fifteen being two hundred and twenty-five, the product of two hundred and twenty-five into the square of AG is equal to the product of GB into sixty. But the product of GB into sixty is equal to the product of AB into sixty less the product of AG into sixty. The product of AB into sixty being sixty, since AB is one, sixty less the product of AG into sixty is equal to the product of AG into itself taken two hundred and twenty-five times. Thus it is known that the products of AG into itself, taken two hundred and twenty-five times, and AG into sixty are sixty. From which it follows perforce that the products of AG into itself, taken once, and into a fifth and a third of a fifth are a fifth and a third of a fifth.[1884] I shall now draw a line AD which shall be a fifth and a third of a fifth. Thus the products of AG into itself and DA into AG are a fifth and a third of a fifth. But the products of AG into itself and AG into DA are equal to the product of DG into AG. Thus the product of DG into AG is a fifth and a third of a fifth while DA is a fifth and a third of a fifth.[1885] Let then DA be bisected at the point H. So the products of DG into AG and HA into itself will be equal to the product of HG into itself. But the product of DG into AG is a fifth and a third of a fifth, and the product of HA into itself is four ninths of a fifth of a fifth. Thus the product of HG into itself is a fifth and a third of a fifth and four ninths of a fifth of one fifth; therefore HG is two fifths and two thirds of a fifth. Since

---

[1882] B.320 (see also B.316e).

[1883] Geometrical treatment.

[1884] Equation of the form $x^2 + bx = c$.

[1885] Difference and product of DG and AG thus known. Therefore *Elements* II.6 will now be applied.

LIBER MAHAMELETH

HA is two thirds of a fifth, this leaves as *AG* two fifths, and so much does he receive from the money at fifteen nummi to the morabitinus. Hence he receives from the money at fifteen two fifths of the morabitinus, namely six (nummi), and from the money at sixty three fifths of the morabitinus, namely thirty-six: whereby he receives from the money at fifteen the root of what he receives from the money at sixty, and the value of the morabitinus is made up. This is what we wanted to demonstrate.

Consider accordingly the numerous other problems which can be formulated and which I have not included. Bearing well in mind what has been set forth before, you will easily understand anything which may be proposed.[1886]

B   G   A H D

(B.324)

AGAIN ON THE SAME TOPIC.

(**B.325**) There are two moneys such that ten nummi of one of them are given for a morabitinus and an unknown number of the other, and a man exchanging a morabitinus for nummi of both moneys receives two nummi from the money at ten to the morabitinus and twenty from the other, which thus make up the value of the morabitinus. What is the unknown number of nummi given for a morabitinus?

(**a**) You will do the following. It has been supposed that he receives two nummi from the money at ten to the morabitinus. Thus he receives a fifth of the morabitinus, and four fifths remain to be received from the nummi in unknown number. But you know that, two nummi being received from the money at ten to the morabitinus, a fifth of the morabitinus is received, which leaves four fifths of the morabitinus at ten nummi, thus eight nummi. It is clear, then, that the ratio of eight to ten is the same as the ratio of twenty to the unknown money.[1887] There are therefore four proportional numbers. Thus the product of the first into the fourth is equal to the product of the second into the third. So multiplying ten by twenty and dividing the product by eight will give the unknown number of nummi. Or otherwise. Find the number which multiplying eight produces ten; this is one and a fourth. Multiplying it by twenty produces twenty-five, and so much is the unknown number of nummi.

(**b**) Or otherwise.[1888] Put, as the unknown number of nummi, a thing. Four fifths of it, which is four fifths of a thing, equal twenty. Thus the thing equals twenty-five, and so much is the unknown number of nummi.

[1886] This might have been intended to close this section and thus follow B.329.

[1887] See B.308.

[1888] Here the two parts of the morabitinus are assumed to have been calculated (above, *a*).

1034   PART TWO: TRANSLATION, GLOSSARY

AGAIN ON THE SAME TOPIC.

(B.326) Someone asks: There are three moneys such that ten nummi of the first are given for a morabitinus, twenty of the second, and an unknown number of the third, and a man exchanging a morabitinus for nummi of the three moneys receives two of the nummi at ten to the morabitinus, four of those at twenty, and thirty of those in unknown number. What is the unknown number of nummi given for a morabitinus?

(*a*) You will do the following. Multiply the two nummi received from those at ten to the morabitinus by twenty, which produces forty. Divide this by ten, which gives four. Adding it to what is received from those at twenty to the morabitinus makes eight. Then it is like saying: 'From one money twenty nummi are given for a morabitinus and from another an unknown quantity, and a man exchanging a morabitinus for nummi of both moneys receives eight of the nummi at twenty to the morabitinus and thirty of those in unknown number.' Proceed as I have taught in the previous chapter.[1889] That is, subtract eight from twenty, which leaves twelve. Put this as the principal. Next, multiply twenty by thirty, which produces six hundred. Dividing it by the principal gives fifty, and so much is the unknown number of nummi given for a morabitinus.

The reason for multiplying two by twenty and dividing the product by ten is the following.[1890] You know that two nummi received from those at ten to the morabitinus are a fifth of them, and so also the four received from those at twenty to the morabitinus. Thus it is clear that taking two from ten is the same as taking four from twenty. For the ratio of two to ten is the same as the ratio of four to twenty. So if the product of two into twenty is divided by ten, the result will be four. [*It has thus been proved that receiving two from those at ten is the same as receiving four from those at twenty; for the two are a fifth of the morabitinus and the four likewise. Clearly then, receiving two from those at ten and four from those at twenty is the same as receiving two fifths of twenty, which is eight.*]

Or otherwise. Look for the number which multiplying twelve produces twenty; you will find one and two thirds. Multiply one and two thirds by thirty; the result is fifty, and so much is the unknown number of nummi[1891].

(*b*) Or otherwise. You know that when two are received from those at ten to the morabitinus, a fifth of the morabitinus is received. Likewise also, when four are received from those at twenty to the morabitinus, a fifth of the morabitinus is received. This leaves to be received three fifths of the morabitinus. It follows that thirty is three fifths of the unknown number

---

[1889] B.325.

[1890] How to convert nummi at ten into nummi at twenty is known from B.307 (and B.325*a*). Furthermore, the subsequent explanations are unclear (whence the early reader's comments).

[1891] '*required* number of nummi' would have been more appropriate (see note 902, p. 790). The same above (B.325) and below, several times.

LIBER MAHAMELETH   1035

of nummi, therefore fifty is the unknown number of nummi given for a morabitinus.

(**c**) Or otherwise. Let the unknown number of nummi be a thing. Three fifths of it, which is three fifths of a thing, equal thirty. Thus the thing equals fifty, and so much is the unknown number of nummi.

(**B.327**) Someone asks: Of an unknown quantity of morabitini each is exchanged for three solidi, of an equal quantity of other morabitini each is exchanged for four solidi, of another, likewise equal and unknown quantity of other morabitini, each is exchanged for five solidi, and from the exchange of all in the end a quantity of sixty solidi is obtained. How many are all the morabitini? [1892]

You will do the following. Add three, four and five, which makes twelve. Divide sixty by it; the result is five, and such is the first unknown quantity of morabitini, and such are the second and the third.

This is proved as follows. Let all the morabitini be AB. Thus the products of AB into three, four and five are sixty. But the products of AB into three, four and five are equal to the product of AB into twelve, as explained in the chapter on premisses. [1893] Thus the product of AB into twelve is sixty. Therefore divide sixty by twelve; the result will be that AB is five. This is what we wanted to demonstrate.

A   B

(B.327)

(**B.328**) Someone asks: Of an unknown quantity of morabitini each is exchanged for three solidi, of an equal quantity plus four of other morabitini each is exchanged for four solidi and of a quantity equal to the second plus five of other morabitini each is exchanged for five solidi, and the sum resulting from the whole exchange is a hundred solidi. What is the unknown number of morabitini?

You will do the following. It is clear that since the third quantity exceeds the second by five and the second exceeds the first by four, the third exceeds the first by nine. Then you are to know what corresponds to nine morabitini at five solidi each, which is forty-five solidi, and likewise what corresponds to four morabitini at four solidi each, which is sixteen solidi. Adding this to forty-five makes sixty-one. Subtract it from a hundred; this leaves thirty-nine. Then it is like saying: 'Of an unknown quantity of morabitini each is exchanged for three solidi, of an equal quantity of other morabitini each for four solidi, of a further equal quantity of other morabitini each for five solidi; from the whole exchange results a sum of

---

[1892] 'all the morabitini' (here and below, in the proof), rather: 'each equal quantity of morabitini'.

[1893] Premiss $PE_1 = $ *Elements* II.1.

1036 PART TWO: TRANSLATION, GLOSSARY

thirty-nine solidi.' Then proceed as I have taught above.[1894] The result will be the first quantity. Adding to it four will give the second quantity. Adding to it five will give the third quantity. The proof of all this is clear from the foregoing.

AGAIN ON THE SAME TOPIC.

(B.329) Someone asks: A man exchanges a hundred morabitini, consisting of melequini at fifteen solidi each and baetes at ten solidi each, and he obtains one thousand two hundred solidi from the hundred morabitini. How many melequini and baetes were there?

If more than one thousand five hundred solidi or less than one thousand were obtained, the problem would not be solvable. It cannot be solvable unless, after multiplying the price of the morabitinus of lesser value, namely ten, by the quantity of morabitini, in this case a hundred, and the price of the morabitinus of greater value, namely fifteen, by the quantity of morabitini, the quantity of solidi obtained altogether from the hundred morabitini falls between the two products. This will be demonstrated by means of the proof.

(a) You will do the following. Subtract ten from fifteen, thus leaving five, which you take as the principal. When you want to know how many baetes there are (do the following). Multiply the value of one melequinus, namely fifteen, by the quantity of morabitini, namely a hundred; this will produce a number larger than the number of solidi resulting from all the morabitini. Subtract then from this product the quantity of solidi and divide the remainder by the principal. This will give sixty, which is the quantity of baetes. When you want to know how many melequini there are (do the following). Multiply the value of one baetis, namely ten, by the quantity of morabitini, namely a hundred; this will produce a number smaller than the number of all solidi resulting from the hundred morabitini. Subtract it from the latter and divide the remainder by the principal. This will give forty, which is the quantity of melequini.

The proof of this is the following. Let the hundred morabitini be AB, the baetes, AG, and the melequini, GB. Thus the products of AG into ten and GB into fifteen are one thousand two hundred. But the product of GB into fifteen is equal to the products of GB into ten and into five. Thus the products of GB into ten, AG into ten and GB into five are one thousand two hundred. But the products of GB into ten and AG into ten are equal to the product of the whole AB into ten. Thus the products of the whole AB into ten and GB into five are one thousand two hundred. Now the product of the whole AB into ten is one thousand. Therefore subtract one thousand from one thousand two hundred, thus leaving, as the product of GB into five, two hundred.[1895] Then divide two hundred by five; this gives as GB forty. When you want to know AG, you will do the following. As you know, the products of AG into ten and GB into

---

[1894] B.327.

[1895] Whence the first part of the condition.

fifteen are one thousand two hundred. Put in common the product of AG into fifteen. Thus the products of AG into ten and fifteen, and GB into fifteen, will be one thousand two hundred plus the product of AG into fifteen. But the products of AG into fifteen and GB into fifteen are equal to the product of the whole AB into fifteen. Thus the products of the whole AB into fifteen and AG into ten are one thousand two hundred plus the product of AG into fifteen. The product of AB into fifteen being one thousand five hundred, one thousand five hundred plus the product of AG into ten is one thousand two hundred plus the product of AG into fifteen. Subtracting then the product of AG into ten from its product into fifteen will leave the product of AG into five, plus one thousand two hundred, equal to one thousand five hundred. Therefore subtract one thousand two hundred from one thousand five hundred, thus leaving, as the product of five into AG, three hundred.[1896] Then divide three hundred by five; this gives as AG sixty. This is what we wanted to demonstrate.

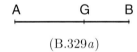

(B.329a)

(**b**) Or otherwise. Multiply ten by a hundred, thus producing a thousand, and multiply fifteen by a hundred, thus producing one thousand five hundred. Then it is like saying: 'From one money are given a thousand nummi for a morabitinus and, from another, one thousand five hundred; a man exchanges a morabitinus for nummi of the two moneys and receives one thousand two hundred.' Proceed as I have taught above.[1897] It will be found that what he receives from those at one thousand five hundred[1898] is forty, which is the quantity of melequini, and that what he receives from those at one thousand[1899] is sixty, which is the quantity of baetes. The proof of this is clear.

If there are more than two kinds of morabitini, you will do as was explained before; such problems will be indeterminate unless rendered determinate by some additional condition.[1900]

AGAIN ON THE SAME TOPIC.

(**B.330**) Someone asks: From a morabitinus exchanged for fourteen nummi a piece is cut which, together with the nummi it is worth according to the

---

[1896] Whence the second part of the condition.
[1897] B.314.
[1898] 'one thousand' in the text.
[1899] 'one thousand five hundred' in the text.
[1900] As already known. B.316–318 are about a single morabitinus exchanged for various kinds of nummi; here, a number of identical solidi is exchanged for various kinds of morabitini.

1038 PART TWO: TRANSLATION, GLOSSARY

given rate of exchange, makes four nummi[1901]. How much does this piece weigh?

Suppose the morabitinus weighs two nummi.[1902]

(*a*) Since it is exchanged for fourteen nummi, the morabitinus[1903] plus its value must make sixteen nummi. As it has been supposed that the piece plus its nummi makes four nummi, it is clear that the ratio of the piece to itself and its nummi is the same as the ratio of the morabitinus to itself and its nummi altogether. But the ratio of the morabitinus to itself and its nummi is an eighth since it has, itself, a weight of two nummi. It follows that the piece is an eighth of four nummi, that is, half a nummus. Such is the weight of the piece, and this is what you wanted to know. Or otherwise. Multiply two by four and divide the product by sixteen; this will give half a nummus, and such is the weight of the piece.

(*b*) Or otherwise. Put as the weight of the piece a thing; this leaves as its value four nummi minus a thing. As it has been supposed that the morabitinus weighs two nummi and is exchanged for fourteen nummi, it is clear that the ratio of the piece, which is a thing, to its value, which is four minus a thing, is the same as the ratio of the morabitinus, with a weight of two nummi, to its value, which is fourteen. Therefore the product of a thing into fourteen is equal to the product of four minus a thing into two nummi. The multiplication thus ends with fourteen things equal to eight nummi minus two things.[1904] Then complete eight by adding the two lacking things and add as much to fourteen things; this gives sixteen things, which are equal to eight nummi. Thus a thing equals half a nummus, which is the weight of the piece.

(*c*) You may want to verify this problem. As you know,[1905] the piece is a fourth of the morabitinus (in weight); then its value is a fourth of fourteen, which is three nummi and a half. As you further know, the weight of the piece is half a nummus. Therefore the piece and its value make four nummi.

[1901] That is, weight plus value makes (*ponderat*) four nummi, the nummus here being understood as both coin and unit of weight.

[1902] This supposition is valid for both subsequent treatments.

[1903] Here and in what follows this means, when not otherwise specified, its weight. Same for the piece.

[1904] Equation of the form $b_1 - a_1 x = a_2 x$.

[1905] Result just found.

Chapter (B–XIX) on Cisterns

(B.331) For instance, three pipes feed one cistern which the first fills in one day, the second in half a day and the third in a third of a day. If the three begin to run at the same time, what fraction of the day will they take to fill the cistern?

You will do the following. You know that the first pipe fills one cistern in just one day, that the second pipe, which takes half a day, fills two in one day and that the pipe which takes a third of a day fills three in one day. Thus it follows that the three pipes running together fill six cisterns in one day. Thus it follows that they fill one cistern in a sixth of a day.[1906] For one cistern is a sixth of six (cisterns). Therefore all the pipes fill one cistern in a sixth of a day.

(B.332) It is now asked thus: the first pipe fills a cistern in one day, the second in half a day and the third in a third of a day, but there is an outlet below whereby a full cistern will be emptied in a third of a day. What fraction of the day will it take to fill it when the three pipes are running together and the outlet is open?

You already know that[1907] the three pipes running together fill six cisterns in one day and that[1908] the outlet below, which when open empties a full one in a third of a day, will empty three full ones in one day. Therefore subtract three cisterns from six, thus leaving three, and denominate from it one cistern, which gives a third. [*Since three empty cisterns cancel out three full ones and three full ones are left, it follows that the three pipes running together fill one of these three in a third of a day.*][1909] This is what you wanted.

(B.333) Someone asks thus: There are above a cistern three pipes the first of which fills it in two days, the second in three, and the third in four. In how many days will the three running together fill it?

You will do the following. You know that the pipe filling a cistern in two days fills half a cistern in one day and that the other filling it in three days fills a third of a cistern in one day and that the pipe filling a cistern in four days fills a fourth of a cistern in one day. Therefore these three pipes running together fill in one day one cistern and half a sixth. Thus consider what fraction one cistern is of one cistern and half a sixth; this is twelve thirteenths. In that fraction of a day, namely twelve thirteenths, do the three pipes fill one cistern.

[1906] What follows, being superfluous, may be an addition.
[1907] From the previous problem.
[1908] From the data.
[1909] This is repetitive and looks like an addition.

1040 PART TWO: TRANSLATION, GLOSSARY

AGAIN ON THE SAME TOPIC [1910]

(**B.334**) Into a cistern ten cubits long, eight wide and six deep, containing a thousand measures of water, someone throws a stone which is four cubits long, three wide and five thick.[1911] How much water overflows from the cistern?

(**a**) You will do the following. Find the volume of the cistern as follows. Multiply its length by its width and the result by its depth; this produces four hundred and eighty. Take this as the principal. Next, find the volume of the stone in the same way, that is, multiplying its length by its width and the result by its thickness; this produces sixty. Multiply it by a thousand measures; this produces sixty thousand measures. Dividing this by the principal gives a hundred and twenty-five, and so many measures of water overflow from the cistern.

The reason for this is the following. We know that the ratio of the volume of the cistern, which is four hundred and eighty, to a thousand measures, which it contains, is the same as the ratio of the volume of the stone to the quantity of water which overflows. There are therefore four proportional numbers. Thus the product of the first, which is four hundred and eighty, into the fourth, which is the unknown quantity of water overflowing, is equal to the product of the second, which is a thousand measures, into the third, which is sixty. Then dividing the product of a thousand into sixty by four hundred and eighty will give the fourth, which is the unknown quantity of water.

(**b**) Or otherwise. You know that the ratio of the stone's volume to the cistern's volume is the same as the ratio of the water overflowing to the whole quantity of water the cistern contains. But the ratio of the stone's volume to the cistern's volume is an eighth. Therefore the quantity of water overflowing is an eighth of a thousand, which is a hundred and twenty-five.

(**c**) Or otherwise. Look for the number by which four hundred and eighty must be multiplied to produce a thousand; you will find two and half a sixth. Thus multiply sixty by two and half a sixth; this produces a hundred and twenty-five, which is what you wanted.

AGAIN ON THE SAME TOPIC.

(**B.335**) Into a cistern ten cubits long, eight wide and six deep, containing a thousand measures of water, someone throws a stone which is three cubits wide and four long. One hundred and twenty-five measures of water overflow. How thick is the stone?

(**a**) You will do the following. Multiply three by four, which produces twelve. Multiply it by a thousand, which produces twelve thousand. Take this as the principal. Next, multiply the volume of the cistern, which is four hundred and eighty cubits [1912], by a hundred and twenty-five and divide

---

[1910] The common topic is the cistern.

[1911] Usually the length is the largest dimension. But see B.335.

[1912] This value (in cubic cubits) is known from the previous problem.

LIBER MAHAMELETH   1041

the result by the principal; this gives five cubits, and so many cubits is the stone thick.

The reason for this is the following. We know that the ratio of the cistern's volume, which is four hundred and eighty, to the thousand measures of water it contains is the same as the ratio of the stone's volume to the quantity of water overflowing, which is a hundred and twenty-five measures. There are therefore four proportional numbers. Thus the product of four hundred and eighty into a hundred and twenty-five is equal to a thousand multiplied by the volume[1913] of the stone is found by multiplying its length by its width, that is, four by three, and the result by its thickness. Thus it is clear that the product of four into three multiplied by the thickness of the stone, and this in turn by a thousand, gives a result equal to the product of four hundred and eighty into a hundred and twenty-five. Thus it follows that if the product of four hundred and eighty into a hundred and twenty-five is divided by the result of multiplying the product of a thousand into three by four, this will give the thickness of the stone.

(*b*) Or otherwise. Denominate a hundred and twenty-five from a thousand; this gives an eighth. Take such a fraction, that is, an eighth, of the cistern's volume, which is four hundred and eighty, namely sixty, and divide this by the surface area[1914] of the stone, which is twelve; the result will be the thickness of the stone, which is five.

(*c*) Or otherwise. Denominate four hundred and eighty from a thousand; this is two fifths and two fifths of a fifth. Take this fraction of a hundred and twenty-five, namely sixty. Dividing it by twelve gives five, and this is what you wanted.

AGAIN ON THE SAME TOPIC.

(**B.336**) Into a cistern ten cubits long, eight wide and six deep, containing a thousand measures of water, someone throws a [*square*][1915] stone similar to the cistern —that is, such that the ratio of the cistern's width to its length is the same as the ratio of the stone's width to its length— and three cubits thick. A hundred and twenty-five measures of water overflow. How long and wide is the stone?

(*a*) You will do the following. Multiply a thousand by the thickness of the stone, which is three, thus producing three thousand; take it as the principal. Next, multiply the volume of the cistern, which is four hundred and eighty, by a hundred and twenty-five, thus producing sixty thousand. Divide it by the principal; this gives twenty, which is the stone's surface area.[1916]   If you want to know its length, multiply the cistern's length,

---

[1913] *tota magnitudo* here (instead of *magnitudo*). See next note.

[1914] *magnitudo superficiei*, in this case the product of the two known dimensions.

[1915] Rather: rectangular. The upper section (length × width) is meant.

[1916] That with the required dimensions.

1042   PART TWO: TRANSLATION, GLOSSARY

which is ten, by the stone's surface area, which is twenty, and divide the result by the width of the cistern; this gives twenty-five, and its root, which is five, is the length of the stone.   But if you want to know the width of the stone, multiply the cistern's width by twenty and divide the result by the cistern's length; the root of the result, namely four, is the width of the stone.

(**b**) Or otherwise. Consider what fraction of a thousand measures are the hundred and twenty-five that have overflowed; it is an eighth. Taking such a fraction, namely an eighth, of the cistern's volume gives sixty, which is the stone's volume. Now you know [1917] that this volume is the result of multiplying the stone's length by its width and the product by its thickness. Thus dividing sixty by the thickness of the stone will give twenty, which is the stone's surface area, obtained by multiplying its length by its width. [*Now you know that the cistern's surface area is eighty.*][1918]   It has also been supposed that the ratio of the cistern's width to its length is the same as the ratio of the stone's width to its length. But the ratio of the cistern's width to its length is four fifths; so multiplying the length of the stone by its four fifths produces twenty. Therefore, multiplying the length by itself produces twenty-five. Consequently, the root of twenty-five is the length of the stone, which is five, and the width is four fifths of the root, which is four.

The reason for all these ways [1919] is that indicated in the (problems on) drapery; those who have understood them will therefore easily understand these.[1920]

AGAIN ON THE SAME TOPIC.

(**B.337**) Into a square [1921] cistern, ten by ten cubits, and likewise ten in depth, containing a hundred measures of water, someone throws a stone, which is also square, four by four cubits, and the same in thickness. How many measures of water will overflow?

(**a**) You will do the following. Find the volume of the cistern as indicated above,[1922] by multiplying its length by its width and the result by its depth; this will produce a thousand. Take it as the principal. Next, multiply the volume of the stone, found in the same way, which is sixty-four, by a hundred, and divide the product by the principal; the result will be the quantity of water which has overflowed, namely six measures and two fifths. The reason for this is that which we have previously indicated.[1923]

[1917] B.334a, if need be.

[1918] This is irrelevant.

[1919] Just two.

[1920] B.129 (& B.132).

[1921] That is, the upper section is square (see note 1661, p. 977). In fact, both cistern and stone are cubic.

[1922] B.334 (& B.336b).

[1923] The ratio of the volume to the quantity of water it contains remains constant (B.334).

LIBER MAHAMELETH 1043

(**b**) Or otherwise. Consider what fraction a hundred is of a thousand; it is a tenth. Taking such a fraction, namely a tenth, of sixty-four will give what you wanted.

AGAIN ON THE SAME TOPIC.

(**B.338**) Into a square[1924] cistern, each side ten cubits, and the same in depth, containing two hundred measures of water, someone throws a square stone, having (thus) equal sides, but four cubits in thickness, and twenty measures of water overflow. What is the amount in cubits of each of the stone's (two equal) sides?

(**a**) You will do the following. Multiply two hundred by four, thus producing eight hundred; take this as the principal. Next, multiply the cistern's volume, found as shown above[1925], which is a thousand, by twenty and divide the product by the principal; this gives twenty-five. Its root, which is five, is the size of each of the stone's (two equal) sides.

(**b**) Or otherwise. Find the number which multiplying two hundred produces a thousand; this is five. Multiplying it by twenty produces a hundred. Dividing this by four gives twenty-five. Its root is the size of each of the stone's (two equal) sides. The reason for this is as we have shown above.[1926]

(**B.339**) Into a cistern ten (cubits) long, eight wide and five deep, containing a thousand measures of water, a stone is thrown and a hundred measures overflow. By how much will the (level of the) water drop when the stone is removed?

You will do the following. Multiply the hundred which have overflowed by the cistern's depth, which is five, and divide the product by what the cistern holds, namely a thousand; the result will be how far the water drops.

The proof of this is the following. We know that when the stone is removed, there is a part in the cistern[1927] which remains without water but taking on the shape of the cistern.[1928] It is thus as if the surface of the water divided the cistern into two parts, a lower one with water and an upper one without. Now Euclid has said that when a body is thus divided by a surface, as here, the ratio of one part to the other will be the same as the ratio of the base of the first to the base of the second.[1929] Now the base of the part without water is a (rectangular) area, and the base of the part with water is likewise a (rectangular) area. Since these two areas have an equal height, the ratio of these two areas is the same as the ratio

---

[1924] Refers to the upper section. Same for the stone.

[1925] B.334.

[1926] B.336b.

[1927] *aliquis in cisterna*, the text says.

[1928] In other words: the volume of the empty part equals the volume of the stone, but with the same cross section as the cistern.

[1929] *Elements* XI.25. The 'bases' of these two parallelepipeds together form one vertical wall of the cistern.

1044   PART TWO: TRANSLATION, GLOSSARY

of their bases.[1930] But the base of the area which is itself the base of the part without water is the upper part of the cistern's height, and the base of the area which is itself the base of the part with water is the lower part of the cistern's height. [*For the cistern's height is any one of the four vertical corner edges, since the sides of the cistern are parallel and the surfaces containing it vertical, according to the hypothesis of the problem.*][1931] The cistern is therefore divided into two parts in such a way that the ratio of the part without water to the part with water is the same as the ratio of the section of the corner edge visible above the water to the hidden section of it which is immersed. By composition, the ratio of the part without water to the whole cistern is the same as the ratio of the visible section of the corner edge to the height of the whole cistern. ⟨But the ratio of the volume of the part to the volume of the cistern is the same as the ratio of the quantity of water overflowed to the quantity of water in the whole cistern.⟩ [*But the part without water is equal to the quantity of water which has overflowed, while the cistern is equal to its whole content.*][1932] Thus the ratio of the whole content of the entire cistern to the quantity of water which has overflowed is the same as the ratio of the height of the whole cistern to its visible section, which is required. This is why the treatment consisted in multiplying the second, namely a hundred, by the third, namely five, and dividing the product by the first, namely a thousand; this gave the fourth, namely by how much the (level of the) water drops when the stone is removed.

(**B.340**) Someone asks: Into a cistern ten cubits long, eight wide and five high,[1933] containing a thousand measures of water, a stone three long, two wide and one high is thrown. By how much will the water (level) drop when the stone is removed?

You will do the following. You are to determine according to what we said above[1934] how much water overflows; this is fifteen measures. Thus it is like saying: 'Into a cistern ten (cubits) long, eight wide and five deep a stone is thrown and fifteen measures of water overflow; by how much will the water (level) drop after removal of the stone?' Do as I have taught above[1935], and the result will be what you wanted.

AGAIN

(**B.341**) Someone asks: How many measures of wine does a cask with a diameter of ten cubits and a depth of eight contain?

[1930] *Elements* VI.1. The 'equal height' is the width of the vertical wall considered and the 'bases', the vertical edges.

[1931] This specification looks like an addition. In all these problems the cisterns are parallelipipedal.

[1932] Gloss to make up for the lacuna. The author himself would not have identified water content (in measures) with volume (in cubits).

[1933] *in altum* instead of the usual *in profundum*. Same for the stone.

[1934] B.334.

[1935] B.339.

LIBER MAHAMELETH 1045

You will do the following. Multiply the cask's diameter by itself and subtract from the product its seventh and half a seventh; the remainder is the size of the cross section of the cask.[1936] Multiply this by the cask's depth and double the result —for it is said that a vessel one cubit long, wide and deep contains two measures. The result is the content of the cask.

Or otherwise [*in order to find the size of the cross section of the cask*][1937]. Always multiply the diameter by three and a seventh, and the product will be the length of the circumference: multiplying the diameter by three and a seventh produces the length of the circumference, for it has already been proved by the Ancients that any circumference is thrice its diameter plus a seventh of the diameter.[1938] Having thus found the circumference, multiply its half by half the diameter; the result is the area of the bottom. Multiply this by the depth of the cask, double the result, and this will give what you require.

But if the cask is square[1939], find its volume as we have shown above[1940], and twice the result will be what you wanted.

[1936] Literally: 'the size of the inner area of the whole cask.'

[1937] The only difference is indeed found in this calculation.

[1938] Archimedes (*Dimensio circuli*, prop. 3) is in fact more precise: 'plus *less than* a seventh of the diameter'. If this reference originated with the author, he is less cautious than elsewhere (p. 868).

[1939] That is, parallelipipedal.

[1940] B.334.

Chapter (B–XX) on Ladders

(**B.342**) A ladder ten cubits long standing against an equally high wall is withdrawn from the bottom of the wall by six cubits. By how much will it come down from the top?

(*a*) You will do the following. Multiply six by itself, and ten by itself, and subtract the lesser product from the greater. Subtract the root of the remainder, namely eight, from ten; this leaves two. By so much does the ladder come down from the top of the wall.

This will be demonstrated using the following figure. Let the wall be line AB and the ladder, line DG. Hence line DG is ten, and line GB is six. Now it is known that the triangle DBG is right-angled. Thus the products of line DB into itself and BG into itself are equal to the product of DG into itself.[1941] Therefore it is clear that subtracting the product of line BG into itself from the product of GD into itself will leave, as the product of DB into itself, sixty-four. Therefore line DB is the root of sixty-four, which is eight. But line AB is ten. Thus line AD is two, and by so many cubits does the ladder come down from the top of the wall. This is what we wanted to demonstrate.

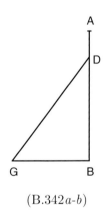

(B.342 *a-b*)

(*b*) Or otherwise. Put as line AD one thing; this leaves as line DB ten minus a thing. Multiply this ten minus a thing by itself, thus producing a hundred and one square minus twenty things. Next, multiply line BG by itself, thus producing thirty-six. Add together these two products; this makes one square and a hundred and thirty-six minus twenty things; this is equal to the product of DG into itself, which is a hundred. Complete then the hundred and thirty-six and the square by adding the lacking twenty

---

[1941] *Elements* I.47 (used throughout in what follows).

things, and add as many things to a hundred. Then remove a hundred with a hundred. This leaves a square and thirty-six, which are equal to twenty things.[1942] So multiply half of the things, which is ten, by itself; this produces a hundred. Subtracting thirty-six from it leaves sixty-four. Subtract its root, which is eight, from half of the things. This leaves two, which is the value of the thing, thus line AD. This is what you wanted.

AGAIN.

(**B.343**) A ladder ten cubits long standing against a wall of the same height is withdrawn and comes down by two cubits from the top of the wall. By how much is it withdrawn from the bottom of the wall?

(**a**) You will do the following. Subtract two from ten; this leaves eight. Multiply this by itself, and ten by itself, and subtract the lesser product from the greater; the remainder is thirty-six. Its root, which is six, is how far the ladder moves away from the bottom of the wall.

This is demonstrated by means of a figure like the following one. Let the wall be AB, the ladder, DG, and let AD be two. Thus DB is eight. But the triangle DBG is right-angled. Therefore the products of DB into itself and GB into itself are equal to the product of DG into itself [*as Euclid declares in the first book*][1943]. Thus, subtracting the product of DB into itself from the product of DG into itself will leave, as the product of line BG into itself, thirty-six. Then line BG is the root of thirty-six. Therefore it is six, and by so many cubits is the ladder withdrawn from the bottom of the wall. This is what we wanted to demonstrate.

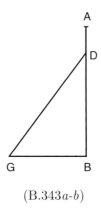

(B.343 *a-b*)

(**b**) Or otherwise. Let line BG be a thing, which is by how much the ladder is withdrawn from the bottom of the wall. But you know that DB is eight. So multiply DB by itself, and BG by itself, and add the two products; this makes one square and sixty-four nummi, which are equal to a hundred. Continue then as has been shown before in algebra[1944]; the value of the

---

[1942] Equation of the form $x^2 + c = bx$.

[1943] *Elements* I.47. Here Euclid is mentioned, although the theorem was already used in B.342.

[1944] Equation $x^2 + c - bx = x^2$, thus reducing to $bx = c$.

1048   PART TWO: TRANSLATION, GLOSSARY

thing will be six, and by so many cubits is the ladder withdrawn from the bottom of the wall.

AGAIN ON THE SAME TOPIC.

(**B.344**) A ladder of unknown length standing against a wall of the same height is withdrawn by six cubits from the base of the wall and comes down two cubits from the top of the wall. What is its length?

(*a*) You will do the following. Multiply six by itself, and two by itself, and subtract the lesser product from the greater, thus leaving thirty-two. Divide its half, which is sixteen, by two cubits; this gives eight. Adding two cubits to it makes ten, and so much is the height of the ladder or the wall.

This is demonstrated as follows. Let the wall be AB and the ladder, DG. Then it is evident that line AB is divided at the point D into two unequal parts.[1945] Therefore the products of AD into itself, DB into itself and AD into DB, twice, are equal to the product of AB into itself, as Euclid said in the second book [1946]. But you know that line AB is equal to line DG. So the product of DG into itself is equal to the products of AD into itself, DB into itself and AD into DB, twice. Now, as you already know,[1947] the product of DG into itself is equal to the products of DB into itself and BG into itself. Thus it has been established that the products of AD into itself, DB into itself and AD into DB, twice, are equal to the products of DB into itself and BG into itself. Removing what is common, which is the product of DB into itself, leaves the products of AD into itself and AD into DB, twice, equal to the product of GB into itself. Subtract the product of AD into itself, which is four, from the product of BG into itself, which is thirty-six, thus leaving, as twice the product of AD into DB, thirty-two. Thus once the product of AD into DB is sixteen. Dividing sixteen by line AD, which is two, will give as line DB eight. But line AD is two. Therefore line AB is ten, and so much is the height of the ladder and of the wall. This is what you wanted to know.

(*b*) Or otherwise. Let line AB be a thing. Now line AD has been supposed to be two, so line DB is a thing minus two. Then multiply a thing minus two by itself, thus producing one square and four minus four things. Multiply also line BG by itself, thus producing thirty-six. Adding these two products makes one square and forty minus four things, which are equal to the product of DG into itself, that is, a square. Do then as we have taught above in algebra.[1948] The value of the thing will be ten, and so much is the height of the ladder and of the wall.

[1945] No real need to specify that the parts are unequal here. Same remark for B.351.

[1946] *Elements* II.4 = PE_4.

[1947] *Elements* I.47, already used in the two previous problems.

[1948] Equation reducing to the form $ax = b$.

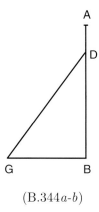

(B.344a-b)

(**B.345**) Someone asks: A ladder ten cubits long standing against an equally high wall is withdrawn from the base of the wall by a distance which, when added to that by which the ladder comes down from the top of the wall, makes eight. By how much is it withdrawn and by how much does it come down?

You will do the following. Subtract eight from the height of the wall; this leaves two. Multiply it by itself, thus producing four. Subtract it from the product of the length of the ladder into itself; this leaves ninety-six. Keep half of this, which is forty-eight, in mind. Next multiply half of two by itself, and add the result, which is one, to forty-eight; this makes forty-nine. Keep its root, which is seven, in mind. When you want to know by how much the end of the ladder comes down from the top of the wall (do as follows). Add seven to half of two; this gives eight. Subtract it from the height of the wall; this leaves two, and by so much does the end of the ladder come down from the top of the wall. But when you want to know by how much it is withdrawn from the bottom of the wall (do as follows). Subtract half of two from seven; this leaves six, and by so much is the ladder withdrawn from the base of the wall.

The proof of this is the following. Let the previous figure remain as it was.[1949] Then adding AD and BG gives eight. Let (a line) equal to BG, say DH, be cut off from DB. Thus the whole AH is eight. AB being ten, this leaves two as HB. But we know[1950] that the products of DB into itself and HD into itself are a hundred, for they are equal to the product of DG into itself, and DG is ten. Now the products of BD into itself and HD into itself are equal to the products of BD into DH, twice, and BH into itself.[1951] So subtract the product of BH into itself from a hundred; this leaves, as twice the product of BD into DH, ninety-six. Thus once the

---
[1949] *Maneat precedens figura sicut erat.* This merely means that the signification of the segments will not be repeated. (The figure is redrawn, as it will be each time.)
[1950] *Elements* I.47.
[1951] *Elements* II.7 = PE_7.

product of BD into DH is forty-eight.[1952] Let then BH be bisected at the point Z. So BH will be a bisected line to which is added, in prolongation, some (line), namely DH. Therefore the products of BD into DH and HZ into itself will be equal to the product of ZD into itself, as Euclid said in the second (book)[1953]. So ZD is seven. ZB being one, the whole BD is eight. But AB is ten. This leaves as AD two, and by so much does the end of the ladder come down from the top of the wall. And since DZ is seven and HZ is one, this leaves as HD, which is equal to BG, six; and by so much is the ladder withdrawn from the base of the wall. This is what we wanted to prove.

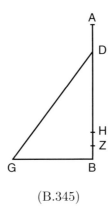

(B.345)

(**B.346**) Someone asks: A ladder ten (cubits) long standing against an equally high wall is withdrawn from the base of the wall by a distance which, when diminished by the distance by which it comes down from the top, leaves four.[1954]

If it were said here that the ladder had been withdrawn from the base of the wall by a distance which, when subtracted from that by which it had come down from the top, left four,[1955] or three, or something else, the problem would not be solvable. For that by which it comes down from the top is always less than that by which it is withdrawn from the bottom.

This is proved as follows. Let the figure remain the same. Then I say that BG is always longer than AD. This is proved as follows.[1956] We know[1957] that the products of BD into itself and BG into itself are equal to the product of DG into itself. But DG is equal to AB. Thus the products of BD into itself and BG into itself are equal to the product of AB into

---

[1952] Difference and product of BD and DH thus known.
[1953] *Elements* II.6 = PE_6.
[1954] Question: How far is it withdrawn and how far does it come down?
[1955] Same value as in the formulation.
[1956] *Sic* (repetitive). In fact the result we shall arrive at ($BG^2 - AD^2 = 2\,AD \cdot DB$) has been already found in B.344a.
[1957] *Elements* I.47.

itself. But the product of AB into itself is equal to the products of AD into itself, DB into itself, and AD into DB, twice.[1958] Thus the products of DB into itself and BG into itself are equal to the products of DB into itself, AD into itself, and AD into DB, twice. Removing what is common, that is, the product of DB into itself, leaves the products of AD into itself and twice into DB equal to the product of BG into itself. Thus the product of BG into itself is larger than the product of AD into itself by twice the product of AD into DB. Therefore the square of BG is always greater than the square of AD. Thus BG is always greater than AD. This is what we wanted to demonstrate.

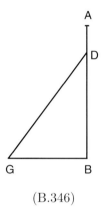

(B.346)

This being demonstrated, let us return to the proposed problem and indicate how to treat it. You will then do the following. Multiply four by itself, thus producing sixteen. Keep its half, which is eight, in mind. Next, subtract four from ten; this leaves six. Multiply its half, which is three, by itself; this produces nine. Subtracting from it the eight kept in mind leaves one. Add its root, which is one, to three; this makes four. Subtract it from the height of the wall; this leaves six, and by so much is the ladder withdrawn from the base of the wall. But you may want to know by how much the end of the ladder comes down from the top of the wall. Subtract four, which is proposed as remainder in the problem[1959], from six; this leaves two, and by so much does the ladder come down from the top of the wall.

This is proved as follows. Let the former figure remain the same. Thus subtracting AD from BG leaves four. Let AH be equal to BG. Thus DH is four. We have already demonstrated[1960] that the product of BG into itself is equal to the products of AD into itself and twice into DB. But BG is equal to AH. Thus the product of AH into itself is equal to the products of AD into itself and twice into DB. But the product of AH into itself is

---

[1958] *Elements* II.4 = PE_4.
[1959] Not the 'four' found above.
[1960] In the proof of the condition.

equal to the products of AD into itself, DH into itself, and AD into DH, twice.[1961] Therefore the products of AD into itself, DH into itself, and AD into DH, twice, are equal to the products of AD into itself and twice into DB. Removing what is common, which is the product of AD into itself, leaves the products of AD into DH, twice, and DH into itself, equal to twice the product of AD into DB. But twice the product of AD into DB is equal to the products of AD into DH, twice, and into HB, twice. Then the products of AD into DH, twice, and into HB, twice, are equal to the products of AD into DH, twice, and DH into itself. Removing what is common, which is the product of AD into DH, twice, leaves the product of AD into HB, twice, equal to the product of HD into itself. But the product of HD into itself is sixteen. Then twice the product of AD into HB is sixteen. Therefore once the product of AD into HB is eight. Let BZ be equal to AD, and let BD be (added in) common. Therefore the whole ZD is equal to AB. AB being ten, ZD is ten. But HD is four. This leaves as HZ six. Now this HZ is divided at the point B into two ⟨unequal⟩ parts the product of which is eight.[1962] So let ZH be bisected at the point K. Then [1963] the products of HB into BZ and KB into itself will be equal to the product of KZ into itself. But the product of KZ into itself is nine. Subtract from it the product of HB into BZ, which is eight; this leaves, as the product of KB into itself, one. Thus KB is one. KZ being three, BZ is two. BZ being equal to AD, AD is two. But DH is four. Therefore AH is six. AH being equal to BG, BG is six. This is what we wanted to demonstrate.

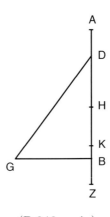

(B.346 again)

(B.347) Someone asks: A ladder ten (cubits) long standing against an equally high wall is withdrawn from the base of the wall by thrice that by which it comes down from the top. By how much is it withdrawn and by how much does it come down?

[1961] *Elements* II.4 = PE_4.
[1962] Sum and product of HB and BZ thus known.
[1963] *Elements* II.5 = PE_5.

Let the above figure remain the same. Then BG will be thrice AD. Now we have already demonstrated[1964] that the products of AD into itself and twice into DB are equal to the product of BG into itself. But the product of BG into itself is nine times the product of AD into itself. Let once the product of AD into itself be (added in) common. Thus the products of AD into itself, twice, and AD into DB, twice, are equal to ten times the product of AD into itself. But the products of AD into itself and into DB are equal to the product of AD into AB. Thus the products of AD into itself, twice, and AD into DB, twice, will be equal to twice the product of the whole AB into AD. So ten times the product of AD into itself will be equal to twice the product of AB into AD. But twice the product of AB into AD is equal to the product of AD into twenty, for AB is ten. So the product of AD into twenty is equal to ten times the product of AD into itself. Therefore the product of AD into two will be equal to once the product of AD into itself. Thus AD is two. This is what we wanted to demonstrate.

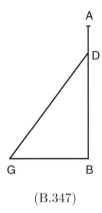

(B.347)

Consequently, the way to treat this and similar problems is the following. You multiply three by itself; this gives nine. You always add one to it, making ten, which you keep in mind. Next, always double the length of the ladder. Divide the result by the ten kept in mind. The quotient is the distance by which the ladder comes down from the top of the wall. Taking thrice this will give the distance between the lower end of the ladder and the base of the wall.

If it were said in such a problem that the distance between the lower end of the ladder and the base of the wall is two thirds, three fourths, or any (other fraction) less than one, of the distance the ladder comes down from the top, or that it is equal to it, this problem would not be solvable. Indeed, we have already established[1965] that the distance the ladder drops down from the top is always less than the distance it is withdrawn from the base.

[1964] B.346, condition.
[1965] B.346, preliminary condition.

1054 PART TWO: TRANSLATION, GLOSSARY

[(**B.347′**) [1966] *Someone asks: A ladder of unknown length standing against an equally high wall is withdrawn from the base of the wall by six cubits and comes down from the top by two cubits. What is its length?*

You will do the following. Multiply six by itself, thus producing thirty-six. Multiply two by itself; this gives four. Subtracting this from thirty-six leaves thirty-two. Divide its half, which is sixteen, by two; this gives eight. Adding two to it makes ten, and so many cubits is the ladder long.

The proof of this is the following. Let the above figure remain the same. We have already established [1967] *that the products of AD into itself and AD into DB, twice, are equal to the product of BG into itself. Now the product of AD into itself is four and the product of BG into itself, thirty-six. So subtract the product of AD into itself from the product of BG into itself; this leaves, as twice the product of AD into DB, thirty-two. Thus once the product of AD into DB is sixteen. AD being two, DB is eight. Since AD is two, the whole AB is ten. This is what we wanted to demonstrate.*]

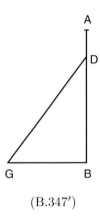

(B.347′)

(**B.348**) Now someone asks: A ladder of unknown length standing against an equally high wall is withdrawn from the base of the wall by a distance which, when added to the distance it comes down from the top of the wall, makes eight, while multiplying one distance by the other produces twelve. What is its length?

Let the above figure remain the same. Adding AD to GB gives eight. Let GH be equal to AD; thus the whole BH is eight. But it was supposed that the product of BG into GH is twelve.[1968] Let therefore BH be bisected at the point Z. Thus [1969] the products of BG into GH and ZG into itself are equal to the product of ZB into itself. But the product of ZB

[1966] Repetition (not verbatim) of B.344.
[1967] B.346, in the demonstration of the condition (also in B.344, source of the present problem).
[1968] Sum and product of BG and GH thus known.
[1969] *Elements* II.5 = PE_5.

into itself is sixteen. Subtracting from it the product of BG into GH will leave, as the product of ZG into itself, four. Thus ZG is two. BZ being four, BG is six; GH will be two, and it is equal to AD. Then it is like saying: 'A ladder of unknown length standing against an equally high wall is withdrawn from the bottom of the wall by six cubits and comes down from the top by two cubits; what is the length of the ladder?' Proceed then as I have taught above;[1970] this will give what you wanted.

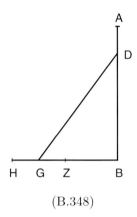

(B.348)

(B.349) Someone asks: A ladder of unknown length standing against an equally high wall is withdrawn from the bottom of the wall by a distance which, when diminished by the distance by which it comes down from the top of the wall, leaves four, while multiplying one distance by the other produces twelve. What is its length?

Let the figure remain the same. Thus the product of AD into BG is twelve, and subtracting AD from BG leaves four. Then let (a line) equal to AD, say HG, be cut off from BG. Thus BH is four. Now the product of BG into HG is twelve.[1971] Let then BH be bisected at the point Z. So the products of BG into HG and HZ into itself are equal to the product of ZG into itself, as Euclid said[1972]. But the product of BG into GH is twelve. Adding to it the product of ZH into itself, which is four, the product of ZG into itself will be sixteen. Thus ZG is four. BZ being two, the whole BG is six. Likewise again: ZG being four and ZH two, this leaves as HG two. HG being equal to AD, AD is two. Therefore it is like saying: 'A ladder of unknown length standing against an equally high wall is withdrawn from the bottom of the wall by six cubits and comes down from the top by two.' Proceed as I have taught above,[1973] and you will find what you require.

AGAIN, ON ANOTHER TOPIC

[1970] B.344 (or B.347'), with the same data.
[1971] Difference and product of BG and HG thus known.
[1972] *Elements* II.6 $= PE_6$.
[1973] B.344 (& B.347'), with identical data.

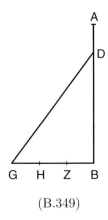

(B.349)

(**B.350**) Someone asks: A tree thirty cubits high bends to the earth at a height of ten cubits. How far from its root does its top touch the ground?

You will do the following. Subtract ten from thirty, leaving twenty. Multiply it by itself, and ten by itself, and subtract the lesser product from the greater; the remainder is three hundred. Its approximate root, which is seventeen and eleven thirty-fourths, is what you wanted.

This is demonstrated by means of a figure like this. Let the tree be line AG, the ten (cubits) to the bending point, line DG, the place where it bends being the point D. This leaves as line AD twenty. AD being equal to line DB, line DB is twenty. Now it is known that an angle of the triangle BGD, namely DGB, is a right angle. Thus [1974] the products of DG into itself and GB into itself are equal to the product of DB into itself. Subtracting the product of DG into itself from the product of DB into itself leaves as the product of GB into itself three hundred. Thus GB is the root of three hundred. This is what we wanted to demonstrate.

The treatment by algebra is as we have shown in the chapter on ladders.[1975]

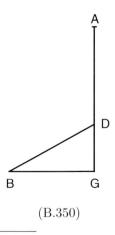

(B.350)

[1974] As usual, *Elements* I.47.
[1975] B.343*b* (*section* on ladders, thus B.343–B.349).

LIBER MAHAMELETH   1057

Again.

(**B.351**) A tree thirty cubits high bends in such manner that its top is ten cubits away from its root. How much of it remains upright to the bending point [1976]?

(**a**) You will do the following. Multiply thirty by itself, ten by itself, and subtract the lesser product from the greater; this leaves eight hundred. Divide its half, which is four hundred, by thirty; this gives thirteen and a third, and so much is its upright part to the bending point.

This is demonstrated as follows. Let the tree be line AG, ten, line GB, the part beyond which the tree bends, line GD, the bending point, point D. It is then evident that line AG is divided into two unequal [1977] parts at the point D. So [1978] the products of AD into itself, DG into itself and AD into DG, twice, are equal to the product of AG into itself. Now you know that line AD is equal to line DB. Therefore the products of DB into itself, DG into itself, and GD into DA, twice, are equal to the product of line AG into itself. But the product of DB into itself is equal to the products of DG into itself and GB into itself. Thus it is clear that the products of GB into itself and GD into itself, twice, increased by twice the product of GD into DA, are equal to the product of AG into itself. Subtract the product of GB into itself, which is a hundred, from the product of AG into itself, which is nine hundred; this leaves, as twice the product of DG into itself and twice the product of GD into DA, eight hundred. Thus once the product of DG into itself, increased by once the product of the same into DA, is four hundred. But the products of GD into itself and into DA are equal to the product of DG into GA. [*For if a line is divided into two unequal parts, the products of one part into itself and into the other are equal to the product of this part into the whole line, as Euclid declares in the second book.*][1979] Thus it is clear that the product of DG into GA is four hundred. Therefore dividing four hundred by line AG, which is thirty, will give for line DG thirteen and a third, and so much is the upright part, beyond which the tree bends. This is what we wanted to demonstrate.

(**b**) Or otherwise. Multiplying ten by itself produces a hundred. Dividing it by thirty gives three and a third. Add it to thirty; this makes thirty-three and a third. Subtracting three and a third from half of this, which is sixteen and two thirds, leaves thirteen and a third, and so much is the part beyond which it bends.

This is demonstrated in the following way. Let the tree be line AG, the part beyond which it bends, line DG, the place where it bends, point D, ten, line GB. Next, let B and D be joined. So line AD is equal to line DB. I shall then take point D as the centre of a circle encompassing the distances between D, A and D, B, say the circle ABH. Next, let G, H

[1976] Literally: from where it is bent.

[1977] Not a condition (see p. 1048, note 1945).

[1978] *Elements* II.4 $= PE_4$.

[1979] *Elements* II.3 $= PE_3$.

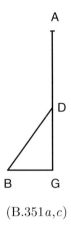

(B.351a,c)

and G, K be joined. Hence AK is the diameter of the circle. Thus it is known that the product of KG into GA is equal to the product of BG into GH, as Euclid said[1980], as follows: 'If two straight lines intersect within a circle, then the (rectangle) contained by the two parts of one of the lines is equal to the (rectangle) contained by the two parts of the other'. Now the multiplication of GB by GH produces a hundred: line BG is equal to line GH for, as Euclid said[1981], 'if a line falls within a circle without passing through the centre and another one, drawn from the centre, stands on it at right angles, the second will divide the first into equal parts'; now, as you know, DGB is a right angle, so line BG is equal to line GH. Thus if a hundred is divided by line AG, which is thirty, this will give for GK three and a third. Hence the whole AK is thirty-three and a third. But you know that line AD is equal to line DK; so line DK is sixteen and two thirds. Line KG being three and a third, line GD is thirteen and a third, and so much is the part beyond which it bends. This is what we wanted to demonstrate.

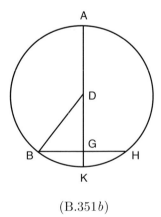

(B.351b)

[1980] *Elements* III.35.
[1981] *Elements* III.3.

(c) Or otherwise. Let the part beyond which the tree bends be a thing, and it is line DG. This leaves as line AD thirty minus a thing, equal to line DB. Therefore multiply DB by itself; this gives one square and nine hundred minus sixty things, which are equal to the products of DG into itself and GB into itself, which are one square and a hundred. Proceed then as has been shown in algebra.[1982]; this will give as the thing thirteen and a third. This is what we wanted to demonstrate.[1983]

(**B.352**) Someone asks: A tree of unknown height bends beyond six cubits and its top touches the ground at a distance of eight cubits from its root. What is its height?

You will do the following. Multiply eight by itself, thus producing sixty-four. Next, multiply six by itself, thus producing thirty-six. Add the two products; this makes a hundred. Adding to its root six will give what you wanted, namely sixteen, and so much is the height of the tree.

This is proved as follows. Let the figure remain as it was.[1984] Thus BD will be six and BG, eight. Now the products of BD into itself and BG into itself are equal to the product of DG into itself. So the root of the sum of DB multiplied by itself and BG multiplied by itself will be DG, which is equal to AD. Add DB to AD; the whole will be AB, which is the height of the tree.

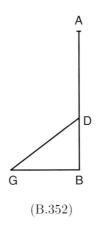

(B.352)

(**B.353**) Someone asks: A tree of unknown height bends beyond its three eighths and the place where its top falls is eight cubits away from its root. How high is it, and where is the point at which it bends[1985]?

If it were said that it bends beyond two thirds of itself, or three fourths, or a half, or more than a half, the problem would not be solvable, since the top would not fall to the ground.

[1982] Equation reducing to $b_1 - a_1 x = b_2$.
[1983] Geometrical computation. See note 584, p. 724.
[1984] It is the same as in the *subsequent* problem, for we find the places of B and G inverted in the two previous problems.
[1985] This second part is not answered.

Let the tree be AB, and let it bend at the point D. So BD is three eighths of AB. This leaves as AD five eighths of AB. Let it descend as DG. So DG is five eighths of AB. Now the place where its top falls is the point G, so BG is eight. Now the products of BG into itself and BD into itself are equal to the product of DG into itself. But the product of DG into itself is three eighths and an eighth of an eighth of the square of AB and, further, the product of BD into itself is an eighth and an eighth of an eighth of the square of AB. This therefore leaves as the product of BG into itself a fourth of the square of AB. Therefore a fourth of the square of AB is sixty-four, thus AB is sixteen. This is what we wanted to demonstrate.[1986]

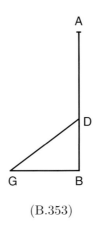

(B.353)

(B.353′) If it were said that it bends beyond three eighths of its height and that the distance between the place where its top falls and its root is half its height, the problem would be indefinite[1987]: taking any number whatsoever, the products of its three eighths into itself and of its half into itself are equal to the product of its five eighths into itself. [*This is what we wanted to demonstrate.*]

AGAIN [1988]

(B.354) The top of a tree thirty cubits high bends each day by one cubit. After how many days will it fall to the ground?

You will do the following. Multiply half of thirty by three and a seventh and divide the product by the distance (its top) descends in one day; this will give the number of days whereafter it will fall to the ground, which is forty-seven and a seventh.

The reason for this is the following. As we know, the tree will descend no more than a fourth of a circumference. [*This is proved as follows.*][1989]

[1986] Geometrical computation.

[1987] *multiplex*, thus with 'numerous' answers. Same sense as *interminata* in B.187 (note 1313, p. 890).

[1988] Now the whole tree falls progressively to the ground.

[1989] What will be proved is the formula.

Let the height of the tree be line AG. Thus it is known that line AG will descend until it becomes like line BG. Let then the point G be the centre of a circle meeting the extremities of lines GA and GB, say the circle ABT. Thus it is clear that the figure ABG is a quadrant and the arc AB, a fourth of the circumference. If you want to know the whole circumference, double line AG; this gives line AK, which is the circle's diameter; then multiply it by three and a seventh to obtain the whole circumference.[1990] Now if you want to know arc AB, take a fourth of the whole circumference. But you know that doubling AG, multiplying the result by three and a seventh, and taking a fourth of the product is the same as multiplying half of line AG by three and a seventh. This is what we wanted to demonstrate.

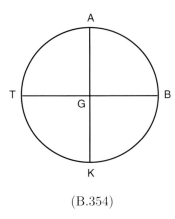

(B.354)

If it were said that it descends each day by two, or three, cubits, you would take half of forty-seven and a seventh, or its third, or more, according to what was proposed.

(B.355) The top of a tree seventy cubits high descends each day by three cubits and rises each day by one cubit; for how many days will it be falling towards the ground?

You will do the following. Multiply half of its height, which is thirty-five, by three and a seventh, thus producing a hundred and ten.[1991] Then add one cubit to three cubits, which makes four cubits; keep it in mind. Next, subtract one from three cubits, which leaves two. Then subtract one cubit from a hundred and ten; this leaves a hundred and nine. Divide it by two; this gives fifty-four, and one remains. Add it to one cubit by which the tree rises each day; this gives two. Denominate it from the four kept in mind; this gives a half. Adding it to the fifty-four resulting from the division makes fifty-four and a half, and for that many days will the tree be falling towards the ground.

(B.356) (The top of) a tree of unknown height descends each day by two cubits and falls towards the ground for forty-four days. What is its height?

[1990] As seen in B.341.
[1991] Arc calculation seen in the previous problem.

1062   PART TWO: TRANSLATION, GLOSSARY

Multiply the two cubits by forty-four, thus producing eighty-eight. Always divide it by three and a seventh; this will give twenty-eight. Always double it; this will make in this case fifty-six, and so much is the height of the tree.

OTHER TOPIC [1992]

(B.357) How many trees, with a distance between them of two cubits, can be planted on a plot twenty cubits long and ten wide?

Divide the length of the plot by two cubits and always add one to the quotient; in this case, the result will be eleven. Then divide the width of the plot by two cubits and always add one to the quotient; in this case, the result will be six. Multiply it by eleven; this produces sixty-six, which is the number of trees which can be planted.

AGAIN.

(B.358) Along the length of a certain plot eleven trees are planted and along its width, six, all with a distance between them of two cubits. What is the length and the width of this plot in cubits?

Subtract one from eleven; multiply the remainder by two, thus producing twenty, which is its length. Likewise, subtract one from six, thus leaving five; multiply it by two cubits, thus producing ten, which is its width.

AGAIN

(B.359) Of two towers, one is thirty cubits high and the other twenty, and their bases are eight cubits apart. What is the distance between their tops?

You will find the answer as follows. Multiply the difference between twenty and thirty, which is ten, by itself, and eight by itself, and add the two products; this will give a hundred and sixty-four. Its (approximate) root, which is twelve and five sixths, is the distance in cubits between their tops.

This will be demonstrated as follows. Let one of the towers be line AB, the other, line DG, and let eight be line GB. I shall draw from the point D a line parallel to line BG, say line DK. Then line DG is equal to line KB.[1993] Line DG being twenty, line KB is twenty. But line AB is thirty, thus line AK is ten. And line BG is equal to line DK. Now it is evident that the triangle AKD is right-angled. Thus the products of AK into itself and KD into itself are equal to the product of AD into itself. Therefore the product of AD into itself is a hundred and sixty-four, and AD is its root. This is what we wanted to demonstrate.

AGAIN.

[1992] Problems B.357–358 are found only in MS $\mathcal{D}$, among a collection wholly taken from the *Liber mahameleth*. The same holds for B.362.

[1993] *Elements* I.33 (if need be). Also used in B.360–361.

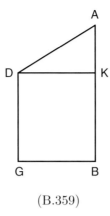

(B.359)

(**B.360**) The tops of two towers, one thirty cubits high and the other twenty, are twelve and five sixths cubits apart. What is the distance between their bases?

You will find the answer as follows. Multiply the difference between thirty and twenty, namely ten, by itself and subtract the result from the product of twelve and five sixths into itself. The root of the remainder is the distance between the bases.

The reason for this is the same as that indicated in the previous (problem), namely the following. Let the figure remain as it was. Thus line AB is thirty, line DG, twenty, and line AD, twelve and five sixths. We want to know the length of line GB. I shall draw from the point D line DK parallel to line GB. Hence KB is twenty. AB being thirty, AK is ten. Now the products of AK into itself and DK into itself are equal to the product of AD into itself. So subtracting the product of AK into itself [*which is ten*] from the product of DA into itself will leave the product of DK into itself. The root of the remainder is then DK, which is equal to GB. This is what you wanted to know.

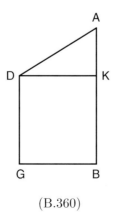

(B.360)

(**B.361**) Someone asks: Two towers, the height of one of which is unknown and the height of the other, eighteen cubits, have a distance between their tops of ten cubits and between their bases, of six. What is the height of the tower which is unknown?

This problem is subject to two interpretations: the tower of unknown height may be the higher or the lower one.

(*i*) Assume it to be the higher one. You will then do the following. Multiply the distance between the bases by itself, subtract the result from the product of the distance between the tops into itself, and add the root of the remainder to the height of the tower which is known. The result is the height of the tower which is unknown.

The proof of this is the following. Let the known tower[1994] be AB and the unknown one, DG. So BG is six, AD is ten, and AB is eighteen. I shall draw the perpendicular AH, parallel to BG. Then HG will be equal to AB. AB being eighteen, HG is eighteen. But BG is equal to AH. Thus AH is six. Now AD is ten, and AHD is a right angle. Therefore ⟨the products of AH into itself and DH into itself are equal to the product of AD into itself. Subtracting then the product of AH into itself from the product of AD into itself will leave as the product of DH into itself sixty-four. Therefore⟩ DH is eight.[1995] But HG was eighteen. Thus the whole DG is twenty-six. This is what we wanted to demonstrate.

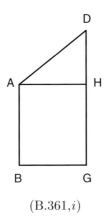

(B.361,*i*)

(*ii*) Assume it to be the lower one. Multiply the distance between their tops by itself and subtract from the result the product of the distance between the bases into itself. Subtract the root of the remainder from the height of the tower which is known; the remainder is the unknown height of the second tower.

The proof of this is the following. Let the figure remain the same. Let (now) the known tower[1996] be DG and the unknown one, AB. Then AD is

---

[1994] 'tree' in the MSS (corrected in 𝒜). See note 1996.
[1995] The computation of DH, although banal, must have been in the original text since it is found in the (similar) proof below.
[1996] 'tree' in the MSS.

ten, BG is six, and DG is eighteen. I shall draw from the point A line AH parallel to BG. Thus AH is six. Now it is clear that the products of AH into itself and HD into itself are equal to the product of AD into itself. Therefore multiply AD, which is ten, by itself and subtract from the result the product of AH, which is six, into itself; this leaves, as the product of DH into itself, sixty-four. Therefore DH is eight. DG being eighteen, this leaves ten as HG. But it is equal to AB; thus AB is ten, and so much is the unknown height of the tower. This is what we wanted to demonstrate.

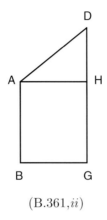

(B.361,*ii*)

(*iii*) If it were said in one of these problems[1997] that the distance between the bases is greater than the distance between the tops, this would not be solvable. Likewise also, the distance between the tops cannot be said to be equal to the distance between the bases if one of the towers is to be higher: if the distances are equal, the towers will perforce be equal. The proof of all this is evident.

AGAIN. SCIENCE OF FINDING THE HEIGHT OF A TOWER OR A TREE

(B.362) If you want to know the height of a tower or a tree (do the following). Take two sticks, one longer than the other by one or two cubits. Then place each of the two sticks upright on flat ground, (thus) parallel one to the other, and both aligned with the tower or the tree. Then direct the visual ray[1998] in such a way that, beginning at the top of the smaller stick and traversing the top of the longer, it reaches the top of the tower (or the tree). This being done, subtract the length of the smaller stick from that of the other and denominate the remainder from the amount of the distance between the two sticks.[1999] Multiply this fraction by the distance between the smaller stick and the tree or the tower and add the product to the length of the smaller stick. The result will be the height of the tower or the tree.

[1997] B.359–361.
[1998] According to the Ancients, the eye emits this ray.
[1999] We may also have to divide if the distance is less than the 'one or two cubits' mentioned above. See note 1790, p. 1009.

1066 PART TWO: TRANSLATION, GLOSSARY

Again, (B–XXI) on
Another topic

(B.363) A rope of four cubits surrounds a bundle of a hundred rods. How many identical rods are surrounded by a rope of ten cubits?

(*a*) You will find the answer as follows. Multiply four by itself, thus producing sixteen, which you take as the principal. Next, multiply ten by itself, thus producing a hundred. Multiply it by the number of rods, which is a hundred; this will produce ten thousand. Divide it by the principal; this gives six hundred and twenty-five, and such is the required number of rods.

(*b*) Or otherwise.[2000] Multiply ten by itself, thus producing a hundred. Divide it by sixteen; this will give six and a fourth. Multiply it by the number of rods, which is a hundred, thus producing six hundred and twenty-five. This is what you wanted.

The reason for this is the following: the ratio of one quantity of rods to the other is the same as the ratio of the square of (the length of) one rope to the square of the other.[2001]

AGAIN.

(B.364) A rope of four cubits surrounds one bundle and another rope of twelve cubits surrounds another bundle. How many times is the smaller bundle contained in the larger?

You will find the answer as follows. Multiply four by itself, thus producing sixteen. Next multiply twelve by itself, thus producing a hundred and forty-four. Dividing this by sixteen gives nine, and so many times is the smaller contained in the greater, namely nine times.

(B.365) Likewise also, if it were said that a rope of four palms surrounds a bundle of harvest the price of which is half a nummus, what would be the price of a bundle surrounded by a rope of twelve palms?

You will find the answer as follows. Multiply four by itself and take the result as the principal. Next, multiply twelve by itself, thus producing a hundred and forty-four. Divide it by the principal; this will produce nine. Multiplying it by a half will give four and a half, and so much is its price.

We have multiplied nine by a half for the following reason. We wanted to know how many times the bundle with the four-palm rope is contained in the bundle with the twelve-palm rope,[2002] and, having found nine times,

---

[2000] The usual 'other way'.

[2001] Incomplete justification.

[2002] Which is B.364.

LIBER MAHAMELETH 1067

and having attributed to each bundle half a nummus, to all of them correspond four nummi and a half. [*The reason for this is what Euclid said in the twelfth book.*][2003]

(**B.366**) Someone asks: A rope ten cubits long surrounds a thousand rods. What is the length of a rope surrounding two hundred and fifty of them?

We know [2004] that the ratio of a thousand to two hundred and fifty is the same as the ratio of the square of ten to the square of the required quantity. Proceed then as explained above, [2005] and twenty-five will be found as the square of the required quantity. Therefore the required quantity is five.

(**B.367**) Someone asks: A rope of three cubits surrounds a bundle the price of which is eighteen nummi. What is the length of a rope surrounding a bundle the price of which is two nummi?

We know [2006] that the ratio of the square of three, which is nine, to the square of the required quantity is the same as the ratio of eighteen to two. Thus proceed as I have taught above, [2007] and the required result will be one.

Treat all similar problems accordingly, and you will succeed.

[2003] If at all, this should have been mentioned in the justification found in B.363*b*.

[2004] B.363.

[2005] Plain proportion (thus rule of three, p. 777).

[2006] Probably alluding to B.365.

[2007] Again, plain proportion.

Chapter (B–XXII) on Messengers

(**B.368**) For instance, a messenger is sent to a town and advances daily by twenty miles. In how many days will another messenger, sent five days later and advancing daily by thirty miles, overtake him?

(***a***) You will do the following. Take the difference between twenty and thirty, namely ten, as the principal. Next, multiply five by twenty and divide the product by the principal; this will give ten, and so many days did the second messenger walk. The first has been on his way that many days plus five, which is fifteen, when the second messenger met him.

This will be demonstrated by the following proof. Let the days the first messenger walked be line AB. You know that the days the first walked outnumber the days the second walked by five. Thus let a line of five, say line GB, be cut off from line AB. This leaves line AG as the days the second messenger walked. Now you know [2008] that the numbers of miles are equal for each messenger and that the product of the number of days each messenger walked into the number of miles he covers daily is the total number of miles each one covers until they meet. Therefore it is clear that the product of line AG into thirty is equal to the product of line AB into twenty. But the product of AB into twenty is equal to the products of each of AG and GB into twenty. Now you know that multiplying line GB by twenty produces a hundred, for line GB is five. Thus the product of AG into twenty plus a hundred is equal to the product of AG into thirty. Subtract then the product of AG into twenty from the product of the same into thirty; this will leave, as the product of AG into ten, a hundred. Dividing then a hundred by ten will give line AG, which is ten.

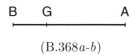

(B.368 *a-b*)

(***b***) Or otherwise. You know [*it has been demonstrated*][2009] that the product of line AG into thirty is equal to the product of AB into twenty. Therefore the ratio of AG to AB is the same as the ratio of twenty to thirty. Twenty being two thirds of thirty, AG is two thirds of AB, and therefore line AB will perforce be thrice line GB. Thus line AB is fifteen, and line AG is ten, and so many are the days the second messenger walked until he overtook the other, the first having walked fifteen days. This is what we wanted to demonstrate.

[2008] Inferred from the hypothesis.
[2009] See part *a*.

LIBER MAHAMELETH **1069**

(*c*) Or otherwise. Put, as the number of days after which they met,[2010] a thing, this being the number of days the second messenger walked; multiplying the thing by the number of miles this second messenger covers daily produces, as the number of miles he covered (altogether), thirty things. But the number of days the first messenger walked will be a thing and five days; multiplying this by the number of miles he covers daily, which is twenty, will give twenty things and a hundred miles. This is equal to thirty things. Proceed then according to algebra.[2011] The amount of the thing will be ten, and so many are the days after which the second messenger overtook the first.

(*d*) You may want to verify this. You know that the first messenger walked fifteen days, covering daily twenty miles; therefore the miles he covered are three hundred, for multiplying fifteen by twenty produces this result. The second messenger walked ten days, covering daily thirty miles; therefore the miles he covered are three hundred. Thus the numbers of miles are equal when the second messenger overtakes the first.

(**B.369**) A messenger, sent from one town to another at a distance of four hundred miles, advances daily by twenty miles. Another messenger is sent fifteen days later and is to overtake him at the entry of the town so that they walk in at the same time. How many miles must he cover daily?

You will do the following. Multiply fifteen by twenty; this produces three hundred. Subtract it from four hundred; this leaves a hundred. Divide it by twenty; this gives five. Dividing four hundred by this gives eighty, and so many miles does the second need to cover daily so as to overtake the first after five days.

(**B.370**) A ship moves from one place to another at a distance of three hundred miles. It sails daily twenty miles but is driven back five miles daily by the wind. In how many days will it reach this place?

You will do the following. Add five to twenty; this makes twenty-five. Next, subtract five from three hundred; this leaves two hundred and ninety-five. Divide it by the difference between five, by which it goes back, and twenty, which it sails, that is, by fifteen; this gives nineteen and a remainder of ten. Add this ten to five miles, thus making fifteen. Denominate it from twenty-five; this gives three fifths of a day. Adding these three fifths of a day to nineteen will make nineteen and three fifths of a day, and in so many days will it reach the intended place.

(**B.371**) A snake comes out of a hole daily by a third of its length and returns daily by a fourth of its length. In how many days will it be out completely?

You will do the following. Multiply the denominators of a third and a fourth, thus producing twelve. Add its third and its fourth, which are three

---

[2010] See subsequent specification.

[2011] Equation of the form $a_1 x + b_1 = a_2 x$.

and four [2012]; this gives seven, which you keep in mind. Next, subtract three from twelve; this leaves nine. Divide it by the difference between three and four, which is one; this gives nine. Next, denominate three from seven; this gives three sevenths. Adding it to nine makes nine and three sevenths, and in so many days and fraction of a day will it be out completely.

(B.372) A snake seven cubits long comes out of a hole daily by one cubit and returns daily by a third of a cubit. In how many days will it be out completely?

You will do the following. Add a third to one; this makes one and a third, which you keep in mind. Next, subtract a third from seven cubits; this leaves six and two thirds. Divide it by the difference between a third and a cubit; this gives ten. Then denominate a third from one and a third; this gives a fourth. Adding it to ten makes ten and a fourth, and in so many days and fraction of a day will it be out completely.

(B.373) A snake comes out daily by its third and returns by an unknown amount, and is completely out after nine days and three sevenths. What is the unknown fraction by which it returns?

You will do the following. Always subtract one from a third of nine days and three sevenths, thus leaving two and a seventh. Next, subtract three sevenths from nine, thus leaving eight and four sevenths. Denominate two and a seventh from it; this gives a fourth. So much is the unknown fraction, namely a fourth.

(B.374) A snake seven cubits long goes out daily by one cubit and returns daily by some unknown fraction of a cubit, but comes out completely in ten days and a fourth. What is this fraction?

You will do the following. Multiply the cubit by ten and a fourth, thus producing ten cubits and a fourth. Subtract from it the length of the snake, which is seven; this leaves three and a fourth. Next, subtract the fourth added to the ten days from ten, thus leaving nine and three fourths. Denominate three and a fourth from it; this gives a third. Thus it returns daily by a third of a cubit.

[2012] *Sic* (not respectively).

Chapter (B–XXIII) on Another topic

(B.375) [2013] There were three men. The first said to the other two: 'Receive as much from what I have as you have each.' Likewise the second said to the first and the third: 'Let each of you receive as much from what I have as he has.' The third said the same to the first and the second. This being done, they were found to have the same amount. How much did each one have?

You will do the following. Always add one to the number of men, thus making in this case four; so much did the third have. Double it and subtract one from the result, thus leaving seven; so much did the second have. Double it and subtract one from the result, thus leaving thirteen; so much did the first have. This will be done whatever the number of men.

(B.376) There were four men. The first said to the other three: 'Let each of you receive from what I have as much as he already owns.' This was done. Likewise the second said to the other three: 'Let each of you receive from what I have as much as he already owns.' This was done. The third said the same to the other three, and so did the fourth. This being done they were found to have the same amount. How much did each one have?

Always add one to the number of men, making in this case five; so much did the fourth have. Double it, and subtract one from the result, thus leaving nine; so much did the third have. Double it again, and subtract one from the result, thus leaving seventeen; so much did the second have. Double it, and subtract one from the result, thus leaving thirty-three; so much did the first have.

(B.377) Three men had between them seventy-two nummi. The first said to the other two: 'Let each of you receive from what I have as much as he already owns.' The second said the same to the other two, and so did the third. This was done and they were found to have each the same amount of the seventy-two. How much did each one have?

Find by the above rule how much each man had; you will find that the third had four, the second, seven, the first, thirteen. [2014] Adding them all makes twenty-four. Dividing seventy-two by it gives three. Multiply this by what each one has and you will find what you require. Thus the first had thirty-nine, the second, twenty-one, and the third, twelve.

[2013] MS 𝒜 gives this chapter a heading which relates it to the subject of sharing, which is B–III (probably on account of B.378, end: *participatio*, 'sharing-out'). However, what we have here is a section on money-lending among a fixed number of partners.

[2014] See B.375 (rule and result).

1072   PART TWO: TRANSLATION, GLOSSARY

(B.378) There were three men. The first said to the other two: 'Let each of you receive from what I have as much as he already owns.' The second said the same thing, so also the third. This being done, the first was found to have as much as the second plus two nummi and the second as much as the third plus one nummus. Then how much did each of them have?

Put as what one of them had after all had received any number.[2015] For instance, let the third have five; thus the second will have six and the first, eight. I shall now determine what each has according to almencuz, thus 'inversely' —to mean: beginning inversely from above[2016]. Add to the property of the third half the property of the second and half the property of the first; this gives twelve, and the property of the second becomes three and that of the first, four. Next, add to the property of the second half of twelve and half of four; the property of the second thus becomes eleven, that of the third, six, and that of the first, two. Then add to the property of the first half of six and half of eleven. Therefore the first will have ten and a half, the second, five and a half, and the third, three; and so much did each of them have before the sharing-out.

There can be devised innumerable problems dealing with this subject.

(B.379) Three men wanted to buy a certain horse, each for himself. The first said to the second: 'If you give me half of what you have, it will, together with what I have, make up the price of the horse.' The second said to the third: 'If you give me a third of what you have, I shall have, together with what I (already) have, the price of this horse.' The third said to the first: 'If you give me a fourth of what you have, I shall have, together with what I (already) have, the price of this horse.' Then how much did each of them have and what was the price of the horse?

This problem is indeterminate. In it, you will do the following. Multiply the denominators of a half, a third and a fourth; this gives twenty-four. Always add one to it if the number of men is odd, but if it is even always subtract one [*from the denominating number*][2017]; the result after this addition or subtraction of one will be the price of the horse: twenty-five in this case.   If you want to know what the first has, subtract one from the denominator of a half, which is two; this leaves one. Multiply it by the denominator of a third, which is three, thus producing three. Add one to it, which makes four. Multiply it by the denominator of a fourth, which is four, thus producing sixteen. So much does the first have.   Subtract it from the price of the horse and multiply the remainder by the denominator

---

[2015] The problem is indeterminate.

[2016] That is, computing backwards: *secundum almencuz* corresponds to *e converso* (with *almencuz* transcribing the Arabic *al-mankūs*).

[2017] That is, from the product of the denominators (see note 345, p. 660). This must be an early gloss prompted by the ambiguity of a rule expressed generally ('always') but applied to a specific case ('to it', referring to twenty-four).

LIBER MAHAMELETH

of a half, which is two, thus producing eighteen. So much does the second have. Again, subtract this from the price of the horse and multiply the remainder by the denominator of a third, which is three, thus producing twenty-one. So much does the third have.

(B.380) Four men met together to buy a certain horse, each for himself. The first said to the second: 'If you give me half of what you have, I shall have, together with what I (already) have, the price of this horse.' The second said to the third: 'If you give me a third of what you have and I add it to what I have, I shall have the price of this horse.' The third said to the fourth: 'If you give me a fourth of what you have and I add it to what I have, I shall have the price of this horse.' The fourth said to the first: 'If you give me a fifth of what you have and I add it to what I have, I shall have the price of this horse.' Then how much did each of them have and what was the price of this horse?

This problem is indeterminate. In it, you will do the following. Multiply the denominators of all the given fractions, without omitting any; this will produce a hundred and twenty. Subtract one from it, for the number of men is even,[2018] leaving one hundred and nineteen. So much is the price of the horse. If you want to know what the first has, subtract one from the denominator of a half, thus leaving one; multiply it by the denominator of a third and add one to the product, thus making four; multiply this by the denominator of a fourth and subtract one from the product, thus leaving fifteen; multiply it by the denominator of a fifth, thus producing seventy-five. So much does the first have. Subtract it from the price of the horse and multiply the remainder by the denominator of a half, thus producing eighty-eight. So much does the second have. Subtract it from the price of the horse and multiply the remainder by the denominator of a third, thus producing ninety-three. So much does the third have. Subtract it from the price of the horse and multiply the remainder by the denominator of a fourth, thus producing a hundred and four. So much does the fourth have. This is what you wanted to know.

(B.381) Four men met together to buy a certain horse, each for himself. The first said to the other three: 'If you give me half of what you have and I add it to what I (already) have, I shall have the price of this horse.' The second said to the other three: 'If you give me a third of what you have and I add it to what I have, I shall have the price of this horse.' Likewise the third asked that a fourth be given to him, and the fourth, a fifth. Then how much does each of them have and what is the price of the horse?

This problem is indeterminate. I shall present for it a treatment involving algebra, but not following Abū Kāmil.

Let what the first has be one and what the three (others) have be a thing. Add then one to half of the thing; this gives one and half a thing. So much is the price of the horse. Multiply it by the denominator of a

---

[2018] Rule in B.379.

third; this produces three and one and a half things. Subtract from it one and a thing, which is their total amount; this leaves two and half a thing. Half of it is one and a fourth of a thing. So much does the second have. Next, multiply the price of the horse by the denominator of a fourth; this produces four and two things. Subtract from it one and a thing, which is their total amount; this leaves three and one thing. A third of it is a third of a thing and one. So much does the third have. Next, multiply the price of the horse by the denominator of a fifth; this produces five and two and a half things. Subtract from it one and a thing; this leaves four and one and a half things. Its fourth is one and three eighths of a thing. So much does the fourth have. Next, add what the second, the third and the fourth have; this makes three and five sixths of a thing and three fourths of a sixth of a thing, which are equal to what the second, the third and the fourth have after another manner[2019], namely a thing. Then do as has been said in mucabala, which is to remove what is common, that is, what is repeated on the two sides[2020]. This will leave a fourth of a sixth of a thing, which equals three. Thus a thing equals seventy-two. But the price of the horse was one and half a thing; thus the price of the horse will be thirty-seven. The second had one and a fourth of a thing; thus he will have nineteen. The third had one and a third of a thing; thus he will have twenty-five. The fourth had one and three eighths of a thing; thus he will have twenty-eight. The first had one. So we know what each of them had.

There can be devised numerous other problems dealing with this subject.[2021]

[2019] Namely as set initially.

[2020] Use of the word *mucabala* (Arabic *muqābala*) alone is new, but not the application of this second of the two main algebraic operations (note 921, p. 794).

[2021] Similar remark at the end of B.378.

Glossary

In providing more than the mathematical terms, this glossary may be useful for comparison with other occurrences, either in the present text or elsewhere. The references are to lines in the Latin text, or to a figure. A reference in ( ) means that it is our addition or correction; in [ ] that it belongs to a gloss; when preceded by *, the word referred to is by a reader's, or former reader's, hand, in which case the reference is followed by n since it will be found in the critical notes. A reference number followed by (2) means that there are two occurrences in the line. Finally, Tn followed by a number will refer to a note in the translation. Dictionaries of mediaeval Latin consulted are the *Mittellateinisches Wörterbuch*, the *Dictionary of Medieval Latin*, Niermeyer's *Lexicon* as well as du Cange's *Glossarium*. Klamroth's study of the Arabic Euclid lists Greek and Arabic geometrical terms, some of which appear in this glossary. Abbreviations used here are: abl(ative), absol(ute use), accus(ative), adj(ective), adv(erb), ant(onyme), dat(ive), f(eminine), gen(itive), imperf(ect tense), indic(ative), inf(initive), inv(ariable), m(asculine), n(euter), opp(osite), perf(ect tense), pluperf(ect), pl(ural), *post(erius)*, prep(osition), pres(ent tense), *pr(ius)*, *q(uod)* *v(ide)*, subj(unctive), syn(onym).

— A —

abscindere: 8374, 8377, 8394, 8397, 8410, 8427, 8441, 8452-53, 8460, 8476, 8496, 8508, 8527, 8541, 8563, 8571, 8597, 8630 (2), 8635.

absque: 7, 9 (2), 781n, 1090, 1736.

Abū Kāmil: (*Avoquemel*) 254, [277], 278, 15758; 4935 (*Avochemel*); 9940 (*Abuquemil*).

ac: (*talis est ac si*, *est ac si*; syn. *est quasi*) 9654, 9677, 11687, 13433.

accedere: [441].

acceptio: 1867 (*partis*), 15687.

accĭdere: 1635, 1637, [1646], 1646, 3049, 4196, 4353 (2), 4357, 4358, 4514 (2), 4523, 4524, 4904, 4905, 5694, 6892, 8015, 8016 (2), 8024, 8025, 8036 (2), 8093, 8094, 8125, 8127, 8143, 8144, 8223, 8224 (2), 8225 (2), 8226, 8235, 8236, 8237, 8248, 8249 (2), 8250, 8263 (2), 8264. *Quod accidit uni* (*l.* 1635) = ما يصيب الواحد, see Abū Kāmil's *Algebra*, fol. 54^r, 14–15 or

Rebstock, *Muʿāmalāt Traktat*, pp. 106–107.

accipere: 197, 205, 384, 385, 390, 391, 445, 483, 496, 580 (2), 581 (2), 582, 585 (2), 586, 592, 604, 609, 949, 955, 966, 967, 1020, [1073], [1074], 1381, 1458, 1567 (*fractionem*, syn. *sumere*), 1573, 1582, 1588, 1591, 1602, 1603, 1627 (2), 1629, 1668, 1686, 1724, 1735, 1743, 1746, 1754, 1773, 1798, 1809, 1917 (2), 1921, 1922 (2), 1923, 1925, 1926, 1927, 1931 (2), 1932, 1936 (*nichil*), (...), 5501 (*radicem*), (...); 1773 (*de*), 2170 (*ab*), 8897 (*pro*), 1735 (*per se*); (*numerum*, e.g.:) 4922, 11217, 11231 (syn. *ponere*, cf. 11253).

accomodare: 32.

accrescere: 9211.

activus: 21.

actor: [1057], [1482], *2246n.

actus: 8878 (*in actu*), 8882 (*ad actum*).

adaequare: See *adequare*.

adaptare: 7455 (*aptare*, 𝒜), 11087 (Tn1495).

addere: (to add) 54 (2), *211n (2), 525, 529, 559, 564, 592, 603, 607, 783, 784, 808, 810, 811 (2), 816, 819, 822, (...); (*addere superficiem linee*; syn. *adiungere*, *agregare*) 5671, 5719; (to append —names; syn. *adiungere*, *agregare*, *apponere*) 1355, 1362, 1366, 1395, 1397, 1400, 1571, (...); (***additus***, noun; زائد) 526, 527 (2), 561, 562, 603, 604, 605, 1192 (3), 1193, 1194, 1195, figure p. 48 (2), [1536], [1537], [1562], [1565], 3485, *3677n (2), 4078, (...).

additio: (addition) 55, 68, *3553n, *3649n, *4881n, 5666, *6180n, 9826, 10233, 13743, 13832, 15713; (appending) 1835.

adequare: 9726, 9739; (*adequari*) 4112, 4113, 6794, 6796, 6797, 6798, 6801, 6849, 6851, 6852, [6854], 7925, 7926, 9083, 9084, 9102 (2), 9120, 9137, 9138, 9629, 9722, 9723, (...), 11326, 11339, 11354, (...); (absol.) 11323, 14347, 15586.

adherere: 9651.

adhuc: 2037, 4960, 4964, 14108.

adicere: 179, [199], *211n, 437, 746, 10031, 10065, 10089, 10101, 10118, 10139, 10193, 10206, 10262, 10268, 10303, 10319, 10338, (...).

adiectio: 10001, 13984, [14157-58], 14713. Syn. *additio*.

adinvenire: 436, [3070], 8130-31, 9625, 10705, 12011, 12463.

LIBER MAHAMELETH 1077

adiungere: (to add; syn. *addere*) 846, 862, 2343, 2752, 2894, 6419, 6546, 6549, (6556); (to append —names; syn. *addere*) 1358, 1388; (*scala adiuncta parieti*) 15003, 15025, 15091, 15153, [15182], 15197, 15209-10, 15213, 15228.

adiutorium: 432.

admiscere: See *amiscere*.

aeq–: See *eq–*.

afferre: 8997, 9157.

agere: 623 (*de*), 632, 981, 985, [1076], 1143, 2508, 2511, 4900, 4932, 4934, 4938, 5773 (*in*), 5786, 5803, 5902, 6437, 7951 (*nichil agimus*), (...), 12205 (*secundum sensum questionis*), 13840, 14940; (*secundum algebra*) 15248, 15757; (*modus agendi*) 1641-42, 1657, 3074, 3143, 4024, 6638, 7196, 7279, 8018, 8027, (8174), 8255, 8485, 8699, 8892-93, 8978, 9041, 9375, 9584-85, 9813, 10518, 10519, 10520, 11009, 11765, 12488, 12623, 12713-14, 13709, 14412, 15113, 15171; (*regula agendi*) 5771.

agitare: 379, 380, (388).

agregare: 27, [39], 177, 180, 183-84, 185, 191 (2), 202, 203, *211n, 380, 388, 660, 687, [773], 906, 914, 926, 936, 950, 962, 977, 1021, 1022, 1023, 1024, 1025 (2), 1201, 1204, 1207, 1235, 1248, (...); (*cum*) 3081, 3104, 3200, 3205, 3793, 5143, (...); (*et*) 6988, 7074, 14281, (...); (*nomina*) 1383; (*agregatum,* noun) 1333, 1440, 1507, 1740, 1917, 1926-27, 1949, 1971, 1984, 1989, 2038-39, 2305, 2330, 2497, 3061, 3067, 3082 (2), 3099, 3100, 3118 (2), 3121, 3122, 3123, 3125-26, 3126, 3128 (3), 3129, 3134 (2), (...). Syn. *addere*. See Part I, p. lxxx (*aggregare*).

agregatio: *211n (2), 630, 631, 906, [1202], 1248, 1284, 1326, 1512, 2874, 3193, 3203, 3213, 3270, 3316, [3328], 3382, 3386, 3582, 3587, 3595, 3600, 3608, 3616, 3764, (...); (*nominum*) 1354. Syn. *additio*.

Alcorizmi: 19.

alcoton (القطن)**:** 8508, 8519, 8527, 8536, 8540, 8563, 8570.

algebra: 255 (*gebla et mugabala*), 4935 (*gebra et muchabala*), 5725 (*age-bla, A*), [6371] (*elgabre,* Tn934), 6791 (Tn1000), 6846, 6857, 7930, 9454, 9510 (*angebla, B*), 9598, 9737, 9740, 9746, 9755, 9798, 9835, 9865, 9966, 10086, 10128, 10156, 10160, 10214, 10240, 10247, 10275, 10294, 10330, 10346, 10518, 10520, 10572-73, 10745, 10894, 11327, 11340, 11355, 11480, 11577, 11625, 12011, 12019, 12135, 12488, 12500, 13975, 14419, 15022,

1078   PART TWO: TRANSLATION, GLOSSARY

15056 (*algebla*, *B*), 15248, 15310, 15579-80, 15758 (*gebla*, *D*). MS *A* mostly has *algebra*; *B*, *agebla*.

alias: 11918 (*iam probatum est alias*).

aliquando: 635, 636, 637.

aliquantulum: 10774.

aliquid: 788, 1901, 2035, 2037, [2070], 2517 & 2519 (*aliquid unum*), [3072], [7063], [7502], 8338, 8876, 10664 (2), 11092, 11105, 13987, 15084, 15178; (*x*, شى) 6265n, 9082n, 9120n, 9135 (2), 9136, 9137, 9138, 9193n; see also 9084n (omission) and 9099n, 9102n, 9191n (n. *B*). Tn1250, Tn1260, Tn1266.

aliqui(s): 70, 221, 256, *256n, [262], [263], 293, 312, 350, 389, 411, 466, 477, 491, 506, 525, 542, 559, 572, 602, 802, 814, 932, 998, 1001, 1002, 1006, 1007, 1008, 1043, 1093, 1103, 1105, 1113, 1114, (. . .); 1839 (*si aliquis diceret*, لو قال, ان قال), 1856 (*si aliquis querat*; syn. *quis*, cf. 1862).

aliquot: 4882, 4889, 7749, 10348, 10771, 10795, 12504.

aliquotiens: 701.

aliquotus: 638.

aliter: 781n, 1089, 1759, 1766, 1809, 1821, 2166, 2192, 2195, 2232, [2246], 2262, 2267, 2285, 2308, 2333, 2387, 2404, 2442, 2474, 2481, 2484, 2580, 2647, 2649, 2713, 2731, 2776, 2795, 2827 (. . .); (*aliter non*) 1918, 1928, 1950, 1971, 1984, 1990.

alius: 26, 33, 211n, 212-14n, 222, 223, 227n, 229, [263], 293, 312, 351, 380, 411, 412, 444, 470, 525, 564, 602, 607, 655, 656, 663, 691, 705, 707, (. . .); (= *alter*, *secundus*, e.g.:) 5400, 5437, 6499, 14265, 14374, 14390, 14434; (*alius . . . alius*) 4, 5, 6-7, 8-9, 22-23, 24-25 (4), 27 (3), 28 (3), 31-33 (4), [37], [38] (4), *256n, 643-44 (3), 645-46, [647-48] (3), 650, (. . .).

aliusmodi: 8694.

almencuz (المنكوس): 15690 (Tn2016).

almodi (m. & n., pl. inv. *almodis*; المكّ, dry measure): 7259, 7261 (n.), 7262, 7263, 7448, 7449, 7451 (m.), 7451, 7454, 12996 (m.), 12998, 12999, 13001, 13004.

alter: 178, 211n, 227n, 258, 259, 260, [263], [265], 482, 495, 508, 545, 562, 636 (2), 802, [1049], [1050], [1070], [1078], [1203], 1272, 1332, 1636, 1638, 2397, 2418, 2435, 2663, (. . .); (*alter . . . alter*) 352, [355], 479-80, 493, 758,

LIBER MAHAMELETH

1353, 1687, 2381, 2495, 2849, 3025, 4020-21, 4409, 4578-79, 5183, 5550, 5755, (...).

alteruter: 543.

altitudo: 12581 (syn. *profunditas*), 14912, 14922 (2), 14925, 14926, [14927], 14934, 14939, 15004, 15025, 15031 (syn. *longitudo*), 15048, 15057, 15353, 15371, 15381, 15383, 15480, 15481.

altrinsecus: 11969.

altus: 12557 (syn. *profundus*), 12615, 12617, 14945 (Tn1933), 14946, 14961, 15232, 15251, 15347, 15368, 15387, 15405, 15423, 15478.

ambo: 5118, 5419.

amiscere: 8338.

amittere: 10641, 10988.

amodo: 1030, 2508, 7170.

amplus: (*in amplum*) 8617, 8637, 8638, 8669, 8670, 8681, 8682, 8702; (*amplius*) 4030, 7613, 10776, [11637], 14041, 15329, 15367.

an: 904 (+ subj.), 1919 (+ ind.), 1929, 1951, 1960, 1972, 1978, 2011, 8334, 8337.

angelus: 6.

angularis: (*linea*) [14927], 14931, 14933.

angulus: 10696, 10698 (2), 11974 (3), 14981, 15011, 15242, 15243, 15299, 15418, 15455.

animadvertere: 8129.

animal: 12972, 12974-75, 12976, 12977, 12981, 12982, 12984, 12986, (...), 13311.

annona: 6860, 6862, 6863, 6870, 6872, 6876, 6877, (...), 7670, [8921], (9048), 9076, (...), 9636, 10553.

annus: 12996, 12998, 12999, 13001, 13002.

anser: 13313, *13313n, 13325.

ante: [870], 9285, 14038, 15698; (*paulo ante*) 3115, 3129, 3866, 7162, 8666, 12156, 12377, 12466.

antea: 2238.

antecedere: 1735, 1754, 11898, 12014, 14142.

antequam: 1997.

Antiqui: 14965.

apertus: 4936 (*aperte declaravit*), 6982 (*questio aperta*), 14760, 14762.

apparere: 3890, 7328, 8255, 10520, 14931, 14933, 14939.

appellare: 4784, 4787, 4790.

apponere: [50], [144], 987, 1309-10, 2722, 2832, 2870, [3327], 4936, 5741, 7197, 7280, 7606, 7640-41, 9650, 14555; (*nomina*) 1430, 1434, 1574, 1592, 1602, 1611, 1619, 1725, 1737, 1744, 1755, 2135; (*determinationem*) 14108, 14278; (syn. *conferre*) 6890, 6892, 7970, 7990, 8024, (...), 8109.

apportare: 12036.

appositio: 14092 (syn. *additio, adiectio*).

aptare: 7187 (*ad*), 7381 (+ dat.), 7455n.

apud: [60].

aqua: 8299 (2), 8300, 8302 (3), 8303 (2), 14781, 14783, 14791, 14794, 14796, 14800, 14802 (2), 14804, 14812, 14813, 14821, (. . .), 14951, 14952.

Arabes: 632.

arbor: *15231n, 15232, 15239, 15251, 15258, 15280, 15286, 15306, 15312, (. . .), 15387, 15390, (. . .), 15489, 15491.

arca, archa: *12614n, 12615, 12616, 12619, 12620, 12621, 12621-22, 12624, 12632, 12633, 12641, 12642, 12646.

Archimedes: See *Azemides*.

arcus: 8869, 8871, 15357, 15360.

argentum: *5801n (6), 8258, 8264, 8290, 8291, 8295, 8302, 8304, 8305, 8307, (. . .), 8365, (8368).

arimethica: 16 (*Nicomachi*), 19 (*Alcorizmi*); 22 (2), 175, 624, 627, 780; (*arithmetica*) *24n, 175n (*A*).

arimethici: 2900.

arrova (الربع): (liquid measure, for oil) 12661 (2), 12667, 12669, 12670, 12673, 12674, 12676, (. . .), 12964, 12970; (dry measure) 13375-76, 13379, 13380, 13381, 13382, 13385, 13386, (. . .), 13589, 13601, [13602].

ars: (*arimethica*) 21, 624, 627, 780; 8297, 8309, 8339. Syn. *scientia*.

articulatus: 59.

articulus: [38], 80 (2), [106], 121, [127] (2), [131], [137], [142], 639, 640, 643, 644, 689, [694] (2), (. . .), 1983, [1985], [1986], 1988, 2122.

artificium: 4, 5.

aspicere: 15484.

assequi: 704.

assignare: [168], 437, 438, 622, 776, 786, 972, 3046, 5252, 5424, 5531, 5659, 5710, 5740, 5787, 7108, 7123, 7835, 9554, 9618, 9642, 9659, 9684, 13502, 14262, 14443, 14457, 14629, 14877, 14890, 15113, 15430, 15757; (syn. *formare*) 10889.

attendere: 15, 17, 20, 2710, 3048, 4734, 4747. Syn. *considerare*.

LIBER MAHAMELETH 1081

attribuere: 12673, 13706, 13707.

auferre: See *afferre*.

augere: [3313].

augmentare: (9049), 9053, 9067, 9144, 9148, 9162-63, 9176, 9178, 9182, 9187, 9197, 9207, 9210, 9690.

augmentatio: 9142.

auricalcum: 8258, 8265.

aurum: *5801n (5; *aureus*), 8257, 8263, 8290, 8291, 8294, 8298, 8304, 8307, 8312, 8317, 8319, 8324, 8326, 8334, 8336, figure p. 314, 8343, 8363.

Avoquemel, Avochemel: See *Abū Kāmil*.

Azemides (Archimedes): 8820.

— B —

baetis (coin): 14654 (2), 14656, 14665, 14670, 14671, 14676, 14709.

basis: 14920 (3), 14921, 14922-23, 14924 (3), 14925 (2), 14926, 15406, 15425, 15429, 15442, 15447, 15462, 15475, 15477.

bene: 1486, 2293, 6599, 7515, 9718, 11231, 14555.

bimediale (ἐκ δύο μέσων, ذو الموسطين): (*primum*) 5649, 5651, 5760, 5764; (*secundum*) 5653, [5769].

binarius: 53, 54, 82, [95], [750], [751], [755], [772], 11456, 11485, 11526.

bini: 656, 658 (3), 665, 674, 675, 680, [691], [692] (2).

binomium (ἐκ δύο ὀνομάτων, ذو الاسمين): (alone) 5237, 5238, 5253 (2), 5291, 5292, 5297, 5494, (5555), 5579, 5589, 5593, 5595, 5602, 5608, 5624, 5644, 5647, 5667, [5692], 5695; (*primum*) 5602, 5603 (2), 5605, 5607, 5609, 5644, 5646, 5662, 5667, 5740, 5745, 5767, 5772, 5776, 5787; (*secundum*) 5648, 5650, 5763, 5772, 5776, 5787; (*tertium*) 5653, 5767-68, 5768, 5776, 5787; (*quartum*) 5654, 5774, 5777, 5786, 6541, 6545; (*quintum*) 5655, 5774, 5793; (*sextum*) 5657, 5774, (5795-96). Tn741.

bis: 493, 497, 505, 544, 548, 550, 552, 553 (3), 554, 557, 568, (...), 601, 758, 1109, 1129, 1257, 1313, 1318, 1337, (...).

brevis: 13269 (*brevis solutio*); (*brevius*) [1057], [1482], 4779; *3138n (*breviter*).

— C —

1082   PART TWO: TRANSLATION, GLOSSARY

cacumen: *15231n, 15233, 15251-52, *15312n, 15313, 15326, (15334), 15342, 15347, 15407, (. . .), 15476, 15476-77.

cadere: 5, 6, 15298.

caficius (قفيز): 7199, 7203, 7205, 7221, 7226, 7232, 7236, 7242, (. . .), 7948, 8482, 8483, 8896, (. . .), 9225, 12976, (. . .), 13479, 13481. Syn. *modius*, see 7147-74n (12).

cambire: 13610, 13611, 13617, 13624, 13634, 13636, 13638, 13640, 13644, 13673, 13675, 13684, 13700, (. . .), 14719, 14731.

cambium: 13604, 13716, 14621, 14635, 14647, 14717.

camera: 8859.

canalis: 14746, 14750, 14751, 14753, 14755, 14757, 14760, 14761, 14768, 14771, 14774, 14778.

candela: [12902].

capacitas: 36.

capere: 14913, [14937], 14938, 14961. Syn. *continere*.

capitale: 7035, 7039, 7040, 7043, 7044 (2), 7045, 7046, 7047, 7048, 7054, 7058, (. . .), 7900, 7961, 7962, (. . .), 8246, 8260, 8263, 8264, 8265, 8485, 11120, (. . .), 11142, 11144.

capitulum (باب): (main chapter) 174, (626), 864, 1586, 1633, 2139, 2983, 3193, 3354, 3694, 3783, 3945 (half-chapter), 4194, 4928, (. . .), 6516, 6525, (. . .); (section of a chapter) 781, 900, 972, 1173, *1443n, 1567, 1759, 1764, 1812 (section of a section), (. . .); (case, kind, subject; Tn1778) 777, 826, 876, 2691, 2897, (. . .), 15700, 15783; (procedure, operation) 379, 380, 388, 1478 (Tn235), 1480.

caput: 15068 (*caput scale*), 15330 (syn. *cacumen*).

carere: 1885 (2), 1887 (2), 1888 (2), 1889 (2), 6177.

carpentarius: 12614.

*carta: 168n (*charta*), 8689n (3), 14972n, 15346n, 15404n, 15492n.

*casa: *casa scazi*, see *scazus*.

causa: 704, 864, 972, 976, [1057], 2158, 2219, 9204, 9244, 9269, 9563, 9568, 9589, 9659, 9662, 9684, 9706, 9711, 9727, 10171, 10444, 10652, 10684, 10801, 10884, [10966], [10974], 11465, 11542, 11565, 11667, 11708, 11733, (. . .); 12051 (syn. *probatio*; Tn1601).

causalitas: 12.

LIBER MAHAMELETH 1083

causare: [8342].

cavea: 15610 (*caverna*, $\mathcal{A}$), 15621.

census (مال): (amount, syn. *pecunia*) 15762, 15766; (x^2) 6800 (Tn1005), 6801 (2), 7924, 7926, 7927 (2), 7929, 7935 (2), 7946, 7947, 8041 (3), 9083, 9084, 9120, 9121, 9504 (2), 9506 (3), 9507 (2), 9509, 9510, 9963, 9970-71, 9973 (2), 9976, 9999, 10001, 10004, 10006 (2), 10008, 10027, 10030, 10034, 10036 (2), 10039, 10515, 10516 (2), 10523 (2), 10524 (2), 11623, 11922, 11923, 11924, *11924n, 11925, 11938, 11939, 12133-34, 12170, (. . .), 14991, (. . .), 15310. In the sense of x^2, *census* sometimes appears as neuter, see 8041n, 10027n, 10030n, 10039n, 10524n ($\mathcal{A}$: *unus census equale est*), 11922n, 11925n, 11938n (2), 12171n, 12173n, 12175n, 13282-83n, 13283n, 13284n, 13301n (2), 15020n, 15054n. Sometimes followed by *habitus* (*q.v.*) in MS $\mathcal{B}$: Tn1260, Tn1736.

centenarius: [1177], [1179], 1926, 1931, 1940, 1954.

centeni: 87, 95-96, 97, [101], [109], 121, [136], [137], figure p. 6, [155], [159], 641, 805 (2), 857, 866, 868, [870], 874, 889, (. . .), [1986].

centesimus: 13327, 13329, 13331, 13333, 13334, (13335), (13336), 13339.

centies: 117, 118, 119, [155], [156], [160], [161], [165], figure p. 42, 1362, 1937, 1942, 1945, 1948, 1963.

centrum: 8866, 15289, 15298 (2), 15355.

certus: 58, 67, *5801n, 8848 (*certissimum*); (*certe*) 8817, 8823; (*certissime*) 8825, 8828.

ceteri: 15, 18, 82, 83, (86), 89, [95], [100], 139, [146], 169, (875), [1033], 1088, 1284, [1304], 1523, (. . .); (*cetera questionis*) 6504, 6630, 6739, 11384, 14223.

cibus: 13314.

circulus: (circle) 8803, 8804, 8806, 8807, 8817, 8818, 8824 (*pr.*), 8825, 8826, 8833 (2), 8834, 8835, 8852, 8854, 8865 (*post.*), 8866-67, 8870, 8871, 8875, 8876 (2), 8876-77, 15289, 15290, (. . .), 15357-58 (*circumferentia circuli*), 15359; (circumference; syn. *circumferentia*, *rotunditas*) 8692, 8824 (*post.*), 8851 (2), 8865 (*pr.*), 15353, 15360.

circumdare: 15493, 15494, 15507, 15508, 15515, 15516, 15526-27, 15527, 15533, 15534-35.

circumferentia: 8819, 8821, 8823, 8864, 12579, 14964, 14965, 14966, 14967, 15357, 15358, 15361, *15492n, *15526n.

circumpositus: 743-44, 745, 746 (2).

cisterna: 14745, 14746, 14750, 14751, 14753, 14754, 14754-55, 14755, 14757, (...), 14944, 14949-50.

civitas: 15541, 15588, 15591.

clarus: 140 (*clarius*), *2246n.

coalternus: [10697] (*coalternatus*, A), [11975].

coctio: 9377.

cogere: 66, 68.

cogitare: 7955.

cognitio: *3553n, *3677n, 6500.

cognoscere: 36, [152], [167] (4), 172, [441], 800, 802, 803, 813 (2), 826, 827, 842 (2), (...), 1883, (...), 5713, (...), 15463, 15465.

collectio: *3095-96n (2).

colligere: *227n, 871, 6942.

comedere: 12976, 12977, 12981, 12982, 12984, 12986, 12987-88, 12990, 12991, (...), 13596, 13597.

comestio: 13027, 13516, 13551.

commendare: (*commendare memorie*) 706, 2677.

*commentarium: 434n.

commiscere: 11.

commixtus: 8257.

commodus: 5185 (*commodius*).

communicans ($\sigma\acute{v}\mu\mu\epsilon\tau\rho\sigma\varsigma$, مشارك): 5089, 5089-90, 5090, 5256, [5346], 5353 (*in longitudine*; $\mu\acute{\eta}\kappa\epsilon\iota$, فى الطول), 5354 (*in potentia*; $\delta\upsilon\nu\acute{a}\mu\epsilon\iota$, فى القوّة), 5358, 5363-64, 5366, 5367, 5373, 5375, 5376-77, 5399, 5418, 5419, 5420, 5443, 5445 (2), 5550, 5560, 5663-64, 5666.

communicare: 5400, 5422.

communis ($\kappa o\iota\nu\acute{o}\varsigma$, مشترك): *227n (3), 1690 (*fractio communis*), 4185, 9479; (*numerus communis*, see Tn335, Tn361) (2208), 2210, 2258, 2259-60, 2282-83, 2304, 2306, 2317, 2319, 2329, 2331, 2383, 2385, 2453, 2673, [4455], [4457], *4678n, 8217; (*commune ponere* by adding; $\kappa o\iota\nu\grave{o}\nu \pi\rho o\sigma\tau\iota\theta\acute{\epsilon}\nu\alpha\iota$, جعل مشتركاً) 551, 616, 1551, 4163, 10428, 10433, 10952, 11035, 11049, 11075, 11084, 13875, 14689, 15141, 15160; (*commune ponere in multiplicatione*; جعل مشتركاً فى الضرب) 11876, 11899, 12449 (Tn1645); (*commune* *reicere*; κοινὸν ἀφαιρεῖν, المشترك ... القى ،اسقط) 599, 15041, 15106, 15131, 15137, 15774-75.

commutare: 7801-02, 9305, 10754, (11240), (11244), 11931.

commutatio: 6494.

comparare: (form the ratio) 8404, [11764], 12013-14.

comparatio: (ratio) 222 (2), 224, 225, 242 (2), 244 (2), 245, 246 (3), 284 (2), 286 (2), (...); (*'comparatio = divisio'*) [1715], [6773]; (*'comparatio = numerus'*, Tn597) 3547, 5272, [5284], [10106], 11933; (*comparationem accipere*) 2341, 2375. *Comparatio composita, geminata, duplicata: q.v.*

*compendiosus: 3649n.

comperire: 8348.

competere: 1647, 4857, 4866, 4870, 4880, 4883, 4890, 4898 (2), 6091, 6101, 8250, 8251, 9994, [10051], [10074], 10087, [10146], 10147, (...), 12673, 13039, 13230, 13715, 13718, 14639, 15524. Syn. *accidere*.

complere: 2721, 5686, [10190], 10368, 11112-13, 11179, 11273, 11374, 11899, 13758, 13829, 13921, 13952, (14060), 14140, 14141, 14296, 14329, 14330, 14348, [14371], 14472, 14491 (2), 14501, 14552, 14561, 15021; (eliminating the subtractive terms; syn. *restaurare*) 6301, 6355, 7925, 9826, 10001, 10030, 10065, 10089, 10101, 10117, 10139, 10193, 10206, 10233, 10262, 10268, 10303, 10318-19, 10338, 10573, 10715, 10745, 10851, 12005, 12073, 12492, 13900, 13967, 14028, 14737, 14994; (changing the coefficient of x to 1; syn. *restaurare*) 10005.

completio: (*completio linee*) 5672, 5720; 6341 (removing the subtractive terms).

componere: 42, [47], 53, 5254 (*binomium*), *5801n (5); (*questio*) 9342, 9358, 9375; 8873 (*componetur, scilicet parificabitur*); (*comparatio*) [1038], 7084, 7107, 7856, 7877, 8166, 8989, 9427, 9875, 10220, 10676, 10700-01, 10842, 10999, (11235), 11977, 11983, 12054, 14400, 14932; ***compositus*** (by addition) [38], [141], [142], 639, 642, 643, 646, [646], [647], 689-90, [695] (2), [696] (2), [696-97], [698] (2), [775], 901, 902, 1092, 1196, 1210, 1221, 1241, (...), 2185 (*fractio*), (...); (by multiplication) [1038] (Tn170).

compositio: 60, 864-65; (*comparatio*) 9873, 10218.

comprehendere: [6370-71], 12624; 7952, 9735.

computare: [1191].

concavitas: 12565-66, 12567-68, 12574 (2), 12578-79, 12581, 12621 (Tn1664). See Tn1658.

concedere: 788.

concludere: 12622 (syn. *continere, comprehendere*).

concurrere: 1879, 1880.

conducere: 32, 9644, 9646-47, 9648 (*diebus*), 9652 (*per mensem*), 9675, 9689, 9699, 9700, 9763, (...), 13428, 13433.

conductor: 10556, 10642, 10984.

conferre: 7958, 7961, 7967, 7988, 8004, 8005, 8013, 8015, 8016, 8025, 8122, 8125, 8126, 8141, 8222.

conficere: 3128. Syn. *efficere*.

congruus: [11763].

coniungere ($\curlywedge$): 24-25, 27, 30, 627-28, 628, 629, 630, 1470 (*cum*), 5706 (+ dat.), 5709, 5755 (*inter se*), *5801n, [14069-70].

consequens: (*in proportione*) 1000, 1001, 1017.

consequenter: 81, 83, [100], [104], [105], [124], 998, 4847.

consequi: 994 (*proportionaliter*), 15543, 15549, 15556, 15572, 15580, 15587, 15590, 15596.

conservare: 33.

considerare: 13, 14, 16-17, 635, 875, 899, 1088, 1632, 1913, 1978-79, 2027, 2029, 2046, 2048, 2051, 2074-75, 2077, 2507, 2793, (...), 14554, 15539; (*considerans*) 6070, 7106, [10934], [10966], [13122].

consideratio: 20, 2104.

consimilis: (*questio*) 1121, 1143, 1268, 1350, 2414, 5301, 5742, 10518, 10526, 11614, 14503, 15171; (*figura*) 1331; (*probatio*) 2472, 8204; (*causa*) 15430; (*fractio*, syn. *similis*) 2610; (*numerus*) 4927; (*numeri superficiales et consimiles*) 5076, 5076-77; (*superficies*) 8477, [8479], 8572, 8664, 14846; (*virga*) 15494.

consors: 6890, 8013, 8141, 8260.

constare: (*constat*) 6478, 6966, 7246, 7347, 7808, 7940, 8038, 8096, 8151, 8343, 8600, [8744], 8822, (...), 11804 (Tn1574), (...), 15291, 15354; (to cost) 9222, 9228.

constituere: 75, 787, 1995, 15482.

consumere: 9239, 9248, 9250, 9293, 9299, 9302, 9316, 9326, 9329, 9330, 9332, (...), 9543, 12660, 12661, (...), 12970, 13481, (...), 13550.

consumptio: 9236, 9237, 9289, 9339, 9352-53, 9353, 9355, 9371, 9372, 9397, 9412, (...), 9487, 12764.

contendere: 706.

continere: 29, [72], 76, *227n, 624, [1031], 1762, 1826, 1842, 1865, 5357, 5646, 5649 (Tn814), 5699, *5801n (2), [7469], 8869, 8870-71, 9729, 9732-33, 10696, 12634, (...), 13706, 14781, (...), 14956, 14961, 15295 (2; *contineri sub lineis*), 15509, (...), 15522.

LIBER MAHAMELETH

contingere: 1479, 3045, 5774, 5852, 5876, 5952, 5987-88, 6111, 7605, 7638, 7960, 7973, 9379, 9648, 13006.

continuare: (*per comparationem*) 7137, 7139.

continue: 3569, 3574, 3583, 3588, 3596, [3638], 3678, 9244-45.

contrarius: 826, 876, 1797, [4080].

conveniens: (*convenientius*) [146-47], 789, 2512.

convenire: 4846 (syn. *competere*), [7062], 8173, 8177, 8183, 8193, 8194, 8195, 8197, 8199, 8202, 8210, 9530 (*in*), 10782 (*pro*), 11322 (*ut*), 14641, 15574, 15724, 15750 (*convenire supra*, اجتمع على).

conventio: 8895, 8897.

conversio: (*fractionum*) 2508, 2512, 2513, 2515, 3054-55, [3069]; (*mensurarum, monetarum*) *12555n, *13623n, 13824.

conversus: 829, 1828, 1861, 4418, 7054, 15690 (2); (*e converso*) 36, [162], 212-14n, [241], [264], 802, 1791, 2237, 3217, 4385, 4414, 5096, [7064], 7185, 7859, 8416, 13637, [13701].

convertere: 631-32; (*mensuras, monetas*) 1877 (2), 1878, figure p. 75 (2), 7377, 13461, 13563, 13570, 13627, 13647, 13648-49, 13816, 13825, (. . .); (*comparationem*) 339-40, [345], 364, 373, 398, 406, 11005, 12045, 12053; (*fractiones*) 2528-29, 2529-30, 2539, 2558, 2571, 2586, 2587, 2590, 2591, 2593, 2616, 2665, 2669, 2724, 2725, 2750, (. . .), 3251; (*questionem*; syn. *reducere*) 14082, 14094.

cooperire: 8299.

coquere: 9234, 9235, 9236, 9238, 9242, 9245, 9247 (2), 9248, 9250, (. . .), 9542, 9543.

corda: 15493, 15494, 15505 (2), 15507, 15508, 15514, 15516, 15522, 15523, 15526, 15527, 15533, 15534.

corpus: 8294, 8296, 8304, 8309, 8312, 8314, 8315, 8321, 8340 (2), 8343, 8350, 8352, 8354, 8357-58, 8358, 14919.

cortina: 8371, 8372, 8381, [8385], 8387, 8389, 8393, 8396, 8398, 8399, (. . .), 8596, 8601, 8659, 8663, 8751, 14878. Tn1153.

costa: 8869, 8871.

creare: 5922.

crescere: 56, 93, 641-42, 784, 9173.

1088 PART TWO: TRANSLATION, GLOSSARY

cubitus: 8372, 8373, 8374 (2), [8385], 8393, 8395, 8396, (. . .), 8569, 8596, 8598, 8616, 8617, (. . .), 8861, 8875, 8879, 8881, 8885, 8887 (2), 12556, 12558, 12566, (. . .), 14960, 14973, (. . .), 15482, 15493, (. . .), 15533, 15621, (. . .), 15644. Tn1155, Tn1159.

cubus: (of a number) 3649 (3), *3649n, 3655 (3), 3665, 3671 (2), 3672; (of the unknown, thus x^3) 12188, 12189, 12194, 12195, 13306.

cunctus: 11100.

cupa: 14955, 14957, 14959, 14961, [14962], 14969. MS $\mathcal{A}$ has *cuppa* except in the last instance.

currere: 15599, 15605.

custodire: 12510, 12511, 12528, 12530, 12531, 12539, 12540; (*custodia*) *12510n.

— D —

dampnum: 7572, 11107.

dare: 5862, 5878, 5891, 5897, 5904, (. . .), 8115, 8150, 8236, 8951, 12088, 12139, 12308, (. . .).

debere: 63, *90n, [147], 174, 432, *440n, *993n, 1881, figure p. 75, 2108, 2130, 2135, 2909, 5475, (. . .); (*dies*) 11288, 11289; (*deberi*, syn. *competere*) 5879, 5959, (. . .), 7978, (. . .), 11685, 11695, (. . .), 13757.

decem: [49], [51], 76, 81 (2), 92, 94, 98, [102], 116, [128], [129], 139, [152], figure p. 6, (. . .), 739 (*decem multiplicatus*), (. . .).

decenarius: [1176].

deceni: 80, 96, [127], [136], figure p. 6, [155], [158], 640, 805, 831, 852, 866, 868, [870], 873, (. . .).

decidere: 15326, 15330, (15334), 15342, 15348 (*in*), 15351 (*ad*), 15370, 15380, 15382.

decies: [107], 116, 117-18, 119, [155], [156], [159], [160], [164], [165], figure p. 42, [1063], [1283], 1539, 1941, (. . .).

decimus: [110-11], [153], [249], 798, 847, 1028, 1172, (. . .); (*pars*) 1886, 1889, 1890, 1913, 1915 (2), 1995, 1996, 1997, (. . .).

declarare: (*figura declarat*) 140, 2065, 10392, 10708; (*Avochemel declaravit*) 4936.

decoctio: 9285.

decoquere: 9281, 9282, 9285-86, 9289, 9290, 9316, 9339, 9353, 9355, 9357, 9370.

decuplare: 760, [762].

decuplus: 80, 81, 82, 83, 93-94, 94, [94], 96, 97 (2), 98, (. . .), 13292, 13293; (*ad*) 5248, 5249.

deducere: 7273.

deesse: 1218 (*duo, que desunt*), 1257, 1545, 1547, 3164, 3466, 3471, 5672, 5720, 10031, 10446, 10470, 10491, 10716, 13968, 14028, 14738, 14995, *15118-19n.

deferre: 9164.

definitus: 59, 62.

defluere: 14746.

deinceps: [134], [649], 904, 983, 986, [1032], 1630, 1943, 1985, 3489, (11408).

deinde: [100], [104], 257, 436, [441], 528, 783, 906, 907, 914, 926, 936, 948, (...).

demere: 964, figure p. 48 (2; syn. *diminutus*), (1227), 1233 (*unum demptum*), 1234 (2), 1235 (2), 1236, 1237, 1238, 1258, 1260 (2), 1292 (2), (...), *3520n, 3522, 3523, (...), 9815, 10178, 10235; 6302 (*deme quod est iteratum*, Tn921).

demonstrare: 291, 624, 6012, 6413, 6435, 6721, 6790, [6803], 6978, 7088, 7094, 7144, 7714, (...); (concluding a computation, see Tn584) 8613, 9314, 9486. Syn. *monstrare* (more common in $\mathcal{B}$).

demonstratio: 5241n, *15280n.

denarius: 77, 701 (2), 702, 760, [762], 969, 1925, 1932, 1940, 1953.

denominare (سمّى): 847, 863, 904, 913, 4697; (fractions, with *de* or *ab*) 1667, 1671-72, 1682, 1883, 1895,1909, 1910, 2022, 2025, 2026, 2029, 2033, 2040, 2061, [2072], 2080, 2087, 2090, 2095, 2098, 2106, 2107, 2112, 2116, 2123, 2125, (...); (instead of *dividere*) Tn552, Tn1483, (Tn1790, Tn1999); (**denominandus** *numerus*) 2025-26, 2042, 2054, 2064, [2073], figure p. 80; (**denominans** *numerus*) 2042, 2046, 2065, [2073], figure p. 80.

denominatio (تسمية): 759-60, [762]; (fractions, see Tn328 & Tn345) *227n (5), 1660-61, 1882, [1993], 2066, 2077, 2136, 2150, 2155, 2171, 2207 (...).

***denominator:** 227n.

deprehendere: 740, 3501-02, 6527, 8816 (*deprehendi in effectu*), 8818, 8828-29.

deprimere: 1879.

descendere: 8117, 8120, 14911, 14914, 14942, 14947, 14952, 14974, 14977, (...), 15215, 15229; (syn. *oriri*) 7177.

1090   Part Two: Translation, Glossary

describere: 1268, 1280.

desiderare: 706.

designare: 57, 68.

desinere: 3571, 3576, 3590.

determinare: [14158], 14712-13.

determinatio: 14108, 14152, 14155, 14249, 14261, 14262, 14268, 14277, 14278, 14491.

Deus: 5, 432.

dexter: 1996-97 (*versus dextram*).

diameter, diametrum (διάμετρος, قطر): (*diameter*) 8266, 8267-68, 8277, 8279, 8692, 8798, 8799, 8801, 8802, 8804, 8805 (2), 8806, 8813, 8819, 8822, 8823, 8830, 8832, 8835, 8836 (3), 8837, 8843, 8849, 8850, 8851, 8852, 8859, 8860, 8865, 8874, 8881, 12580, 14955, 14957, 14963, 14964, 14966, 14967, 14968, 15291, 15359; 5729 (*diametrum superficiei quadrate*).

dicere: 20, 21, [44], [49], [51], 74, 79, 87, 90, *90n, [123], [125], [128], [129], [131], [132], [141], 170, 183, (...); 190 (*dicere quod*), 215 (*quia*), 804 (*quoniam*), 3111 (*ut*); 976 (*causam*), 6472 (*probationem*); 6543 (*radix, dico*; Tn962); (*quod dico*) 7618, 8307, 8308; [8342] (*dicimus gratia exempli*); 9381 (*questiones dicte*); 15477 (*non potest dici*).

dies (m. & f.): 9125 (2), 9126, 9646 (2), 9647 (2), 9653, 9657, 9658, (9667), 9681, (...); (f., e.g.:) 9689, 10641, 11101; (m., e.g.:) 9647, 9720, 10907.

differentia: (decimal place; منزلة, وضع) 71, [100-01], [104], [106], [109], [111], 114, 120, [122], (...), 820, 842, (...), 1172, 1423, 1996, (...), 2017; (difference) 176, 183, 190, *211n, 358, 359, 746, [748], [750], [751], 760, [762], 768, 1174, 1481, 1680, 1682, 2180, 2198, 4851 (2), (...), 4868, 4949, (...), 4976, 6916 (+ gen.), 6918 (*inter*), (...), 15616, 15625; (term in a mathematical expression) 2713, 2827, 2860. See Tn193, Tn421.

differre: 2472-73, 4665, 5278, 5767, 6638, 8205, 9129, 9551-52, 9645, 12973, 14413.

digitus: (finger) 1877, 1878, 1879; (digit) [38] (2), 74, 81-82, 89, *90n, 121, [122] (2), [123], (...), figure p. 6, [155], (...), 638-39, 639-40, 643, 644, 689, (...).

diligenter: 2710, 3048.

dimidia (noun): [2006], 2311, 2577n, 2620, 2846, 2855, 2863, 2864, 2865, 2868, 2893 (2), 2894, 2896, 3255, 3724, (...).

dimidiare: (*lineam*) 9440; (*res*; Tn1366) 9977, 10010, 10040.

dimidium: *211n, 526, 527, 605 (2), 1619, 1620, 1621, 1653, [1654], 1655, 1683, 1747, 1799, (...).

dimidius: 2306 (adj.), 2320, 2324, 2325, 2471, 2479, 2480, 2486, 2488, 2546, 2548, 2551, 2562, (...), 3544 (2), (...); (noun, e.g.:) 3495, 4659.

diminuere: (syn. *subtrahere*) 28, 1471 (*de*), 1872 (*iterationem*, syn. *minuere, reicere*), 3699, 3813, 3982n, 5177 (*ex*), 5179, 6208, (...), 15095; (*diminutus*, adj.; ناقص) 1219, 1226, 1561 (2), 3475, 3476, *3520n (2), (...); (noun) 1193 (2), 1194 (2), 1195 (2), [1536], [1537], [1562] (2), [1565] (2), 6355, 6365.

diminutio: 1452, 1476, 3695, 3716, 3719, 3729, 3733, 3742, 3747, 3765, (...), [7421], 15713; (*iterationis*) 1866.

dimittere: 1310, 1430, 1452, 1469, 1737, 3798, 3900.

dinoscere: [144-45], 786.

in directum: 1879, 1879-80, 11969.

discedere: 15008.

disiungere (διαιρεῖσθαι, فرق): (separate, ant. *coniungere*) 25, 28, 30, 629, [14070]; (separate a ratio) 344, 7770, 10480, 10502.

dispergere (قلب, see Tn1282): (subtract, remove a common part) 9312; (separate *or* convert a ratio) 7107, 7865, 9713, 10251, 10253, 10462, 10511, 10530-31, 12046.

disponere: 71, 785, 1003, 3588, 3596, 4781, 7065, 9966.

dissimilis: 9648, 12474.

distantia: *15231n, *15312n, 15446-47, 15447, 15461, 15462, 15475, 15476 (2), 15477, 15478, 15488.

distare: [149] (2), 178, 185, *211n, 754, [756], 795-96, 796, 1019 (2), 1021, 15175, 15177, 15233, 15252, 15313, 15326, 15342, 15390, 15398, (...), 15588, 15598, 15601.

distincte: 7387.

diversitas: 3075.

diversus: *227n, 350 (see 350n), 354, 1221, 1241, 3553, [3946], *5801n, 6859, [8738], *13343-44n, *13556n, 13557, 13601-02, 13932, *15492n, *15526n.

1092 PART TWO: TRANSLATION, GLOSSARY

dividere: (arithmetical operation) 18, 28, 43, [43], *43n, [44], [45], [46], 256, 262, [263], [265], 266, 268, 272, 291, 293 (2), 297, [300], 310, (...); (a surface by a segment of line) 10386, (10388), 10390, 13957; (instead of *denominare*) Tn368, Tn1571, Tn1828, Tn1833; (*dividere denominando*, Tn789) 5515-16, 6459, 6522, 8083; (sharing) 1636 (*per*), 1645, 1649, 4837, 4858, 4871, 4882, 4889, 4896, 4897, 6891 (*inter*), 8015, 8093, 8116, 8124, 8134 (+ dat.), 8142, 8145, 8147, 8151, (...), 8248; (***dividendus***; noun, syn. *divisus*) 1638, 1644, 1650, 1657, 1671, 1672, 1675, 1676, 1752, 4364, 4377, 4380, 4392, 4406, 4408, 4412, (...); (***dividens***; noun) 256n, 260, [301], 314, 353, 391, 413, 417, figure p. 19 (2), 1638, 1643-44, 1650, 1656, 1670-71, 1672, (...); (***divisus***; noun) *256n, 259, [301], 314, 413, 414, 416, 418, figure p. 19 (2), 445, 5809, 11917.

divisibilis: 835, 897.

divisio: 169, 257, 258, 267, 268, 280 (2), 282, 294, 296 (2), 298, 301, 313, [315], [316], 317, (...), 14452, 15379; (sharing) 1650, 8132; (section of a book, قِسم) 2897.

*divisor: 4675-76n.

docere: *docere* (like *predocere*) being the main expression for references, we indicate all occurrences. 16, 19, 26, [1029], [1030], 1880, 2482, 2777, 2915, 3130, 3527, 3533, 3738-39, 3788, 3902, 4107, 4119, 4346, 4645, 4736, 5004, 5011, 5022, 5052, 5056, 5059, 5064, 5204, 5388, 5405, 5459, 5472, 5488, 5725 (Tn1350), 5749, 5776, 5900, [5923], 5961, 6175n, 6246, 6516, 6598, 6610, 6616, 6625, 6630, 6675, 6857, 7196, 7218, 7230, 7239, 7290, 7467, 7476, 7496, 7499, 7582, 7726, 7737, 7858, 7910, 8458, 8486, 8602, 8666, 8944, 8966, 9016, 9159, 9224, 9349, 9454, 9598, [9852], 9940, 9952, 10041, 10045, 10156, 10346, 10434, 10572, 10745, 10889, 10924, 11084, 11137, 11145, 11169, 11173, 11305, 11385, 11462, 11899, 12019, 12209, 12305, 12500, 12582, 12607, 13128, 13183, 13348, 13363, 13369, 13543, 13563, 13567, 13573, 13592, 13645-46, [13771], 13975, 13996, 14131, 14177, 14192, 14223, 14235, 14240, 14286, 14300, 14315, 14319, 14419, 14502, 14514, 14591, 14648, 14707, 14952, 15056, 15212, 15229, 15538.

dominus: (*Dominus*; syn. Deus) 4751n, *15009n; (*dominus*) 8126, 8127, 8140 (2), 8142, 8143 (2), 8144, 8152 (2), 8154, 8156, 8157, 8163 (2), 8164, (...), 8250, 8337.

domus: (*domus pecunie, domus pecuniosa*) 1812, 1814, 1815, 1817, (...), 1873, 1875, figure p. 75 (7). Syn. **casa*.

donec: 7717, 8137, 15556.

LIBER MAHAMELETH   1093

dragma: (درهم, name of a second unknown; see Tn1002) 6792, 6793, 6795, 6796, 6798 (2), [6799], 6802 (2), 6847, 6847-48, 6850, 6852 (2).

ducenti: 86, [102], [132], [133], [143], 970, 1169, 1235, 1248, 1273, 1274, 1366, 1367, 1387, 1388, 1397, 1398, 1412, 1414, 1437, (. . .).

ducenties: 14533-34, 14535.

ducere: *211n (3), [426], 710, 713, 714, 717, 721 (2), 722 (2), 723 (2), 724, 726, 727 (2), 728, 729, 730, 732 (2), 733 (2), 735 (2), (. . .). Syn. *multiplicare*.

ductus (noun): 212, 212-14n (2), 213 (2), 214, 216, 217 (2), 218 (2), 219, 220 (2), 221, 226 (2), 228, 230, 234, 235, 236, 237, (. . .).

dum: 14759.

duodecies: 967, 970, 2103, 7272.

duodecimus: 13720, [15525].

duodecuplum: 1037, 1039-40.

duodenarius: [2005-06], 3183.

duplare: 27, 180–81, 204, 1044, [1086], 1768, 1853, 1854, figure p. 75 (2), 2014 (2), 3680, 3686, 4948 (2), 4977, 5120, 5132, 5160, 5334, 6999, 11670, 12841, 13021, 13027, 13410, (. . .).

duplatio: 631.

duplex: 15444.

duplicare: 1767, 1770, 1789, 1852, 1854, 1858, figure p. 75 (2), 2012, 2018, 4058, 4952, 6195, 6220, 7001, 7825, 10167, 10353, 12257, 12773, 13019, 13027, 13416, 13991 (2), 14084, (. . .); [1050] (*proportio duplicata*, see Tn122, syn. *geminata*; $\lambda\acute{o}\gamma o\varsigma$ $\delta\iota\pi\lambda\alpha\sigma\acute{\iota}\omega\nu$, نسبة مثّاة بالتكرير).

duplus: 207 (*ad*), 208, 209, 210, 211n, 604 (+ dat.), 608, 613, 619, [756], [1032] (+ gen.), 1040 (*de*), 1053, [1653], 1776 (noun), 2120, 3549, 3668, 3675, 4854, 4863, 4957, 6353, 6354 (2), 6484, 6487, 6624, 6625, 6626, 6723, 6733, 6782, (. . .).

duum (= *duorum*) : 8294, 10695.

— E —

effectus: (*in effectu*) 8816, 8829, 8878; (*ad effectum*) 8811, 8863, 8877.

1094   PART TWO: TRANSLATION, GLOSSARY

efficere: 710, 714, 718, 739, [754], 757, 759, [761], [763], 769, 770, [771], 3126, 3487, 3492, 3498, 3616, 4704, 5806, 6121, 6168, 8483, [8736], [8737], 8873, 9254.

effluere: 14783.

effundere: 9237, 9240-41, 9245, 9249, 9283, 9284, 9290, 9294, 9300, 9317, 9321, 9323, 9327, 9330, 9332, 9333, 9334, 9335, 9339, 9340, 9344, 9348, 9354, 9356, 9361, 9370, (. . .), 14948, 14951.

effusio: 9247, 9377, 9412, 9419.

egere: [3072].

ego: 5852, 5876, 7955, 14555; (*quantum me contingit*) 5852, 5876, 5952, 5987, 6111. See *michi*.

egredi: 15610, 15612, 15619, 15620, 15622, 15628, 15629, 15630, 15637, 15638.

elaboratus: 8356.

emere: 32, 5844, 5845, 5846, 5889, 5951, 6114, 6120, 6148, 6167, 6175n, 6181, 6207, 6233, 6288, 6319, 6332, 6344, 6373, 6742, 6750, 6759, 6769, 6793, 6806, 6821, 6848, 6862, 6870, 6882, (. . .). Other occurrences: see *vendere*.

emina: 13559, 13560, 13563, 13570, 13571, 13572, 13573, 13574, (. . .), 13598.

enim: 14, 42, [43], [44], [46], [48], 58, 81, 94, [101], [122], 193, *211n, 435, 439, [554], (. . .); (beginning a proof, syn. *nam, q.v.*) 192, 204, 220, 240, 298, 319, 360, 394, 419, 449, 470, 497, 513, 531, 549, 566, 579, (. . .).

enumerare: [126], [130], [134], [136], 4683, 4684.

equalis: 190, 192, 193 (2), 194 (3), 195 (2), 196, 197, 198, [199] (2), 200, 205 (2), 206, 209, 390, 464 (2), 471, 474, 484, (. . .); 15483 (*terra equalis*).

equalitas: 8297, *13343-44n; (*secundum proportionalitatem equalitatis*, syn. *secundum equam proportionalitatem*; see Tn60, Tn1097) 343, 347.

equaliter: 6879, 6951, 6981, 6986, 8114, 11138, 14279, 14298, 14375, 14384, 14410, 14436, 14486, (. . .), 14514, 15346n, 15650, 15662, 15673 (*equaliter tantum*).

equare: (*equari*) [1073], *3520n, *3533n, 6255 (2), 6301, 6303, 6341 (2), 6355, 6356, 6364, 7929, 7946, 8041, 9505, 9509, 10523, 11150, 11553, 11576, 11675, 12073, 13345, (. . .); (absol.) 14472.

equidistans: 10695-96, 10698, 11973, [14928], 15414, 15434, 15452, 15467, 15483. MS *B* has *equistans*, except in 15434.

equipollere: 6256, 6273, 6275, 6325, 6328.

equistans: 11410, 11411, 11412, 11414.

equivalere: (being equal, for algebraic expressions) 6276, 6366, 9193, 9578, 9971, 9976, 9996, 9998, 10000, 10005, 10009, 10022, (. . .).

equus (adj.): 213, 216, 219, 226, 229, 235, 239, 248, 251, 258, 274, 289, 294, (. . .); (*eque distant*) 1021; (*equa proportio*) 754; (*equa proportionalitas*) 7710, 11239.

equus (noun): [51], 15701, 15703, 15705, 15707, (. . .), 15768, 15777 (2).

ergo: 13, 196, 202, 225, 268, 270, 272, 274, 284, 324, 347, 419, 422, 475, (. . .).

erigere: 15369, 15377.

etenim: 4260, 13159.

etiam: [48], 76, 117, 118, 179, 195, 200, 201, 203, 208, *211n (4), 243, 254, 271, 273, (. . .).

Euclides: 223, 227, 381, 386, [425], 433, 435, 437, 439, [440], *440n, 458, 622, 1003, 1018, [1341], 1710, [2070], [2654], [2742], [2794], [2876], [3394], 4938 (2), 5077, 5100, [5173], 5356, 5402, 5416, 5419, 5421, 5423, 5447, 5455, 5467, [5483], 5491, 5603, 5645, 5687, 5710-11, 5731, 5765, [5769], [6371], 9442, 10694, 10841, 11413, 11976, 14918, [15013], 15035, 15085, 15222, [15276], 15293, 15297, [15525].

evacuare: 14759, 14762, 14763.

evadere: 10643, 10760, 10768, 10771, 10807, 10868, 10910.

evenire: 5081, 5181, 5223, 5484, 7955, 8130, 9380, 11689.

evidens: 621-25n.

evidentia: 2078, 5358, 5385-86, 5402-03, 5424, 5456, 5468, 7197, 7280, 7386.

examinatio: 8291-92.

excedere: 509, 4024-25, 9387, [10188], [10189], 10197, 10201, 10202, 11321, 11443, 11446, 11452, 11456, 11526, 11540, 11560, 11581, [11618], 15551.

excellere: 10095 (*in*), 10099, 10108, 10113, 10135 (⊦ abl.), 10136, 10142, 10148, 10150.

exceptus: 901, 902, 4851, 6540, 6571, 13834, 14254, [14257], [14259].

excrescere: 712, 1326, 1423, 1537, 1701, 3216, 3365, 3375, 3377, 3833.

exemplificare: 7147.

exemplum: *182n, 2079, 2254, 2269, 5359, 5369, 5386, 5403, 5424, 5456, 5469, 5659-60, 5859, 5860, 5877, 5889, 6118, 6146, 6165, 7171, 7197, 7280, 7387, 7458, 11742; (*gratia exempli*) 8308, [8343]; (*exempli gratia*) 12720.

exercere: [6371].

exire: 257 (*de*), 258 (*ex*), 262, [263], [265], 267 (2), 268, 269, 272, 280, 281, 283, 291, 293, (. . .); (*a centro*) 15298; (***exiens***, noun; *de divisione*) 351 (2), 6142.

*exoriri: 90n.

expeditus: 172.

expendere: 33, 13044, 13074, 13075, 13076, 13077, 13095, 13097, 13118-19, 13134, 13135, 13145, 13148, (13149), 13150, 13152, 13239, (. . .), 13327, 13458.

1096   PART TWO: TRANSLATION, GLOSSARY

expensa: 12972n, 13373.

experientia: 1674, 7481, 7550, 10896.

experimentum: 7397.

experiri: 6286, 7306, 7630, 7637, 7679, 7739, 8308, 9086, 9104, 9756, 9983, 10307, 10579, 10610, 10760, 10868, 11128, 14468, 14741, 15582.

explere: 432.

exsurgere: 8603.

extractio: 14942.

extrahere: 8300, 14911, 14915, 14946, 14951.

extremitas: 15355.

extremus: 177, *211n (4), 747, 1021, 1022.

— **F** —

faber: 8336, 8337.

facere: *90n, 727, 734, 738, 761, 891, 1036, 1284, 1303, 1314, 1331, 1370, (...); (*numerum*) 3087; (*lineam*) 9437; (*superficiem*) 5716; (*questionem*) 5793.

facilis: (*facile*) 3049, [3070], 3501, 4944, 6599, 7956, 9736, 12150, 14556, 14878; (*facilior*) 278, 2691, 4744; (*facilius*) 1480, 1483-93n, [1491], 5144, 5177-78, 5779, 5789, 5794, 5799, 5989, 11230.

falsus (ant. *verus*): 3417, 4025, 6914, 6945, 7009, 7327, 7336, 7338, 7502, 7605, 7614, 7618, 9381, 9385, 9522, 9531, 10544, 10546, 10550, 10773, 10775, 10966, (10970), 12621, 13825, 13853, 13861, 14082, 14088, 14144, 14148, 14151, 14152, 14270, 14277 (2), 14658, 15096, 15179, 15330, 15476. *Falsus* may variously mean 'not solvable' (problem, mostly leading to a negative or zero solution; Tn1434, Tn1030), 'absurd' or 'meaningless' (contradicting the hypotheses; Tn1464), 'inappropriate' (concerning the choice to render determinate an indeterminate problem; Tn1839).

fasciculus: 15493, 15507, 15508, 15509, 15515, 15515-16, 15522, 15522-23, 15524, 15533, 15535.

fieri: 54, 55, 140, 212, 213, *212-14n, 216 (2), 217, 218 (2), 219, 226, (...).

figura: (sign, symbol; Tn185) [60], 1094 (2), 1105, 1106, 1115 (2), 1125 (2), 1135 (2), 1145 (2), 1149 (2), 1156 (2), 1166 (2), [1189] (2); (illustration, diagram) 140, [1030], 1331 (Tn217), 1876, 2065, 10368, 10392, 10456, 10686, 10708, [10967], 11505, 13942, 14050, 14979, 15010, 15073, 15098, 15123, 15156, [15189], 15201, 15217, 15239, 15319, 15431, 15465; (geometrical figure) 8695, 8697-98, 8869, 8870, 8872, 15357.

finire: [431].

finis: *1443n, *1658n, 3075, *8689n, 9691, 13315.

fixus: 15234, 15313.

fodere: *8689n, 12556, 12557, 12584, 12586, 12600, 12602.

LIBER MAHAMELETH   **1097**

foramen: 14758, 14759, 14762.

foris: 705 (*in foribus*).

forma: *5801n, [9990].

formare: 59; (*questionem*) 7561, 9377, 9737.

fortasse: 7956n.

forum: (price set) 6120, 6148, 6167, 6181, 6207, 6233, 6442, 6507, 6588, (...), 7841, 7939.

fovea: *8689n, *12555n, 12556, 12558, 12566, 12568, 12574, 12576, 12577, (...), 12608 (2), 12618.

fractio (كسر): [37], *227n, 637, 638 (*fractio fractionis*), 650 (2), 651, (...), 669 (*fractio et fractio fractionis*; Tn116, Tn351, Tn421), 686 (2), 1567, 1686 (2), 1690 (*fractio communis*), 1895 (*fractio denominata*), 1909, 2067 (*minores fractiones*), 2140, (...); 2185 (*fractio composita*), 2627 (*fractiones similes*), 2665 (*ultimum genus fractionum*), 2874 (*fractio minima, maxima*), 2899 (*fractio irregularis*), 4709 & 4728 (*genus ultime fractionis*).

frust(r)um: 8298, (8300), 8302.

fundus: 14968.

fustis: 15482, 15485, 15487, 15488, 15489, 15490.

— G —

gausape: 8617, 8621 (2), 8622, 8623, 8626, 8637, 8639, 8640, 8642, 8644 (2), 8646 (2), (...), 8826, 8827. See Tn1186.

gebla: See *algebra*.

geminare: 77-78, 79, 85, 89, [648], 711, 782, 1819-20, 1821, 1825, 1836, 1837, 12764, 13516, 13551; (*comparatio geminata repetitione*, نسبة مثنّاة بالتكرير, Tn948; syn. *duplicata*) 3546, 6481, 6483, 6623; (*comparatio geminata per alteram*) 11770, 11771, 11774, 11775, 12255, 12261, 12262.

generalis: (*regula, modus agendi*) 3565, 6339, 6352, 9736, 9813; (*generaliter*) 5808.

generare: 139.

gens: [60].

genus: 1276 (*genus numerorum*), 1281, 1421, 1636, 1639, figure p. 75, 2665 (*genus fractionum*; Tn397), 2722 (*genus numeri*), 2818 (Tn419), 2833,

1098 PART TWO: TRANSLATION, GLOSSARY

2851, 2871, 3054, 3271, 3317, 3325, 3796, *3809-11n, 4238 (*huius generis*, syn. *huiusmodi*), 4377, 4708, 4728, 4745, [7063], 8682, 8703.

gradus: [1031] (Tn166).

gratia: See *exemplum, verbum*.

— H —

habere: 8 (*habent esse*), 10, [44], [45], 63, 65, 70, 73, 138, 267 (*habere quod*), 327 (*sic se habet*), 328, 330, 331, 337, 341, 433, (...), 703, 747, 865, [870], 1014, (...), 1815, 1890, (...), 2826 (*probatio habetur ex*), 2859, [2875], 3074 (*habent se*), (...), 5110 (*numeri habent inter se*), (...), 5853 ('*quantum habebo?*'), 5877, (...).

habitus (مال, syn. *census* in MS *B*): 9121n (Tn1260), 9509n, 9974n, 13281n (Tn1736), 13282-83n, 13283n (2).

hemina: See *emina*.

homo: 4, 5, [51], 63, 68, 1636-37, 1645, 4837, 4840, 4846, 4852, 4853, 4857, 4858, 4870, 4871, 4879, (...), 4905, 5030 (*non esse possibile homini*), 5902, 8112, (...), 8227, 11394, (...), 11671, 13374, (...), 13599, 15646, (...), 15750.

hora: 14748.

hucusque: 2511.

huiusmodi: 12, 16, 18, [52], [143], 1133, 1164, [1304-05], 1477, 1632, 2507, 2737, 2793, 3057, 3075, (...), 13987, 14155.

humanitas: 7, 9.

humanus: 18.

hypothesis: See *ypotesis*.

— I —

iam: (with perf., قد) 303, 432, 621-25n, 1480, [1564], 1846, 3115, 3119, 3123, (...), 6271 (with pluperf.); (with pres.) 1486, 1581, 1702, [1777], 2242, 2966, 3136, 4055, 4111, 5050, 5054, 5136, 5270, (...); (with imperf.) 245, 287, 2701, 3523, 5044, 6324, 15075.

ibi: 908, 909, 920, 1880, 1917, 1918, 1926, 1927, 1933, 1935, (...), 2298, 4277.

idcirco: 62, 71, 690, 777, 817, 9731, 10474, 11439.

LIBER MAHAMELETH   **1099**

***idem*:** 42, [45], 176, 183, *211n (2), *227n (2), [324], 452, 454, 466, 469, 484, 544, (...), 6038 (*eedem*), (...).

***ideo*:** 22, [150], 170, [440], 621-25n, 629, 866, [1075], 1772, (1801), 1958, 2518, [4041n], 4225, (...).

***igitur*:** [45], [47], [52], [145], 183, 190, 194, 195, 198, 206, 207, 209, 210, 215, (...).

***ignis*:** 8292 (2), 9293, 9413.

***ignorare*:** 1010, [2071], 5808, 5924, 5990, 6110, 6111, 6115, 6116, 6117, 7035, 7055, 7068, 7095, 7118, (...), [7472], 8947, 8965.

***ignotus*:** 3678, 4896 (2), 5809, 5810 (2), 5811, 5824, (...), 6128 (Tn898), 6152 (Tn900), 6172 (Tn902), (...), 12930 (Tn1696), (...), 12965 (Tn 1699), (...), 14613; (*questiones de ignoto*) 6114 (Tn893), 6373, 7067, 7163-64, 7273 (Tn1059), 7776 (Tn1103), 7950 (Tn1122), 8027, 8130 (Tn1122), 9648, 9649, 9699 (Tn1336), 12974, (13220).

***immunis*:** 10768.

***impar*:** 15, 179, 201, *211n (2), 1895, 1897, 1911, 1946, (1947), 1948, 1965, 1992 (2), 3583, (...), 3685 (2), 15711; (syn. *inequalis*) 6766.

***impedire*:** 64.

***impensa*:** 12648, 12972, *12972n, 13230, *13313n, *13373-74n, 13536.

***implere*:** 14747, 14748-49, 14750, 14751 (2), (...), 14778.

***implicitus*:** 6176 (*questio implicita*).

***impositio*:** 781.

***impossibilis*:** [46], 70, 5554, 5561, 7611, 8396, 9379-80, 9388-89, 9390 (Tn1291), 9523, 13856, 13857, 13860.

***imum*:** 14974, 15005, 15009, 15016, 15018, 15071, 15097, 15181, 15210, 15214, 15228.

***inaeq–*:** See *ineq–*.

***incedere*:** 15547.

***inchoare*:** 115, 634.

***incĭdere*:** 381, 388, 7161.

***incīdere*:** (*partem*) 8292, 8637, 8657, 8669, 8715, 8747, 8755, 8799, 8802, 8808, 8814, 8880, 13948 (*superficiem*), 14716; (*lineam de*) 4179, 5256, 5537, 5593, 6709, 8875, 9438, 10477, 10500, 12041, 15074, 15218, 15552; (*per medium*) 6390, 6427, 6714.

***incipere*:** 1, [90], 121, 633, 1001, 2027, 2030, 2046, 2049, [2075], 2508, 2874, *3520n, 3578, 3592, [3644], 3660, 5801, 14748, 15485, 15690.

***incisio*:** 6394, 6405, 6492, 13951.

***incisor*:** 12198.

***inclinare*:** 15347 (*cacumen*), 15368 (*verticem*); (*lineam*) 14055, 14057; (*inclinari*) 15287, 15304, 15350, 15352, 15354, 15365, 15381.

incognitus: 3397, 3399, 3401, 3402, 3992, 5812, 5814, 5816, 5817, 5850, 7048, 7049, 7108, 7979, 8002, 8391, 9447, 12369, 14625, 14797, 14800, 15464, 15466.

incommunicans (ἀσύμμετρος, غير مشارك): 5664.

incurvare: 15233, 15240 (2), 15251, 15253, 15257, 15259 (2), 15279, 15285, 15287, 15306, 15312, 15326, 15327, 15328, 15331, 15332, 15341.

inde: 213, 214, 216, 217, 303, 304, 306, 307, 308, 760, 863, 905, 1008, 1537, 1603, 1757, (...).

indivisus: 445.

inducere: 278, 379, 434-35, 439, 2678, 2711, 4937, 5491, 13371, 14262.

inequalis: 350n (syn. *diversus*), 507, 508, 509, 511, 572, 573, 575, 577, 612, *3553n, 3554, 3565, 4867, 6758, 6805, 6820-21, 10553, 10592, 10619, 10834, 10890, 14163, 15033, (15144), 15260, [15274].

inferior: 2046, 12625, 14917, 14926.

infinitus: 62, 64-65, 66, 9520n; (syn. *innumerabilis*; see Part III, p. 1634) 14249, 15700; (*in infinitum*) 56, 93, 113, 120, [134-35], [156-57], [166], 641, [649-50], 784, 8155-56.

influere: 14748, 14753, 14760, 14761, (14767), 14770, 14774-75.

ingredi: 15591.

innumerabilis: 7950-51, 8129.

inquirere: 799, 2151, 3084, 3101, 3427, 4600, 7185 (*de*), 9097, 9567, 9593, 9685, 9845, 10322, (...), 14606, 14806.

insistere: 15298.

instar: (*ad instar*) 75, 86, 140.

instituere: 73, 84, 87, 91, 790, 791, 793.

instrumentum: 58.

***insulse:** 227n.

insuper: 744, [1654], 4091, 4105, 4147, 4162, 4165, 4169, 4172, 4175, 4178, 4181, 4184, 5778, 6515, 6524, 7557, (...), 15684, 15685.

integer: (whole = entire, adj.) 6798, 6852, 9928, 10174-75, 10474, 10497, 10680, 14465; (whole, for a number; adj.) [37], 900, 902, 1089, [1994], (...), 2674, 2745, 2748, (...), 9507 (*census*), 11229; 7024 (integral part of a fractional number); (whole number, noun m. or n.) [37], 636, 638, 650, 652, 661, 662, 664, 666 (2), 667, 668 (2), 669, 670, 672 (2), 673, 674, 675 (2), 676, 678, (...); 4201 (*quodlibet integrum*).

intelligere: 9 (2), [314], 875, 899, 1875, [1993], 3047, [3069], 3136, 3502, 5769, 5770, 5843, 5888, 6599 (2), 7514-15, 7618, 7953 (2), [8384], 9018, 9736, 11898, 12150, 12378 (2), 12508 (2), 13064, 13199, 14556, 14878 (2).

intendere: 387, 1634, 1635, 1637 (2), [1646], 4523, 12678.

intentio: 776, 1651, 1657.

LIBER MAHAMELETH   **1101**

inter: [141], 627, 630, 654, 778, 779, 1016, 1017, 2900, 3325, 4787, (. . .); (*differentia inter*) 4976, 10532, 10797, (. . .); (*agere inter se*) 5902; (*inter diem et noctem*) 9125 (2), 9126; (*complere inter se*) 11112, 11178, 11274, (. . .); (*servire inter se*) 11291, 11293, 11296; (*multiplicare* or *ducere inter se, inter ipsos*) 71, 653, 901, 903, 905, 913, 940, 977, 978, 982, [1029], [1058], [1072], [1077], 1122, (. . .); (*agregare inter se*) 3296, 3301, 4779, 4930, 5083, 5407, 5489, 5491, (. . .); (*dividere inter se*) 4195, (4198), 4264, 4779, 4930, 5187, 5495; (share) 6891, 8015, 8093, 8116, 8124 (. . .); (*minuere inter se*) 4779, 4930, 5147, 5486, 5492, 5493; (*proportio* or *comparatio inter*, *habere inter se*) *227n (4), [1038], 5075, 5084, 5088, 5089, 5092, 5110; (*conversio fractionum inter se*) 2508, 2512, 2873.

interiacere: [141].

interior: 14959, [14962].

interminatus (سِيَّال): 3557, 4900, 4922, 9520 (Tn1313; syn. *multiplex*), 9525, 13832, 13984, [14157], 14712, 15709, 15734, 15757.

interrogare: 3445, 3457, 3458, 3461, 3464, 3473, 5797.

interrogatio: 5185, 5775, 5794, 5798.

intervenire: 1015, 1017.

intra: (*intra te*) 6321, 6361; (*intra circulum*) 15293, 15297.

introducere: 704, 6113n, [6372].

introitus: 15590-91.

invenire: 26, 29, 30, 34, 35, *69n, 687, 829, 953, 964, 1594, 1605, 1726, 1913, 1916, 1925, 1939, 1969, (. . .), [2075], 2094, 2401, 3049, 3342, 3346, (. . .); (*probatio inveniendi*) 6472, 6473; (*sic invenies* instead of *sic facies*; cf. 7956, *invenire questionem*) 3447, 3460, 3466, 3475, 4000, 15408, 15426, (. . .), 15517; (*invenies ita esse*) 4440, 4510, 5082, 5146, 5186, 5523, 5583, (. . .); (*et ita invenies*) 1088, 1632, 3443, (5234), 6113, 13622, 13888; (*invenies quod per* or *ex* or *de* or −) [4455], 4520, 4539, 4551, 4560, 5519, 6874, 11349, 12331, (. . .).

inventio: *3609n, *4881n, 4928, 4933, 4940, *4940n.

invicem: 4312, 15294, *15404n, *15440n.

ipse: 42 (2), [44], [141], [149], *211n (2), 279 (*Avoquemel*), 435 (*liber Euclidis*), 466, 479 (2), 483, 489, 492, (. . .).

ipsemet: 2159.

1102   Part Two: Translation, Glossary

ire: 11779, 11785 (2), 12089, 12146, 12168, 15542, 15543, 15548, 15550, (...), 15591, 15596.

irrationabilis: 5140.

irregularis: (*fractio*) 2899.

ita: 55, 82, (86), 93, [95], [99], [102], [108], 113, 116, 118, 120, [122], (...).

item: 757, 759, 767, 770, 854, 882, [1057], 1089, 1229, [1482], 1599, 1730, 1741, (...).

iterare: [91], 814, 819, 820, 822, 824, 832, 834, 837, 840, 844, 855, 881, 885, 886, (...), 2460, [2654], 6302 (*deme quod est iteratum*).

iteratio (تکرار): *90n, [91], 808 (2), 810, 811, 812 (Tn134), 816, 817, 845, 851, 856, 861, 862, (...).

iterum: 277, 2245, 2409, 2410, 2445, 2518, 2531, 2868, 2909, 4897, 5018, 5684, *5801n (2), 7532, (...).

iungere: *211n, [1071] (2), *3520n, *5801n, 15288, 15291.

iuxta: 9691, [9990], 12172, 12210.

— L —

labor: 1522.

laboratus: 8340, 8345.

lampada: (*lampadarum*) 12649, 12651, 12653, 12655 (2), 12707, 12744, 12756, 12884, 12887, 12937, 12940, 12947, (...), 12973.

lampas: *12648-49n, 12660, 12662, 12669, 12670, 12671, 12672 (2), 12675, 12676-77, 12678, 12683, (...), 12970, 13311-12, 13502-03.

lapis: 12198, 12199, 12200, 12203, 12211, 12214, 12219, 12220, 12221, 12222, (...), 12373, 14781, 14787, (...), 14951, 14952.

latere: 7516, 14932.

latitudo: 36, 8376, 8398, 8400, 8401, 8411, (...), 8612, 8619, (...), 8890-91, 12603, (...), 12612 (2), 12625, 14785, (...), 14875, 15392, 15398.

latus (noun): (side of a rectilinear surface, or of a number: $\pi\lambda\epsilon\nu\rho\acute{\alpha}$) [1039], [1049], [1050], 6480, 6481 (2), 8500, 8501, 10373, 10374, 10695, 10696, 10698, 10699, 11414, 11973, 11974; (side of a parallelepiped or a cube) 14895, 14897, 14899, 14904, 14908, [14927]; (place where an expression is written; Tn667) 1314 (Tn215), 2354, 2380, 2381, 2396, 2398, 2417,

LIBER MAHAMELETH

2418 (*post.*), 2434, 2435, 2492, (. . .); (expression involved in an arithmetical operation; Tn360) 2395, 2397, 2418 (*pr.*), 2669, 2818, 2820, 3261, 4377, 4666, 4728, 4744, (. . .); (side of an equation) 6325, 6326, 6327 (2), 10234, 15775.

latus (adj.): (*in latum*) 8373, 8375, [8385], 8394, 8395, (. . .), 8597, 8617, (. . .), 8887, 12557, 12558, 12585, (. . .), 12639, 14780, (. . .), 14960, 15389, 15399, 15403.

lector: 12473.

legere: [441], 5402.

liber: (*liber mahameleth*) 1, *440n, [442], 624, 864, 7955 (Book B), 12784 (Book A); (*liber Euclidis* = the *Elements*, thus كتاب) 433, 435, *440n, [6371]; (*liber Euclidis* = one of its Books, thus βιβλίον = مقالة) 224, [323], 381, [425], 437, 439, 440, 1004, [2654], [2742], [2794], [2876], [3395], 4938 (2), 5077, 5099, 5356, 5402, 5417, 5447, 5491, 5688, 9442, 10695, 10842, 11413, 11976, [15013], 15035, [15276], 15293, [15525]; (*liber, auctore Abū Kāmil, sc.* كتاب الجبر والمقابلة) 255 (*liber gebla et mugabala*); (*liber de taccir,* كتاب فى التكسير) 8695.

libra: 7148.

limes: 67; (عقد) [38], 75, 77, 84, 88, 95, 96 (2), 97 (Tn24), (. . .), 138, [141], [142], 639 (2), 642, (. . .) , [698] (2), 768, 769, [774], [1029], [1031], (. . .), [1286], 1917 (Tn294), 1921, 1970, 1975, 1988.

linea: [199], 438, 622, 3312 (2), 3314, 3778 (2), 3779 (3), 3780, 3785, 5254, 5357, [5444], 5596, 5597, 5646 (2), (. . .), 5795, 6736, 6840, (. . .), 15571 (3); 10836 (*linea de novem*); 10838 (*linea multiplicata in lineam*); 11410 (*linea equistans linee*), 15414 (*linea equidistans linee*); 11969 (*protrahere duas lineas altrinsecus in directum*); [14927] (*linea recta*); 14931 (*linea angularis*).

lingua: 58.

linteus: 8615, 8616, 8619, 8622, 8626, 8629, 8631, (. . .), 8879, 8880, 8893; (*linteus:*) 8680, 8681, 8701, 8802, 8830, 8831, 8843.

linum: 8508, 8519, 8527, 8535, 8541, 8563, 8571.

liquescere: 8292.

littera: [60].

locare: 32, [2076].

1104 PART TWO: TRANSLATION, GLOSSARY

locus: 1452, 1880, 2055, 2057, 8299, 8303, 15240, 15259, 15287, 15326, (15333), 15341, 15598, 15600, 15609; (*numeri denominati suorum locorum*) 905, 913, 925, 935; (*loco* + gen. or *de*, syn. *pro*) 944, 946, 6295, 6889.

longitudo: 36, 8376, 8397-98, 8398, (...), 8613, 8619, (...), 8845, 8890, 12603, (...), 12624, 14785, (...), 14875, 15027 (syn. *altitudo*), (...), 15490 (2), 15527, 15534, 15610, 15611, 15641; (*communicans in longitudine*) [5346], 5354, 5366, 5377.

longus: [8385], 15025, 15099 (*longius*), 15478n (*longior*); (*in longum*) 8373, 8374, 8393, 8395, 8408, 8410, 8428, 8442, 8451, 8475, 8507, 8509, 8526, (...), 8596, 8616, (...), 8887, 12557, (...), 12638, 14780, (...), 14950, 14960, 14973, (...), 15441, 15526, 15620, 15637.

loqui: 58, 439, 777, 2510, 2512, 3046, 5903.

lucrari: 6891, 7037 (2), 7057 (2), 7058, 7070, 7071, 7097, (...), 7942, 7959-60, (...), 8123, 8142, 8222, 8247, 8261, 8484, 11032, 11121, 11143.

lucrum: 7033, 7035, 7040, 7042, 7043 (2), 7045, 7046, (...), 7954, 7957, (...), 8109, 11107.

— M —

magis: 2677, 5799, 14249, *15440n.

****magister***: 2099n, 6113n.

magnitudo: (length: *magnitudo circumferentie*) 14965; (area; Tn1155) 8379, [8384], 8387, 8388, 8389, 8390, 8400, 8402 (3), 8404 (2), 8405, 8406, 8411, 8412, 8417, 8419, 8421, 8423, 8454, 8455, 8457, 8463, 8466, 8473, 8474, 8481, 8487, 8489-90, 8513, (...), *8689n, (...), 12625, 12626, (...), 12642-43, 14064, 14836-37, 14855, (...), 14958, [14962]; (volume; Tn1658, Tn1913) 8295, 8297, 8305, 8310, (8340-41), 8344, 12576, 12577, 12588 (2), 12589, 12591, 12605-06, 12606, 12607, 12608, 12619 (2), 14784, 14787, 14792-93, 14794, (...), (14935), 14970.

magnus: 779, 1521-22, 5843, 8302-03, 8512. See also: *maior*, *maximus*.

mahameleth (معاملات): 2, 19, 434, [442], 624. Also written *mahamelet* (3n, 434n, 442n), but the first has been adopted as *scriptio difficilior*.

maior: [150], 351, 632 (*maiores Arabum*), 794, 1002, 1659, 1660, [1662], 2034, 2036, 2055, 2057, 2059, 2066, 2078 (2), [4030], 4031, (...), 15513; 5654 (*linea maior*; μείζων, أعظم); (*res maior*, first unknown) 6764, 6769, 6791, 6803, 6819, 6821, 6846, [6854], [6855].

manere: (*questio* or data) 7585, 9525, 10935, 12538; (*figura*; Tn1949) 15073, 15098, 15123, 15156, [15189], 15201, 15217, 15319, 15431.

maneria: 5548, 5554.

manifestare: 6473, 10686.

manifestus: 864, 1670, 1883, 2373, 2402, 2642, 2907, 3136, 3936, [4432], 4530, 4701, 4937 (*manifestior*), 5074, 5099, (...); 5858 (*manifestius*); (*manifestum est*) 206, 208, 249, 429, 1034, 1043, [1073], [1079], [1561], [1564], 2242, 2520, 2978, 3449, 3516, 3846, 4005, 4016-17, (...).

manus: (*retinere* or *agregare in manu*) 949, 961, 1202, [1202], [1203], 1245, 1269, 1279; (*habere ad manum*) 5799.

*marcha: 5801n (10).

marmoreus: 8886.

massa: *5801n (8), 8253, 8257, 8266, 8267, 8272, 8276, (...), 8331, figure p. 314 (3).

materia: [7], 9, 10 (2), 14, 16, 21, 8506, 8510, 8514, 8516, 8521, 8528, 8532, 8536, 8538, 8543, 8550.

maximus: figure p. 314, 11586, 11597, 11598, 11639, (13335); (*fractio*) 2874.

medialis (μέσος, موسّط): 5356, 5357, 5358, 5363, 5372, 5378, 5400 (2), 5401 (Tn765), 5421, 5422 (2), 5422-23, 5455, 5467, 5656, 5658, 5702, 5704, 5707, 5709 (2), 5795.

mediare: 28.

medietas: *211n (2), [540], 610, 611 (2), 1682-83, 1798, 1801, 1802, 1867, (...), 15760, 15763.

medius: 178, 180, *211n (3), 747, 5847, 5848, (...), 10608; (*dividere per medium*) 528, 606, 612, 4186, 5259, 5596, 5670, (...), 6390, 6714 (*incidere per medium*), (...), 15204, 15220.

melequinus: 14653, 14654, 14656, (...), 14677, 14708. Same form in Niermeyer's *Lexicon*.

melius: 1875, 3324, *3809-11n.

memoria: (*commendare memorie*) 706, 2678.

memoriter: 702.

mensis: 9652, 9658 (*dies mensis, scilicet triginta*), 9675, 9689, 9691, (...), 11372, 11374, 12996, (...), 13326.

mensura (قدر): (size) *8689n, 8818, *12578n, 12581, *13343-44n,

*13556n, 13557, *14745n, 15008, *15404n; (amount) 9571, 9572; (amount of a ratio, *πηλικότης*) 9248, 9250; (quantity of a liquid, namely: *musti*) 9235, 9237, 9238, (...), 9545, (9547); (*olei lampadarum*) *12648-49n, 12651, 12653, 12654, 12655-56; (*aque*) 14781, (...), 14951; (*vini*) 14956, 14961.

mensurare: [8386].

mercari: (*mercando*) 7611, 7622, 8092.

mercator: 8246.

merces: 9653, 9659, 9662, 9664, 9667, [9670], 9673, 11387, 11396.

messis: 15515.

***metallicus:** 5801n.

metallum: *5801n (2), 8259.

michi: *2246n, 5879, 5959, 5966, 5976, 6864, 6929, 15702, 15704, (...), 15753.

miliarium: 11678-79, 11679, 11681, 11683, (...), 12180, 12187-88, 12188-89, 12207, 12208, 12475, 12476, 13434, 13435, 15542, (...), 15601, 15606.

milies: (*milies mille*) [106], [107], [108], [109], [110], 117, 118 (2), [155], [156] (2), (...), 993 (2), figure p. 42 (3), [1035], (...), 2128, 2133; (*milies mille iterata*) 814, 832 (*decem milies mille iterata*), (...), 1600 (*decem milies mille iteratum*), (...); (*milies milia*) [105], [107], [110], 819, 820, 822, 824, 834, 837, 967, 1308, 1312, 1313, (...), 1858, 1859; (*milies milies mille*) [112], 118-19, 119, 119-20, [156], [161], (...), 994, figure p. 42 (2), 1172, 1302, (...); (*milies milies mille iteratum*) 1445, 1577, 1604; (*milies milies milia*) [111], 1307, 1859.

millenarius: 1935, 1936, 1936-37, 1941, 1955.

milleni: 90, 96-97, figure p. 6 (7), [155], [159], 641, 806, 990, 1114, 1123, 1144, 1154, (...) , 1390, 1405, 1406.

minimus: 54, 2874 (*fractio*; Tn424), 11585, 11598, 11600, (...), 11646, 13334.

minor: *211n, 352, 509, 575, 783, 1002, [1662], 1679, 1680, (...), 4947, (...), 5664, (...), 15490, 15512; (*fractio*) 2067, 3325, 4377, 4745; (*linea minor*; ἐλάσσων, اصغر) 5705; (*res minor*, second unknown) 6759, 6763, (...), [6853].

minuere: 31, 379, 687, 839 (*de*), 972, 976, 1045, [1058], [1074], [1076], (...), 3824 (*ab*), 3829 (*de*), 3831 (*ab*), (...), 3893 (*ab*), (...), 4803 (*ex*), (...); (*iterationem*) 1788, 1868, figure p. 75 (3).

LIBER MAHAMELETH   **1107**

minus: 838, 1215, 1216, 1224, 1231 (2), 1255, 1256, 1290 (2), 1490, [1490], 1492 (2), (. . .); 1493-500n (*multiplica in minus uno; minus uno in minus uno facit unum additum*, here *minus uno* is −1); 4197-98 (*divisio per minus uno*, here *minus uno* is 'less than 1').

mittere: 8302, 8303, 15541, 15542, 15588, 15590.

modius: 5965, 5969 (2), 5975 (2), 6119, 6120, (. . .), 6178, 6249, (. . .), 6299 (2), 6318, (. . .), 6363, 6521, (. . .), 6531, 6741, (. . .), 6750, 6859, (. . .), 6948, 7010-21n (9; instead of *sextarius*), 7108, 7123, 7148, (. . .), 7257, 9553, (. . .), 9641, 9655, 9656, 9678, 13344, (. . .), 13369.

modulatio: 59.

modus: 13, 22, 30, 276, 655, 656, 657 (2), 658, 659, 660, *661-63n, *663-65n, (. . .); *modus agendi*, see *agere*; (*modus solvendi*) 9617; (*modus solutionis*) 6252; (*tres modi*) 5864 (Tn852); 6115 (*tres modi 'de ignoto'*), 7273 (*tres modos ignoti*); (syn. *species*) 7161 (Tn1051).

molaris: 8894 (Tn1227).

molendinarius (noun): 8896, 8897, 8951, 8985, 9036.

molendinum: 9124, 9131, 9132, 9134.

molere: 8902, 8903, 8906, 8910, [8913], [8914], [8916], 8920, (. . .), 9227, 9228.

moneta: 13617, *13623n (2), 13755, 13756 (2), 13757, 13761, 13763, (. . .), 14705, 14707.

monstrare: 192-93, 195, 200, 201-02, 210, 227 (*demonstrare*, C), [243], 252, 254, 275, (. . .), 620, 828, 1028, [1078], 1345, 1349, 1566, 1589, (. . .), 15550, [15566]; (computation) 4145 (Tn584); (*causa*) 12911, 13111. See *demonstrare*.

morabitinus: 1760, 1762, (1785), 1792, 1794, 1805, 1813-14, 1815, 1821, (. . .), 1876, figure p. 75 (6), 13605, 13607, 13608, 13610, (. . .), 14723, 14730, 14733, 14742.

motus: 5, 6, 7, 8, 11.

movere: 15598.

mucabala (مقابلة): 255 (*mugabala*), 4935 (*muchabala*), 15774 (*mucabala*; Tn2020).

multi: 24, 26, 33, 434, 3027, [3071], 3073, 3501, 5844, 7578, 7665, 7716, 7724, 7728, 7937, 7950, 7983, 8128, 9232 (more than one), 12618, 12652, 14155, 14554, 15783.

1108   PART TWO: TRANSLATION, GLOSSARY

multiplex: *4812n, 8506, 15343 (Tn1987).

multiplicare: 18, 27, 31, 71, 78, 79, 85, 89, *211n, 212-14n, *227n (4), 257, [262-63], [264], [301], 312, 325, 360-61, 376, 395, 408, 416, (...); (*lineam in lineam ut fiat superficies*) 6475, 6477, 6620, 6621, 10365, 10689, 10691, 10837, 10838, 13921, 14054, 14056, 14058; (*multiplicans*; multiplier: b in $a \cdot b$) 314, 769, [1061], 1116, 1119, 1146, 1150, 1152, 1167, 1171, 1525, 2233, 2375, (...), 4767-68, 4781, (...), [8740], [8741] (*multiplicans, quod*), (...), 11918; (both factors a and b) 3414, 3430, 3455, 4008-09, 4020, 4041-42, 4052, 5833-34, 5869 & 5870, (...), 9001; (*multiplicatus, multiplicandus*; multiplicand: a in $a \cdot b$) 769, 1060-61, 1314, 1524, 2375, 2664, 2666, 2737, 4752-53, 4754, 4768, 4781-82, 7834; (both factors a and b) 223 (2), 1710, 5938, 6003.

multiplicatio: 169, *211n, 228-31n (2), *227n (2), 259, [315], [316], (626), 631, 633, 653, 702, 704, 707, 740, 760, 766, [774], 900, 921-22, 940, (...).

multiplicitas: 3903.

multitudo: 65.

multo: 278.

multotiens: 702, 5803.

multum: 1478.

mustum (عصير): 9234, 9235, 9288, 9298, 9315, 9325, 9337, 9343, (...), (9549).

mutare: 2008, 2017, 10685.

mutequefia (ἀντιπεπόνθασιν, متكافئة): (10374; Tn1414), 10697, 10699, 11975.

mutuare: 32, 9550, 9553, 9556, 9570, 9581, 9599, 9611, 9636.

— **N** —

nam: [145], 196, *227n, 614, 615, [944], [946], [1536], [1653], [1715], [1937], 2226, 2240, 2405, 2511, 3346, 3417, [4956], (...); (beginning a proof, syn. *enim*; Gr. γάρ) 261, 338, 1670, 2212, 2642, 4481, 5114, 5136, 5139, 5314, (...),

namque: 2373.

nasci: 78, 79, 85, 89, 1422.

LIBER MAHAMELETH   **1109**

natura: 14.

naturaliter: 59.

navis: 15598.

necessarius: 171, [277], [435], 438, 621-22, 2508-09, 14249, *15280n; (*probatio necessaria*) 625, 987, 5468; (*necessario*) 629, 5340, 6467, 6470, 6579, 11211, 12997, 13240, 14536, 15478.

necesse: 66, 70 (2), 653, 690, 784, 1019, 1268, 4937, 5142, 5435, 5441, 5477, 7482, 7640, 10757, 13292, 13294, 13296, 15570.

necessitas: 66.

negotiari: 25, 31, 35, 7585, 7590, 7595, 7600, 7609, 7959, 7989, (...), 8079, 8247, 11120-21, 11143.

nemo: 11434.

*nequire: 43n.

nescire: 8654; (*nescio quot*) 7778, 9050, 9326, 11559, 11580, 11627, 11857, 11944, 12023, 12476, (...), 14645; (*nescio quantum*) 9299, 9317, 15630; 15025 (*nescio quam longa*).

nichil: 628, 1603, 1619, 1935, 1985, 2038, 3418, 4912, 4918, 4925, 7516 (*nichil latebit*), 7951 (*nichil agimus*), 8137, 8155, 9245, 9249, 10643 (2), 10760 (2), 10768 (2), 10771 (2), 10773, 10780, 10910 (2), 10923 (2), 10928 (2), 10985 (2), 10990 (2), 13828; (*nichil aliud vis*) [1062], 2340; (*nichil aliud est*) [2376], 2917, 2969.

Nicomachus: 16.

nimis: 64, 1485, 4934.

nisi: 10, 440, 628, 708, 709, 710, [1062], [1085], 1140, 1192, 1193, 1195, 1208, 1626, 1637, 1640, [1646], 1661, (...), 9524, 10968, 11313, 13833, 13834, (...), 15352, 15522.

nocuplus: 1048, 11451, 11452, 11510, 15159 (*ad*).

nomen: 57, *57n, 61-62, 62, 65, 67, 69, 70, 74, 138, [162] (2), (...), [166], 1306, 1352, (...), 1366, 1384, 2343; (math. term, ὄνομα) 5254 (*duo nomina*), 5587 (*tria nomina*), 5590; (*comparatio geminata repetitione nominis*; Tn948) 6482, 6483, 6623.

nominare: 7145, 7185, 7264, 8896, 9943, 13831, 13833.

nonagies: [109], 1295.

nonaginta: 81, 83, 92, [100], [129], [130], 738, 790, 1101, 1102, 1253 (2), 1280, 1288, 1298, 1447, 1448-49, 1450, 1453, 1454, (...).

noncenties: 1295n, 1298n. See *nongenties*.

nondum: [1070].

nongenti: 87, [104], [132], [134], 791, 1288, 1448, 1449, (...), 1540, 2436, 3663, 9079, 9084, 9951, 9963, 9968, (...), 15269, 15308.

nongenties: [110], 1295, 1298.

nongies: 1446, 1453, 1496, 1499.

nonus: [109], [249], 1003, 1590, 1591 (2), 1593 (2), 1595 (2), 1596 (2), 1598, 1887, 1891, (...).

nos: 204, [277], 386, [435], 436, 438, 633, 776, 816, 1371, 1504, 1547, 2219, 2362, 2685, 3074 (*placuit nobis*), 4469, 4936, 7606, 7745, 8185, (9649), 12473, 12677, 14262.

noscere: 1846, [3070], 5490, 5491, 6100, 6101 (2), 7096, 7119, 7178, (7367), 7373, 7445, 7522, 8117, 11167, 11897, 11997, (...), *15312n; (*notum est*) 9092, 9111, 9839, 12907.

nota (أُتَس): [144], [145], (...), [167] (2), 777, 778, 781, 786, 787, (...), 1090.

notare: 2031, 5339, 6109, 12650, 13375; ('*nota*') *41n, *434n, *440n (2), *878n, *942n, *1659-61n, *1882n, *2166n, *2274n, *2339n, *2513-14n, *2585n, *3503n, *3982n, *4266n, *5804n, *5847n, *5868-69n, *7147n, *7254n, *8289n, *15280n.

novenarius: [2006], 2118, 9929, 14639.

novies: [106], [112], 1100, 1539, 1540, 2102, 11439, 15513, 15523.

novus: 68.

nox: 9125 (2), 9126, 12651, 12652, 12654, 12656, (...), 12970, 12977, (...), 13222, 13426, (...), 13480, 13591.

nullus: 1895, 1903, (...), 1992, [4798], 4938, [5232] (pl.), 10641, 11098, 11103, 11179, 11374, 11443, [11618], 15735.

numeralis: 65.

numerare: [48], [50], [52], 66, 68, *227n, 361, 362, 374, 375 (2), 376, 395, 396, 1002, 1012, 1013, 1686, 1690, 1896, 1900, (...), [2006], 4684n, 11181-82, 11228. See Tn64.

numeratio: 64.

numerus: 3, 12, 13, 14, 20, 22, 25, 26, (...); [37], 41 (*unitas non est numerus*), 42 (*integer*), 71 (*ordines*), 635, 643 (*compositus*), 1421 & 2871 (*genus numeri*), 4759 (*integer et fractio*, cf. 4768), 5030 (*surdus*), 5076 (*superficialis*). See also Tn328, Tn335, Tn345.

nummus (درهم): [51], 1636, 1645, 1647 (2), 3459, 3465, (...); (unit in an algebraic expression, Arabic درهم) Tn1120, Tn1456, Tn1527, Tn1593, Tn1597, Tn1738; (unit of weight, Arabic درهم) 14717, 14718, 14724, 14725 (2), 14728, (...), 14743. See Part I, p. lxxx (*numus*).

nunc: 976, *5801n.

nunquam: 703, 711, 8810, 8828, 8873, 8877, 8878, 8882.

nuntius: 15540, *15540n, 15541, 15547, (15549), 15553, 15554, 15555, 15572, 15575, 15577, 15581, 15582, 15585, 15588.

— O —

ob: 1509, [2348], 2919, 9027, 9041, 11009, 11688, 11835, 13006, 13161, 13443, 13471, 13518, 13551, 13672, 13702.

obicere: 14556.

obolus: 4106 (2), 4131, 5976, 5979, 6867, 6869, 6909, 6911, 6914, 6923, 6926, 8247. See Tn877.

observare: [1189], [1662-63], 3383.

occīdere: 13314.

occultus: 26, 34.

occupare: 15289, 15355.

occurrere: 2032.

octavus: [106], [1018] (2), [1039], [1050], 1616, 1618, 1622, 1885, 1888, 1892, 1900, (...).

octies: 849, 1295, 1540, 1541, [1938].

octingenti: 1232-33, 1236, 1248, 1249, 1277, 1393, 1399 (2), 1400, 1401, 1402, (...), 15271.

octonarius: [751], [755], 1934, 1937.

octuplus: 11438.

oculus: 1876, 5764, 15485.

oleum: *12648-49n, 12649, 12660, 12661, 12676, 12678, 12751.

omnino: 9128.

omnis: 4, [39], 42, [46], 53, 58, [60], 61, 63, 69, (...), 176, 190, *211n (3), 212, (...).

operari: (3766), 5888, 5924, 12199, 12200, 12203, 12204 (2), 12214, 12218-19, 12219-20, 12221, (...), 12504.

operarius: 7010-21n, 11110, 11177, 11272, 11294, (...), 11675, 12199, (...), 12503, 13330, 13332, 13333, 13338.

oportet: 2236, [2346], 2907, 3340, 3350, 3940, 4475, 4530, 7049, 8645, [8722], 9380, 9395, 9777, 9786, (...), 13587, 13597.

*opponere: 3520n.

oppositus: 3088, (11405), 11405 (Tn1525), 11414, 11427-28, 11429, 12626, 12628, 12630, 12635, 12637-38, 12640.

opus: 12292, 12293.

ordeum: 6927, 6928, 6930, 6933, (...), 6948, 7778, (...), 7812.

ordinare: 140.

1112 PART TWO: TRANSLATION, GLOSSARY

ordinatio: 5485.

ordo: 61, [103], [105], [108], [125], [129], (...), 2066, 3964; (*ordo nu-merorum*, مرتبة) *69n, 71, [72], 73, 75, 77, 78, 80, 84, 85, (...), 811, 815, (...), [1082]; (syn. *latus*) 1270 (Tn206), 1271, 1272, 1314 (Tn215), 1332, 1333, 1524, 1526, 2602, [4798], 9966, 9971, 9973, 9974, 10038.

origo: 41, [147], 435, 779, 782, 7513, 9282, 9283, 14261.

oriri: 7169 (syn. *descendere*, cf. 7177), 7463, 7516.

orthogonaliter: 15298.

ostendere: [151], 303, 408, [425], [471], 474, 797, 953, 984, 1011, 1023, 1485, 1521, [2351], 2607, 2678, 2983, 2999, 3054, 3056, 3059, 3065, 3354, 3945, 4493, (...), 15248, 15310.

ovis: 8112, 8114, 8126, 8127, *12509n, 12510, 12518, 12519, (...), 12530, 12539.

— P —

paene: See *pene*.

palatium: 8885.

palmus: 8266-67, 8268, 8277, 8279, 15514, 15516, 15522, 15523.

panis: 13374, 13376, 13378, 13380, 13381, (...), 13554, 13559, (...), 13594, 13597 (2).

par: 15, 177, 183, *211n (4), 1895, 1896, 1898, (...), 1992, [2006], 3596, 3600, (...), 3689, 15712, 15736.

paries (m. & f.): 12621, 12623, 12627, (...), 12645, 14973, 14974, (...), 15215, 15228 (2).

parificare: 8873, 8874.

pars: [44] (2), *211n, 443, 445, 447, 466, 467, 468, (...), 563, 637 (3), (...), 771, [772], 1667, 1672, [1897], [1993], 1997, (...), 2089, 2170, 2518 (3), (...), 3447 (*denominatio partis*), 3472, 3994, 4080, 4257, (...), 4747, 5444, 5724, 5753, (...), *5801n (2), 5969, 6309, (...), 8258, (...), 8635, 8822, (...), 15706; (subdivision of a book, of a chapter) 254, 5801 (*pars secunda* = Book B), 7381, 7455, 8895, 8899, 8901, (...), 9008; (syn. *latus*) 2610, 2769, 6366.

particeps: 7957, 7958, 7983, 7988, 8034, 8045, 8054, 8061, 8070, 8078, 8091, 8104, 8118, 8121, 8122, 8221, 11119, 11142, 15645n.

participatio: 8130, 8254, 9128, 15699.

participium: 8113-14.

partim... partim: 10642-43, 10778-79, 10878-79, 10923, 10927-28, 10984, 11019, 11060, 11091-92, 11104, 14653-54.

parvus: 779.

pastor: 12510, 12528, 12538-39.

patere: *90n, 994, 998, 2168, 2212, 2342, 2440, 2648, 2653, 2676, (...), 15260, 15418.

LIBER MAHAMELETH **1113**

***pauci*:** 67, 7983; (*paucior*) 6456, (6552), 6661, 7009, 7617, 13840, 14657.

***paulo*:** 2957 (*paulo supra*); (*paulo ante*) 3115, 3129, 3866, 7162, 8666, 12155, 12377, 12466.

***pavimentum*:** 8859, 8864, 8866, 8867, 8886, 8888.

pecunia (مال): (*domus pecunie*, بيت المال; syn. *domus pecuniosa*) 1812, 1824, 1826, 1829, 1834, 1837, 1839-40, 1843, 1844, 1845, 1846, 1848, 1849, 1862, 1865, figure p. 75 (5); (amount, مال) 3444, 3445, 3451, 3453-54, 3456, 3458, 3461, (...), 3551, 4033, 4034, (...), 4192 (2), 8208, 8209.

***pecuniosus*:** (*domus pecuniosa*; syn. *domus pecunie*) 1814, 1815, 1817-18, 1822.

***pendēre*:** 2343 (Tn357), 2511 (Tn380).

***pene*:** 64.

***per*:** 26, 34, 55, 68, 71, [103], [105], (...); (*multiplicare *per*) *211n, *227n (4), (...); (*dividere per*) 256, 259, 262, [263], [265], (...); (*per se*) [15], 20, 635, 1451, 1724, (...).

***percipere*:** 13314.

***peregrinus*:** 7744.

***perfecte*:** [441], 3047, 7514.

***perficere*:** (*questionem*) 11612.

***permixtus*:** 8290, 8316, (8317), 8319, 8321, 8335, figure p. 314, 8348, 8525, 8539, 8561, 8569.

***permutare*:** (ratio; ἐναλλὰξ λόγος, اذا بدّلنا) 9259, 10531.

***perpendĕre*:** 7606, 14156.

***perpendicularis*:** 15452 (noun).

***perpendiculariter*:** 15483.

***persolvere*:** 8903, 8907, 8925, [8928], [8929], 8954, 8962, 8963, 8964, 8965, 8985, 9034, 9050, (...), 9205, 9552-53, 9557, (...), 9639, 10568, (...), 11105.

***pertinere*:** 7951.

***pertingere*:** 5492, 8300, 8303.

***pertractare*:** 4936.

***pervenire*:** 121, 1444, 4416, 9207, 9245, 9377, 15486, 15600, 15609.

***pes*:** 15023 (syn. *radix*, *imum*).

1114   Part Two: Translation, Glossary

petere: 15755.

pistare: 9218, 9219, 9221 (2), 9226, 9227, 9228.

placere: 3074.

plantare: 15390, 15395, 15397.

planum: 15234.

plenus: 14759, 14762, 14763, [14765], [14766]; (*plene*) 699, 703, 2710.

pluralitas: 12657, 13383.

plures: 2143 (Tn325), 2174, 2185, 2529, 2585, *2585n, 3045, 3074, *3649n, 4023, 4783, (...); (*plures quam*, e.g.:) 5228, 6905, 13160, 13824, 13833, 13983, [14157], 14248, 14412, 14657, 14711.

plurimum: [6370] (*quam plurimum*).

plus: 4029, 4178, 5663, 6946, 8301 (2), [8342], [9388], 9473, (...), 13856, 13859.

ponderare: *5801n (2), 8290, 8293, 8297, 8300, 8301 (2), 8305, (...), [8342], 8347, 14717 (2), (...), 14730.

pondus: 35, *5801n (2), 8258, 8294, 8296 (2), 8298, 8306 (2), (...), 8358 (2), 8373, 8375, (...), 8597, 8598, 14726, 14729, 14733, 14740, 14743.

ponere (جعل): [43], 113, [145], [148], *182n, 470, 484, 787, 788, 790-91, 792, (...), 1116 (*pone in differentia*), 1119, 1146, (...), 15688; (*questionem*) 1270 (syn. *disponere*), 7745, 9554 (syn. *apponere*); (*probationem*) 279, 386, 5687; (*numerum*) 4755, 11253 (syn. *accipere*); (*lineam*) 9424; (*punctum centrum*) 15289; (*rem*) 9448, 9496; (*prolationem*) 12473; (syn. *proponere*) [5233]; (*commune: q.v.*).

portare: [8930], 8942, 8946, 8951, 8955, (...), 9209, 9214, 11679, 11687, (...), 12187, 12188, 12206, 12207, 12305, 12383, 12410, 12415, 12431, 12472, 12474, 12475, 12487, 12545, 13433, 13434.

portio: 8133, 9291, 9296, 14458. Tn1135.

positio: (*ex positione*) 6766, 12699.

posse: 6, 7, 8, 9, 11, [46], 57, 63, 113, *211n (2), *227n, 687, 704, 860, 1444, 2008 (2), 2427, 2510-11 (Tn379), 2678, 2711, 3027, 3049, (...), 9523 (*questio non potest fieri*), (...), 13828 (*hoc esse non potest*), (...), 14556, 15477; (*posse supra*, Tn777; δύνασθαι + acc., قوى على) 5655, 5657, 5698, 5700, 5701, 5703, 5705, 5708; (*potens*; δυναμένη) 5455, 5467, 5646, 5650, 5706, 5732, 5795.

LIBER MAHAMELETH **1115**

possibilis: 5030, 5548, 5558, 5754.

post: 114, 120, [135], 436, 811, 816, 867 (2), 940, 1282 (syn. *postea*), (...), 15687, 15713.

postea: 960, 1204, 1282, [1283], 1320, 1414, (...), 13901, 14057.

posterior: 4899, 14298.

posterius: 2294, 2413.

postponere: 61, 2428, [11884].

postquam (بعد ان، بعد ما): 276, 427, 621, 807, 974, 4242, 4358, 4513, 4523, 4932, (...), 5862, 5878, 5891, (...), 7315, 7374, 7539, 8412, (...), 15112, 15570.

potentia: (*in potentia*; δυνάμει, بالقوّة & فى القوّة) 5092, 5255, 5354, 5357, 5363, 5366, 5372-73, 5374-75, 5573, 5589, 5593, 5596, 5611; (potentially) 8862.

potius: *90n, 4437.

practicus: (*practica arimethica*) 21, 23, 24, 175, 623.

precedere: 409, *440n, 474, 795, 818, 872, [1006], 2252, 2307, (...), 15073; (*precedens*, noun) [108], [110], 114, [151], 642, [1041], [1080], [1083], (...), 13199; (*in precedenti, ex precedenti*) 1521, [4432], 4509, 5385, [6015], 6089, 6361, (...), 15430.

precellere: 10060.

precipue: 63, 4934.

precognoscere: 5613.

predicere: 34, 623, 627, 650, 848, 853, 859, 918, 921, 953, 969, 1141, [1191], 1213, 1239, 1278, 1281, 1403-04, 1419-20, 1440, 1451, 1480, 1522, 1586, (...), 15675, 15774.

predocere: (see *docere*) 1370, 1409, 1495, 1807, 1817, 2251, 2866, 3119, 3135, 3322, 3782, 3953, 3980, 4061, 4071, 4077, 4089, 4095-96, 4616, 4963, 5274, 5333, 5521, 5566, 5636, 5941, 6269, 6845, 7066, 7207-08, 7249, 7267, 7328, 7370, 7376, 7490, 8103, 8108, 8124, 8455, 8463, 8481, 8550, 8555, 8663, 8675, 9345, 9746, 9755, 9865, 10214, 10239-40, 10865, 11122, 11131, 11134, 11479, 11482, 11823, 12767, 12783, 12971, 13083, 13582.

prefatus: 8657.

prelatus (إمام): 2355 (Tn361), 2357-58, 2399, 2401, 2420, 2422, 2437, 2439, 2468, 2470, 2495, 2498, 2639, 2641, 2664, 2767, 2772, 2817, 2824,

1116   PART TWO: TRANSLATION, GLOSSARY

2850, 2857, 2884, 2887-88, 2903, 2905, 2932, 2935, 2951, 2955, 2961, 2965, 3009, 3013, (...), 3449 (Tn493), (...), 4361 (Tn610), (...), *4679n, (...), 8718 (Tn1202), (...), 15546, 15547. Common in C is the occurrence of the pair *communis* and *prelatus*, with the first found in the text and the other as a variant (e.g. *ll.* 2357–58, 2468, 2887–88, 2935, 2951, 2955 (...), but: 3037; B: 2850), though not always (e.g. *ll.* 1468, 2399, 2401, 2420, 2422, 2437, 2439).

premittere: 171, 173, [306], 349, 974, 1034, 2775, 3272, 4701, [5619], 5660, 6520-21, 7954, 8237, 8882, [9990], 12193, 14555; (*ex* or *in premissis*; see Tn395) 828, 1023, 1043, 2169, 2212, 2242, [2351], 2403, 2441, 2648, 2653, 3366, 3963, 5112, 5873, 6070, (...).

premonstrare: 203, [270], 1630, 1763-64, 1786, 2896, 3123, 3251-52, 3489, 3500, 3520, 3764, 4132, 6177-78, 7186, 9510, 9797, 9834-35, 9900, 10246, 10540, 11327, 11339, 11354-55, 13213, 15022.

preostendere: 1048.

preponere: 61, 169, 175, 388, 432, 621, 1008, 1043-44, 1883, 2067, 2293, 2427, 7044, [11884], 13751, 15123.

prepositio: (a theorem in A–II) [431], 1586, 2983, 3354, 3783, 3946, [12719], 14629.

prescire: 9380.

prescribere: 705.

presens: (*ad presens*) 7951.

preter: [72], 788, (1979), 1980, [3072], [3137], 11230, 11231, 11253, 12974, 13676, 15297.

pretermittere: 386-87, 897, 898, 1398, 1400, 2603, 2610, 2627 (*pretermittere cum*, syn. *reicere pro*), 7955, 10397, 10644, 10677, 10679, 10771, 10795, 10867, 10871, 10892, 10898, 14252, 15735.

pretium: (price) 5846, 5860, 5863, 5865, 5880, 5890, 5892, (...), 15776, 15777; (value of a coin) 13677, 13758, 13764, 13784, 13793, 13871 (2), 13873, (...), 14734, 14745; (wage, syn. *merces*) 9647 (2), 9707, 9712 (2), 9728 (2), 9730, 9734, (...), (13336), 13337.

prevalere: 10592, 10593, 10620 (2).

primus: [39], 52, 53, [72], 73, 75, 79, 89, [126], [131], [134], 178, 180, (...), 777, 783, 786, 789, 804, (...); (*primum*, adv.) 2028, 2030, 2033 (2), (...); (*in primis*) 5887, 6020; (*prior*) 561, 628, 629, 630, 812, 826, (...); (*prius*, adv.) 78, 115, 172, [199], [440], 632, 634, 706, 708, 742, 2371, 2390, 2413 (*prius vel posterius*), 2443, (...).

principium: 41, 74, 77, 84, 88, *90n, [91], 137, 139, [146], [147], 778, 1633n, 2998, 5903, [6490], 7954 (*in principio libri*), 8652, 10751, 10806, 14155-56, 14250.

probare: 170, 192, 204, 219, 239, 283, 298, 360, 387, 434, (438), [1070], 1350, 2711, 5250, 6329, 7517, [8744], 9719, 10963 (*et hoc est quod probare voluimus*, syn. *demonstrare*), [14602], 14965.

probatio: 170-71, 172, 278, 319, 338, 386, 394, 419, 434, 436, 449, (...), 2691 (*probatio facilior*), (...), 6472 (*probatio inveniendi*), 6473, (...), 12794 (Tn1679), 12864 (Tn1687), (...), 13873 (Tn1804), (...), 15479, 15550.

procedere: 69, 113, 2921, 3951, 4492, 4499, *5801n.

procreare: 3142, 6105.

producere: 3402, 8863 (*producere ad effectum*).

productum, productus: (of a multiplication) *211n, 212-14n (3), 213, 214, 216, 217, 222 (2), 229 (2), 230, 231, (...), 258 (*productum ex multiplicatione*), 259 (*productum ex ductu*), (...), 15740, 15741; (*productus*) [263], 417, 519, [773], 980, 984, 999, 1010, (...), 14963, 14968; (of an addition) 1562, 3673; (of a division) 8589, 8595, 8644, 11298.

profecto: 192, [1069], 4062, 6748, 6756, 7633, 7634.

***profluere**: 211n, 11924n.

profunditas: 35, 12609-10, 14785-86, 14897, 14959, 14969.

profundus, profundum: 12592, 12594, 12596, 12598, 12599, 14885; (*in profundum*) 12585, 12587, 12601, 12603, 14781, 14812, 14845, 14910, 14950, 14956; (*in profundo*) 14880-81, 14896.

progressio: 67; (mathem.) *211n, *3568n, *3609n, *3649n, *3677n, *11386-87n.

prohicere: 14781, 14812, 14845, 14881, 14896, 14910, 14945-46, 14950-51.

prolatio: 5485, 12472, 12473 (Tn1647).

prolixus: 1485-86, 3143, 3501.

***in promptu**: 65, 703.

pronuntiare: 5104, [5227], [5232].

propinquus: (*numerus*) 2177, 2196, 4946, 4955, 4975, 8322, 8359, 8828, 8858; (*comparatio*) 8821; (*superficies*) 8826, 8854; (*radix*) *4940n, 4945, 4950, 4953, 4959, 4960, 4963-64, 4964, 4967, 4971, 4972, 4979, 6177, 8457-58, 12147, 15237; (*corpus propinquius in pondere*) 8321, 8322, 8358 (2).

proponere: 170, 211, 227, 253, 275, 276, [277], 292, 623, 860 (*propositus numerus*), 1468, 1864, 2735 (*propositi*, noun), 2825, 3203, 3419, 3451, (...), 12800 (*propositum questionis*), (...), 15609, 15735.

proportio: *211n, *227n (4), 382 (2), 383 (2), 384, 385, 754 (*equa proportio*), [756], *995n, 1000, 1017, [1049] (2), 3990, (...), 10843, 12173.

***proportionabilis**: 2520n.

proportionalis: *227n (2), 247, 323, [426], 1009, 1015, 1016, [1065], 2152, *5801n, 5802, 5804, 5820-21, 5866, 5882, 6480 (*latera*), 7085, 9211-12, 9253-54, (...), 10380 (*tres termini*; Tn1418, Tn1568), (...); (*proportionalia*, syn. *numeri proportionales*) 225, 12445.

1118   PART TWO: TRANSLATION, GLOSSARY

proportionalitas : 1003; (*proportionalitas equalitatis*; δι' ἴσου λόγος, نسبة المساواة) 343, 347; (*equa proportionalitas*; Tn1097) 7711, 11239.

proportionaliter : 995, 1002, (*reducere proportionaliter*) 7928, [9974], 9975, 10037, 10038, 11924, 13284.

propositio : *182n, 379, 387.

proprietas : 14-15.

proprius : 9, 10, 57, *57n, 138, 1660 (*proprie*); (*de proprio*) 15658, 15660, 15671, 15682.

propter : 972, 976, [1057], 1586, 8174, 8990, 11515, 12757, 13073, 13806, *15280n.

prosequi : 1630, [2076], 3863, 4968, [5008], 6504, 6598, 6610, 6615, 6630, 6739, 6844, 10434, 11084, 11384, 12882-83, 14223.

protrahere : (*lineam*) 192, 204-05, 5729, 6736, 6840, 8874, 9434, 10688 (*a puncto*), 10690, 10836, 11409 (*de puncto*), 11411, 11412, 11505, 11968, 12041, 13919, 13944, 13946, 14053-54, 14537, 15414, 15433, 15451, 15467.

prout : [17], 15367.

provenire : (*ex ductu*) 220 (2), 221, 240 (2), 241, 244, 258, 261, [264], 266, 269, 270, 271 (2), 273, (. . .), 15765, 15768.

punctum, punctus : 510, 511, 528, 546, 563, 576, 577, 606, 612 (2), 4186, 5259, 5597, 5670, 5676, 5718, 6391, 6392, 6395, 6405, 6427, 6489, 6714, 6844, 8607, 9433, 9434, 9440, 9480, 10688, 10690, 10835, 10836, 11410, (. . .); (*punctus*) 15240, 15260, 15287, (15334), 15355.

purus : *5801n, 8334 (*purissimus*), 8335, 8343, 8345 (2), 8346, 8350 (2), 8352, 8354.

putare : 12473, 12618, 12619.

— Q —

quadragies : [107-08].

quadratura (تربيع) : 7, 10.

quadratus (noun & adj.; τετράγωνον, τετράγωνος, مربّع) : (figure) 5672 (*superficies quadrata*), 5683, 5684, 5721, 5728, 8453, 8563, 8658, 8817, 8893, 12582 (Tn1661), 14880 (Tn1921), 14882, 14895 (Tn1924), 14897 (*quadratus equalium laterum*; مربّع متساوى الاضلاع), 14970 (Tn1939); (square power) 579 (2), 580 (2), 581 (2), 582, 583 (2), 584 (2), 585 (2), 586,

LIBER MAHAMELETH 1119

(. . .), 1045, 1048, [1049], (. . .), 3609 (2), 3610, 3616, (. . .), 4952, (. . .), 5665, 5666, 5671, (. . .), 15536 (2); (*radix*) *4940n.

quadringenti: 957, 959, 962, 1100, 1102, 1255 (2), 1256 (2), 1258, 1266, 1272, 1273, 1392, 1396, 1397, 1401-02, (. . .).

quadruplare: 14272, 14273, 14282 (2).

quadruplus: [1032], 1035, 1036-37, 2115, 6297 (*comparatio*), 6298 (+ gen.), 6483 (*ad*), 6624, 11436 (*superatur in quadruplo*), 11647.

quae–: See *que–*.

qualis: (*talis . . . qualis*) 222, 224, 242, 284, 286, 322, [945], [947], [1064], 1671, 1709, 2150, 2341, 2375, 2696, 2701, 3396, (. . .).

qualiscumque: 5808, 7616, 13509.

qualiter: 981, 984, 985, 3046, 3123, 3991 (*taliter . . . qualiter*), 7640, 10760 (syn. *quomodo*).

quamvis: 1486, 2067, 5803.

quandoque: 95n, 96n.

quandoquidem: 407, 799, 1018.

quantitas: 4043, 5355, 5363, 5372, 5444, *5801n (2), 7961.

quanto: (*tanto . . . quanto*) 4967.

quantum: 3983-84, 4196, 4353, 4832, 5431, *5801n (2), 5846, 5852 (2), 5860, 5863, 5876 (2), 5879, 5890, (. . .); 7951 (*quantum pertinet*); 8301 (*quantum plus*); 14375 (*quantum ad*); (*tantum . . . quantum*) [149], 177, 178, 180, 184, 185, 191, 202, 204, *211n (4), [427], 743, 745, [748], (. . .); (without *tantum*) 13251, 13256-57, 13290, 13295, 13826, 13989; (*tantum . . . quantum . . . et quantum*) 14170-71, 14195-96, 14198-99, (. . .).

quantumlibet: 2271.

quantus: 3445, 3458, 3464, 3468, 3473, 3555, 3570, 3575, (. . .); (*tantus . . . quantus*) *227n (2), 1305, 2170-71, 13508, 14004, (. . .); (*totiens . . . quantus*) 894, 896.

quantuslibet: 14121.

quare: [10974], 12833.

quartus: 87, 98, [122], [124], 229, 230, 233, 250, 252, 328, 329, 331, 334, 335, (. . .), 15780.

quasi: 115, 120, 705, 944, 946, 959, 960, 1224, 1231, 1255, 1290, 1802, 2895, 2914, 3042, 3462, 3467, 3476, 3486, (. . .), 4213, 4235, 4236, (. . .), 6178 (*quasi numerus*; 2), 6698 (*radix, quasi unus*), 6725, (. . .), 8459 (*quasi radix*), (. . .), 10313 (*sunt quasi*), (. . .), 15333.

quater: 562, 566, 570, 582, 586, 590, 591, 814, 819, 820, 822, 824, 832, (. . .), 2137, 2297.

quaternarius: 55, [748], [754-55], [756], [772], 5822, 5823, 7012, 14638.

quaterni: [692], [693] (4).

quemadmodum: 7147-74n, 9663.

1120 PART TWO: TRANSLATION, GLOSSARY

querere: *227n, 800, 814, 828, 847, 863, 872, [1065], [1066], [1075], [1085], 1207, *1443n, [1502], 1665, 1671, 1678, 1717, 1765, (...), 14389, (...), 14406 (syn. *accipere*, see Tn1864, Tn1865), (...), 15537, 15679.

questio (مسألة): (theorem) [277], 408; (problem) 1268, 1270, 1478, [1850], 2232-33, 2517, 2731, 2921, 3053, [3070], 3073, 3416, 3470, 3501, *3503n, 3557, 3927, 3951, 4023, 4025, 4028, *4108n, 4352, 4492, 4499, 4840, 4900, 4904, 4921, 5181, 5223, 5484, 5643, 5694, 5741, 5774, 5793, 5797, *5801n, 5922, 5967, *5991n, 6092, 6101, 6109-10, 6110, 6176 (*questio implicita*), 6286, [6370], [6490], 6504, 6512, 6586, 6598, (...); 6912, 6914 (*questio vera*); 6945 (*questio falsa*); 6982 (*questio aperta*); 6997 (*questio non terminata*), 9519 (*questio interminata*); 7605 (*questiones false*); 7744 (*questio peregrina*); 8130 (*questio de ignoto*); 8396 (*questio impossibilis*); 8810 (*questiones huiusmodi nunquam possunt exire ad effectum*); 9342 (*questio composita*); 9377 (*formare questionem*); 12202 (*hec questio habet se ad duos sensus*); 15343 (*questio multiplex*); (...), 15783.

quia: 30, 98, [145], [149], 216, 218, *227n, 233, 238, 318, 336, 447, 469, [471], 482, (...).

quicquid: 2222, 3048, [3313], 4267, 4268, 4788, 4793, [8384], 8520, 11338, 13283, 14556.

quicumque: [265], 1772, 1892, 1893, 1894, 2500, [4456], [5650], 7516, 7641, 8697, 8797, 8853, 11404, 13688, 14159, 14162, 14251, 14254.

quidam: 170, 439, 1478, 1883, [5923], 7605, 7744, 7829, 7830, 7840, 8336, 11389, 15397, 15638, 15701, 15724, 15750.

quidem: 120, 433, 4479, (8648), (10970), 12621.

quilibet: [152], 178, 185, *211n, 350, 389, 654, 753, 842, 985, 999, [1042], 3342, 3559, *3609n, 3795, 4200, 4755, 4768 (2), (...), 15687-88.

quinarius: [748], 7023, 14638, 15552.

quincuplus: 1054 (+ gen.), 1056, [2347], 7938, 7941, 10755-56, 10862 (+ dat.), 11437, 13222, 13226, 13226-27.

quindecuplus: 1055, [2347-48].

quindenarius: 13705.

quingenti: 931, 933, 934, 938, 956, 1138, 1159, 1163, 1220, 1225, (...), 14708.

quinquagesimus: 13327-28, 13329, 13331, 13333, 13334-35, (13335), (13336), 13339.

LIBER MAHAMELETH

quinquies: 844, 890, 1140, 1160, 1169, 1171, 1311, 1323, 1328, 1431, 1435 (2), 1457, 1458, 1459 (2), (. . .), 14535.

quintus: *90n, 91, 97, [101], [127], [129], 328, 329, [330], 334, 336, figure p. 16, 368, (. . .), 15768.

quisquis (*cognoverit, intendit, intellexerit* etc.): 171, 702, 705, 1634, 2710, 3047, [3069], 3502, 4196, 4513, 5490, 5769, 7514, 7953, 8129, 9718, 12150, 12508, 14878.

quomodo: 801, 1337, 2678, 5491, 5712, 5804-05, 6891, 8116, 8124, 8142, 8223, 8248, (. . .), 11144.

quoniam: 21, 24, 29, 69, 70, 169, 386, 632, [694], 776, 789, 804, 865, 1170, 1619, 1772, (. . .).

quoque: 633, [1083], 4208, 4217, 4220, 6588, 7448, [9153], 9349, 10554, 10593, 11505-13n, 12591, 13788, 13882, 14079, 14209.

quot: *227n, [1186], 1761, 1785, 1792, 1805, 1814, 1824, 1829, 1840, 1844, 1845, (. . .), 15673.

quotiens: 362, 374, 375, 376, 396 (2), 1012, 1013, 1935, 2162, 2459, 4683, 4684, 9732, [13497], 13652, 13704, 15508, 15522; (syn. *cum*) 654, 4024, 4751, 5181 (cf. 5223), 7571, 7955, 8129, 11915. See also *totiens*.

quotienslibet: 1114, 1145, 1166.

quotlibet: 443, 466, 4901, 12769.

quotquot: 176, *211n, 2585, 15656.

quotus: 903, 911, [2071], 4257, 4390, 4600, 4614, 4628, 4715, 4734, 4747, 7966, 7968, 9263, 9278, 9661, [9669], 9682, 11734, 11740, 12062, 12117, 12157, 12269, (. . .), 14891.

quotuslibet: 8912, [8918].

quousque: 990, 1444, 2032, 2038, 7666, 8155, [8230], 8299, 9249, 9251, (. . .), 9368, 9377, 9382, (. . .), 9542, 14712, 15251, 15354.

— R —

radius: 15485 (*radius oculi*).

radix: (mathem.) 29, 30, 3531, 3682, 3686, 3692, 4156, 4887, 4894, 4898, 4905, 4906, 4928, 4933, 4938, 4940, *4940n, 4941, 4942 (2), 4943, 4944, 4945 (*radix propinqua*), 4946 (*radix rationabilis*), 4948, 4949, 4950, 4952 (2), 4953, 4954, (. . .), 5089 (*radices communicantes*), (. . .), 5140 (*radix irrationabilis*), (. . .), 5302 (*radix radicis*), (. . .), 6177 (*numerus carens*

1122   PART TWO: TRANSLATION, GLOSSARY

radice, عدد غير مجذور; Tn904), (...), 6543 (*radix, dico;* Tn962, Tn964), (...), 8457 (*numerus non habens radicem*), 8458, 8459 (*quasi radix*), (...), 15067, 15117, (...), 15463; (syn. *pes, imum*) 15026 (*radix parietis*), 15059, 15072, 15090, 15092, 15094, 15119, 15154, 15175, 15176, 15177, (...), 15234 (*radix arboris*), (...), 15342.

ratio: (*secundum rationem*) 8151, 12089; (reasoning) 13519.

rationabilis: 4946, 4955, 5068, 5073, 5079, 5141, 5455, 5664 (2), 5762.

rationalis (ῥητός, منطق): 5091, 5238, 5253, 5255, 5258, 5264, 5265 (2), 5267, 5272, 5291, 5298, 5356, 5358, 5368, 5573, 5589, 5593, 5595, 5611, 5646, 5649-50, 5656, 5699, 5707, 5795.

recipere: 4028 (*hec questio non recipitur*), 7749, 8113, 8115, 8118, 8121, 8861, 8862, 8868, 8886, 11096, 14249.

rectus: [14928], 15253, 15257, 15279; (*linea*) 8820 (2), [14927], 15294, 15484; (*angulus*) 14981, 15011, 15243, 15299, 15418, 15455; (*recte,* adv.) 2297, 2459, 3403.

recuperare: 7532, 7557, 7560, 7563, 7645, 7652, 7659, 7666, 7682, 7717, 7721, 7722, 7725, 7729, 7732, 7733-34, 7736, 7741.

recuperatio: 7568.

*redactio: 13556n.

reddere: 177, 180, 184, 191, 202, 203, *211n (3), [427], 743, 745, 753, [764], 3570, 3575, 3584, 3589, 3597, (...), 3672, 10556, 10581, *15280n.

redigere: 9243, 9249, 9251, 9253, 9257, 9281-82, (...), 9542, 9544, 14916.

redire: 622-23 (*ad propositum*), 1675, 2851, [10189], 15112, 15600, 15604, 15611, 15621, 15630, 15631, 15638, 15644.

reducere: (*fractiones;* syn. *convertere*) *227n, 2818 (*ad*), 3054, *3809-11n, 4242 (*in*), 4377, 4406, 4638, 4654, 4666, 9059, 9061; (*monetam,* syn. *convertere*) 13686 (*in*), 13709; (equation; اكمل) *3520n, 7927, 9232, 9893, 9973, 9974, 10036, 10037, 10225, 11923, 11924, 13282, 13283; (*questiones*) 7639 (*ad*), 7640, 14117, 14161, 14253, 14270-71, 14332.

*reductio: 227n.

regio: 13602, [13603]. Syn. *terra.*

regula: [166], [262], [299], (381), 386 (syn. *propositio*), 388, 740, [774], 776, 799, 802, 813, 826, 975, 1034, 1091, 1174, [1189], 1208 (syn. *modus agendi*), 1594, 1605, [1662], 1880, [2005], 2014, 2047, [2072], [2074], 2170,

LIBER MAHAMELETH
1123

2176, 2243, 2510, *2520n, 2957, 3044, 3136, 3223, 3272, 3394, 3463, 3469, 3477, 3565 (*regula generalior*), 3990, 4265, 4275, 4287, 4293, 4302, 4325, 4354, 4778, 4848, 4850, 5138, 5173, 5424, 5659, 5740, 5745, 5768, 5770, 5770-71, 5771 (*regula agendi*), 5772, 5786, *5801n, 5987, 6284, 6337, 6339, 6350, 6352, 6521, 7042, 7065, 7571, 7685 (*probatio regule*), 7835, 7974, 8000, 8011, 8387, 8419, 8517, 8676, 8797, 8945, 8961, 9002, 9736, 11433, 15675.

reicere: 598 (*reiecto communi*), 951, 1352, 1361, 1365, 1618, 1733 (2), 1752 (2), 1767, 1769, 1819, 2013, 2017, 2127 (2), 2606 (*pro*; syn. *pretermittere cum*), 3682, 9478, 15041, 15105, 15131, 15136, 15774.

reintegrare (اكمل): 4199-200, 4202, 4203, 4207, 4214, 4216, 4265, 4277, 4280, 4283, 4286, 4290, 4292, 4297-98, 4300, 4307, 4538-39.

relinquere: [14765], 14996 (القى).

reliquus: [147], 1451, 1757, 2820, 3456, 5588, 5592, 5741, 5809, 6961, 7170, 7990, 8115, 8149, 8906, 9355, 9840, 10175, 10176, 10854, 10905, 10977-78, 11427, 13757, 13761, 13929, 13935, 13944, 13972, 13976, 14160, 14163, 14228, 14561, (15646), 15657, (. . .), 15753.

remanere: 599, [764], 830-31, 833, 839, 879, 884, 888, 893, 895, 907, 916, 927, (. . .), 15769, 15775; 1361 (*remanebit multiplicare*); 1733 (*remanebit dividere*); 2038 (*quousque nichil remaneat*); 6781 (*ut*); (*remanens*, noun) 1777, 3620, 9237, (. . .).

repercutere: 15600.

repetere: [649], 850 (syn. *iterare*), [1060], 1145, 1431-32, 1437, 1439 (2), 2098, 2099, *2099n, 2101, 2102, 2103, 2104, 2110, 2132, (. . .), 2677 (*probationem repetam*), *3138n, 5802 (*hic repetimus*), 7482, 8501, 15775.

repetitio (تكرير): 1430, 2105, 3547, 6481-82, 6483, 6623.

reponere: 1877.

reportare: 8337, 8903-04, 8911, [8914], [8916], 8921, 8925, [8931], 8937, (. . .), 9198, 9202.

representare: 61.

requirere: 1299, 1356, 1538, 1609, 1615, *1658n, 1771, 1804, 2024, 2277, 2906, 2941, (. . .), 3228 (*summa requisita*), (. . .), 3396 (*comparatio requisiti*), (. . .), 9259 (*quas requiris*, syn. *quas scire voluisti*, cf. 9260), 9378, 12419.

res: 35, [48], [50], [52], 61, 64, 864, 1335, 1635, 1639, 4265, 4353 (2), 5074, 5099, 5112, 5155, 5277, 5286, 5307, 5358, 5385, [5483], [5618], *6180n

1124   PART TWO: TRANSLATION, GLOSSARY

(2), 6386, 6438, [7063], 8803, 9662, 10652, 11542, 11667, (...), 15430, [15525]; (object of unknown value, unknown quantity, thus our x; see Tn1001) *3520n, *3533n (2), 6248, 6250 (2), 6251, 6253, 6254 (2), 6255, 6257 (Tn912), (...), 6856, 9081, 9082-84 (Tn1250), (...), 9120, 9191, (...), 9448, (...), 9510, 9573, (...), 9630, 9699, (...), 12495, 13235, (...), 13308, 13890, (...), 14739, 14989, (...), 14739, 14989, (...), 15780; 6757, 6804 (*due res inequales*); 6766 (*due res impares*); (*res prima*) 14424; (*res secunda*) 12135, 14425; (*res minor*) 6759, 6763, 6769 (2), 6791 (*rem minorem pone rem*), 6793, 6801, 6802, 6810, 6818, 6846 (*minor res sit res*), [6853], [6854]; (*res maior*) 6764, 6769, 6791 (*rem maiorem pone unam dragmam*), 6803, 6819, 6846, [6854], [6855]. See *aliquid*.

residuum: (remainder) 897, 899, 1867, 1869, figure p. 75, 2014, 2017, 3417, 3421, 3425, 3480, 3483, 3487, (...), 4058 (+ gen.), (...), 5176 (*ex*), 5179 (*de*), (...), 15487; (apotome; ἀποτομή, منفصل) 5238, 5238-39, 5253, 5257, 5270, 5281, 5290, 5493, 5527, (...); (*primum*) 5698, 5700, 5740, 5770; (*secundum*) 5701, 5771; (*tertium*) 5703, 5773; (*quartum*) 5705, 5797; (*quintum*) 5706, 5797; (*sextum*) 5708, 5798; (*residuum mediale primum*) 5702; (*residuum mediale secundum*) 5703-04.

residuus: 9772.

resolvere: 42, 7517n.

sine respectu: 635-36.

respicere: 8501 (*latus ... se respiciens*).

respondere: 4205.

responsio: 5185, 5775, 5794, 5798.

restare: 1476, 1753 (*ad*), 2509, 2611 (*ut*), 4932, 6275, 6302, 6356, 9840, 10004, 10177, 13784, 13793, 13814, 13815, 13868, 13904 (+ inf.), 13955, 14311, 15553.

restaurare: (Tn930) 4117, 6365; 9507 (Tn1309).

retinere: 905, 908, 914 (Tn147), 916, 919, 1531, 2179, 2181, 3065, 3067, 3149, 3153, 3375, 3377, 3619, 3621, [3643], [3647], 3659, 3747, 3839, 3974, 3979, (...), 15615, 15623; (*in manu*) 948, 961, 1245, 1269.

retrahere: (*numerus retrahitur una differentia*) 1996, 1997 (2), 1999, 2003, 2011; (*scala retrahitur*) 14974, 15004, 15005, 15016, 15018, 15023, 15026, (...), 15214, 15228.

retro: 1999, 2003, 2008, 2012, 2017.

LIBER MAHAMELETH 1125

revolvere: (7517).

rotunditas: 8777, 8778, 8788, 8789, 8793, 8796.

rotundus: 8266, 8267, (8277), 8278, *8689n (2), 8691 (noun), 8694, 8776, 8778, (8788), 8797, 8798, 8799, 8810 (*questiones de rotundo*), 8813, 8830, 8832, 8843, 8846, 8849 (noun), 8859, 8860, 8880, 12578 (Tn1660), *12578n, *14954n.

— S —

sacellus: 1761, 1762, 1765, [1783], 1785, 1786, 1792, 1794, 1805, 1844, 1845, 1847, 1848, 1849, [1851], 1855, 1856, 1860, 1862, 1864, 1868, 1870, 1874, 1875, figure p. 75 (5).

saepe: See *sepe*.

satis: *90n, [11763].

scala: 14972, *14972n (2), 14973, 14977, 14979, 14987, 15003, 15009, 15010, 15016, 15018, 15023, (. . .), 15227, 15249.

*scazus: (*casa scazi*, 'Schatzkammer'; بيت المال) 1443n. See du Cange's *Glossarium* (*Scatz*), Niermeyer's *Lexicon* (*scattus*, 'pecuniary resources').

scientia: 24, 25 (2), 28-29, 31, 34, 433 (*scientia mahameleth*), 436, 703, 778, 779, 1173, *1443n, 5492, [6370], 8334, 8694-95, 15480.

scilicet: 13, 30, *168n, [198], 201, *211n, 255, 387, 446, 510, 511, 623, 637, 645, 652, (. . .), 15644, 15774.

scire: 271 (*quod*), 635, 782, 799, 801, 816, 844, 849, 855, 860, 878, 883, 887, 891, (. . .), 5429 (*quoniam*), (. . .), 5446 (*patet scienti*, cf. *patet legenti* 5402), (. . .), 11982 (*de*), (. . .), 15749, 15781. See Tn754, Tn1585.

se: See *sibi*.

secare: 8876, 15294.

Secobia: 13366, 13559, 13564.

Secobiensis: 13344, 13572, 13575.

secundum: 15, 17, 60, 89, 343, 347, 434, 777, (875), 901, 903, 975, 1002, [1057], (. . .), 15690, 15758 (2); (regarding) 5793.

secundus: 76, 86, *90n, 115, 116, 117, 119, [126], [128], (. . .), 15773, 15778.

sed: 41, [47], 62, 188, 194, 273, 320, 433, 463, 474, 517, 536, 614, 617, 631, 632, 689, 746, (. . .).

1126 PART TWO: TRANSLATION, GLOSSARY

semel: [555] (2), [556], 659, 685, 701, 920, 929, 1110, 1130, 1140, 1161, 1554, 1555, 1574, 1743, (...), 15271, 15272.

semper: 55, 65, 69, 93, 115, 120, [135], [142], 642, 816, 845, 861, 892, [1076], 1103, 1113, 1303, 1305, (...), 2232 (Tn340, Tn677), 2528, 3538, 3603, 3611, 3614, 3617, 3619, (...), 15663, 15711, 15712.

senarius: [748], [761], [763], [764], 1977, 5822, 5824.

sensus: 2517, 3074, 4352, 7748, 8136, 8147, 10807, 10987, 12202, 12205, 15444.

separatim: 2717, 2733, 6528-29.

sepe: 1166; (*sepius*) [649], 2298.

septemcuplus: 11437-38.

septenarius: [749-50], 1101, 1111, 1957, 1967, 2091.

septies: 954, 955, 956, 958, 959, 962, 1382, 1383, 1384, 1541, (1542), [2069].

septimus: [104], 223, [227], [243], [249], [322], [425], 439, 440, [453], 801, 809, [1068], 1624, 1627, (...).

septingenti: 1266, 1272-73, 1276, 1280, 2319, 2673, 3298, 3301, 3308, 3770, (...), 12841.

sequi: 33, [43], [90-91], *90n, 93, [101], [106], [108], [109], [111], 113, [135], [168], 171, (...), 14874, *14972n, *15346n, *15404n.

sericum: 8507, 8518 (2), 8526, 8535, 8540, 8562, 8570.

serpens: 15610, 15620, 15629, 15637, 15641.

servatum (محفوظ): 4790 (Tn670), 4790-91, 4794, 4795, 4796, 4797, [4799], [4801], 4801, 4803.

servire: 9646, 9647, 9653, 9657, 9668, 9676, 9681, (...), 9693 (*diebus*), 9696, 9701 (*dies*), (...), 12414, 12434.

servitium: 9984, 10047, 10174, 10364, 10579, 10584, 10589, 10600, 10613, 10627, 10861 (2), 10917.

sesquitertius: 7703, 7704.

sexcenti: 1217, 1219, 1244, 1246, 1325, 1329, 1411, 1419, 1784, 1788, (...), 14593.

sexcuplus: 1054 (+ gen.), 11437, 14482 (*ad*), 14497 (+ dat.).

sexies: 855, 1322, 1328, 1431, 1514, 1523, 1525, 1528, 1531, 1532, 1544, 2099, 2101, 2131, [2576].

LIBER MAHAMELETH   **1127**

sextarius: [51], 1640 (2), 1641, 1652-53, 1653, 5862, 5863, figure p. 224 (2), 5864-65, 5878, 5880, (. . .), 7032, 7108, (. . .), 7664, 9217, (. . .), 9231, 9552, (. . .), 9638, 10553, (. . .), 10637, 11678, (. . .), 12476, 13313, (. . .), 13433, 13434. Written *sexstarius* initially (according to the disordered text) by the copyist of $\mathcal{B}$.

sextuplus: see *sexcuplus*.

sextus: [100], [105], [132], 328-29, 330, [331], 335, 337, (. . .).

sibi: 177, 180, 977, 1015, 2726, 3570, 3573, 3574, (. . .), 3689, 4174, 6546, (. . .), 6569, 6573, (. . .), 6873, 6946, 6947, (. . .), 15751, 15755.

sic: 95, [103], [125], [129], (. . .); (*sic facies*) 1036, 1314, 1486, 1678, 1696, (. . .), 15734; (*sic invenies*) 3447, 3460, (. . .), 3475, 4000, 15408, 15426, 15495, 15510, 15517, 15517n.

sicut: 78, 94, [101], 115, [121], [123], [126], [128], [131], [199], 203, 223, (. . .), 15714, 15774.

signare: 8299, 8303.

significare: [49], [50], 880, 898, 3075, 10060, 10098, 10136.

significatio: [151], *figure p. 6, 796.

significativus: 885, 889, 894, 895.

signum: [144].

similis: *211n, 380 (*propositio*), [409], 438, 1912, 2627 (*fractiones*), 2669, 2724 (Tn407), 2873, *3609n, *3649n, 5583, 5639, 5800, 6479 (*superficies*), 7106 (*probatio*), 7334, 7863, 8427, 9128 (*questio*), 9380, 9390 (Tn1291), 9892, 12618, 13371, 13622, 13835, 14247, 14334, 15539.

similiter: 97, 117, 179, 200, *211n (2), 243, 266, 285, 365, [380], 399, 458, 792, (. . .), 15683, 15755.

similitudo: *90n, 114, 8755.

simplex: 645, 1305.

simul: [45], 197 (*simul accepti*), *211n (*simul iuncti*), 384, 385, 390, 391, 444, 448, 483, 496, 499, 604, (. . .), 14774, 15591.

sin: 5184, [5227], 6097, 6099, 10770, 10778, 10794, 10908, 10922, 10926, 11095.

sine: 6, 8, 11, 635, 1724, 2178, 2197, 4438, 4694, 6016, 6065, 6113n, 8291, 11229, 14363, 14452, 14916, (. . .), [14936].

singularitas: 12657, 13384.

singuli: 57, 63, 256, 445, 655 (3), 656, 657, 661 (2), 663 (2), 665, 671, 691 (2), [691], [692] (2), 990, 1332, 2722, 2819, 2821, 4847, 6880, 6981, 6986, 7386, 7457, [8385], 8637, 8899, (. . .), 14022, (. . .), 14457.

siquidem: 13291.

sive: 71, 74, 75, 77, 80, 84, 88, 779 (2), 986 (3), 1970, 1988, 2122 (2), 2123, 2529, 2585 (2), 2590, (. . .), 14162, 15031.

1128   PART TWO: TRANSLATION, GLOSSARY

socius: (*numerus*; مقترن) 5807, 5809 (2), 5811; (syn. *consors*) 8150.

solidus (coin): 7183, 7184, 7188, 7191-92, 7192, 7194, 7195, 7269, 7272, 7365, 7366, (...), 7378 (2) , 7380, 13625, 13629, (...), 14668, 14673.

solus: [39], [49], 1911, 4113, 4693, *5801n (2), 7611.

solutio: 3053, 6252, 10894, 13269.

solvere: (to split up) 1444, 1445, 1446, 1448 (2), 1449, 1450, 1460, 1461, 1464, 1465, 1466; (to solve) 5922, 5967, 8180, 9617-18, 10218, 10251.

sors: 7987, 8239.

spatium: 15008, 15290, 15488, 15489; (*spatium concavitatis*; syn. *magnitudo*, see Tn1658) 12565, 12567, 12578, 12581.

species: 24, *24n, 35, 170, 623, 627 (2), 629 (2), 630, 653, 687 (syn. *modus*, cf. 660), 7034, 7054, 7068, 7095, 7118, 7147, 7149, 7155, 7158, 7161, 7168 (2), (...), 7849 (2), 12652, 12657, 13378, 13383, 14711. Tn1051.

speculativus: 20-21.

spissitudo: 14828, 14829, 14833, 14837, 14848, 14851, 14866, 14867.

spissus: (*in spissum*) 14782, 14788, 14814, 14819; (*in spisso*) 14882.

statim: [50].

sub: 5, 6, 29, 2026, 2028, 2031, 2042, 2048, 2051, 2052, 2053, [2073], [2076], 2196, 2354, (...).

subicere: (*oculis*) 1876, 5764; (*exempla*) 2079, 5386, 5403, 5424, 5456, 5468, 7387, 7457.

subiectus: [17], 140, [1030], 2065, 10392, 10708.

sublatus: 7098, 7593, 7603, 9777, 9782, 9783, 9788, 9789, 9871, 10254.

substantia: 8279.

subtractio: 4057, 4059, 4064, 4069, 4075, 4079, 4088, 5778, 5781, 5788-89, 5791.

subtrahere: 760, [763-64], 768, 906, 915, 927, 937, 1026, 3513, 3693, 3998, 4011, 4012, 4026, 4060, (...).

subtus: 4787, 14758, 14762.

sufficere: 4694, [6372], 11614, 14263, 14301.

suffragare: 19.

sumere: 206 (*ex*), 433 (*de*), 591, [1072], 1569 (*fractionem*, syn. *accipere*), 1587, 1620, 2970, 2975, 3102 (*ab*), 3103, 3106, 3394, 3472, 3990, 6565 (*radicem*), 7042 (*regulam, ex*), 7974, 8000, (...), 8676 (*regulam, ab*), (...), 13892.

summa: (result of a multiplication) 759, 768, 905, 955, 969, 1207, 1261, 1297, (...); (of an addition; cf. 2874-75) 1248, 1279, 1326, 1354 (*nominum*), 1403, (...), 3570, (...); (of a subtraction) 1238, 1261, 1454, 1475, (...); (of a doubling) 1044, [1086]; (of a division) 1727, 1740, 1770, (...); (of a conversion) 4745, 4746; (*proposita summa*) 3460, 3466, 3470, 4769, (...); 4784 (Tn668), 4785, 4786, (...); 4856 (*summa nummorum*), 4869, 8211, 8233, (...); 6090 (*summa verborum*).

LIBER MAHAMELETH   1129

summitas: (*summitas scale*, syn. *caput*) 15070, 15088, 15120; (*fustis*) 15485, 15485-86; (*turris*) 15486.

summum: 14974, 14978, 14987, 15004, 15026, 15060, 15068, 15070, 15088, 15093, 15095, 15096, 15120, (...), 15215, 15229.

super: 215, 232 (2), 233 (2), 592 (*addere super*), 1877, 1878, 5728, 6292, 8872, 11469, 14768, 14931, 15279, (...), 15341, 15355, 15483, 15750.

superare: 176, 183, 187 (2), 188, 189 (2; *in*), 190 (+ abl.), *211n (2), 329, [329], (...), 575, [1006], [3948], [3949], 4848, (...), 4877, 4949, 7014, 7023, 7025, 11073, (...), (11663), 13331, 14638. Syn. *vincere, transcendere*.

superesse: 3471, 3482, 3495, 4093, 4100, 4103, 4127, 8137, [8231].

superexcellere: 10056, 10059 (+ abl.), [10061], [10062], 10070 (*in*), 10077, 10078.

superficialis: 5075, 5076.

superficies: 5357, 5646, 5649, 5669, 5671, 5672, 5681, 5682, 5683 (2), 5684 (2), 5685 (2), 5686 (2), 5687, 5688, (...), 5735, 6479 (*superficies similes*), 8496, (...), 14968.

superhabundantius: 207.

superior: 1879, 3495, 6813, 12624, 12634, 14918, 14925.

superius: 968, [1029], 3027, 3621, [4309], 4827, 4932, 4964, 5803, 15691.

superponere: 8299, 8872.

supplere: 3524, 13784 (syn. *complere*).

supra: 953, 1011, 1020, 1327, 1485, 1836 (*additio supra*), 1876, figure p. 75 (2), 2482, 2607, 2675, 2752, 2755, 2774, 2777, 2801, 2808, 2840, 2888, 2897, 2915, 2957, (...), 15257, 15326 (adv.), (...); 3832, 7654 (syn. *insuper*)

surdus (اصمّ, or, rather: *non quadratus*, see Tn689): 4943, 4945, 4950, 4951, 4953, 5031.

— T —

tabula: *figure p. 6, 8860, 8865, 8867-68, 8870, 8871, 8872, 8886, 8890, 8892, *12614n.

taccir (تكسير; Tn1197): 8696.

talis: 1672, 1674, 2064, 2079, 2588 (2), 3223, 4325, 4352, 4354, 5386, 5403, 5468, 5483, *5801n (2), 6252, 7258, 7733, (8174), 8208, 8211, 9041, 9551, 9654, 10216-25n, 11009, 11687 (*ac si*), 11719, 12623, 12864, 13432, 13916, 13942, [14258], 15010, 15171, 15239.

taliter: 3990, 7585, 10935, 12538, 13161, 15258.

tam: (... *quam*) 68.

tamen: 434, 1009, 1680, 2677, 4024, 5402, 5423, 5468, 6177, 7279, 7386, 8457, 8877, 15758.

tamquam: [1780], 2239.

tantillum: [13642].

tanto: (... *quanto*) 4967.

tantum: (syn. *solum*) 15, 138, 700, 701, 1170, 1211, 4784-85, 5229, 5255, 5350, 5357, (...), 5548, 5812, (...); (... *quantum*) see *quantum*; (that much) 3595, 3608, 4064, 4835, 6140, (...), 7272, (...), 7945, 7946, (...).

tantumdem: 1733, 1752, 2127, 6365, 7005, 8220, 8241-42, 9333, 9565, 9975, 11329, 12174, 13284, (...), 14409, 14430.

tantus: (such) 1667, 3453, 3587, 3615, 3994, 4756, 6186, 6212, 6214, 6295, 6309, (...), 8236, 8243, (...); (... *quantus: q.v.*).

tempus: [45], 9127, *12648-49n.

ter: 881, 942, 944, 950, 969, 1302, 1312, 1316, 1321, 1323, 1324, 1329, 1335, 1368, 1371, 1372, 1374, 1376, (...), 1832, 1934, 2098, (...), 2136, 14074, 14075.

terminare: 58, 67, 6997 (*non terminatus*, syn. *interminatus*), 13832, 13833, 13834, 13984.

terminus: (term of a proportion) 3396, 3992, 5915, 5982, 10380; (end) 9387.

ternarius: 54 (2), 82, [749], [752], [771], 1990.

terni: 657, 658, 659 (2), 671, 680, 684 (2), [692] (2), [693] (2).

terra: 15233, 15313, 15330, 15348, 15351, 15370, 15380, 15382, 15389, 15391, 15393, 15397, 15399, 15483; (syn. *regio*) 13557.

tertius: 83, 114, 116, 117, 118, 119, 120, [131], [135], [136], [137], [148], 213 (2), 229, (...).

testari: 11413, [15013], [15276].

theorema: [224], *227n, 386n, 388n, [425], [1004], [2654], [2741], [2794], [2875-76].

theoricus: 20, 23.

Toletanus: 13571.

Toleto: 13345, 13367 (2), 13558, 13569.

tollere: See *sublatus*.

LIBER MAHAMELETH 1131

tot: [1176], [1177], [1179], 3382, 4846, 4880, 4903, 6194, 6218, 6461, (. . .);
(*tot vel tot*; syn. *tantum vel tantum*) 2588 (2), 7188, 7190, 7266, 7275,
(. . .).

totidem: 686, 7749, 8526, 8562, 9327, 10319, 10339, 10384, 10574, 11293,
11307, 11309, 12006, 12074, (. . .); (*totidem quot*) 14620, 14634.

totiens: 919, 920, 930, 1099, 1101, 1110, 1111, 1130, 1131, (. . .); (with
quotiens) *227n (2), 361, 837, 898, 908, 909, 917 (2), 929, 967, [1060],
[1062], 1095, 1096, 1106, (. . .); (with *quot*) [1187]; (with *quantus*) 894,
896.

totus: 196, 197, 198, 208, 209 (3), 210, 393, 394, 404, 405, 462, 463, 478,
[478], 492, 543, 544, 560, 603, (. . .).

tractare: 22, 34, 688, 699, 4932.

tractatus: [5923].

tradere: 8336.

transcendere: 14638. Syn. *vincere, superare*.

transilire: 805, 807, 809, 810.

transire: 1423, 15485.

transmutare: 10372. Syn. *permutare*.

tredecimus: 820, 2089, 2187, 2188, 2188-89, 2191, 2193, 2195, 2197 (2),
2198, (. . .).

trescenti: 86, [103], [132], [133], 1108, 1111, 1159, 1162, 1243, 1244,
1245, 1253 (2), 1319, (. . .).

triangulatus: 8694.

triangulus: *8689n, 14981, 15011, 15242, 15418.

tricesimus: 15238.

trigies: [107].

triplicare: 5054, 5332, 7000, 10632, 14172, 14173, 15175; (*comparatio
triplicata*; λόγος τριπλασίων, نسبة مثلّثة بالتكرير) 12255, 12257, 12261,
12263.

triplum: 1035, 1037, 1047 (2), 1053, 1054, [2346], 4905, 6340, 6706,
7915, 7937, 7940, 8473, 9459, 9460, (. . .).

triplus: *211n, [1032], 6340 (*comparatio*), 6487, 6696 (+ gen.), 6698,
6781, 8472 (+ dat.), 8661, 8821, 8823, 9927 (2), (. . .).

triticum: 6927, 6928, 6930, 6932, (. . .), 6948, 7777, 7779, 7781.

1132   PART TWO: TRANSLATION, GLOSSARY

tu: [1083], 1485 (*om.* *A*), 1802, 2032, 2081, 2088, 2091, 2096, 2113, 2118, 2543, 3050, 3477 (*om.* *A*), 3513, 3521, 4013 (*om.* *AB*), 4131, 4694, 5114, 5157, 5237, 5290, 5893 (*om.* *A*), 5899 (*om.* *A*), 5931, [5950], [5955], 5960, 5962, 5978, 6140, 6175n, 6297, 6321, 6612, 6917, 6932, 7075, 7306, 7365, 7452, 7897, 7963, 7992, 8007, 8546, 8898, 8940, 8954, 8966, 9220, 9883, 9915, 10461, 10585, 10628, 10730, 11433, 11614, 12986, 12990, 13133, 13179, 13225, 13396, 13410, 13537, 13734, 13774, 13812, 14368, 14468n, 14609, 14741, 14750, 14771, 14801, 14865.

tunc: 258, 294, 313, 329, 351, 373, 384, 389, 406, 407, 412, 428, 444, 491, 507, 526, 542, (. . .), 15732, 15756.

turris: *15404n, 15405, 15413, 15423, 15440, 15442, (. . .), 15489, 15491.

tuus: (*in manu tua*) 961, 1202; (*in manibus tuis*) 1279; (*tua responsio*) 5184, 5775, 5794, 5798.

$$\text{— U —}$$

ubi: 1876, 2008, 4850, 5645, 6123, 6151, 10216-25n, (10686), 13557, 13601, [13602], 15326, 15333, 15341.

ubicumque: 1422.

ulterior: 705 (*ad ulteriora*), *5801n (*ulterius*).

ultimus: 179, 180, 184, 186, *211n (2), 412, 995, 2246, 2274, 2502, 2590, (. . .); (*ultimum genus*) 2665, 2818, 2851; (*ultima fractio*, syn. *minor fractio*) 4709, 4728; (*ad ultimum*) 1283, 2447, [2456], 5176, 7781, 7928, 9363, 9378, 10034, 10606, 10635, 11823, 11925, 12303, 13284, 13305, 13316, 14621, 14736.

ultra: 1898, 1901, 1904, 10164, [10186].

uncia: *5801n (5), 8257, 8259, 8291, 8311, 8313, 8336, 8342, (. . .), 8347, 8373, (. . .), 8571, 8597.

unde: (syn. *igitur*) 53, 57, 59, 66, 137, [152], *227n, 376, 632, 704, 796, 870, 872, [1079], 1269, 1486, (. . .); (*unde denominatur*) 2022, 2145, 2167, 2417, 2629, 2681, 2902, 3144 (2), 4394, 4395, (. . .); (*locus unde = locus a quo*) 15240, 15253, 15259, 15306.

undecies: 965, 966, 971, 1538.

undique: 14880, 14882.

uni (pl.): 74, [158], 789, 804, 866, 867, [869], 872, 885, 894, 1917. Syn. *digiti, unitates*.

unitas: 11, [39], 41 (*unitas non est numerus*), *43n, [47] (2), [48], 53, 54, 55 (2), 75, 79, 82, 94, 115, [122], [135-36], [145], [146], [147], *227n (2), 362 (3), 645, 746, (. . .), 1911, 2032.

unquam: 14658 (*non . . . unquam nisi*).

unusquilibet: 2245, 3414, 3430, 6384, 14251.

unusquisque: [72], [144], [148], 785, 795, 954, 965, 978, 982, 988, [1005], [1175], [1178], 1271 (2), 1271-72, 1275, 1280, 1332, (. . .); (syn. *uterque*) see Tn1850, Tn1878.

LIBER MAHAMELETH **1133**

usque: [105], [108], [110], [112], [134], 791, [2075]; (*usque ad*) [39], 73, 76, 81, 83, 84, 86, 88, 92, [95], [100], [104], 121, [124], [125], [129], (...), 15382, 15486.

usus: 18-19.

utcumque: 9892-93.

uterque: 13, 22, 29, 30, [197], 257, 390, 391, 492, 573, 758, [773], 903, 906, (...), 15554, 15556, 15775; (pl., e.g.:) 1434, 3025, 3381, 6510, 6592.

uti (verb): 64.

utilis: 433; (*est utile*) 1478, 4198, 4264, 4934.

utilitas: 5843.

utpote: 1007.

*utrobique: 995n.

$$— \text{ V } —$$

'vacat': (text indicated as superfluous by copyists) 781n ($\mathcal{A}$), 1084-88n ($\mathcal{A}$), 2095-105n ($\mathcal{C}$), 2183-84n ($\mathcal{C}$), 3359n ($\mathcal{C}$), 4276-311n ($\mathcal{C}$), 4953n ($\mathcal{A}$), 5810-11n ($\mathcal{B}$), 6429n ($\mathcal{A}$), 11372n ($\mathcal{A}$).

vacuare: [14765].

*vacuitas: 12614n.

vadere: 15555, 15556.

valere: *2221-22n, *3520n, 6257, 6265, 6283, 6284, 6303, 6306, (...), 15056, 15580; (equation, syn. *equivalere*) 10159, 10302, 10305, 10310; (*regula*) 4848, 4849, 4850, 6285, 6337, 6350.

variare: 12657, 13383.

varietas: 11372n, 11386.

varius: 62.

vas: 8298, 8336, 8337 (2), 8352, (...), 8365, (8368), 14960.

-ve: [8342].

vector: 11677, 11715, 11830, 12096, 12150, 13428.

vehere: 11678, 11683, 11686, 12507.

vel: 14, 15, [17], 20, 21, 35, 36 (3), [107] (3), (...).

velle: [50], 171, *227n, 433, (...), 844 (*si volueris scire*), 849, (...), 15738, 15749; (*hoc est quod monstrare voluimus*) 311, 326, 348, 378, 410, 430, (...).

velut: 3318.

veluti: 637, 1636, 1639, 1645 (syn. *verbi gratia*), 1652, 1839, 1856, 1870, 2343, 2604, (...), 14659, 14660 (2).

vendere: 31, 5844, 5845, 5846, 5889, 5951, [5954], 6114, 6253, (...), 8118, 8120, 9551, 9564, 9585-86, 9642, 9645, 9649, 9660, 9684, 10731, 11170, 12396, 12621, 13349, 13370, 13372.

1134   Part Two: Translation, Glossary

venditio: 6942, 6946, 7732, 7733.

venire: 706, 7202 (syn. *oriri, descendere*), 8877 (*venire ad effectum*), 8882 (*venire ad actum*).

ventilare: 2900.

ventus (*– i*): 15599.

verbum: *2246n, 6090, 8136, 8147, 10807; (*secundum verba questionis*) 2921, 3951, 4492, 4499; (*secundum ordinem verborum*) 3964; (*verbi gratia*; Tn1795) 182, 215, 232, 261, 280, 296, 317, 332, (...), 606, 910, 1097, 1108, (...), 15541, 15688.

veritas: 10686.

vero: 16, 21, 28, 31, 34, [37], 42, [50], 54, 83, 96, 139, [141], [148], [153], 179, (...).

versus (prep.): 1996.

vertex: 15369. Syn. *cacumen*.

verus: (*questio*; solvable, صحيح, ant. *falsus*) 3418, 6913, 6915, *7329-30n, 7331, 7614, 8399, (9531), 10460, 10547, (10971), 10974, 13862, 13995, 13996, 14092, 14094, 14099, 14102, 14107, 14108, 14115, 14117, 14118, (14122), 14154, 14160, 14164, 14171, 14177, 14197, 14199, 14201, 14253, 14255, 14278, 14281, 14300, 14303, 14332, 14333, 14658; (*radix*; Tn704) 5030.

via: 11756.

vicesimus: 853, 858, 10119-20, 10120, 13590, 13594.

vicies: 14534, 14535. See also *vigies*.

vicis: 13707 (2), 13708; (*vice eius*, syn. *pro eo*) 10794, 10927.

videlicet: (138), *227n, 800, 861, 1232, 1256, 1383, 1396, 1399, 1429, 1433, 1492, 1529, (...).

videre: (*vide*, syn. *scias, considera*) 903, 907, 911, 964, 1685, 1690, 1991, 2010, 5205, 5223, 5272, 6872, (...); (*videri*) 622; (*videtur nobis*) 12473; (*nobis visum est* or *fuit*) 436, 7606, 7744, 9649, 14261.

vigies: [107], 1937, 8805, 8806.

vincere: 196, 1020, [1080], [1083], 4838, 4859, 7011, 8598, 10553, 10554, 11289, 14637. Syn. *superare, transcendere*.

vinum: 14956.

virga: 15493-94, 15494, 15497, 15499, 15501-02, 15504 (2), 15527.

vocare: *57n, 793.

vos: (*unusquisque* or *quisque vestrum*) 15647, 15648, 15649, 15658, 15660, 15670, 15682.

vox: 59.

— Y —

ypotesis: [220], [240], [261], [299].

Printed by Books on Demand, Germany